APPLIED RELIABILITY ENGINEERING AND RISK ANALYSIS

WILEY SERIES IN QUALITY & RELIABILITY ENGINEERING
*and related titles**

Applied Reliability Engineering and Risk Analysis:
Probabilistic Models and Statistical Inference
Ilia B. Frenkel, Alex Karagrigoriou, Anatoly Lisnianski and Andre Kleyner

Electronic Component Reliability:
Fundamentals, Modelling, Evaluation and Assurance
Finn Jensen

Measurement and Calibration Requirements
For Quality Assurance to ASO 9000
Alan S. Morris

Integrated Circuit Failure Analysis:
A Guide to Preparation Techniques
Friedrich Beck

Test Engineering
Patrick D. T. O'Connor

Six Sigma: Advanced Tools for Black Belts and Master Black Belts*
Loon Ching Tang, Thong Ngee Goh, Hong See Yam and Timothy Yoap

Secure Computer and Network Systems: Modeling, Analysis and Design*
Nong Ye

Failure Analysis:
A Practical Guide for Manufacturers of Electronic Components and Systems
Marius Bâzu and Titu Băjenescu

Reliability Technology:
Principles and Practice of Failure Prevention in Electronic Systems
Norman Pascoe

APPLIED RELIABILITY ENGINEERING AND RISK ANALYSIS

PROBABILISTIC MODELS AND STATISTICAL INFERENCE

Editors

Ilia B. Frenkel
SCE – Shamoon College of Engineering, Beer Sheva, Israel

Alex Karagrigoriou
University of Cyprus, Nicosia, Cyprus

Anatoly Lisnianski
The Israel Electric Corporation Ltd., Haifa, Israel

Andre Kleyner
Delphi Electronics & Safety, Indiana, USA

WILEY

Registered office
John Wiley & Sons Ltd, The Atrium, Southern Gate, Chichester, West Sussex, PO19 8SQ, United Kingdom

For details of our global editorial offices, for customer services and for information about how to apply for permission to reuse the copyright material in this book please see our website at www.wiley.com.

Library of Congress Cataloging-in-Publication Data

Applied reliability engineering and risk analysis : probabilistic models and statistical inference / Ilia B. Frenkel, Alex Karagrigoriou, Anatoly Lisnianski, Andre Kleyner. – First edition.
1 online resource.
Includes bibliographical references and index.
Description based on print version record and CIP data provided by publisher; resource not viewed.
ISBN 978-1-118-70189-8 (ePub) – ISBN 978-1-118-70193-5 (MobiPocket)
ISBN 978-1-118-70194-2 – ISBN 978-1-118-53942-2 (hardback) 1. Reliability (Engineering)
2. Risk assessment – Mathematical models. I. Frenkel, Ilia, Ph.D., editor of compilation.
II. Gnedenko, B. V. (Boris Vladimirovich), 1912-1995.
TA169
620′.00452 – dc23

2013025347

A catalogue record for this book is available from the British Library.

ISBN: 978-1-118-53942-2

Set in 10/12pt Times by Laserwords Private Limited, Chennai, India
Printed and bound in Malaysia by Vivar Printing Sdn Bhd

1 2014

Contents

Part II NETWORKS AND LARGE-SCALE SYSTEMS

Remembering Boris Gnedenko

Andre Kleyner[1] and Ekaterina Gnedenko[2]
[1]*Editor of Wiley Series in Quality and Reliability Engineering,*
[2]*Granddaughter of Boris Gnedenko, Faculty, Department of Economics, Tufts University*

Boris Gnedenko was one of the most prominent mathematicians of the twentieth century. He contributed greatly to the area of probability theory, and his name is permanently linked to pioneering and developing mathematical methods in reliability engineering. Gnedenko is best known for his contributions to the study of probability theory, such as the extreme value theorem (the Fisher–Tippett–Gnedenko theorem). He first became famous for his work on the definitive treatment of limit theorems for sums of independent random variables. He was later known as a leader of Russian work in applied probability and as the author and coauthor of outstanding textbooks on probability, mathematical methods in reliability and queuing theory.

Boris Gnedenko was born on January 1, 1912, in Simbirsk (later Ulianovsk), a Russian city on the Volga River. He was admitted to the University of Saratov at the young age of 15 by special permission from the Minister of Culture and Education of the Soviet Union.

After graduation from the university, Gnedenko took a teaching job at the Textile Institute in Ivanovo, the city east of Moscow, which for many years was the center of the Soviet textile industry. While lecturing at the university, Gnedenko simultaneously was involved in the solution of some practical problems for the textile industry. This is when he wrote his first works, concerning queuing theory, and became fond of the theory of probability. That triggered his later works on the applications of statistics to reliability and quality control in manufacturing.

In 1934, Gnedenko decided to resume his university studies at the graduate level. He was awarded a scholarship which allowed him to undertake research at the Institute of Mathematics at Moscow State University. He became a graduate student under the direction of Alexander Khinchin and Andrei Kolmogorov. The latter became one of the most famous mathematicians of the twentieth century (the Kolmogorov-Smirnov test in statistics, the Kolmogorov-Arnold-Moser theorem in Dynamics, the Kolmogorov Complexity, and other landmark achievements). As a graduate student, Gnedenko became interested in limiting theorems for the sums of independent random variables. In 1937, he defended his dissertation on "Some Results in the Theory of Infinitely Divisible Distributions," and, soon after the defense, he was appointed a researcher at the Institute of Mathematics at Moscow State University.

Years later, Kolmogorov would say:

> Boris Gnedenko is recognized by an international mathematics community as one of the most prominent mathematicians who is currently working in the area of probability theory. He combines an exceptional skill and proficiency in classical mathematical methods with deep understanding of a wide range of modern probability problems and a perpetual interest to their practical applications.

In 1937, during the infamous Stalin's Purges, Gnedenko was falsely accused of "anti-Soviet" activity and thrown in jail. The NKVD (the Soviet secret police at that time) were trying to coerce him to testify against Kolmogorov, who was not yet arrested but was under investigation by the NKVD for running a conspiracy "against the Soviet people," a very common bogus charge at the time. However, Gnedenko survived brutal treatment at the hands of the NKVD and refused to support the false accusations against his mentor. He was released six months later, though his health suffered.

He returned to Moscow University as an assistant professor in the Department of the Theory of Probability in 1938 and as a research secretary (an academic title in Russia) at the Institute of Mathematics. During this period at Moscow State University, he solved two important problems. The first problem involved the construction of asymptotic distributions of the maximum term of the variation series and defining the nature of limit distributions and the conditions for convergence (Gnedenko 1941b). The second problem involved the construction of the theory of corrections to the Geiger–Muller counter readings used in many fields of physics and technology (Gnedenko 1941a). This paper is a landmark in what later became "the theory of reliability".

In 1939, Gnedenko married Natalia Konstantinovna and subsequently they had two sons.

During World War II, Gnedenko continued his research work at Moscow State University, although for two years, together with all the university colleagues, he had to temporarily relocate to Turkmenistan and later to the Ural mountains because of Moscow's proximity to the front lines. Some of his work during the war was of national defense nature, including quality and process control at military plants. During that time Gnedenko continued working on a variety of mathematical problems, including the limit theorem for the sums of independent variables, and discovering the classification of the possible types of limit behavior for the maximum in an increasing sequence of independent random variables. The Weibull distribution, one of the most popular time-to-failure distributions in reliability engineering, is occasionally referred to as 'Weibull-Gnedenko' (see, e.g. Pecht 1995, or Chapter 19 in this book). Gnedenko developed this model at about the same time as Waloddi Weibull in Sweden, however, due to the relative isolation of the Soviet Union at that time, this was not common knowledge. Only in 1943, two years after Gnedenko had published his results in Russian, due to the warming relationship between the USA and the Soviet Union during World War II, did Gnedenko receive an invitation to publish his hallmark paper on extreme value limit theorem in the American journal, *Annals of Mathematics*. His research into limit theorems continued and in 1949 this resulted in a monograph with Kolmogorov, entitled "Limit Distributions for Sums of Independent Random Variables". This monograph was awarded the Chebyshev Prize in 1951 and was translated into many languages. It was later published in English by Addison-Wesley publishing (Gnedenko and Kolmogorov 1954) and underwent a second edition in 1968.

Boris Gnedenko and his wife Natalia, 1978

During World War II, the western part of the Soviet Union was devastated by the German occupation, therefore after the war in 1945, Gnedenko was sent to Lviv, the largest city in Western Ukraine, to help rebuild Lviv University and undertake the restoration of the overall Ukrainian system of higher education. He accepted this challenging job with great energy and enthusiasm.

Building on the works of Kolmogorov and Smirnov establishing the limit distributions for the maximum deviation of an empirical distribution function from the theorctical,

Gnedenko developed effective methods to obtain the exact distributions in the case of finite samples in these and other related problems. This work received worldwide recognition, because it served as the basis for compiling tables which were very valuable in applied statistics at that time.

In 1948, Gnedenko was elected a full member of the Ukrainian Academy of Science and in 1950 he was transferred to Kiev, the capital of the Ukraine, to become Head of the recently created Department of the Theory of Probability at the Institute of Mathematics of the Ukrainian Academy of Sciences and also Head of the Physics, Mathematics and Chemistry Section of the Ukrainian Academy of Sciences. At the same time he served as the Chair of the Department of Probability Theory and Algebra at Kiev State University. Later he became Director of the Kiev Institute of Mathematics.

His work in Kiev followed several directions. Besides mathematics and statistics, his contribution was instrumental in developing computer programming and setting up a computing laboratory and encouraging his younger colleagues to study programming. His earlier efforts at the Ukrainian Academy of Science helped to create, in 1951, one of the first fully operational electronic computers in continental Europe.

In 1958, Gnedenko was a plenary speaker at the International Congress of Mathematicians in Edinburgh with a talk entitled, "Limit Theorems of Probability Theory". One of Gnedenko's most famous books is called *Theory of Probability*, which first appeared in 1950. Written in a clear and concise manner, the book was very successful in providing an introduction to probability and statistics. It has undergone six Russian editions and has been translated into English (Gnedenko 1998), German, Polish and Arabic. Earlier, in 1946, Gnedenko also co-authored, with Khinchin, the book, *Elementary Introduction to the Theory of Probability*, which also has been published many times in the USSR and abroad.

In 1960, Boris Gnedenko returned to Moscow State University and later, in 1966 became Head of the Department of Probability holding this post until his death in 1995. He took over from Andrei Kolmogorov, who became Head of the Interdepartmental Laboratory of Probability and Statistics in Moscow.

During the sixties, Gnedenko's interests turned to mathematical problems with industrial application, namely, the queuing theory and mathematical methods in reliability. In 1961, with several of his students and colleagues, he organized and chaired the Moscow Reliability Engineering Seminars. This was a very successful undertaking with around 800 participants: academics, engineers and mathematicians. Many attendees traveled from other cities, and besides academic activities, it resulted in a number of practical consultations helping engineers in various industries. This seminar was also a big promoter of reliability engineering, which at that time was in its infancy.

As Professor Vere-Jones, then a British graduate student at Moscow State University, remembers: "It was into this seminar that I strayed in 1961. I was much impressed, not so much by the academic level, which varied from excellent to indifferent, as by the strong impression I received that this was an environment in which everyone's contribution was valued" (Vere-Jones 1997). This "owed a great deal to Boris Vladimirovich Gnedenko's own personality and convictions, and the influence he had on his colleagues". Later, at the end of 1965, Vere-Jones managed to arrange for Gnedenko a two-month trip "down under". Gnedenko was invited by the Australian National University in Canberra, which had a special exchange agreement with Moscow State University. Unfortunately, Gnedenko was not able to take his family with him on this exciting trip – the Soviet government, afraid of losing Gnedenko to the foreign capitalistic country, did not allow his wife and sons to accompany him. According to Vere-Jones, Gnedenko lectured on two themes: one was reliability theory (estimation and testing of the life-time distribution); and the second was mathematical education in the Soviet Union, which generated the greatest interest. Throughout his visit, Gnedenko displayed great interest in everything: the people, the birds and animals, the scenery, the universities and schools, shops, etc. He was particularly fascinated by the koala in Australia and by the kiwi and its huge egg in New Zealand. “His interest in matters ‘down under’ continued well past this visit. I believe he became a president or patron of the USSR side of the New Zealand-USSR friendship society”.

Most people who met Gnedenko remember their personal interactions very fondly and pay tribute to his disposition and personal qualities. Vere-Jones says:

> I consider myself extraordinarily fortunate in having happened to drift into his seminar in October 1961. Not only was it a chance to step inside the legendary world of Russian probability theory, it was a chance to come to know a rare human being, to see him at home and with his family, and to work briefly alongside him.
>
> (Vere-Jones 1997)

In 1991, Gnedenko visited the USA. Professor Igor Ushakov, his longtime colleague, who was his host during this trip, remembers:

> I was blessed in life to have an opportunity to work closely with Boris Gnedenko. I accompanied him on various academic business trips and spent many evenings at his house and in the company of his family. It would not be an exaggeration to say that I’ve never met another person with more zest for life and more willingness to share his kindness and help others in need.
>
> (Ushakov 2011)

During that trip Gnedenko, accompanied by his son Dmitry, visited the University of North Carolina, where he lectured and had several research meetings. Then he went to Washington, DC, to give a lecture at the Department of Operations Research of George Washington University. Upon learning about Gnedenko’s arrival, the university photographers took a lot of photos and the university newspaper published an article about his visit. While in Washington, DC, Gnedenko met quite a few local mathematicians. Many invited him to their homes and he was always the center of attention. Since this was an international crowd, his knowledge of German and French (in addition to English) came in very handy.

During that visit he was also interviewed by Professor Nozer Singpurwalla at George Washington University. Answering questions, Gnedenko recalled many important events in his life, his work with Kolmogorov and other prominent Russian mathematicians, as well as other significant milestones (Singpurwalla and Smith 1992). During his professional life he held a number of high administrative positions, both at a university and the Academy of Science levels, however, his heart was clearly in academic work and research. Gnedenko said: “I prefer scientific work, lecturing and writing. I enjoy working with students. I have had over a hundred doctoral students, of whom 30 are professors in my country or in other countries”. Seven of his students became members of the Academy of Science, the highest academic distinction in Russia and the former Soviet Republics.

His teaching activities extended beyond academia, Boris Gnedenko had also contributed greatly to popularizing math and science. Besides the book on history of mathematics mentioned earlier, he also wrote for primary and secondary school. Gnedenko said: "This year I have also written a short book for school children on mathematics and life", and later, "The second book I plan is for school children – a trip into a mathematical country".

Boris Gnedenko and his colleague and friend, Igor Ushakov, at Gnedenko's home, 1970

Besides studying and doing research in mathematics, Gnedenko also took a keen interest in the history of this discipline, which he considered very important to a future development of mathematics. According to O'Connor and Robertson (2000), Gnedenko's interest in the history of mathematics extended well beyond his text aimed at secondary school pupils. He published much on this topic, including the important *Outline of the History of Mathematics in Russia* which was not published until 1946, although he wrote it before the start of World War II. It is a fascinating book which looks at the history of mathematics in Russia in its cultural background.

Later in 1993, Gnedenko visited the USA again, now by invitation from the MCI Corporation, at that time a telecommunication giant, where Professor Ushakov was a consultant at the time. Gnedenko was 81, but despite his health problems he put together a rigorous plan to visit all the technical and academic centers he was invited to. The first visit was to the MCI Headquarters in Dallas, where he lectured to a large audience about statistical problems in the telecommunications industry. Introducing him to the audience, Chris Hardy, the MCI Chief Scientist, said: "I did not have any difficulties inviting Prof. Gnedenko, I just said to our CEO that for us hosting Professor Gnedenko would be like for Los Alamos Labs hosting Norbert Wiener."

Next stop was at Harvard University, hosted by Eugene Litvak, professor at the School of Public Health. For his lecture topic Gnedenko chose, "Probability and Statistics from

Middle Ages to Our Days". Gnedenko always had a sixth sense and a feel for his audience; therefore, because this time he was not speaking to expert mathematicians, he chose one of his favorite subjects: the history of mathematics.

During his lifetime Gnedenko produced a remarkable number of published works. One of the most complete lists of his publications can be found in Gnedenko D.B. (2011). Interestingly enough, all his life, even after the introduction of word processors, Gnedenko still used a typewriter. When asked during one interview how many drafts it took for a paper or book, Gnedenko answered, "one draft only". It was difficult to believe one draft would have no errors or need for improvement. Gnedenko replied, "it is necessary to think first and only then to write. At this stage I am almost finished" (Singpurwalla and Smith 1992).

Many of the facts and events of Gnedenko's life that are presented here are also recounted in greater detail by the great man himself in his memoirs (Gnedenko 2012). More information about Boris Gnedenko can be found on a dedicated website, Gnedenko Forum, an informal association of specialists in reliability (Gnedenko Forum 2013). The Forum, created by Gnedenko's followers, I. Ushakov and A. Bochkov, is designed to commemorate his legacy and also to promote contacts between members of the global reliability community. It contains the latest professional news in the areas of probability theory, statistics, reliability engineering, risk analysis, mathematical methods in reliability, safety, security and other related fields. Many contributors of this book are members of the Gnedenko Forum and were inspired or influenced in some way by the lifelong work of this great mathematician. This book commemorates the centennial of his birth and pays tribute to his immense contribution to the probability theory and reliability mathematics and also celebrates his legacy.

References

Gnedenko, B.V. (1941a) To the theory of GM-counters, *Experimental and Theoretical Physics* 11: 101–106. (In Russian).

Gnedenko, B.V. (1941b) Limit theorems for maximum term of variational series. Moscow, *DAN; Sc. Sci. of the USSR* 32 (1): 231–234. (In Russian).

Gnedenko, B.V. (1943) Sur la distribution limite du terme maximume d'une série aléatoire, *Annals of Mathematics* 44 (3): 423–453.

Gnedenko, B.V. (1998) *Theory of Probability*, 6th edition. Boca Raton, FL: CRC Press.

Gnedenko, B.V. (2005) *Essays on History of Mathematics in Russia*, 2nd ed. Moscow: KomKniga. (In Russian).

Gnedenko, B.V. (2012) *My Life in Mathematics and Mathematics in My Life: Memoirs*. Moscow: URSS. (In Russian).

Gnedenko, B.V. and A.Ya. Khinchin (1962) *Elementary Introduction to the Theory of Probability*. New York: Dover Publications, pp. 1–130.

Gnedenko, B. and Kolmogorov, A. (1954) *Limit Distributions for Sums of Independent Random Variables*. Cambridge, MA: Addison-Wesley.

Gnedenko, B.V. and Kolmogorov, A. (1968) *Limit Distributions for Sums of Independent Random Variables*, 2nd ed. Cambridge, MA: Addison-Wesley.

Gnedenko, B.V. and Ushakov, I. (1995) *Probabilistic Reliability Engineering*. New York: John Wiley & Sons.

Gnedenko, D. B. (2011) Gnedenko's bibliography. *Reliability: Theory and Applications* 6 (4). (December). Accessible at: http://gnedenko-forum.org/Journal/2011_4.html. (In Russian).

Gnedenko Forum (2013) http://gnedenko-forum.org/

O'Connor, J. and Robertson, E. (2000) Boris Vladimirovich Gnedenko. Available at: http://www-history.mcs.st-and.ac.uk/Biographies/Gnedenko.html

Pecht, M. (ed.) (1995) *Product Reliability Maintainability Supportability Handbook*. Boca Raton, FL: CRC Press.

Singpurwalla, N. and Smith, R. (1992) A conversation with Boris Vladimirovich Gnedenko, *Statistical Science* 7 (2): 273–283.

Ushakov, I. (2011) The 100th anniversary of Boris Gnedenko birthday. *RT&A* # 04 (23), Vol. 2 (December). Available at: http://www.gnedenkoforum.org/Journal/2011/042011/RTA_4_ 2011-01.pdf

Vere-Jones, D. (1997) Boris Vladimirovich Gnedenko, 1912–1995: A personal tribute. *Australian Journal of Statistics* 39 (2): 121–128.

List of Contributors

Vilijandas Bagdonavičius, Department of Mathematical Statistics, Faculty of Mathematics and Informatics, Vilnius University, Lithuania
Vilijandas.Bagdonavicius@mif.vu.lt

N. Balakrishnan, Department of Mathematics and Statistics, McMaster University, Canada and Department of Statistics, King Abdulaziz University, Saudi Arabia
bala@univmail.cis.mcmaster.ca

Ayanendranath Basu, Indian Statistical Institute, India ayanbasu@isical.ac.in

Ji Hwan Cha, Department of Statistics, Ewha Womans University, Korea
jhcha@ewha.ac.kr

Lirong Cui, School of Management & Economics, Beijing Institute of Technology, P.R. China lirongcui@bit.edu.cn

Emilio De Santis, Department of Mathematics, University La Sapienza, Italy
desantis@mat.uniroma1.it

Yi Ding, Department of Electrical Engineering, Technical University of Denmark, Denmark yding@elektro.dtu.dk

Shijia Du, School of Management & Economics, Beijing Institute of Technology, China
dushijia111@126.com

Maxim Finkelstein, Department of Mathematical Statistics, University of the Free State, South Africa FinkelM@ufs.ac.za

Ilia B. Frenkel, Center for Reliability and Risk Management, Industrial Engineering and Management Department, SCE - Shamoon College of Engineering, Israel iliaf@sce.ac.il

Stylianos Georgiadis, Laboratoire de Mathématiques Appliquées de Compiègne, Université de Technologie de Compiègne, Centre de Recherches de Royallieu, France
stylianos.georgiadis@utc.fr

Ilya B. Gertsbakh, Department of Mathematics, Ben-Gurion University, Israel, elyager@bezeqint.net

Abhik Ghosh, Indian Statistical Institute, India abhianik@gmail.com

Alan G. Hawkes, School of Business and Economics, Swansea University, UK alanhawkes@virginmedia.com

Hong-Zhong Huang, School of Mechanical, Electronic, and Industrial Engineering, University of Electronic Science and Technology of China, China hzhuang@uestc.edu.cn

Kailash (Kal) C. Kapur, Industrial & Systems Engineering, University of Washington, USA kkapur@u.washington.edu

Alex Karagrigoriou, Department of Mathematics and Statistics, University of Cyprus, Cyprus alex@ucy.ac.cy, alex.karagrigoriou@gmail.com

Lev Khvatskin, Center for Reliability and Risk Management, Industrial Engineering and Management Department, SCE - Shamoon College of Engineering, Israel khvat@sce.ac.il

Andre V. Kleyner, Delphi Electronics & Safety, USA andre.v.kleyner@delphi.com, kleyner.consulting@sbcglobal.net

Gregory Levitin, The Israel Electric Corporation Ltd., Israel levitin@iec.co.il

Ruta Levuliene, Department of Mathematical Statistics, Faculty of Mathematics and Informatics, Vilnius University, Vilnius, Lithuania Ruta.levuliene@mif.vu.lt

Yan-Feng Li, School of Mechanical, Electronic, and Industrial Engineering, University of Electronic Science and Technology of China, China lyfmoon@gmail.com

Yan-Fu Li, Ecole Centrale Paris LGI-Supelec, France yanfu.li@ecp.fr, yanfu.li@supelec.fr

Zhaojun (Steven) Li, Electro-Motive Diesel – A Caterpillar Company, USA zjli@uw.edu

Nikolaos Limnios, Laboratoire de Mathématiques Appliquées de Compiègne, Université de Technologie de Compiègne, Centre de Recherches de Royallieu, France nikolaos.limnios@utc.fr

Yan-Hui Lin, Ecole Centrale Paris LGI-Supelec, Paris, France yanhui.lin@ecp.fr

Anatoly Lisnianski, The Israel Electric Corporation Ltd., Israel anatoly-l@iec.co.il

Yu Liu, School of Mechanical, Electronic, and Industrial Engineering, University of Electronic Science and Technology of China, China yuliu@uestc.edu.cn

Avijit Maji, Indian Statistical Institute, India avijit.maji@hotmail.com

Krishna B. Misra, International Journal of Performability Engineering, India kbmisra@gmail.com

Ramin Moghaddass, Reliability Research Lab, Department of Mechanical Engineering, University of Alberta, USA rmoghadd@ualberta.ca

Toshio Nakagawa, Department of Business Administration, Aichi Institute of Technology, Japan toshi-nakagawa@aitech.ac.jp

Mikhail Nikulin, IMB, University Victor Segalen Bordeaux 2, France mikhail.nikouline@u-bordeaux2.fr

Suvra Pal, Department of Mathematics and Statistics, McMaster University, Canada pals3@math.mcmaster.ca

Michael Pecht, Center for Advanced Life Cycle Engineering (CALCE), Department of Mechanical Engineering, University of Maryland, USA pecht@calce.umd.edu

Rui Peng, University of Science & Technology Beijing, China pengrui1988@gmail.com

Mikael Petersson, Department of Mathematics, University, Stockholm, Sweden mikpe@math.su.se

Hoang Pham, Department of Industrial and Systems Engineering, Rutgers, The State University of New Jersey, USA hopham@rci.rutgers.edu

Fabio Postiglione, Department of Electronic and Computer Engineering, University of Salerno, Italy fpostiglione@unisa.it

Pedro O. Quintero, Department of Mechanical Engineering, University of Puerto Rico at Mayagüez, Puerto Rico pquintero@me.uprm.edu

Yoseph Shpungin, Software Engineering Department, SCE – Shamoon College of Engineering, Israel yosefs@sce.ac.il

Dmitrii Silvestrov, Department of Mathematics, Stockholm University, Sweden dmitrii.silvestrov@mdh.se, silvestrov@math.su.se

Nozer D. Singpurwalla, Department of System Engineering and Engineering Management, and Department of Management Science, City University of Hong Kong, Hong Kong nozer@gwu.edu

Fabio Spizzichino, Department of Mathematics, University La Sapienza, Italy fabio.spizzichino@uniroma1.it

Ola Tannous, Reliability and Quality Engineering Department, Electro Motive Diesels, USA ola.tannous@electromotivediesels.com

Kishor S. Trivedi, Department of Electrical and Computer Engineering, Duke University, USA kst@ee.duke.edu

Ilia Vonta, Department of Mathematics, School of Applied Mathematical and Physical Sciences, National Technical University of Athens, Greece vonta@math.ntua.gr

Irene Votsi, Laboratoire de Mathématiques Appliquées de Compiègne, Université de Technologie de Compiègne, Centre de Recherches de Royallieu, France irene.votsi@utc.fr

Min Xie, Department of Systems Engineering & Engineering Management, City University of Hong Kong, Hong Kong, China minxie@cityu.edu.hk

Liudong Xing, University of Massachusetts Dartmouth, USA lxing@umassd.edu

Xiaoyan Yin, Department of Electrical and Computer Engineering, Duke University, USA xy15@duke.edu

Xufeng Zhao, School of Economics and Management, Nanjing University of Technology, China g09184gg@aitech.ac.jp

Shun-Peng Zhu, School of Mechanical, Electronic, and Industrial Engineering, University of Electronic Science and Technology of China, China zspeng2007@uestc.edu.cn

Enrico Zio, Ecole Centrale Paris LGI-Supelec, France and Dipartimento di Energia, Politecnico di Milano, Italy enrico.zio@ecp.fr, enrico.zio@polimi.it, enrico.zio@supelec.fr

Ming J. Zuo, Reliability Research Lab, Department of Mechanical Engineering, University of Alberta, USA ming.zuo@ualberta.ca, mzuo@ualberta.ca

Preface

This book is a collective work by a number of leading scientists, analysts, mathematicians, and engineers who have been working on the front end of reliability science and engineering. This work is dedicated to the memory of Boris Gnedenko and commemorates the 100-year anniversary of his birth. Boris Gnedenko was a remarkable mathematician and one of the pioneers of reliability science, thus most of the material presented in this book in some way have been influenced by his work. More about Gnedenko and his life can be found in the introductory chapter "Remembering Boris Gnedenko".

All chapters in this book are written by the leading researchers and practitioners in their respective fields of expertise and present quite a number of innovative methods, approaches and solutions not covered before in the literature. The amount of research and study in these areas, which represent conventional, contemporary, and recently emerged fields of reliability science have been growing rapidly due to ample opportunities to apply their results in various areas of academic research and in industry.

Despite the large number of contributing authors, this manuscript presents a continuous story of modern reliability engineering, shows the recent advances in mathematical methods in reliability and also celebrates the legacy of Boris Gnedenko. This book has been divided into five logically contiguous parts:

Part I Degradation Analysis, Multi-State and Continuous-State System Reliability

Multi-state and continuous-state system reliability, along with degradation analysis, have undergone vast growth and development in recent years and now is one of the most intensively developing areas in reliability science.

Chapter 1 presents a stochastic process, which models a degraded system behavior, as inhomogeneous continuous time Markov chain. It compares four possible approaches to the problem solution: Monte-Carlo simulation, uniformization, state-space enrichment and numerical differential equations solvers.

Chapter 2 studies multiple independent failure modes for a device under condition monitoring and suggests how to assess the key reliability measures for such a device. In addition, a method for estimation of characteristic parameters associated with the multistate degradation structure of each failure mode is presented.

Chapter 3 covers applications of time series regression with exponentially distributed errors to acceleration test models, degradation monitoring, and financial risk tracking. The chapter also explores inference and prediction modeling under both Bayesian and frequentist methods.

Chapter 4 introduces the inverse L_z-transform for a discrete-state continuous-time Markov process. It shows that the application of inverse L_z-transform to multi-state system reliability analysis significantly expands the range of problems that can be solved by using Ushakov's universal generating operator and corresponding universal generating function (UGF).

Chapter 5 presents a case study of L_z-transform applied to a real world problem: the availability assessment of an aging cooling system of MRI equipment under stochastic demand. The method presented is used to determine the system structure and to achieve the required availability level for complex multi-state aging systems. The proposed solution offers a significant decrease in computational burden compared with conventional Markov methods and also can be used as a tool in engineering decision-making.

Chapter 6 combines clustering and the L_z-transform technique to reduce the computational complexity of multi-state system reliability analysis. The proposed method reduces the computational dimension for a complex system and consequently decreases computational complexity.

Chapter 7 presents three new types of multi-state systems that generalize a linear multi-state sliding window system in the case of allowed gaps (groups of consecutive elements with insufficient cumulative performance). These models can be applied in manufacturing technology, radar and military systems, system integration testing and multi-target tracking.

Chapter 8 presents a new way to model reliability of a component or system using the theory and methods for fuzzy sets. A comprehensive literature survey for traditional fuzzy reliability is also given. This chapter discusses the use of a substitute variable to fuzzify the states of a component or a system. The approach and the methods presented constitute a generalization of traditional binary and multi-state reliability and can also be applied to model the continuous state behavior for systems that degrade over time. The applications of fuzzy reliability models and their advantages in decision-making processes are demonstrated in several case studies.

Chapter 9 explores the relationship between multi-state system reliability and performability. The fast depleting resources of the world with an exponentially increasing population prompted the development of sustainable products, systems and services that should also be dependable. This chapter discusses the necessity of introducing a performance index based on an overall consideration of the problem suitable to be used as a criterion for developing future products, systems and services.

Part II Networks and Large-Scale Systems

Part II covers various aspects of reliability analysis of complex and large systems (both continuous and discrete), including networks.

Chapter 10 discusses monotone binary systems with binary components and its structural invariants as multidimensional parameters. The three invariants are: (1) cumulative D-spectrum, which numerically matches the cumulative signature; (2) the importance spectrum; and (3) the set of border states. The authors present a simple formula to compute the Birnbaum Importance Measure using the second structural invariant, which can be used to approximate the system reliability function.

Chapter 11 presents six hierarchical analytic models to evaluate the performance and availability of mobile networks, especially useful for the next generation of networks based on the Third Generation Partnership Project IP Multimedia Subsystem.

Chapter 12 presents an empirical method to estimate the probability of the first failure, the rate of occurrence of failures, and the steady-state availability for a discrete-time semi-Markov system. A numerical application is given to illustrate the stability of this process.

Chapter 13 presents a system of two finite networks with a certain topology of the network nodes failures. Failure of a randomly chosen node of the first network causes a failure of another randomly chosen node of the second network, which in turn, causes a failure of a randomly chosen node of the first network, and so on. As a result, a random number of nodes will fail in each network. Using the distribution of the failed nodes in each network and the D-spectra (cumulative signature) technique, it generates the failure probability models for both networks.

Part III Maintenance Models

Part III covers repairable systems along with various maintenance models and their applications, placing emphasis on both mathematical rigor and practical applications.

Chapter 14 presents a comparative analysis of four preventive replacement policies with minimal repairs: (1) the unit is replaced at scheduled regular times; (2) the unit is replaced at a planned time or at a random operating time, whichever occurs first; (3) the unit is replaced at a planned time or at a random operating time, whichever occurs last; (4) the unit is replaced upon completion of some operating times over a planned time. Optimal replacement policies of each model are summarized and compared.

Chapter 15 uncovers the similarities between waiting times to occurrences of words in a random sampling of letters from an alphabet and random degradation of devices or systems. It utilizes this "words sequences" approach to model systems with aging, degradation, or reliability-maintenance in discrete time.

Chapter 16 covers distributions of occupancy times in subsets of Markov and semi-Markov processes in terms of Laplace-Stieltjes transforms. It discusses the applications of the well-studied ion-channel theory to system reliability analysis and systems that suffer terminal failure.

Chapter 17 discusses the imperfect maintenance model selection method with a novel GoF test method and a tailored Bayesian model selection method. An example of scheduling the preventive maintenance plan for diesel engines is presented.

Chapter 18 studies the reliability of a warm standby system with consideration of imperfect fault coverage based on binary decision diagrams (BDD). It shows applications of a BDD-based approach to warm standby systems with any type of time-to-failure distribution.

Part IV Statistical Inference in Reliability

The statistical inference Part consists of six chapters which cover important issues associated with goodness of fit tests, divergence measures, cure rate models, perturbed discrete time renewal equations, and extreme shock models.

Chapter 19 offers a comprehensive survey of statistical methods on validation of the two-parameter Weibull-Gnedenko distribution and its connection with the extreme value theory.

Chapter 20 deals with heavy-tailed distributions which are of great interest in reliability systems where the interest lies in the occurrence of rather exceptional and rare events associated with tails of the distributions.

Chapter 21 explores two important families of divergence measures, namely, the power divergence and the density power divergence, which have proven to be useful tools in robust inference analysis.

Chapter 22 investigates cure rate survival models where the competing cause scenario is considered and the exponential and Weibull distributions for the time-to-event are assumed.

Chapter 23 presents asymptotic expansions for solutions of perturbed discrete time renewal equations.

Chapter 24 examines a generalization of the classical extreme shock model where the probability of failure depends not only on the history of the shock process but also on the operational history of a system as well.

Part V Systemability, Physics-of-Failure and Reliability Demonstration

Part V is comprised of several miscellaneous topics to cover other important areas of modern reliability science, such as physics-of-failure (PoF), accelerated testing and operating environments.

Chapter 25 defines the systemability of complex systems and shows how it is affected by the uncertainty of operating environments. A simple systemability function approximation using the Taylor series expansion is presented. Numerical examples are given to illustrate and compare the results of reliability and systemability functions for systems with loglog distribution.

Chapter 26 presents a review of the physics-of-failure (PoF) reliability modeling approach via a step-by-step discussion on how this methodology is applied. The key advantage of the PoF approach is that the root causes of individual failure mechanisms are studied and corrected to achieve the desired lifetimes. Various applications of PoF are also discussed.

Chapter 27 considers the effect of variance in field environmental conditions on the demonstrated reliability. In many cases the acceleration factor for a reliability demonstration test is calculated based on a high percentile field stress level, corresponding to severe user conditions. This chapter presents a mathematical approach to calculating the "true" field reliability, which in those cases will be higher than that demonstrated by the test.

We hope that this book will appeal to a wide range of reliability practitioners including engineers, scientists, researchers, statisticians and postgraduate students.

Ilia B. Frenkel
Alex Karagrigoriou
Anatoly Lisnianski
Andre Kleyner

Acknowledgements

We would like to thank all the contributing authors for sharing their technical expertise, their engineering foresight and their hard work on this book.

We would also like to thank the following individuals for their invaluable help in reviewing the book chapters and their timely feedback and constructive suggestions:

Professor Vilijandas Bagdonavičius (Vilnius University, Lithuania), Professor Maxim Finkelstein (University of the Free State, South Africa), Professor Ilya B. Gertsbakh (Ben-Gurion University, Israel), Professor Gregory Levitin (The Israel Electric Corporation Ltd., Israel), Professor Ying Ni (Malardalen University, Sweden), Professor Dmitrii Silvestrov (Stockholm University, Sweden), Dr Lina Teper (Rafael, Israel), Professor Zhigang (Will) Tian (Concordia University, Canada), Professor Marian Grendar (Slovak Academy of Sciences, Slovakia) and Professor Ilia Vonta (National Technical University of Athens, Greece).

We would like also to thank the SCE–Shamoon College of Engineering (Israel), and its president, Professor Jehuda Haddad, and the SCE Industrial Engineering and Management Department and its dean, Professor Zohar Laslo, for their help and continuous support at all stages of the work.

It was indeed our pleasure to work with Laura Bell, Assistant Editor at John Wiley & Sons.

Acknowledgements

Part One

Degradation Analysis, Multi-State and Continuous-State System Reliability

Part One

1

Methods of Solutions of Inhomogeneous Continuous Time Markov Chains for Degradation Process Modeling

Yan-Fu Li,[1] Enrico Zio[1,2] and Yan-Hui Lin[1]

[1]*European Foundation for New Energy-Electricité de France, Ecole Centrale, France*
[2]*Politecnico di Milano, Italy*

1.1 Introduction

Degradation process modeling has been an active field of research in reliability engineering for some time (Barata *et al.* 2002; Black *et al.* 2005; Hosseini *et al.* 2000; Li and Pham 2005; Chryssaphinou *et al.* 2011). Multi-state modeling (MSM) (Lisnianski and Levitin 2003; Ding *et al.* 2009; Lisnianski *et al.* 2010) is often applied in degradation process modeling, because it offers the possibility of describing the degradation state through a number of consecutive levels from perfect working to complete failure. To model the dynamics of such multi-state degradation process, Markov models have often been used (Black *et al.* 2005; Kim and Makis 2009; Chryssaphinou *et al.* 2011). In doing this, it is typically assumed that the rates of transition among the degradation states are constant, which implies that the degradation process is memoryless. The resulting stochastic process is called a homogeneous continuous time Markov chain (HCTMC). In many realistic situations, e.g. cracking of nuclear component (Unwin *et al.* 2011), battery aging (Cloth *et al.* 2007), cancer patients' life quality (Liu and Kapur 2008), etc., with varying

Applied Reliability Engineering and Risk Analysis: Probabilistic Models and Statistical Inference, First Edition.
Edited by Ilia B. Frenkel, Alex Karagrigoriou, Anatoly Lisnianski and Andre Kleyner.

external factors influencing the degradation processes, the transition rates can no longer be considered as time-independent. Under these circumstances, the inhomogeneous CTMC (ICTMC) is more suited to modeling the degradation process.

One of the drawbacks of ICTMC is that its closed-form solution is difficult, if not impossible, to obtain. Therefore, a number of numerical solution techniques have been proposed. Four representative methods are: numerical solver of differential equations (Telek *et al.* 2004) (e.g. the Runge–Kutta method), uniformization (van Moorsel and Wolter 1998; Arns *et al.* 2010), Monte Carlo simulation (Lewis and Tu 1986), and state-space enrichment (Unwin *et al.* 2011). To the best of our knowledge, very few studies have been carried out to analyze the performances of these numerical approaches with the special application background of degradation process modeling.

This work aims to carry out a comparative study of the above-mentioned methods, from the point of view of accuracy and efficiency. The rest of the chapter is organized as follows. Section 1.2 introduces the formalism of the ICTMC model. Section 1.3 briefly introduces the numerical techniques for its solution. Section 1.4 presents empirical comparisons with reference to two degradation case studies. Section 1.5 analyzes the comparison results and provides guidelines for the use of the methods. Section 1.6 draws the conclusions of the study.

1.2 Formalism of ICTMC

Let $\{X(t), t \geq 0\}$ be a Markov process on a finite state space $S = \{0, 1, \ldots, M\}$. The primary quantity of interest in many applications of CTMC is the state probability vector at any time instant $t, \boldsymbol{p}(t) = \{p_0(t), \ldots, p_M(t)\}$. By the definition of probability, we have $\sum_{i=0}^{M} p_i(t) = 1, \forall t$.

In the case of HCTMC, $\boldsymbol{p}(t)$ is typically found by solving the following system of differential equations:

$$\frac{d}{dt} p_i(t) = \sum_{\substack{k=0 \\ k \neq i}}^{M} p_k(t)\lambda_{ki} - p_i(t) \sum_{\substack{j=0 \\ j \neq i}}^{M} \lambda_{ij} \tag{1.1}$$

where i is the state index ranging from 0 to M, λ_{ij} is the rate which characterizes the stochastic transition from state i to state j. The transition rate λ_{ij} is defined as:

$$\lambda_{ij} = \lim_{\Delta t \to 0} \frac{\Pr(X(t + \Delta t) = j | X(t) = i)}{\Delta t} \tag{1.2}$$

In the case of ICTMC, the transition rate λ_{ij} is dependent on time t: $\lambda_{ij}(t)$. Due to the time dependency, it is in general difficult to obtain the closed-form solutions to the ICTMC differential equations:

$$\frac{d}{dt} p_i(t) = \sum_{\substack{k=0 \\ k \neq i}}^{M} p_k(t)\lambda_{ki}(t) - p_i(t) \sum_{\substack{j=0 \\ j \neq i}}^{M} \lambda_{ij}(t) \tag{1.3}$$

To obtain the state probability vector $\boldsymbol{p}(t)$, Equation (1.3) has to be solved by numerical methods (Arns *et al.* 2010).

1.3 Numerical Solution Techniques

1.3.1 The Runge–Kutta Method

Runge–Kutta methods (Butcher 2008) are an important family of iterative approximation methods used to solve the differential equations of ICTMC. Let $\boldsymbol{Q}(t)$ denote the transition matrix of ICTMC, Equation (1.3) can be rewritten as:

$$\frac{d}{dt}\boldsymbol{p}(t) = \boldsymbol{p}(t)\boldsymbol{Q}(t) = \boldsymbol{f}(t, p) \tag{1.4}$$

The main idea of the Runge–Kutta methods is to compute $\boldsymbol{p}(t + \Delta t)$ by adding to $\boldsymbol{p}(t)$ the product of the weighted sum of S derivatives at different locations within the time interval $(t, t + \Delta t)$. Mathematically, $\boldsymbol{p}(t + \Delta t)$ can be expressed as follows:

$$\boldsymbol{p}(t + \Delta t) = \boldsymbol{p}(t) + h \cdot \sum_{i=1}^{s} b_i \cdot \boldsymbol{k}_i \tag{1.5}$$

$$\boldsymbol{k}_i = \boldsymbol{f}\left(t + c_i \cdot h, \boldsymbol{p}(t) + h \cdot \sum_{j=1}^{s} a_{ij} \cdot \boldsymbol{k}_j\right) \tag{1.6}$$

where $\boldsymbol{f}$ is the first-order derivative of $\boldsymbol{p}(t)$ at $(t + c_i \cdot \Delta t, \boldsymbol{p}(t) + h \cdot \sum_{j=1}^{s} a_{ij} \cdot \boldsymbol{k}_j)$, $h = \Delta t$, and a_{ij}, b_i and c_i are the coefficients which are usually arranged in a Butcher tableau (Butcher 2003):

$$\begin{array}{c|cccc} c_1 & a_{11} & \dots & a_{1s} \\ c_2 & a_{21} & \dots & a_{2s} \\ \vdots & \vdots & \ddots & \vdots \\ c_s & a_{s1} & \dots & a_{ss} \\ \hline & b_1 & \dots & b_s \end{array}$$

The Runge–Kutta method is consistent if $\sum_{j=1}^{s} a_{ij} = c_i$. It is also noted that the Runge–Kutta method is explicit if the Butcher tableau is lower triangular, while if the Butcher tableau is not necessarily lower triangular, then the Runge–Kutta method is implicit, which is more general than the explicit case (Butcher 2003).

The coefficients in Butcher tableau are chosen to match as many as possible of the terms in the Taylor series:

$$\boldsymbol{p}(t + \Delta t) = \boldsymbol{p}(t) + \Delta t \cdot \boldsymbol{p}^{(1)}(t) + \frac{\Delta t^2}{2!} \cdot \boldsymbol{p}^{(2)}(t) + \frac{\Delta t^3}{3!} \cdot \boldsymbol{p}^{(3)}(t) + \dots + \frac{\Delta t^s}{s!} \cdot \boldsymbol{p}^{(s)}(t) + O(\Delta t^{s+1}) \tag{1.7}$$

in order to minimize the approximation error. The vector quantity $\boldsymbol{p}^{(i)}$ can be expressed by $\boldsymbol{f} = \boldsymbol{p}^{(1)}$ and its derivatives, for example, $\boldsymbol{p}^{(2)} = \frac{\partial f}{\partial t} + \boldsymbol{f} \cdot \frac{\partial f}{\partial p}$ and $\boldsymbol{p}^{(3)} = \frac{\partial^2 f}{\partial t^2} + 2 \cdot \boldsymbol{f} \cdot \frac{\partial^2 f}{\partial t \partial p} + \boldsymbol{f}^2 \cdot \frac{\partial^2 f}{\partial p^2} + \frac{\partial f}{\partial p} \cdot \left(\frac{\partial f}{\partial t} + \boldsymbol{f} \cdot \frac{\partial f}{\partial p}\right)$. On the other hand, $\boldsymbol{k}_i$ can also be expressed by $\boldsymbol{f}$ and its derivatives using the Taylor series: $\boldsymbol{k}_i = \boldsymbol{f}(t, \boldsymbol{p}(t)) + c_i \cdot h \cdot \frac{\partial f}{\partial t}(t, \boldsymbol{p}(t)) + \sum_{j=1}^{s} a_{ij} \cdot \boldsymbol{k}_j \cdot h \cdot \frac{\partial f}{\partial p}(t, \boldsymbol{p}(t)) + O(h^3)$. The coefficients in the Butcher tableau can be obtained by setting the right-hand side of Equation (1.5) equal to the Taylor series of $\boldsymbol{p}(t+\Delta t)$ in Equation (1.7). For example, a general form of an explicit two-stage Runge–Kutta method is:

$$\boldsymbol{p}(t+\Delta t)=\boldsymbol{p}(t)+h\cdot\sum_{i=1}^{2}b_i\cdot\boldsymbol{k}_i$$

$$\boldsymbol{k}_1=\boldsymbol{f}(t,\boldsymbol{p}(t))$$

$$\boldsymbol{k}_2=\boldsymbol{f}(t+c_2\cdot h,\boldsymbol{p}(t)+h\cdot a_{21}\cdot\boldsymbol{k}_1) \tag{1.8}$$

where the coefficients are in a lower triangular Butcher tableau. By Taylor expansion, we obtain $\boldsymbol{k}_2=\boldsymbol{f}(t,\boldsymbol{p}(t))+c_2\cdot h\cdot\frac{\partial\boldsymbol{f}}{\partial t}(t,\boldsymbol{p}(t))+a_{21}\cdot h\cdot\boldsymbol{k}_1\cdot\frac{\partial\boldsymbol{f}}{\partial\boldsymbol{p}}(t,\boldsymbol{p}(t))$ and $\boldsymbol{p}(t+\Delta t)=\boldsymbol{p}(t)+(b_1+b_2)\cdot h\cdot\boldsymbol{f}(t,\boldsymbol{p}(t))+\left[c_2\cdot b_2\cdot\frac{\partial\boldsymbol{f}}{\partial t}(t,\boldsymbol{p}(t))+a_{21}\cdot b_2\cdot\boldsymbol{f}(t,\boldsymbol{p}(t))\cdot\frac{\partial\boldsymbol{f}}{\partial\boldsymbol{p}}(t,\boldsymbol{p}(t))\right]\cdot h^2+O(h^3)$. Therefore, $b_1+b_2=1$, $c_2=a_{21}$ and $c_2\cdot b_2=a_{21}\cdot b_2=1/2$.

1.3.2 Uniformization

Uniformization for HCTMC has been known to be the most efficient approach to obtain the state probabilities of HCTMC since it was proposed by Jensen (Jensen 1953). In a later study, van Dijk (van Dijk 1992) was the first to formulate uniformization for ICTMC. Subsequently, van Moorsel and Wolter (van Moorsel and Wolter 1998) have proposed three numerical algorithms to realize the uniformization formulation of ICTMC. More recently, Arns *et al.* (2010) have proposed two novel variations of uniformization for ICTMC which are shown to outperform standard differential equation solvers if the transition rates change slowly.

The idea of uniformization is to represent the behaviors of CTMC by a Poisson process and an embedded discrete time Markov chain (DTMC) (Jensen 1953). Let $\boldsymbol{Q}$ denote the transition matrix of HCTMC and $\alpha\geq max_i(|\lambda_{ii}|)$, then $\wp=\frac{Q}{\alpha}+I$ is a stochastic matrix of the DTMC. Furthermore, let

$$f(\alpha t,k)=e^{-\alpha t}\cdot\frac{(\alpha t)^k}{k!} \tag{1.9}$$

denote the probability of a Poisson process with rate α to have k events in the time interval $[0,t]$. According to Peano-Baker series, the state probability vector $\boldsymbol{p}(t)$ can be rewritten as (Antsaklis and Michel 1997):

$$\boldsymbol{p}(t)=\boldsymbol{p}(0)\sum_{k=0}^{\infty}f(\alpha t,k)\cdot\wp^k \tag{1.10}$$

In the case of ICTMC, the transition matrix $\boldsymbol{Q}(t)$ is dependent on time. Suppose that there are M different time-dependent events/transitions in $\boldsymbol{Q}(t)$ and the rate of the m-th event/transition is described by the function $\beta^{(m)}(t)$. At the state level of CTMC, the m-th event leads to the transitions described by the matrix $\boldsymbol{R}^{(m)}$. Consequently, $\beta^{(m)}(t)\boldsymbol{R}^{(m)}(i,j)$ defines the overall transition rate from state i to state j at time t caused by event m. Moreover, let $R^{(m)}(i,i)=-\sum_{j\neq i}R^{(m)}(i,j)$. By an appropriate scaling of $\beta^{(m)}$, we can achieve $max_i(|R^{(m)}(i,i)|)=1$. Therefore, the matrix $\boldsymbol{Q}(t)$ can be decomposed into M

time-dependent transition rates multiplying their corresponding time-independent DTMC transition probability matrices:

$$\boldsymbol{Q}(t) = \sum_{m=1}^{M} \beta^{(m)}(t)\boldsymbol{R}^{(m)} \tag{1.11}$$

To compute the state probabilities $\boldsymbol{p}(t)$, the total time horizon T has to be discretized into K_u equal time intervals of size Δt, during which the transition rate matrix $\boldsymbol{Q}(t, t+\Delta t)$ is constant and the uniformization formulation of HCTMC can be utilized. After the discretization, starting from (0), the state probabilities at any time instant $t \le T$ can be computed by the following recursive formula:

$$\widehat{\boldsymbol{p}}(t+\Delta t) = \widehat{\boldsymbol{p}}(t)\sum_{k=0}^{\infty} f(\overline{B}(t, t+\Delta t)\Delta t, k) \cdot \left(\sum_{m=1}^{M} \frac{\overline{\beta}^{(m)}(t, t+\Delta t)}{\overline{B}(t, t+\Delta t)} \wp^{(m)}\right)^{k} \tag{1.12}$$

where $\overline{\beta}^{(m)}(t, t+\Delta t) = \frac{1}{\Delta t}\int_t^{t+\Delta t} \beta^{(m)}(t)$ is the average transition rate of event m within the interval $(t, t+\Delta t)$, $\overline{B}(t, t+\Delta t) = \frac{1}{\Delta t}\sum_{m=1}^{M}\int_t^{t+\Delta t} \beta^{(m)}(t)$ is the summation of all the averaged transition rates in $(t, t+\Delta t)$, and $\wp^{(m)} = \boldsymbol{R}^{(m)} + I$. To implement this method, two parameters have to be determined: the truncation point for the summation which can be computed from the Poisson probabilities, and the size of time interval Δt, which can be obtained by minimizing a local error estimate. More detailed descriptions of the theory and implementation of the uniformization of ICTMC can be found in (Arns *et al.* 2010).

1.3.3 Monte Carlo Simulation

In the Monte Carlo (MC) simulation approach, Equation (1.3) is rewritten as (Lewis and Tu 1986):

$$\frac{d}{dt}p_i(t) = \sum_{\substack{k=0 \\ k\neq i}}^{M} p_k(t)q_{ki}(t)\lambda_k(t) - p_i(t)\lambda_i(t) \tag{1.13}$$

where $\lambda_i(t) = \sum_{\substack{j=0 \\ j\neq i}}^{M} \lambda_{ij}(t)$ and $q_{ki}(t) = \lambda_{ki}(t)/\lambda_k(t)$. The quantity $q_{ji}(t)$ is regarded as the conditional probability that, given the transition out of state j at time t, the transition arrival state will be i. To rewrite Equation (1.13) into integral form, an integrating factor $M_i(t) = \exp\left[\int_0^t \lambda_i(t')\,dt'\right]$ is used. Multiplying both sides of Equation (1.13) by the integrating factor, we obtain:

$$\frac{d}{dt}[p_i(t)M_i(t)] = M_i(t)\sum_{\substack{k=0 \\ k\neq i}}^{M} p_k(t)q_{ki}(t)\lambda_k(t) \tag{1.14}$$

Taking the integral of both sides, we obtain:

$$p_i(t)M_i(t) = p_i(0) + \int_0^t \left[M_i(t')\sum_{\substack{k=0 \\ k\neq i}}^{M} p_k(t')q_{ki}(t')\lambda_k(t') \right] dt'$$

$$\equiv p_i(t) = \exp\left[-\int_0^t \lambda_i\left(t'\right)dt'\right]\left\{p_i\left(0\right) + \int_0^t \left[M_i\left(t'\right)\sum_{\substack{k=0\\k\neq i}}^{M} p_k(t')q_{ki}(t')\lambda_k(t')\right]dt'\right\}$$

$$\equiv p_i(t) = p_i(0)\exp\left[-\int_0^t \lambda_i\left(t'\right)dt'\right]$$

$$+\int_0^t \exp\left[-\int_{t'}^t \lambda_i\left(t''\right)dt''\right]\sum_{\substack{k=0\\k\neq i}}^{M} p_k(t')q_{ki}(t')\lambda_k(t')dt' \qquad (1.15)$$

In the MC simulation of the Markov process, the probability distribution function $p_i(t)$ is not sampled directly. Instead, the process holding time at one state i is sampled and then the transition from state i to another state j is determined. This procedure is repeated until the accumulated holding time reaches the predefined time horizon. The resulting time sequence consists of the holding times at different states.

Therefore, the holding time is of interest in the MC simulation. To sample the holding time, the probability density (or total frequency) of departing state i, $\psi_i(t)$, can be obtained by multiplying $\lambda_i(t)$ to both sides of Equation (1.15):

$$\psi_i(t) = \lambda_i(t)p_i(t)$$

$$= p_i(0)\lambda_i(t)\exp\left[-\int_0^t \lambda_i\left(t'\right)dt'\right]$$

$$+\int_0^t \lambda_i(t)\exp\left[-\int_{t'}^t \lambda_i\left(t''\right)dt''\right]\sum_{\substack{k=0\\k\neq i}}^{M} \psi_k(t')q_{ki}(t')dt'$$

$$= p_i(0)f_i(t|0) + \sum_{\substack{k=0\\k\neq i}}^{M}\int_0^t \psi_k(t')q_{ki}(t')f_i(t|t')dt' \qquad (1.16)$$

where

$$f_i(t|t') = \lambda_i(t)\exp\left[-\int_{t'}^t \lambda_i\left(t''\right)dt''\right] \quad t \geq t' \qquad (1.17)$$

is defined as the conditional probability density function that the process will depart state i at time t, given that the process is at state i at time t'. Equation (1.16) indicates that the probability density function $\psi_i(t)$ consists of the sum of contributions from the random walks with transitions passing through all the states (including state i) from time 0 to t. From Equation (1.16), the MC simulation procedure mentioned above can be derived.

The cumulative probability distribution function of the holding time is obtained by integrating (1.17):

$$F_i(t|t') = 1 - \exp\left[-\int_{t'}^t \lambda_i\left(t''\right)dt''\right] \qquad (1.18)$$

Given the current time t' at state i, the holding time t can be sampled through direct inversion sampling, acceptance-rejection sampling, and other sampling techniques (Zio and

Zoia 2009). Following the departure, the sampling of the arrived state j^* can be achieved by choosing a uniformly distributed random number U and selecting the state which satisfies the following condition:

$$\sum_{k=0}^{j^*-1} q_{ik}(t) < U < \sum_{k=0}^{j^*} q_{ik}(t) \tag{1.19}$$

1.3.4 State-Space Enrichment

The state-space enrichment method has recently been applied to model the degradation process of a repairable nuclear component, where the transition rates are dependent on the state holding times (rather than the system time) (Unwin *et al.* 2011). In this method, the ICTMC $\{X(t), t \geq 0\}$ is embedded into another stochastic process $Y(t)$, which is a discrete time Markov chain (DTMC) with an enriched state space. This enriched state space can be described by a tuple $S_e = (S, t_c)$ where S is the finite set of original states and t_c is the discretized holding time at each state. Let T denote the total system time and Δt denote the interval size of discretization. Then $K_c = \frac{T}{\Delta t}$ is the total number of intervals. Let $k = \{0, 1, \ldots, K_c\}$ denote all the possible numbers of time intervals that the process could reside in at each state. S_e can be rewritten as $S_e = (S, k\Delta t)$.

The transition probability of $Y(t)$ is defined as:

$$\rho_{(i,k),(j,k')} = \Pr(X(k'\Delta t) = j | X(k\Delta t) = i) \tag{1.20}$$

which is the probability of transition to state (j, k') from the current state (i, k). In relation to the embedded ICTMC $X(t)$, $Y(t)$ has the following non-zero transition probabilities:

$$\rho_{(i,k),(j,0)} = \Pr(X((k+1)\Delta t) = j | X(k\Delta t) = i) = \lambda_{ij}(k\Delta t)\Delta t \tag{1.21}$$

which is the probability of the one-step transition from state i to a new state j, and

$$\rho_{(i,k),(i,k+1)} = \Pr(X((k+1)\Delta t) = i | X(k\Delta t) = i) = 1 - \lambda_{ii}(k\Delta t)\Delta t \tag{1.22}$$

which is the probability of the process remaining at state i. Given the definitions of the individual transition probabilities, the transition probability matrix of $Y(t)$ can be written as:

$$\boldsymbol{P}_{Y(t)} = \begin{bmatrix} \begin{bmatrix} 0 & \rho_{(0,0),(0,1)} & \cdots & 0 \\ 0 & 0 & \ddots & \vdots \\ \vdots & \vdots & \ddots & \rho_{(0,K_c-1),(0,K_c)} \\ 0 & \cdots & \cdots & 0 \end{bmatrix} & \cdots & \begin{bmatrix} \rho_{(0,0),(M,0)} & 0 & \cdots & 0 \\ \vdots & \vdots & \ddots & \vdots \\ \vdots & \vdots & \ddots & 0 \\ \rho_{(0,K_c),(M,0)} & 0 & \cdots & 0 \end{bmatrix} \\ \vdots & \ddots & \vdots \\ \begin{bmatrix} \rho_{(M,0),(0,0)} & 0 & \cdots & 0 \\ \vdots & \vdots & \ddots & \vdots \\ \vdots & \vdots & \ddots & 0 \\ \rho_{(M,K_c),(0,0)} & 0 & \cdots & 0 \end{bmatrix} & \cdots & \begin{bmatrix} 0 & \rho_{(M,0),(M,1)} & \cdots & 0 \\ 0 & 0 & \ddots & \vdots \\ \vdots & \vdots & \ddots & \rho_{(M,N_c-1),(M,K_c)} \\ 0 & \cdots & \cdots & 0 \end{bmatrix} \end{bmatrix} \tag{1.23}$$

Given that the process starts from state 0 when $t = 0$, the initial state probability vector of $Y(t)$ is defined as follows:

$$\prod_0 = \underbrace{[\underbrace{1, 0 \ldots,}_{K_c+1} \ldots, \underbrace{0, \ldots, 0]}_{K_c+1}}_{M+1} \tag{1.24}$$

By the recursive property of CTMC, we can obtain the state probability vector after k steps as follows:

$$\prod_k = \prod_0 \boldsymbol{P}_{Y(t)}^{(k)} \tag{1.25}$$

Give $\prod_k$, we can obtain the state probability of $X(t)$ as:

$$\boldsymbol{p}_i(k\Delta t) = \pi_{k,i} \tag{1.26}$$

where $\pi_{k,i}$ is the $(i \times (K_c + 1) + k)$ th element of the vector $\prod_k$.

1.4 Examples

To compare the techniques for ICTMC numerical solution, two examples of degradation processes are considered. Each numerical solution technique has been run 20 times on each example. All the experiments were carried out in MATLAB on a PC with an Inter core i5 CPU at 2.67 GHz and a RAM of 4 GB.

1.4.1 Example of Computing System Degradation

The first example is taken from (van Moorsel and Wolter 1998). It concerns a non-repairable computing system with two processors. Both of them have a time-dependent failure rate $\lambda(t) = 0.2t$, where t is the system time. The failure of each processor can lead to a system crash or a safe shutdown of the other processor. The probability of a safe shutdown is 0.6 for both processors. The transition diagram is shown in Figure 1.1.

The corresponding transition rate matrix is:

$$Q(t) = \begin{bmatrix} -0.4t & 0.24t & 0 & 0.16t \\ 0 & -0.2t & 0.12t & 0.08t \\ 0 & 0 & 0 & 0 \\ 0 & 0 & 0 & 0 \end{bmatrix}$$

The numerical solution techniques are applied to this problem. Because the state-space enrichment method was designed to handle the changing transition rate dependent on holding time, it is not applicable to this example. The parameter setting of each technique is presented as follows: in the Runge–Kutta method, the time interval size Δt is 0.1 time unit; in the uniformization method, the truncation point is 100 and the time interval size Δt is 0.1 time unit; for the MC simulation, the number of repetitions is 5000. Table 1.1 summarizes the probabilities of the safe and unsafe failure states at the time steps from 1 to 5, in one experiment run. It is shown that the results obtained by the uniformization and Runge–Kutta methods are very close, that is, the mean absolute error (MAE) between

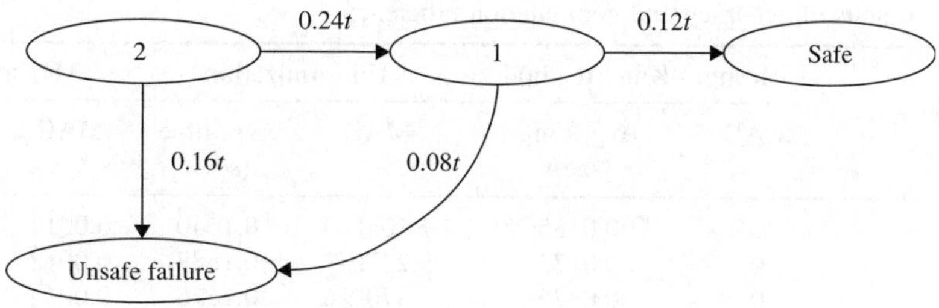

Figure 1.1 Degradation process of a two-processor computing system

Table 1.1 State probabilities obtained by the different solution techniques in one experiment run

System time	Uniformization		Runge–Kutta method		State-space enrichment		MC simulation	
	Prob (safe failure)	Prob (unsafe failure)	Prob (safe failure)	Prob (unsafe failure)	Prob (safe failure)	Prob (unsafe failure)	Prob (safe failure)	Prob (unsafe failure)
1	3.260E-3	0.0747	3.260E-3	0.0747	N.A.	N.A.	3.000E-3	0.0751
2	0.0391	0.2464	0.0391	0.2464	N.A.	N.A.	0.0384	0.2470
3	0.1268	0.4184	0.1268	0.4184	N.A.	N.A.	0.1256	0.4203
4	0.2293	0.5366	0.2293	0.5366	N.A.	N.A.	0.2277	0.5376
5	0.3033	0.5995	0.3033	0.5995	N.A.	N.A.	0.3011	0.6027

them is 1.1035E-6, whereas the results of MC simulation are relatively distant from those of the Runge–Kutta method, with the MAE between them equal to 0.0013.

To investigate the impacts of different variation speeds of transition rates on the techniques' accuracies and efficiencies, we have considered four additional examples in which the transition rates are 2, 4, 8, and 16 times those of the original case, respectively. Table 1.2 summarizes the results, in terms of MAE with reference to the Runge–Kutta method and of average computation time. The results confirm the finding of Table 1.1, that uniformization is the closest to the Runge–Kutta method, this latter being the most efficient followed by uniformization and MC simulation. It is also seen that the accuracy of uniformization might deteriorate if the variation speed of transition rate increases, whereas that of MC simulation does not exhibit any significant tendency. As to the average computation time, the Runge–Kutta method and MC simulation show small-sized steady increases when the variation speeds of the transition rates grow, while uniformization does not have a clear trend in this respect.

1.4.2 Example of Nuclear Component Degradation

The second case study refers to the cracking process in an Alloy 82/182 dissimilar metal weld in a primary coolant system of a nuclear power plant (Unwin *et al.* 2011). Cracks can grow from the inner to the outer diameter of the dissimilar metal welds in one of

Table 1.2 Results of accuracy and computation efficiency

Multiplier of transition rates values	Runge–Kutta method		Uniformization		MC simulation	
	MAE	Avg. time (sec.)	MAE	Avg. time (sec.)	MAE	Avg. time (sec.)
1	0	0.0155	1.104E-6	0.1640	0.0014	0.5534
2	0	0.0173	9.237E-7	0.1687	0.0018	0.5610
4	0	0.0175	1.470E-6	0.1676	0.0015	0.6352
8	0	0.0183	1.749E-6	0.1575	0.0014	0.6553
16	0	0.0188	2.053E-6	0.1634	0.0016	0.6825

the three major morphologies: axial, radial, and circumferential. The latter two types can lead to the rupture of the component. The radial crack mainly grows outward from the initiation site towards the outer diameter; the process can lead to a leak and potentially to rupture. The circumferential crack grows relatively evenly around the circumference, potentially leading to a rupture.

The multi-state degradation model proposed in (Unwin *et al.* 2011; Li *et al.* 2012) to describe the crack growth in the case study of interest, is represented in Figure 1.2. The transition rates φ_1, φ_2, φ_3 and φ_4 are dependent on the state holding time, whereas the other transition rates are taken as constants. Detailed information on these parameters can be found in (Li *et al.* 2012).

It is noted that uniformization and Runge–Kutta methods are not applicable to this example, as they are not able to handle the transition rates dependent on the state holding times. The parameter setting of each applicable technique is presented as follows: in the state-enrichment method, the time step size Δt is 1, 0.5, and 0.1 year, respectively; for the MC simulation, the number of repetitions is 50,000. The numerical comparisons on the state probability values at year 80 are reported in Table 1.3. The differences between the simulation and state-space enrichment methods decrease as the step size is reduced, the MAEs of the three comparisons being 1.843E-4, 1.073E-4, and

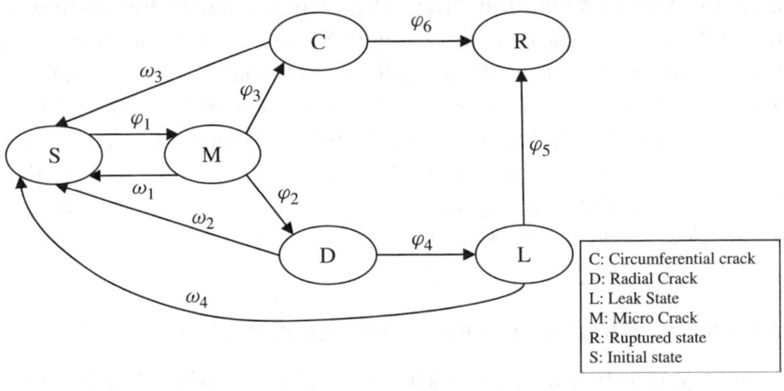

Figure 1.2 Transition diagram of the multi-state physics model of crack development in Alloy 82/182 dissimilar metal welds. *Source*: (Li *et al.* 2012).

Table 1.3 State probabilities at year 80 obtained by the different solution techniques in one experiment run

	Simulation	State-space enrichment Step size = 1 year	State-space enrichment Step size = 0.5 year	State-space enrichment Step size = 0.1 year
Initial state probability	3.656E-3	3.274E-3	3.424E-3	3.559E-3
Micro crack probability	9.958E-1	9.963E-1	9.961E-1	9.959E-1
Circumferential crack probability	2.860E-4	1.943E-4	2.333E-4	2.767E-4
Radial crack probability	7.200E-5	6.377E-5	6.974E-5	7.658E-5
Leak probability	1.000E-5	8.928E-6	1.060E-5	1.249E-5
Rupture state probability	2.080E-4	1.379E-4	1.727E-4	2.116E-4

Table 1.4 Results of accuracy and computation efficiency

Multiplier of transition rates values	State-space enrichment Step size = 0.1 year		MC simulation	
	MAE	Avg. time (sec)	MAE	Avg. time (sec)
1	0	19.80	3.299E-5	216.9
2	0	19.69	3.925E-5	238.2
4	0	19.95	6.247E-5	267.5
8	0	20.42	1.155E-4	352.6
16	0	20.92	2.643E-4	561.7

3.550E-5, respectively. This confirms that the state-space enrichment method is sensitive to the step size, which must then be properly set *a priori*.

Similarly to Section 1.4.1, four additional examples have been considered for testing, with transition rates 2, 4, 8, and 16 times of those of the original case. Table 1.4 presents the testing results. It is seen that MC simulation is generally slower than the state-space enrichment method with step size = 0.1 year, with the difference between their state probability estimations increasing as the transition rates increase. However, it should be noted that the memory requirement of the state-space enrichment is much higher than that of MC simulation.

1.5 Comparisons of the Methods and Guidelines of Utilization

This section summarizes the quantitative results from the two case studies and includes three qualitative attributes for the comparisons of the numerical solution techniques. Table 1.5 presents the comparisons. From the descriptions in Section 1.2, we can observe that Runge–Kutta, uniformization, and state-space enrichment involve the discretization of the time domain. This leads to the need to choose the time step size Δt. Further, uniformization uses infinite series to expand the differentiations so that the truncation point also has to be determined for the computation purposes. MC simulation does not require

Table 1.5 Qualitative comparisons of the four numerical solution techniques

Name of solution technique	Runge–Kutta method	Uniformization	Monte-Carlo simulation	State-space enrichment
Discretization	Yes	Yes	No	Yes
Parameters	Time interval size Δt	Truncation point, Time interval size Δt	Number of replications	Time step size Δt
Transition rate time dependency	System time	System time	Both	Process state holding time
Accuracy	High	High	Medium	Medium
Efficiency	High	High	Low	Low
Memory consumption	Low	Low	Low	High
Influence of transition rate variations on efficiency	Low	Low	Medium	Low

the discretization of time but it requires a number of replications to reduce the estimation variations to the desired level of accuracy.

The two numerical examples in Section 1.4 illustrate two different types of time-dependencies of the transition/degradation rates: dependency on the system time and on the process holding time at a certain state. These two types of time-dependency can describe degradation processes with and without the repairs, respectively. In this sense, MC simulation is more flexible than the other methods, because it is the only method capable of dealing with the two types of transition rates. Table 1.5 also shows that the Runge–Kutta method and uniformization are more accurate and efficient, and less demanding, than other methods when they are applicable. They are also less sensitive to transition rate variations. State-space enrichment is specialized to cope with transition rates dependent on process state holding time.

Given the above evaluations of the four numerical solution techniques considered, the following guidelines for utilization can be helpful:

- When the transition/degradation rates are dependent on system time only, Runge–Kutta and uniformization are good options.
- When the transition/degradation rates are dependent on process holding time at each state, MC simulation and state-space enrichment are good options.
  - When the number of original states is small, state-space enrichment is preferred.
  - When the number of original states is large, MC simulation is preferred.

- When some transition/degradation rates are dependent on system time and some transition/degradation rates are dependent on process holding time at each state, MC simulation is the only option.

1.6 Conclusion

This chapter introduces four numerical solution approaches to ICTMC for degradation process modeling and compares them qualitatively and quantitatively on two case studies. The main findings are: Runge–Kutta and uniformization are more accurate and efficient, less demanding of memory and less sensitive to transition rate variations, than the other methods when the transition rates are only dependent on system time; state-space enrichment is specialized to cope with transition rates dependent on process state holding time; MC simulation is the only method capable of dealing with the two types of transition rates dependence. As a result of the analyses, a number of guidelines are provided for the utilization of these techniques, in the various problem settings.

References

Antsaklis, P.J. and A.N. Michel. 1997. *Linear Systems*. New York: McGraw-Hill.

Arns, M., P. Buchholz, and A. Panchenko. 2010. On the numerical analysis of inhomogeneous continuous-time Markov chains. *INFORMS Journal on Computing* 22 (3): 416–432.

Barata, J., C. Guedes Soares, M. Marseguerra, and E. Zio. 2002. Simulation modelling of repairable multi-component deteriorating systems for 'on condition' maintenance optimisation. *Reliability Engineering & System Safety* 76: 255–264.

Black, M., A.T. Brint, and J.R. Brailsford. 2005. A semi-Markov approach for modelling asset deterioration. *Journal of the Operational Research Society* 56: 1241–1249.

Butcher, J.C. 2008. *Numerical Methods for Ordinary Differential Equations*. Chichester: John Wiley & Sons, Ltd.

Chryssaphinou, O., N. Limnios, and S. Malefaki. 2011. Multi-state reliability systems under discrete time semi-Markovian hypothesis. *IEEE Transactions on Reliability* 60 (1): 80–87.

Cloth, L., M.R. Jongerden, and B.R. Haverkort. 2007. Computing battery lifetime distributions. In *Proceedings of the 37th Annual IEEE/IFIP International Conference on Dependable Systems and Networks, 2007 (DSN '07)*. IEEE Computer Society Press, pp. 780–789.

Hosseini, M.M., R.M. Kerr, and R.B. Randall. 2000. An inspection model with minimal and major maintenance for a system with deterioration and Poisson failures. *IEEE Transactions on Reliability* 49 (1): 88–98.

Jensen, A. 1953. Markoff chains as an aid in the study of Markoff processes. *Scandinavian Actuarial Journal* 36: 87–91.

Kim, J. and V. Makis. 2009. Optimal maintenance policy for a multi-state deteriorating system with two types of failures under general repair. *Computers and Industrial Engineering* 57: 298–303.

Lewis, E.E. and Z.G. Tu. 1986. Monte Carlo reliability modeling by inhomogeneous Markov processes. *Reliability Engineering* 16: 277–296.

Li, W J. and H. Pham. 2005. Reliability modeling of multi-state degraded systems with multi-competing failures and random shocks. *IEEE Transactions on Reliability* 54 (2): 297–303.

Li, Y.F., E. Zio, and Y.H. Lin. 2012. A multistate physics model of component degradation based on stochastic petri nets and simulation. *IEEE Transactions on Reliability* 61 (4): 921–931.

Lisnianski, A., I. Frenkel and Y. Ding. 2010. *Multi-state System Reliability Analysis and Optimization for Engineers and Industrial Managers*. London: Springer.

Lisnianski, A. and G. Levitin. 2003. *Multi-state System Reliability: Assessment, Optimization and Applications*. Singapore: World Scientific.

Liu, Y. and K.C. Kapur. 2008. New model and measurement for reliability of multi-state systems. In *Handbook of Performability Engineering*, ed. K.B. Misra. London: Springer, pp. 431–445.

Telek, M., A. Horváth, and G. Horváth. 2004. Analysis of inhomogeneous Markov reward models. *Linear Algebra and Its Applications* 386: 383–405.

Unwin, S.D., P.P. Lowry, R.F. Layton, P.G. Heasler, and M.B. Toloczko. 2011. Multi-state physics models of aging passive components in probabilistic risk assessment. In *Proceedings of International Topical Meeting on Probabilistic Safety Assessment and Analysis* (PSA 2011), 1. Amercian Nuclear Society, La Grange Park, IL, pp. 161–172.

Van Dijk, N.M. 1992. Uniformization for nonhomogeneous Markov chains. *Operations Research Letters* 12: 283–291.

Van Moorsel, A. P. A. and K. Wolter. 1998. Numerical solution of non-homogeneous Markov processes through uniformization. In *12th European Simulation Multiconference. Simulation – Past, Present and Future*. Saarbrücken, Germany: SCS Europe, pp. 710–717.

Zio, E. and A. Zoia. 2009. Parameter identification in degradation modeling by reversible-jump Markov chain Monte Carlo. *IEEE Transactions on Reliability* 58 (1): 123–131.

2

Multistate Degradation and Condition Monitoring for Devices with Multiple Independent Failure Modes

Ramin Moghaddass and Ming J. Zuo
Department of Mechanical Engineering, University of Alberta, USA

2.1 Introduction

The reliability analysis of multistate systems has attracted considerable research interest in the past decade and numerous analytical models to evaluate the reliability of such systems have been developed. A device with multistate health levels may perform at different intermediate health states between working perfectly and complete failure. With respect to the failure mechanisms associated with a device, the reliability of mechanical devices can be studied in two modes: (1) single failure mode; and (2) multiple failure modes. The degradation mechanism associated with a failure mode may lead to different levels of operational performances and, thus, a multistate structure. Different types of failure modes may exist in a single device. Examples are inner race spall, outer race spall, and cage crack in a rolling element bearing (Sikorska *et al.* 2011). Under the existence of multiple failure mechanisms, a particular combination of physical, electrical, chemical, and mechanical stresses results in multiple types of failures (Mathew *et al.* 2012).

Multiple failure mode reliability analysis has also been referred to as competing risk analysis, particularly in medical research (Andersen *et al.* 2002; Andersen and Keiding 2002, 2012; Putter *et al.* 2007; Beyersmann *et al.* 2012). A review of the connections between competing risk analysis and engineering reliability was carried out by (Ma and Krings 2008). Despite many advances in reliability analysis, the efficient reliability

Applied Reliability Engineering and Risk Analysis: Probabilistic Models and Statistical Inference, First Edition.
Edited by Ilia B. Frenkel, Alex Karagrigoriou, Anatoly Lisnianski and Andre Kleyner.

estimation of systems with multiple failure modes is still an important challenge (Bichon *et al.* 2011). One of the methodologies to address multiple failure modes is to use failure modes, mechanisms, and effects analysis (FMMEA), which aims to identify potential failure mechanisms and models for all potential failures modes, as well as to prioritize failure mechanisms. Modeling failure mechanisms helps in the design and development of a new product (Mathew *et al.* 2008).

As most mechanical devices are subject to multiple failure modes, which may be only indirectly observable through condition monitoring (CM), it is important to develop diagnostic and prognostic methods, which can employ multi-dimensional condition monitoring features to evaluate the health condition of the device and predict future failures of the device, while considering all possible failure modes. Examples of multiple failure modes indirectly observable by condition monitoring are pitting and cracks that occur independently in a gearbox system (Moghaddass and Zuo 2012b). Ignoring multiple failure modes may result in overestimating the device's reliability. Surprisingly, most reported work on multistate degradation models has considered only one degradation process and one failure mechanism at a time. This chapter aims to discuss the challenges of considering multiple independent failure modes for a single device under condition monitoring, and provide details on how to find important reliability measures for such a device. In addition, how to estimate the characteristic parameters associated with the multistate degradation structure of each failure mode is illustrated.

Generally, two approaches are reported in the literature regarding the reliability analysis of devices with multiple failure modes. The most common and conventional approach is to consider multiple-state levels corresponding to the overall health status of the device, irrespective of different failure modes. For example, the overall health states of the devices can be divided into normal, damage level 1, damage level 2, ..., and failure. The main shortcoming of considering the overall health states of the device as opposed to considering all the failure modes is that finding the optimum number of health states and the interpretation of the physical meaning of overall health states are challenging. In addition, the fitted multistate model may be subject to the risk of overestimating or underestimating the reliability.

The second approach, which is more common for devices under condition monitoring, is to consider a single failure mode at a time, usually the dominant failure mode, which is the failure mode most likely to be responsible for the functional failure of a device (Bello 2006). Although finding the dominant failure mode is still difficult, one of the main challenges of considering one failure mode at a time is that the set of condition monitoring features selected to evaluate the degradation level of a failure mode may not perform well when the device is subject to multiple failure modes. For example, if a feature such as Root Mean Square (RMS) is selected as the best feature for monitoring a failure mode, such as a crack in a gearbox system, it may not be a valuable feature for monitoring damage levels when another type of failure mode, such as wear, exists (Moghaddass and Zuo 2012b). Under laboratory experiments, very often only single failure modes are considered and the effects of multiple simultaneous failure modes are not analyzed. The remaining challenge is that considering only one failure mode may result in overestimating the reliability function and, therefore, may cause more unexpected catastrophic failures.

Many examples of failure modes for different types of systems are reported in the literature. A reliability-based robust design method for a structural system with multiple independent failure modes is proposed in (Lu and Zhang 2011). Examples of standard failure mechanisms in avionics systems, such as corrosion, fatigue, and electro-migration, are reported in (Xu and Xu 2011). In all reported works, two types of failure modes are considered: (1) dependent failure modes; and (2) independent failure modes. In dependent failure mode analyses, the failure or the degradation of the physical property of one failure mode may affect the failure or degradation of physical properties of other modes (Mohammadian-Masooly 2009). The main challenge in considering dependent failure modes is the complexities associated with mathematically defining the dependency relationships between failure modes. Several research studies have been carried out to analyze dependent failure modes. Examples can be found in (Bagdonavičius *et al.* 2004; Peng *et al.* 2010; Tan and Raghavan 2010; Chen *et al.* 2011; Liang *et al.* 2011; Wang 2011; Chen and Chen 2010; Wang and Pham 2012).

This chapter focuses only on independent failure modes, in which the damage progression of a failure mode has no effect on the damage progression of other failure modes. Independent failure modes have been studied in many published research papers. In (Li and Pham 2005b), models for evaluating the reliability of multistate degraded systems subject to multiple independent competing failure processes, such as degradation and random shocks, are presented. It was assumed that, similar to a series system, any of the failure modes would cause the system to fail. Some applications are also reported. In (Carr and Wang 2012), individual stochastic filters are used to predict the device's residual life under the influence of each potential failure mode with binary states. Their results indicate that modeling processes that consider multiple failure modes can greatly improve residual life prediction capabilities (compared to the general model with no failure mode considerations). An extension of the reliability analysis of electronic devices with multiple independent competing failure modes involving performance aging degradation and important reliability measures (failure rate, reliability, mean time to failure) is presented in (Huang and Askin 2003). In (Li and Pham 2005a), a generalized condition-based maintenance model subject to multiple independent competing failure processes, including two degradation processes and random shocks, is developed.

To our knowledge, no research work has been carried out on the parameter estimation and time-dependent reliability analysis of a condition-monitored device with multiple multistate independent failure modes. The aim of this work is to propose possible approaches to deal with such a device and introduce details of the calculation of important reliability measures. This chapter is organized as follows: Section 2.2 presents the general model used for a device with multiple competing failure modes. The problem of parameter estimation is briefly discussed in Section 2.3. Important reliability measures are introduced in Section 2.4. A simple numerical example is employed in Section 2.5 for illustration purposes. We offer a conclusion in Section 2.6 and introduce directions for future work.

2.2 Multistate Degradation and Multiple Independent Failure Modes

2.2.1 Notation

The list of notations used in this chapter is as follows:

N: The overall number of states for the device
L: Total age of the device
l: The number of independent failure modes

F^i: The ith failure mode
I^i: The ith condition monitoring indicator
X_n^i: The state of the device after the nth transition of the ith failure mode
T_n^i: The time of the nth transition of the ith failure mode
Z_t^i: The state of the ith failure mode at time t
Y_t^i: The ith condition monitoring indicator at time t
n^i: The number of health states for the ith failure mode
s_j^i: The jth state of the ith failure mode
θ^i: The set of characteristic parameters associated with the ith failure mode
$E^i = \{s_1^i, s_2^i, \ldots, s_{n^i}^i\}$: The finite state space of the ith failure mode
FS_j^i: The set of states to which a degrading transition may occur from state j of the ith failure mode
B^i: The marginal observation probability distribution of the ith failure mode
K: The number of historical (real-time) sequences of the observation process
$O^{(k)}$: The kth sequence of the observation process
$O^{i,(k)}$: The kth sequence of the observation process associated with the ith failure mode
O_p^i: The ith condition monitoring indicator at the pth CM point
Q_p^i: The state of the ith failure mode at the pth CM point
d_k: The number of observation points for the kth sequence of observations
$t_1^{(k)}, t_2^{(k)}, \ldots, t_{d_k}^{(k)}$: The observation time-points of the kth sequence of observations
$o_p^{i,(k)}$: The CM indicator value at time $t_p^{(k)}$ aassociated with the kth observation sequence and ith CM indicator
$q_p^{i,(k)}$: The actual state of the ith failure mode at time $t_p^{(k)}$ associated with the kth observation sequence
$b_j^i(k)$: The probability of observing the kth value of the ith CM indicator when the device is in state j of the ith failure mode.

2.2.2 *Assumptions*

The main assumptions made in this chapter for a condition-monitored device are as follows:

1. The device has l independent failure modes denoted by $F^1, F^2, \ldots, F^l$.
2. The ith failure mode has n^i mutually exclusive states ranging from perfect functioning (s_1^i) to complete failure ($s_{n^i}^i$). For any failure mode i, the device can degrade from its current state to one of its degraded states according to a degradation transition. Therefore, at each state of a particular failure mode, the device is subjected to multiple competing deterioration processes.
3. The degradation transition between two states of a degradation failure mode follows a general and flexible stochastic process called the nonhomogeneous continuous-time semi-Markov process (NHCTSMP). Based on this stochastic process, the transition between two states depends on the states involved in transition, the time that the state has reached, the time spent at the current state, or any combination of these factors. Readers may refer to (Moghaddass and Zuo 2012a; Moghaddass *et al.* 2013) for more details.

4. The device may also be subject to a catastrophic failure mode independent of the degradation failure modes. The transition rates of catastrophic failure mode follow a nonhomogeneous Markov process (otherwise called the homogeneous semi-Markov process). This means that the probability of a catastrophic failure may change with the total age of the device. It should be noted that those sudden failure transitions that depend on the states of a particular failure mode are included in the corresponding failure mode as a degradation transition (see assumption 2). Therefore, state-dependent catastrophic failure modes are also considered.
5. This work considers a general model similar to the one introduced in (Lisnianski and Levitin 2003) for a multistate system in the sense that the overall state of the device at time t (Z_t) is defined through a structure function (ϕ), which is a function of the states of all failure modes at time t, given as $Z_t = \varphi(Z_t^1, Z_t^2, \ldots, Z_t^l)$, where Z_t^i is the state of the device at time t with respect to the ith failure mode. For any type of structural function, the device fails when the first of all possible competing failure modes reaches a failure state. The possible states for the device can be found by calculating the Cartesian product of the states of each failure mode, considering that the device fails when a failure mode reaches its threshold.
6. Apart from the failure state, which is assumed to be self-announcing, the state of the device with respect to each failure mode is directly observable only by inspection and indirectly observable through condition monitoring. A single condition monitoring indicator is extracted at certain points to monitor each failure mode. This single indicator can be the output of a feature fusion (combination) process (Al-Azmi *et al.* 2009), which transforms a set of indicators into a single indicator.
7. The condition monitoring indicator of the ith failure mode is stochastically related to the corresponding states of that failure mode. This relationship is shown by an observation probability distribution with discrete values (B^i). In a general form, the probability that v_k^i is observed when the device is in states $j_1, j_2, \ldots, j_l$ with respect to the ith failure mode can be defined as $b^i_{j_1, j_2, \ldots, j_l}(k) = \Pr(Y_t^i = v_k^i | Z_t^1 = j_1, Z_t^2 = j_2, \ldots, Z_t^l = j_l), \forall t, i, j_1, j_2, j_l, k$. It should be noted that when condition monitoring indicators are independent, this relationship can be rewritten as $b^i_j(k) = \Pr(Y_t^i = v_k^i | Z_t^i = j), \forall t, i, j, k$.
8. The device is not repairable and regardless of the type of failure that occurs first, it is replaced after failure, and therefore, it is as good as new after a failure replacement.

2.2.3 *The Stochastic Process Model*

In this chapter, three approaches are reviewed, which can be used to model a device with multiple independent failure modes. This section is devoted to illustrating these three approaches. Our main focus in this chapter is on approach I. The first two approaches depend on the condition monitoring indicators and the third approach is a general approach, which can be used regardless of the type of the failure modes and condition monitoring indicators. The usefulness of each approach depends on the structure of the failure modes and the condition monitoring indicators used to monitor the health status of a particular device. With respect to the CM indicators used to monitor failure modes, devices can be categorized into two cases with: (1) multiple independent failure modes and independent CM indicators; and (2) multiple independent failure modes and dependent CM indicators. Approaches I and II are related to these two cases.

Approach I: When failure modes are independent, and unique independent condition monitoring indicators are used to monitor each failure mode, the device can be modeled as a system with l multistate subsystems connected in series, where each subsystem corresponds to one failure mode with a multistate structure. The condition monitoring indicators used to monitor multiple failure modes are assumed to be independent, if the following relationship holds true:

$$\Pr(Y_t^1, Y_t^2, \ldots, Y_t^l | Z_t^1, Z_t^2, \ldots, Z_t^l) = \prod_{i=1}^{l} \Pr(Y_t^i | Z_t^i), \quad \forall t \tag{2.1}$$

Under the above condition, each failure mode with its corresponding CM indicator can be considered independently of others. Finding condition monitoring indicators sensitive only to a specific failure mode is one of the challenges of using approach I. In this approach, parameter estimation of an individual failure mode can also be done based on experimental data from a device with an isolated failure mode. Figure 2.1 shows a schematic view of a device with two independent failure modes and two independent condition monitoring indicators.

All reliability results available for a series system with independent components are applicable in this approach. Further discussions of parameter estimation and important measure calculations based on this approach will be given in Section 2.4.

Approach II: When condition monitoring indicators are dependent with respect to each failure mode, failure modes should not be considered independently of others. Under such cases, all possible combinations of states should be considered in a single multistate structure. The possible combinations of states are found from the Cartesian product of the states in each failure mode, considering that the device stops operating as soon as one failure mode reaches its threshold. In Figure 2.2, the possible states of a device when considering the possible combinations of states for two failure modes are presented. The couple (i, j) is used to denote the state of the device, where i and j represent the state of the 1st and the 2nd failure modes, respectively. The failure modes considered here have the same structure as shown in Figure 2.1. It is obvious that, as the number of states $(\prod_{i=1}^{l} n_i - l + 1)$ in this approach is much larger than approach I, the reliability

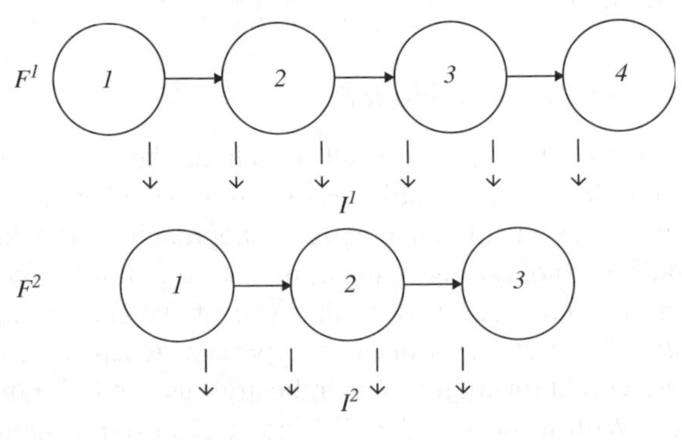

Figure 2.1 Transition diagram of a device with two independent failure modes

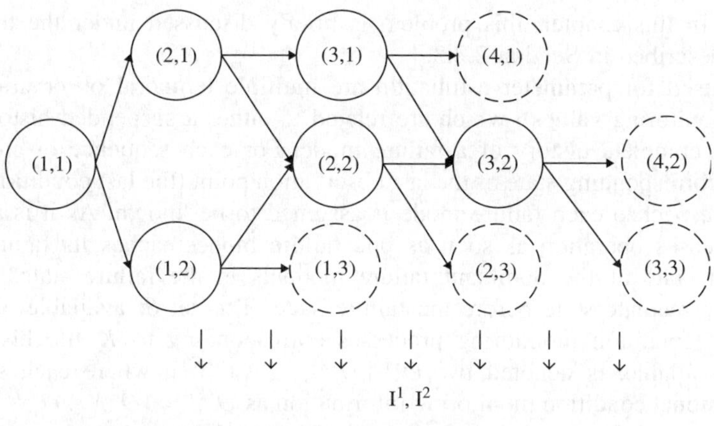

Figure 2.2 Transition diagram of a device with two dependent CM indicators

analysis becomes more complicated. Further discussions of parameter estimation and the calculation of important measures in such cases are given in Section 2.4.

Approach III: The least complicated approach to deal with a device with multiple independent failure modes is to consider multiple levels of health states for the device, regardless of the type of failure modes associated with the device. In other words, a single degradation mechanism with a multistate structure can be considered to model the overall health state of the device. Then either a single or multiple condition monitoring indicators can be used to monitor this failure mode. For example, the overall health states of the device can be divided into perfect functioning, slightly damaged, moderately damaged, severely damaged, and failure. This will avoid all the complexities of simultaneously considering the failure modes. The main disadvantage of considering a general failure mode is the difficulty of finding the best number of states to cover the overall health states of the device. The risk of using such an approach is overestimating or underestimating the reliability of the device. Further discussions of parameter estimation and the calculation of important measures of such cases are given in Sections 2.3 and 2.4. It is important to note that depending on the severity, occurrence, and detection criteria of the possible failure modes, only the dominant failure mode may be considered in reliability analysis. Therefore, approach I can be employed to find the dominant failure mode and then approach III can be employed for further reliability analysis.

2.3 Parameter Estimation

This chapter focuses on a device for which health states are not directly observable and instead condition monitoring indicators providing indirect information regarding the actual health status of the device are used for the health monitoring of multiple failure modes. To be able to employ condition monitoring data for diagnostic and prognostic reliability analysis, the structure of the failure modes and the corresponding observation processes should be fully characterized. The parameter estimation problem deals with estimating the characteristic parameters of the degradation and observation processes using historical data. Surprisingly, no work has been reported for the parameter estimation of a condition-monitored device with multiple independent failure modes and multistate degradation. In this chapter, this problem is briefly discussed under the three possible approaches described in Section 2.2.3.

The data used for parameter estimation are multiple temporal observation processes (condition monitoring values), which are related to either a suspended history (censored history) or a complete history of a failure mode. For each sequence of an observation process, the corresponding state of the last observation point (the last condition monitoring point) with respect to each failure mode is assumed to be known. As it is assumed that the device ceases operation as soon as one failure mode reaches its failure threshold, only the last state of the dominant failure mode is in the failure state, while others are in an intermediate state before the failure state. The set of available data, when K sequences of condition monitoring processes corresponding to K life histories of the device are available, is denoted by $[O^{(1)}, O^{(2)}, \ldots, O^{(K)}]$, where each sequence has multi-dimensional condition monitoring information as $O^{(k)} = [O^{1,(k)}, O^{2,(k)}, \ldots]$. Here, the number of condition monitoring indicators may vary. For each sequence $O^{(k)}$, the last state of each failure mode is also available as $[q_{d_k}^{1,(k)}, q_{d_k}^{2,(k)}, \ldots, q_{d_k}^{l,(k)}]$.

The problem of parameter estimation for a device with a single failure mode and multistate degradation has been addressed previously in (Moghaddass *et al.* 2013). We will use some of the results shown in (Moghaddass *et al.* 2013) and briefly discuss the parameter estimation considering the three approaches presented in Section 2.2.3.

Approach I: As in this approach, independent condition monitoring indicators are used to monitor each failure mode, the parameter estimation method given in (Moghaddass *et al.* 2013) can be applied independently for each failure mode to estimate the corresponding characteristic parameters. It should be noted that for the ith failure mode, the observation processes associated with all the other failure modes involve an incomplete sequence (censored history) of the ith failure mode. Therefore, they can be considered as censored lifetime data in the estimation procedure of the ith failure mode. Let us assume that there are K sequences of observations corresponding to K life histories available to be used for parameter estimation. The corresponding maximum likelihood optimization problem for the estimation of the ith failure mode can be defined as:

$$L(i) = \prod_{k=1}^{K} \Pr(O_1^i = o_1^{i,(k)}, O_2^i = o_2^{i,(k)}, \ldots, O_{d_k}^i = o_{d_k}^{i,(k)}, Q_{d_k}^i = q_{d_k}^{i,(k)} | \theta^i). \quad 1 \le i \le l. \tag{2.2}$$

The set θ^{i*} is defined as the optimum set of characteristic parameters for the ith failure mode as $\theta^{i*} = \arg\max(L(i))$.

Approach II: As in this approach, a single multistate structure is used to demonstrate the relationship between failure modes, a single likelihood function is employed to estimate the characteristic parameters of the device under approach II. This likelihood function is as follows:

$$L = \prod_{k=1}^{K} \Pr\left(\prod_{i=1}^{l}\left(O_1^i = o_1^{i,(k)}, O_2^i = o_2^{i,(k)}, \ldots, O_{d_k}^i = o_1^{i,(k)}\right), Q_{d_k} = (q_{d_k}^{1,(k)}, \ldots, q_{d_k}^{l,(k)}) | \theta^i\right) \tag{2.3}$$

Equation (2.3) is based on the assumption that there are l condition monitoring indicators used to monitor all failure modes.

Approach III: Based on this approach, the overall health state of the device can be modeled as a single multistate device with a single failure mode and N states. In addition, either a single or multiple conditions monitoring indicators can be used for the health monitoring of the device. Based on this scenario, the failure state of the system is equivalent to the failure state of each failure mode. Therefore, the last states of the observation sequences are equivalent to the failure state of the device. The likelihood function can be constructed in the following form:

$$L = \prod_{k=1}^{K} \Pr\left(\prod_{i=1}^{l}\left(O_1^i = o_1^{i,(k)}, O_2^i = o_2^{i,(k)}, \ldots, O_{d_k}^i = o_{d_k}^{i,(k)}\right), Q_{d_k} = N|\theta\right). \quad (2.4)$$

It should be noted here that details on how to calculate a likelihood function in the above form are clearly illustrated in (Moghaddass *et al.* 2013).

2.4 Important Reliability Measures of a Condition-Monitored Device

This section presents important measures, which can be used for the diagnosis and prognosis of a device with multiple failure modes. These measures are conditional in the sense that available condition monitoring data are used to calculate these measures. Our focus here is on approach I (see Section 2.2), where independent condition monitoring indicators are used to monitor independent failure modes. As both approach II and approach III have a single multistate structure as discussed in Section 2.2, they can be considered a special case of approach I. Therefore, the measures developed in this section can also be used for approaches II and III.

The first measure is the probability of being in state j of the ith failure mode at time t, given the condition monitoring data until time t. This important measure can be used as an online diagnostic tool, which identifies the current health status of the device. In (Moghaddass *et al.* 2013), steps to calculate the important forward variables $\alpha_t^i(j) = \Pr(O_1^i, O_2^i, \ldots, O_t^i, Q_t^i = j|\theta^i)$ and $u_t^i(j) = \Pr(O_1^i, O_2^i, \ldots, O_t^i, Q_{t-1}^i \neq j, Q_t^i = j|\theta^i)$, which are respectively the joint probability of observing the sequence of $O_1^i, O_2^i, \ldots, O_t^i$ and being in state j at the tth observation point, and the joint probability of observing the sequence of $O_1^i, O_2^i, \ldots, O_t^i$ and being in state j for the first time at the tth observation point, are illustrated. Using the forward variable α, the probability of being in the jth state of the ith failure mode, given the condition monitoring data up to time t, can be calculated as follows:

$$\begin{aligned}\Pr(Q_t^i = j|O_1^i, O_2^i, \ldots, O_t^i, \theta^i) &= \frac{\Pr(O_1^i, O_2^i, \ldots, O_t^i, Q_t^i = j|\theta^i)}{\Pr(O_1^i, O_2^i, \ldots, O_t^i|\theta^i)} \\ &= \frac{\Pr(O_1^i, O_2^i, \ldots, O_t^i, Q_t^i = j|\theta^i)}{\sum_{j=1}^{n^i}\Pr(O_1^i, O_2^i, \ldots, O_t^i, Q_t^i = j|\theta^i)} = \frac{\alpha_t^i(j)}{\sum_{j=1}^{n^i}\alpha_t^i(j)}\end{aligned} \quad (2.5)$$

The most likely state at time t for the ith failure mode (V_t^i), given condition monitoring data can be found as:

$$V_t^i = \arg\max_j (\Pr(Q_t^i = j | O_1^i, O_2^i, \ldots, O_t^i, \theta^i)). \tag{2.6}$$

As failure modes and conditions monitoring indicators are assumed to be independent, the joint conditional probability of being in different combination of states of all failure modes can be computed by the product of each individual state's probability from Equation (2.5). Other measures, such as average degradation level and its confidence interval, can be computed from Equation (2.5) accordingly.

The second important measure is the conditional reliability function, given the condition monitoring data until a time point t. For the ith failure mode, this measure can be calculated as:

$$\begin{aligned} R^i(t|O_1^i, O_2^i, \ldots, O_p^i) &= \Pr(L^i > t | O_1^i, O_2^i, \ldots, O_p^i, L^i > t_p, \theta^i) \\ &= \frac{\Pr(L^i > t, O_1^i, O_2^i, \ldots, O_p^i|\theta^i)}{\Pr(L^i > t_p, O_1^i, O_2^i, \ldots, O_p^i|\theta^i)} \\ &= \frac{\sum_{j=1}^{n^i-1} \Pr(L^i > t, O_1^i, O_2^i, \ldots, O_p^i, Q_p^i = j|\theta^i)}{\sum_{j=1}^{n^i-1} \Pr(O_1^i, O_2^i, \ldots, O_p^i, Q_p^i = j|\theta^i)}, \end{aligned} \tag{2.7}$$

where $\sum_{j=1}^{n^i-1} \Pr(O_1^i, O_2^i, \ldots, O_p^i, Q_p^i = j|\theta^i) = \sum_{j=1}^{n^i-1} \alpha_{t_p}^i(j)$ and

$$\begin{aligned} \Pr(L^i > t, O_1^i, O_2^i, \ldots, O_p^i, Q_p^i = j|\theta^i) &= \sum_{z=0}^{p} u_t^i(j) \times \prod_{w=z+1}^{p} b_j^i(O_w^i) \\ &\quad \times \Pr(L^i > t, Q_p^i = j|Q_{z-1}^i \neq j, Q_z^i = j, \theta^i) \end{aligned} \tag{2.8}$$

The element $\Pr(L^i > t, Q_p^i = j|Q_{z-1}^i \neq j, Q_z^i = j, \theta^i)$ is the reliability of the device at time t with respect to the ith failure mode, unconditional on the condition monitoring information, as:

$$\begin{aligned} \Pr(L^i > t, Q_p^i = j|Q_{z-1}^i \neq j, Q_z^i = j, \theta^i) &\approx 1 - S_j^i(t_z, t) \\ &+ \sum_{k \neq n^i} \int_{t_p}^{t} \dot{Q}_{j,k}^i(t_z, \tau) \times R^i(t|k, \tau) d\tau, \end{aligned} \tag{2.9}$$

where $S^i j(s, t)$ is the conditional sojourn time, given that state j of the ith failure mode is reached at time s and $R^i(t|k, s) = \Pr(L^i > t|X_n^i = k, T_n^i = s, \theta^i)$ is the solution of the following system of equations:

$$R^i(t|k, s) = 1 - S_k^i(s, t) + \sum_{j \neq n^i} \int_s^t \dot{Q}_{k,j}^i(s, \tau) \times R^i(t|j, \tau) d\tau. \tag{2.10}$$

The detailed steps to estimate a reliability function in the form of Equation (2.10) are given in (Moghaddass *et al.* 2012). Now the conditional remaining useful life of the device, employing the condition monitoring information up to the pth monitoring point can be calculated as:

$$MRL^i(O_1^i, O_2^i, \ldots, O_p^i) = \int_0^\infty R^i(t + t_p | O_1^i, O_2^i, \ldots, O_p^i)\, dt. \tag{2.11}$$

Thus, the conditional reliability of the device can also be calculated as the product of the conditional reliability of its failure modes given by Equation (2.7). Accordingly, the conditional mean residual life of the device can be calculated. In addition, using Equation (2.7), the confidence interval for the remaining useful life can be calculated.

The conditional dominant failure mode can be found at any time point, given the history of the condition monitoring data. A well-known function, entitled the cumulative incidence function (CIF) or crude cumulative incidence function (Putter *et al.* 2007), can be used for this purpose. This function shows the probability of failure (D) from the failure mode i, before time t. Using the available condition monitoring data up to the pth monitoring point, the conditional form of this CIF function can be calculated as:

$$\begin{aligned}\psi^i(t|p) &= \Pr\left(L \le t, D = F^i | \prod_i (O_1^i, O_2^i, \ldots, O_p^i), \theta^i\right)\\ &= \int_o^t \frac{-dR^i(\tau | O_1^i, O_2^i, \ldots, O_p^i)}{d\tau} \prod_{j \ne i} R^j(\tau | O_1^j, O_2^j, \ldots, O_p^j)\, d\tau.\end{aligned} \tag{2.12}$$

Now, at any time point t_p, the probability that the device will finally fail because of the ith failure mode can be calculated as $\psi^i(\infty|p)$. The unconditional value of this index ($\psi^i(\infty|0)$) can be used to find the dominant failure mode unconditional on the condition monitoring data. Such a measure can be used to determine the dominant failure mode of a device.

2.5 Numerical Example

A numerical example is employed to illustrate how online multi-dimensional condition monitoring data can be used for online diagnostic and prognostic indications of a device with multistate degradation. A device with two independent failure modes and independent condition monitoring indicators is considered. It is assumed that failure modes 1 and 2 have respectively four and five levels of health states, ranging from perfect functioning to complete failure. The structure of the degradation process, the transition rate function (homogenous semi-Markov with Weibull distribution) between states, and the distribution of the condition monitoring data for each failure mode are shown below:

$$\alpha_{1,2}^1 = 19,\ \alpha_{2,3}^1 = 13, \alpha_{3,4}^1 = 16, \beta_{1,2}^1 = 6, \beta_{2,3}^1 = 4, \beta_{3,4}^1 = 3, \alpha_{1,2}^2 = 14, \alpha_{2,3}^2 = 11,$$
$$\alpha_{3,4}^2 = 9, \alpha_{4,5}^2 = 10, \beta_{1,2}^2 = 9, \beta_{2,3}^2 = 3, \beta_{3,4}^2 = 6, \beta_{4,5}^2 = 4, n^1 = 4, n^2 = 4,$$

$$B^1 = \begin{bmatrix} 0.55 & 0.35 & 0.1 & 0 & 0 \\ 0.15 & 0.35 & 0.45 & 0.05 & 0 \\ 0 & 0.05 & 0.25 & 0.50 & 0.2 \\ 0 & 0 & 0.05 & 0.15 & 0.8 \end{bmatrix},$$

$$B^2 = \begin{bmatrix} 0.45 & 0.35 & 0.15 & 0.05 & 0 & 0 & 0 & 0 \\ 0.05 & 0.25 & 0.3 & 0.25 & 0.1 & 0.05 & 0 & 0 \\ 0 & 0.05 & 0.1 & 0.4 & 0.25 & 0.15 & 0.05 & 0 \\ 0 & 0 & 0 & 0.05 & 0.2 & 0.45 & 0.2 & 0.1 \\ 0 & 0 & 0 & 0 & 0 & 0.05 & 0.15 & 0.8 \end{bmatrix}.$$

A sample run-to-failure life history is simulated, and the corresponding condition monitoring indicators are recorded (see Figure 2.3). The diagnostic results for each failure mode (using Equation (2.6)) and the actual states of each failure mode are shown in Figure 2.4. Results indicate that estimation results are reasonably close to the actual states of the device.

The conditional mean remaining useful life for each failure mode is shown in Figure 2.5. In addition, the conditional probability that the device will fail according to each failure mode is shown in Figure 2.6. Results shown in Figure 2.6 are consistent with those in Figure 2.5, as after around 15 days, the probability of failure according to the first failure mode exceeds the one for the second failure mode. Figure 2.6 also verifies that the second failure mode is the dominant failure mode at time zero. Similar results can be generated for online diagnostic and prognostic analysis of a device with more than two failure modes.

2.6 Conclusion

This chapter has briefly studied the concept of multiple independent failure modes for a device under condition monitoring. Important diagnosis and prognosis reliability measures

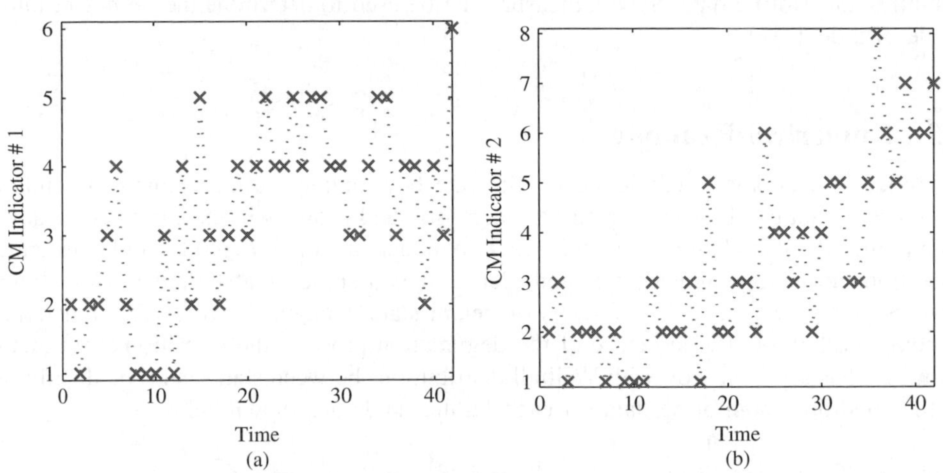

Figure 2.3 Sample condition monitoring data for the two failure modes

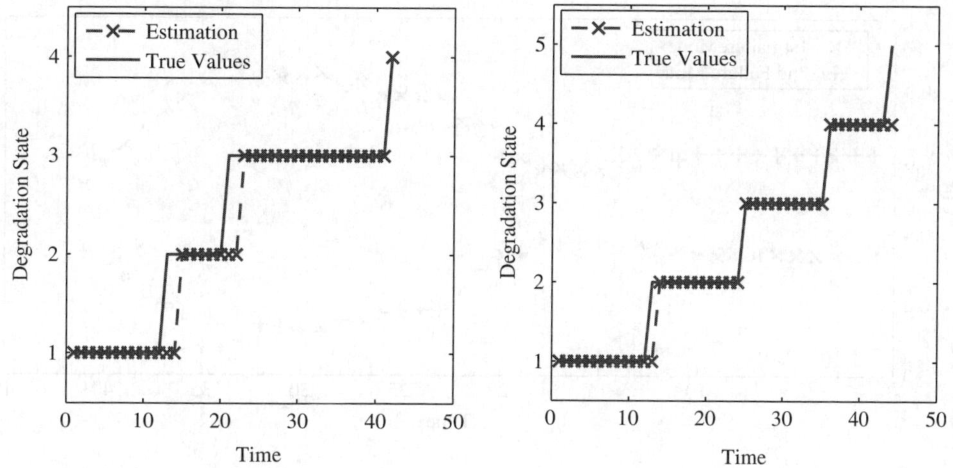

(a) (b)

Figure 2.4 Diagnostic result for each failure mode

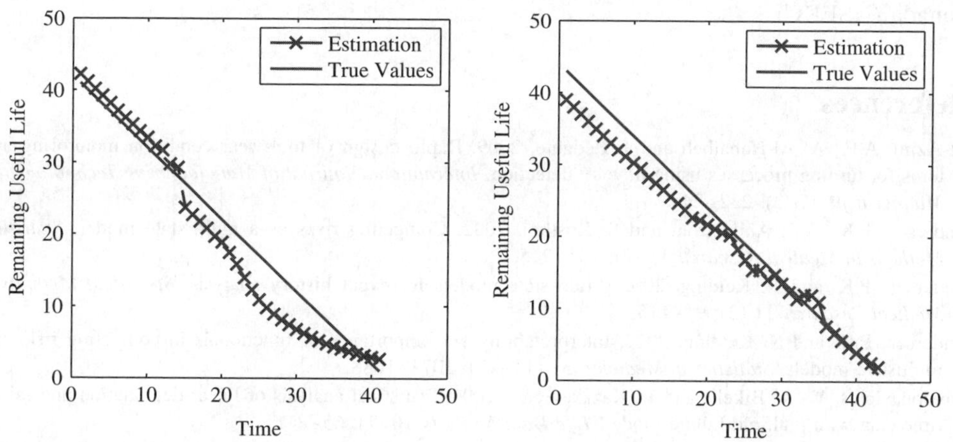

(a) (b)

Figure 2.5 Remaining useful life estimation

are introduced and a numerical example was provided to illustrate the application of these measures. Analyzing multiple failure modes for a condition-monitored device under multistate degradation is still in its infancy. Our direction of future work involves multiple failure mechanisms in a system with more than one device. In other words, how to employ multidimensional condition monitoring information for online diagnosis and prognosis of a system will be the focus of our investigations in our future work.

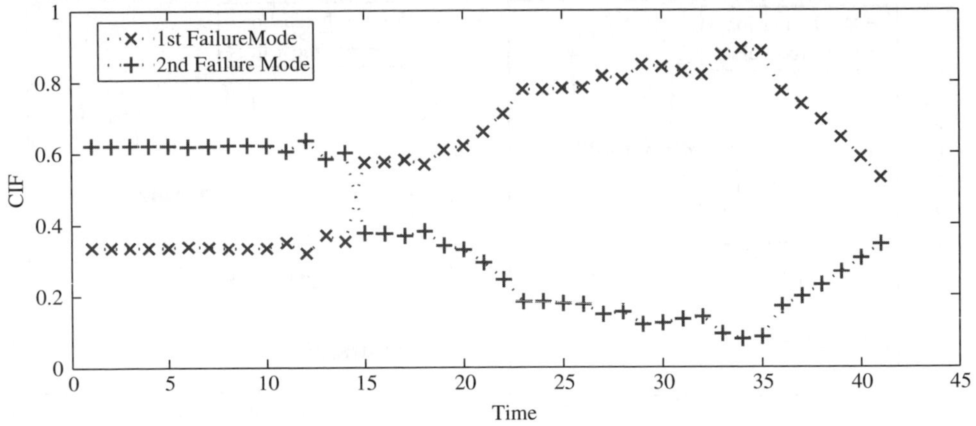

Figure 2.6 Dominant failure probability

Acknowledgements

This work is supported by the Natural Sciences and Engineering Research Council of Canada (NSERC).

References

Al-Azmi, A.F., A. Al-Habaibeh and J. Redgate. 2009. Rapid design of tool-wear condition monitoring systems for turning processes using novelty detection. *International Journal of Manufacturing Technology and Management* 17 (3): 232–245.

Andersen, P.K., S.Z. Abildstrom and S. Rosthoj. 2002. Competing risks as a multi-state model. *Statistical Methods in Medical Research* 11 (2): 203–215.

Andersen, P.K. and N. Keiding. 2002. Multi-state models for event history analysis. *Statistical Methods in Medical Research* 11 (2): 91–115.

Andersen, P.K. and N. Keiding. 2012. Interpretability and importance of functionals in competing risks and multistate models. *Statistics in Medicine* 31 (11–12): 1074–1088.

Bagdonavičius, V., A. Bikelis and V. Kazakevičius. 2004. Statistical analysis of linear degradation and failure time data with multiple failure modes. *Life Data Analysis* 10 (1): 65–81.

Bello, J.H. 2006. Impact of the pump systems availability in the plant maintenance: model development. MSc dissertation, University of Manchester.

Beyersmann, J., A. Allignol and M. Schumacher. 2012. *Competing Risks and Multistate Models with R*. New York: Springer.

Bichon, B.J., J.M. McFarland and S. Mahadevan. 2011. Efficient surrogate models for reliability analysis of systems with multiple failure modes. *Reliability Engineering and System Safety* 96 (10): 1386–1395.

Carr, M.J. and W. Wang. 2010. Modeling failure modes for residual life prediction using stochastic filtering theory. *IEEE Transactions on Reliability* 59 (2): 346–355.

Chen, H. and Y. Chen. 2010. Competing failure with degradation and random shocks based on probabilistic failure threshold. In *Proceedings of the IEEE Conference on the Prognostics and System Health Management PHM 12*, May 23–25, 2012, 1–6.

Chen, L.P., Z.S. Ye and B. Huang. 2011. Condition-based maintenance for systems under dependent competing failures. In *Publications of the IEEE International Conference on Industrial Engineering and Engineering Management*, December 6–9, 2011, 1586–1590.

Huang, W. and R.G. Askin. 2003. Reliability analysis of electronic devices with multiple competing failure modes involving performance aging degradation. *Quality and Reliability Engineering International* 19 (3): 241–254.

Li, W. and H. Pham. 2005a. An inspection-maintenance model for systems with multiple competing processes. *IEEE Transactions on Reliability* 54 (2): 318–327.

Li, W. and H. Pham. 2005b, Reliability modeling of multi-state degraded systems with multi-competing failures and random shocks. *IEEE Transactions on Reliability* 54 (2): 297–303.

Liang, Z., Y. Jun, H. Ling and Z. Su. 2011. Reliability assessment based on multivariate degradation measures and competing failure analysis. *Modern Applied Science* 5 (6): 232–236.

Lisnianski, A. and G. Levitin 2003. *Multi-state System Reliability, Assessment, Optimization and Applications*. Singapore: World Scientific.

Lu, H. and Y. Zhang. 2011. Reliability-based robust design for structural system with multiple failure modes. *Mechanics Based Design of Structures and Machines* 39 (4): 420–440.

Ma, Z.S. and A.W. Krings. 2008. Competing risks analysis of reliability, survivability, and prognostics and health management. In *Proceedings of the IEEE Aerospace Conference*, March 1–8, 2008, Big Sky, MT, 1–21.

Mathew, S., M. Alam and M. Pecht. 2012. Identification of failure mechanisms to enhance prognostic outcomes. *Journal of Failure Analysis and Prevention* 12 (1): 66–73.

Mathew, S., D. Das, R. Rossenberger and M. Pecht. 2008. Failure mechanisms based prognostics. In *Proceedings of the International Conference on Prognostics and Health Management PHM 2008*, October 6–9, 2008, 1–6.

Moghaddass, R. and M.J. Zuo. 2012a. A parameter estimation method for a condition-monitored device under multi-state deterioration. *Reliability Engineering and System Safety* 106: 94–103.

Moghaddass, R. and M.J. Zuo. 2012b. Fault diagnosis for multi-state equipment with multiple failure modes. In *Proceedings of the Annual Reliability and Maintainability Symposium, RAMS*, January 23–26, 2012, 1–6.

Moghaddass, R., M.J., Zuo, Y. Liu and H.Z. Huang. 2012, Reliability analysis of a multi-state deteriorating device using inspection data, submitted to *European Journal of Operation Research*.

Moghaddass, R., M.J. Zuo and X. Zhao. 2013. Modeling equipment with multi-state degradation with non-homogeneous continuous-time hidden semi-Markov process. In *Diagnostics and Prognostics of Engineering Systems: Methods and Techniques*. Hershey, PA: IGI Global, pp. 151–181.

Mohammadian-Masooly, S.H. 2009. A practical contribution to quantitative accelerated testing of multi-failure mode products under multiple stresses, PhD dissertation, Université Laval, Canada.

Peng, H., Q. Feng and D.W. Coit. 2010. Reliability and maintenance modeling for systems subject to multiple dependent competing failure processes, *IIE Transactions* 43 (1): 12–22.

Putter, H., M. Fiocco and R.B. Gekus. 2007. Tutorial in biostatistics: Competing risk and multi-state models. *Statistics in Medicine* 26 (11): 2389–2430.

Sikorska, J.Z., M. Hodkiewicz and L. Ma. 2011. Prognostic modeling options for remaining useful life estimation by industry. *Mechanical Systems and Signal Processing* 25 (5), 1803–1836.

Tan, C.M. and N. Raghavan. 2010. Imperfect predictive maintenance model for multi-state systems with multiple failure modes and element failure dependency. In *Proceedings of the Prognostics and System Health Management Conference PHM' 10*, January 12–14, 2010, 1–12.

Wang, Y. 2011. Dependent competing-risk degradation systems. In *Safety and Risk Modeling and Its Applications*, ed. H. Pham. London: Springer, pp. 197–218.

Wang, Y. and H. Pham. 2012. Modeling the dependent competing risks with multiple degradation processes and random shock using time-varying copulas. *IEEE Transactions on Reliability* 61 (1): 13–22.

Xu, J. and L. Xu. 2011. Health management based on fusion prognostics for avionics systems. *Journal of Systems Engineering and Electronics* 22 (3): 428–436.

3

Time Series Regression with Exponential Errors for Accelerated Testing and Degradation Tracking

Nozer D. Singpurwalla
Department of System Engineering and Engineering Management, and Department of Management Science, City University of Hong Kong, Hong Kong

3.1 Introduction

Our original motivation for considering the material discussed here was one of curiosity. In particular, we were intrigued by the simple problem of predicting the individual future observables in time series regression models. The standard text-book approaches seemed to be, at best, less than satisfactory, and, at worst, a sleight of hand. Thus arose our interest of time series regression with non-Gaussian errors, the exponential distribution being the simplest one to consider. It turns out that the simplest case also happens to be a useful case, in the sense that time series regression with exponential errors is a prototype for several situations involving non-negative random quantities with a non-zero threshold. A few examples, particularly those that are germane to the theme of this volume, are accelerated life testing, degradation monitoring, and tracking financial risk. More generally, time series regression with exponential errors provides a convenient platform for revealing some interesting technical issues which this topic spawns and which may have a broader impact. We have in mind inference for geometric Brownian motion with drift. As stated before, we do not purport to solve a specific problem nor do we endeavor to address any specific issue. The aim is to set out some general matters with the consequence that the assumptions made are not necessarily realistic nor are they motivated by a real application scenario.

Applied Reliability Engineering and Risk Analysis: Probabilistic Models and Statistical Inference, First Edition.
Edited by Ilia B. Frenkel, Alex Karagrigoriou, Anatoly Lisnianski and Andre Kleyner.

3.2 Preliminaries: Statement of the Problem

Suppose that the "true" response at time t, of some phenomenon of interest is η_t, and that η_t is hypothesized as $\eta_t = \beta t$, for some $\beta > 0$; assume that t takes discrete values $t = 0, 1, 2, \ldots$. Let Y_t be the observed response at t, and suppose that Y_t is of the form $Y_t = \eta_t + \epsilon_t$, where ϵ_t is an error. We assume that Y_t (and η_t) are non-negative, and thus it is convenient to suppose that ϵ_t has an exponential distribution with a scale parameter of one. In the interests of clarity, it is useful to note that given β (and t), the Y_ts are independent, but unconditionally (on β), they are dependent.

The choice of a scale parameter other than one, say λ, where $\lambda > 0$ is unknown, will detract from the focus of what we wish to say here by introducing the additional complication of estimating λ. With the above set-up, our model of observations is of the form:

$$Y_t = \beta t + \epsilon_t, \text{ with } \epsilon_t \sim \exp\ (1).$$

Since $E(Y_t) = \beta t + 1$ and $Var(Y_t) = 1$, we have the result that Y_t has an exponential distribution with a threshold βt and scale 1.

The generic problem of time series regression is the following. Having at hand T realizations (observations) of Y_t, say, $y_1, y_2, \ldots, y_T$, what can we say about a future individual unobserved value Y_{T+1}? That is, given as data $y_1, y_2, \ldots, y_T$, what is our (one-step ahead) prediction of Y_{T+1}? Since Y_{T+1} is spawned by η_{T+1}, a related question is, what can be said about η_{T+1} in the light of $y_1, y_2, \ldots, y_T$? Figure 3.1 illustrates this set-up. The dotted line in Figure 3.1 represents (the unknown) βt, and it is to be noted that all the observed y_ts will be greater than or equal to βt.

3.2.1 Relevance to Accelerated Testing, Degradation and Risk

In the context of finance and financial risk, the y_ts could be seen as interest rates at time t, and η_t some hidden parameter of a growing economy which spawns the interest rates.

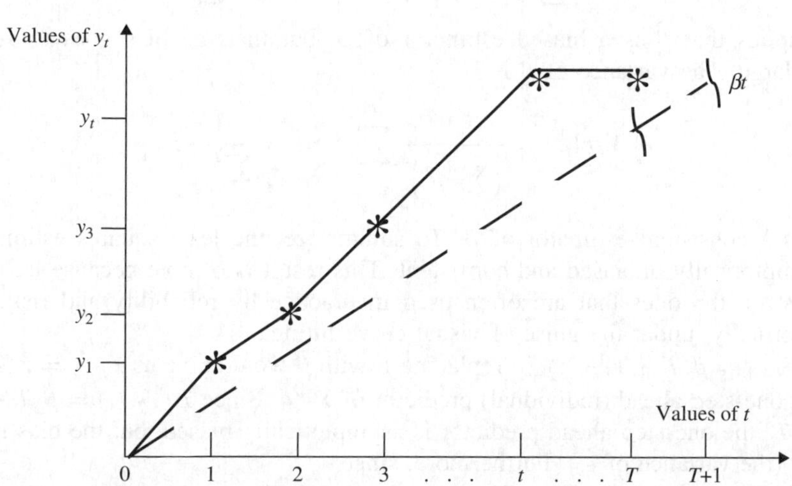

Figure 3.1 An exponential time series regression with support βt

In accelerated testing, t could be conceptualized as the level of an applied stress and η_t an item's propensity to fail under t (see Popper (1959)). Here y_t is the observed lifetime under stress t, where y_t is the realization of an exponentially distributed random variable Y_t with (an arbitrarily chosen) scale 1 and threshold βt. Analogously, η_t can be interepreted as the degradation (which to us, is an undefined notion) which an item experiences under stress t, and y_t an observed manifestation of η_t, such as the incremental size of an observed crack, or the incremental measured level of wear. The usual approach to model degradation and wear is via a compound Poisson process wherein the crack size increases as an exponentially distributed random variable, but the increases occur at random times per the postulates of a Poisson process. In the set-up here the increases occur at fixed times, $t = 1, 2, \ldots$. In principle, this section can be expanded to include a specific application with real data, realistic assumptions, and validation against actuality. However, we refrain from doing so, because our aim is to reveal the basic issues which problems of this type spawn.

3.3 Estimation and Prediction by Least Squares

The method of least squares is the one that is most commonly used in in practice to address problems of regression. Under the assumption of Gaussian errors, it produces results that are optimum in some specified sense. What is the impact of exponentially distributed errors on this method? In what follows it is argued that the method of least squares loses some of its luster under exponentially distributed lifetimes in particular, and skewed distributions in general. To see why, we start by seeking that value of β, say, $\hat{\beta}$ for which $\sum_{i=1}^{T} \epsilon_i^2$ is a minimum. Some elementary algebra shows that the value $\hat{\beta} = \sum^T t y_t / \sum^T t^2$ achieves the required minimization. Thus $\hat{\beta}$ is the least squares estimator of β. To investigate the properties of this estimator, we note that its expectation is

$$E(\hat{\beta}) = \frac{1}{\sum^T t^2} \sum^T t(\beta t + 1) = \beta + \frac{T(T+1)}{2\sum^T t^2},$$

which implies that $\hat{\beta}$ is a biased estimator of β, but that the bias goes to zero as T becomes large. The variance of $\hat{\beta}$ is

$$Var(\hat{\beta}) = \frac{1}{\left(\sum^T t^2\right)^2} \sum^T t^2 * 1 = \frac{1}{\sum^T t^2}.$$

Thus $\hat{\beta}$ is a consistent estimator of β. To summarize, the least squares estimator of β is an asymptotically unbiased and consistent. This result is of note because least squares estimators are the ones that are often used in practice by reliability and risk analysis, albeit informally, under the guise of visual curve fitting.

Since $Y_{T+1} = \beta(T+1) + \epsilon_{T+1}$, replacing $\hat{\beta}$ with β would give us $\hat{Y}_{T+1} = \hat{\beta}(T+1) + \epsilon_{T+1}$ as a one-step-ahead (individual) predictor of Y_{T+1}. Since $E(\hat{Y}_{T+1}) = \beta(T+1) + 1$, for large T, the one-step-ahead predictor is asymptotically biased, but the bias is known, namely 1 (the variance of ϵ_t). Furthermore, since

$$Var(\hat{Y}_{T+1}) = (T+1)^2 Var(\hat{\beta}) + Var(\epsilon_{T+1}) = \frac{(T+1)^2}{\sum^T t^2} + 1,$$

the variance of the one-step-ahead prediction decreases with T, but converges to one when T goes to infinity, suggesting an inconsistency of the estimator. Matters improve if $\hat{\beta}(T+1)$ – rather than $\hat{\beta}(T+1)+\epsilon_{T+1}$ – were to be chosen as a predictor for Y_{T+1}, because this choice leads to consistency. However, such a choice is ad hoc in the sense that it is incoherent vis-à-vis the assumed model of observations. Indeed, $\hat{\beta}(T+1)$ is a consistent estimator of $\beta(T+1)$, the threshold. One way to operationalize the predictor $\hat{Y}_{T+1}=\hat{\beta}(T+1)+\epsilon_{T+1}$ would be to generate a unit exponential deviate and add it to the estimator of the threshold $\hat{\beta}(T+1)$.

The choice of what to use as a predictor of Y_{T+1} is therefore enigmatic. An ad hoc choice provides consistency; a formally defensible choice results in inconsistency.

3.4 Estimation and Prediction by MLE

Let $y=(y_1,y_2,\ldots,y_T)'$; then a likelihood for β with y known is, for all t of the form:

$$\begin{aligned}\mathcal{L}(\beta;y)&=\prod_{i=1}^{T}\exp\,(-(y_t-\beta t)),\ \text{with}\ \beta t\le y_t\\&=0,\ \text{o.w.}\ .\end{aligned}$$

Equivalently,

$$\begin{aligned}\mathcal{L}(\beta;y)&=\exp\left(-\sum_1^T y_t\right)*\exp\left(\beta\sum_1^T t\right),\ \text{for}\ \beta\le\frac{y_t}{t},\\&=0,\ \text{o.w.}\ .\end{aligned}$$

For convenience, we set $A^*=\sum_1^T y_t$ and $A=\sum_1^T t=\frac{T(T+1)}{2}$, and write, that for all t

$$\begin{aligned}\mathcal{L}(\beta;y)&=(\exp(-A^*))*(\exp(\beta A)),\ \text{for}\ \beta\le\frac{y_t}{t}\\&=0,\ \text{o.w.}\end{aligned}$$

Thus for $\beta\le\min\left(\frac{y_t}{t}\right)=y^*$, say, the shape of the likelihood function is of the form shown in Figure 3.2. The likelihood function attains its maximum at y^*, and thus the maximum likelihood estimator of β is $\tilde{\beta}=\min\left(\frac{y_t}{t}\right)$.

3.4.1 Properties of the Maximum Likelihood Estimator

We start by considering the sampling distribution of $\tilde{\beta}$. Specifically, for some $u\ge 0$,

$$\begin{aligned}P(\tilde{\beta}\ge u)&=P(Y_1\ge u,Y_2\ge 2u,\ldots,Y_T\ge Tu)\\&=\prod_{i=1}^{T}\exp\,(-t(u-\beta)),\ u\ge\beta.\end{aligned}$$

The above is so because $P(Y_t\ge tu)=\exp\,(-(tu-\beta t))$. Thus, $P(\tilde{\beta}\ge u)=\exp(-uA)*\exp(\beta A)$, for $u\ge\beta$.

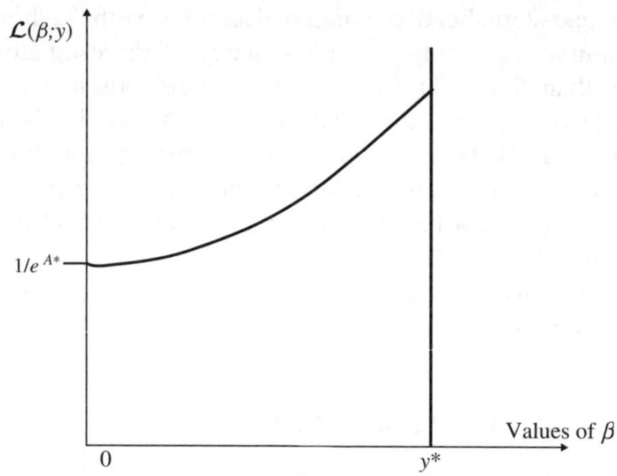

Figure 3.2 The likelihood function of β

Note that $P(\tilde{\beta} \geq u) = 1$, when $u \leq \beta$, and that it decreases to zero as u increases to infinity; see Figure 3.3. Thus $P(\tilde{\beta} \geq u)$ is a legitimate complementary distribution function. The expected value of $\tilde{\beta}$, is the integral of the function, namely

$$E(\tilde{\beta}) = \int_0^\infty P(\tilde{\beta} \geq u)\mathrm{d}u = \beta + \int_\beta^\infty e^{\beta A} * e^{-Au}\mathrm{d}u$$

$$= \beta + e^{\beta A} * \frac{e^{-\beta A}}{A} = \beta + \frac{1}{A} = \beta + \frac{2}{T+1}.$$

Hence for T large $E(\tilde{\beta}) = \beta$ (approximately) implying that $\tilde{\beta}$ is asymptotically un biased.

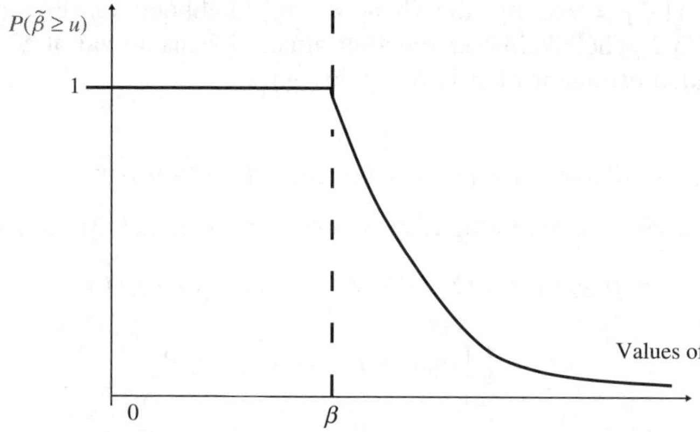

Figure 3.3 The distribution function of the MLE $\tilde{\beta}$

The probability density of the distribution function of $\tilde{\beta}$ is obtained via $-\frac{d}{du}P(\tilde{\beta} \geq u)$ for $u \geq \beta$. Verify that

$$-\frac{d}{du}P(\tilde{\beta} \geq u) = A\exp(-A(u-\beta)), \text{ for } u \geq \beta,$$

which implies an exponential distribution with scale A and threshold β. The variance of $\tilde{\beta}$ is therefore A^{-2}. Consequently,

$$Var(\tilde{\beta}) = \frac{1}{A^2} = \left(\frac{2}{T(T+1)}\right)^2 \downarrow 0, \text{ as } T \uparrow \infty,$$

suggesting that $\tilde{\beta}$ is a consistent estimator of β.

Thus, like the least squares estimator, the maximum likelihood estimator of β is consistent; it is of course asymptotically unbiased.

As was done with the least squares estimator, suppose that the (one-step ahead) predictor of y_{T+1} is $\tilde{Y}_{T+1} = \tilde{\beta}(T+1) + \epsilon_{T+1}$. Now

$$\begin{aligned} E(\tilde{Y}_{T+1}) &= (T+1)\left[\beta + \frac{2}{T(T+1)}\right] + 1 \\ &= \beta(T+1) + \frac{2}{T} + 1, \end{aligned}$$

and

$$Var(\tilde{Y}_{T+1}) = (T+1)^2\left[\frac{2}{T(T+1)}\right]^2 + 1 = \frac{4}{T^2} + 1,$$

which suggests that the variance of the one-step ahead prediction decreases in T, but does not go to zero as T goes to infinity, unless $\hat{Y}_{T+1}$ is taken to be $\tilde{\beta}(T+1)$.

To conclude, we have used prediction in time series regression with unit exponential errors using either the method of least squares or the method of maximum results in a bias, with the size of the bias decreasing in T. By contrast, the error of prediction does not converge to zero under either method. Neither approach is able to provide a predictive distribution of Y_{T+1}.

3.5 The Bayesian Approach: The Predictive Distribution

The Bayesian approach to inference and prediction in time series regression with unit exponential errors is conceptually straightforward and depends on our choice of the prior distribution of β. More important, instead of producing a point and interval (one-step ahead) predictor of Y_{T+1}, it produces an entire predictive distribution of Y_{T+1}. This is a key strength of the Bayesian approach. The matter of bias and inconsistency are no more germane. Besides the choice of a prior, the Bayesian approach requires of a user the specification of a likelihood. The matter of what likelihood to specify has philosophical connotations (cf. Singpurwalla 2006, 2007). The standard text-book approach is to appeal to the "principle of conditionalization" (Howson and Urbach 1989) or (Singpurwalla 2006), and base the likelihood on a transposition of the variable and the parameter of the underlying probability model. In what follows we invoke the aforementioned principle, and use the likelihood and notation of Section 3.4 to write:

$$\mathcal{L}(\beta; y) \propto \exp(\beta A), \qquad \beta \leq y^* = \min(Y_t/t).$$

Since β represents the slope of the assumed linear regression model, β can take values between zero and infinity. A plausible prior for β is therefore an exponential distribution with scale parameter $\lambda > 0$, where λ encapsulates an analysis best guess about the value of β. Thus the proposed prior on β is of the form

$$\pi(\beta; \lambda) \propto e^{-\lambda\beta}, \lambda > 0.$$

Combining the prior and the likelihood via Bayes' Law gives the posterior distribution of β as

$$\pi(\beta; \lambda, y) \propto \exp(-\beta(\lambda - A)), \beta \leq y^*;$$

recall that $A = \frac{T(T+1)}{2}$. Equivalently, by introducing the constant of proportionality, say, C, we have

$$\pi(\beta; \lambda, y) \propto C \exp(-\beta(\lambda - A)), \beta \leq y^*.$$

Since $\int_0^{y^*} \pi(\beta; \lambda, y)d\beta = 1$, it is easy to verify that if $\lambda > A$, then C must be:

$$C = \frac{\lambda - A}{(1 - e^{-y^*(\lambda - A)})}.$$

The requirement that λ must be greater than A could raise a concern, namely, that the prior (parameter) should not depend on the data (the number of observations taken). Thus, the case $\lambda \leq A$ also warrants discussion. This is done later in Section 3.5.2. For now we focus on the case $\lambda > A$, and obtain the posterior distribution of β as

$$\pi(\beta; \lambda, y) = \frac{\lambda - A}{(1 - e^{-y^*(\lambda - A)})} \exp(-\beta(\lambda - A)), \beta \leq y^*,$$

a right truncated exponential distribution with scale parameter $(\lambda - A)$.

3.5.1 The Predictive Distribution of Y_{T+1} when $\lambda > A$

The predictive distribution of Y_{T+1}, after averaging out β with respect to the posterior distribution given above is, for $u \geq 0$:

$$P(Y_{T+1} \geq u; \lambda, y) = \int_0^{y^*} P(Y_{T+1} \geq u|\beta; \lambda, y)\pi(\beta; \lambda, y)d\beta,$$

where

$$P(Y_{T+1} \geq u|\beta; \lambda, y) = 1, \text{ for } u \leq \beta(T+1), \text{ or } \beta \geq \frac{u}{T+1}, \text{ and}$$

$$= \exp\left(-(u - \beta(T+1))\right), \text{ for } \beta < \frac{u}{T+1}.$$

Thus

$$
\begin{aligned}
&P(Y_{T+1} \geq u; \lambda, y) \\
&= \int_0^{\frac{u}{T+1}} \exp(-(u - \beta(T+1))) \frac{(\lambda - A)e^{-\beta(\lambda - A)}}{(1 - e^{-y^*(\lambda - A)})} \mathrm{d}\beta \\
&+ \int_{\frac{u}{T+1}}^{y^*} 1 * \frac{(\lambda - A)}{(1 - e^{-y^*(\lambda - A)})} e^{-\beta(\lambda - A)} \mathrm{d}\beta, \\
&= \frac{(\lambda - A)e^{-u}}{(1 - e^{-y^*(\lambda - A)})} \int_0^{\frac{u}{T+1}} \exp(-\beta(\lambda - A - T - 1)) \mathrm{d}\beta \\
&+ \frac{(\lambda - A)}{(1 - e^{-y^*(\lambda - A)})} \int_{\frac{u}{T+1}}^{y^*} e^{-\beta(\lambda - A)} \mathrm{d}\beta, \\
&= \frac{(\lambda - A)e^{-u}}{(1 - e^{-y^*(\lambda - A)})} \frac{1}{(\lambda - A - T - 1)} \left(1 - e^{-\frac{u}{T+1}(\lambda - A - T - 1)}\right) \\
&+ \frac{(\lambda - A)}{(1 - e^{-y^*(\lambda - A)})} \frac{1}{(\lambda - A)} \left[e^{-(\lambda - A)\frac{u}{T+1}} - e^{-(\lambda - A)y^*}\right].
\end{aligned}
$$

Because A can take the values 1, 3, 6, ..., the case $\lambda > A$ is unlikely to hold when T is large. The more likely scenario is for $\lambda \leq A$, with $\lambda < A$ being the most frequent.

3.5.2 The Predictive Distribution of Y_{T+1} when $\lambda \leq A$

We start with the case $\lambda = A$. Now, the posterior distribution of β is, for some constant C_1, of the form:

$$\pi(\beta; \lambda, y) = C_1 \exp(-\beta(\lambda - A)) = C_1 \text{ for } \beta \leq y^*,$$

Since $\int_0^{y^*} \pi(\beta; \lambda, y)\mathrm{d}\beta = 1$, C_1 must be $\frac{1}{y^*}$, a constant. This implies that the posterior distribution of β is a uniform over $[0, y^*]$. Following the line of reasoning used in Section 3.5.1, the predictive distribution of Y_{T+1} will be:

$$
\begin{aligned}
&P(Y_{T+1} \geq u; \lambda, y) \\
&= \int_0^{\frac{u}{T+1}} \exp(-(u - \beta(T+1))) \frac{1}{y^*} \mathrm{d}\beta + \int_{\frac{u}{T+1}}^{y^*} \frac{1}{y^*} \mathrm{d}\beta, \\
&= \frac{e^{-u}}{y^*} \int_0^{\frac{u}{T+1}} e^{\beta(T+1)} \mathrm{d}\beta + \int_{\frac{u}{T+1}}^{y^*} \frac{1}{y^*} \mathrm{d}\beta, \\
&= \frac{e^{-u}}{y^*} \left[\frac{1}{T+1} e^{\beta(T+1)} \Big|_0^{\frac{u}{T+1}}\right] + \frac{1}{y^*} \left[\beta\Big|_{\frac{u}{T+1}}^{y^*}\right] \\
&= \frac{e^{-u}}{y^*} \left[\frac{1}{T+1}(e^u - 1)\right] + \frac{1}{y^*} \left[y^* - \frac{u}{T+1}\right].
\end{aligned}
$$

For the case $\lambda < A$, the posterior distribution of β is

$$\pi(\beta; \lambda, y) = C_2 \exp(\beta(A - \lambda))$$

for some constant C_2, which must be of the form

$$C_2 = \frac{(A-\lambda)}{e^{y^*(A-\lambda)} - 1};$$

note the symmetric nature of C and C_2. Thus, for $\lambda < A$, the posterior disitribution of β is:

$$\pi(\beta; \lambda, y) = \frac{(A-\lambda)}{(e^{y^*(A-\lambda)} - 1)} \exp(\beta(A-\lambda)), \beta \le y^*.$$

The predictive distribution of Y_{T+1} will therefore be:

$$\begin{aligned}
&P(Y_{T+1} \ge u; \lambda, y) \\
&= \int_0^{\frac{u}{T+1}} \exp(-(u - \beta(T+1))\frac{(A-\lambda)}{(e^{y^*(A-\lambda)} - 1)} \exp(\beta(A-\lambda))d\beta \\
&+ \int_{\frac{u}{T+1}}^{y^*} 1 * \frac{(A-\lambda)}{(e^{y^*(A-\lambda)} - 1)} \exp(\beta(A-\lambda))d\beta, \\
&= \frac{e^{-u}(A-\lambda)}{(e^{y^*(A-\lambda)} - 1)} \int_0^{\frac{u}{T+1}} \exp(\beta(A+T+1-\lambda))d\beta \\
&+ \frac{(A-\lambda)}{(e^{y^*(A-\lambda)} - 1)} \int_{\frac{u}{T+1}}^{y^*} \exp(\beta(A-\lambda))d\beta, \\
&= \frac{e^{-u}(A-\lambda)}{(e^{y^*(A-\lambda)} - 1)} \frac{1}{A-\lambda+T+1} \left[e^{\left(\frac{u}{T+1}\right)(A-\lambda+T+1)} - 1 \right] \\
&+ \frac{(A-\lambda)}{(e^{y^*(A-\lambda)} - 1)} \frac{1}{(A-\lambda)} \left[e^{\left(\frac{u}{T+1}\right)(A-\lambda)} - e^{y^*(A-\lambda)} \right].
\end{aligned}$$

This completes our discussion on Bayesian inference for β, and the predictive distribution of Y_{T+1}. The details tend to be cumbersome but the concept is straightforward.

There could be other strategies for specifying a prior on β. One such alternative is outlined below.

3.5.3 Alternative Prior for β

The exponential prior distribution of β discussed above lacked a constructive motivation. An alternative possibility, based on the notion that $\beta = \tan\theta$, where $0 \le \theta \le \frac{\pi}{2}$, is the angle between the horizontal axis and the relationship $\eta_t = \beta t$, has a constructive feature that could be appealing; see Figure 3.1. If we assume that θ has a uniform distribution over $[0, \frac{\pi}{2}]$. then since $\beta \ge 0$, its distribution function is of the form:

$$P[\beta \le u] = P[\tan\theta \le u] = P[\tan^{-1}(u)] = \frac{2}{\pi}\tan^{-1}(u),$$

for $u \ge 0$. A plot of this distribution function is shown in Figure 3.4.

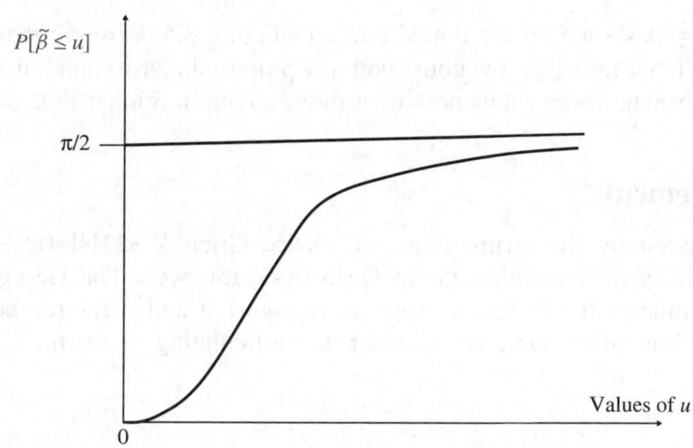

Figure 3.4 The distribution function of β

The probability density of β at $u \geq 0$ is given by

$$\frac{\mathrm{d}}{\mathrm{d}u}\left[\frac{2}{\pi}\tan^{-1}(u)\right] = \frac{2}{\pi}\left(\frac{1}{1+u^2}\right),$$

and this can serve as a prior of β. The bold curve of Figure 3.5 is a plot of this density.

Observe that this prior density parallels an exponential density vis-à-vis its shape, but that it has the advantage of a constructive feature rooted in a uniform distribution for θ. To some, the choice of a uniform distribution is attractive because it conveys an illusion of objectivity. Combining this density with the likelihood function of β discussed before, we have the posterior of β as $\pi(\beta; y) \propto \frac{e^{\beta A}}{1+\beta^2}$, for $\beta \leq y^*$.

It is a challenge to harness this posterior distribution to obtain a predictive distribution of Y_{t+1} in closed form. One way out would be to approximate the density of Figure 3.5 by an exponential distribution with scale parameter $\lambda = \frac{2}{\pi}$. This latter choice is suggested by the fact that the proposed density at $\beta = 0$ is $\frac{2}{\pi}$. The exponential density with scale

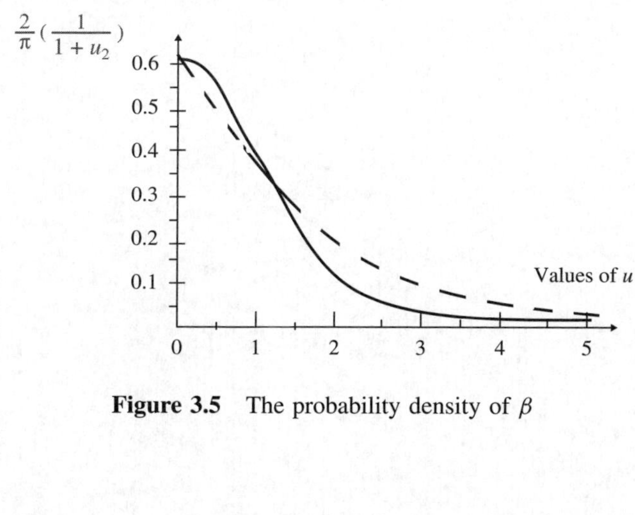

Figure 3.5 The probability density of β

parameter $\lambda = \frac{2}{\pi}$ is shown by the dotted curve of Figure 3.5. Observe that the matching exponential has fatter tails than the constructively proposed density; and distributions with fatter tails encapsulate more robustness than those having a thinner tail.

Acknowledgements

Research supported by the Army Research Office Grant W911NF-09-1-0039 and by The National Science Foundation Grant DMS-09-15156 with The George Washington University. Comments by Professor Alex. Karagrigoriou and a referee helped improve the chapter by eliminating some errors and better articulating its intent.

References

Howson, C. and P. Urbach. 1989. *Scientific Reasoning: The Bayesian Approach*. La Salle, IL: The Open Court Publishing Company.

Popper, K. R. 1959. The propensity interpretation of probability. *The British Journal for the Philosophy of Science* 10 (37): 25–42.

Singpurwalla, N. D. 2006. *Reliability and Risk: A Bayesian Perspective*. Chichester: Wiley.

Singpurwalla, N. D. 2007. Betting on residual life: The caveats of conditioning. *Statistics and Probability Letters* 77 (12): 1354–1361.

4

Inverse L_Z-Transform for a Discrete-State Continuous-Time Markov Process and Its Application to Multi-State System Reliability Analysis

Anatoly Lisnianski[1] and Yi Ding[2]
[1]*The Israel Electric Corporation Ltd, Israel*
[2]*Department of Electrical Engineering, Technical University of Denmark, Denmark*

4.1 Introduction

The main obstacle in Multi-state System (MSS) reliability analysis is what is often referred to as the "curse of dimensionality." In order to overcome this problem, a specific approach called the Universal Generating Function (UGF) technique has been introduced and then successfully applied to MSS reliability analysis. The UGF technique allows one to find algebraically an entire MSS performance distribution through the performance distributions of its elements. The basic ideas of the method were first introduced by Ushakov in the mid-1980s (Ushakov 1986, 1987). Since then, the method has been considerably expanded in numerous research works, see, for example, (Lisnianski *et al.* 1996; Levitin *et al.* 1998; Yeh 2006, 2009; Tian *et al.* 2008; Li *et al.* 2011). In the books (Lisnianski and Levitin 2003; Levitin 2005; Lisnianski *et al.* 2010), one can find a historical overview, a detailed description of the UGF method, and the applications of the method to various important practical cases.

Applied Reliability Engineering and Risk Analysis: Probabilistic Models and Statistical Inference, First Edition. Edited by Ilia B. Frenkel, Alex Karagrigoriou, Anatoly Lisnianski and Andre Kleyner.

Generally the UGF approach allows one to obtain a system's steady-state output performance distribution based on the given steady-state performance distribution of the system's elements and a system structure function. The wide application of the UGF method to MSS reliability analysis is explained by the fact that it avoids an enormous number of states that should be analyzed by using classical straightforward methods.

The main limiting factor of the UGF technique's application to real-world MSS reliability analysis is the fact that the UGF is based on a moment-generating function that is mathematically defined only for *random variables*. As one can see, this fact is the reason for considering the performance of each MSS element as a random variable, in spite of the fact that in reality it is a discrete-state continuous-time stochastic process (Lisnianski and Levitin 2003; Lisnianski *et al.* 2010; Natvig 2011). In practice, this important restriction leads one to consider only steady-state parameters of MSS.

In order to remove this essential restriction a special transform that was called L_z-transform was introduced in (Lisnianski 2012a) for the discrete-state continuous-time Markov process. This transform is similar to the UGF for a discrete random variable which allows Ushakov's universal generating operator Ω_f to be applied. Such important properties as existence and uniqueness were proven for L_z-transform. It was shown that many important MSS's reliability measures such as instantaneous availability $A(t)$, mean instantaneous performance, mean instantaneous performance deficiency, etc., can be found by using the L_z-transform. But the main restriction of this method is that such important reliability measures as, for example, the reliability function $R(t)$ still cannot be found by using the L_z-transform. The reason for this is the fact that by using the L_z-transform and Ushakov's universal generating operator one can find *only* the L_z-transform for the entire MSS output performance stochastic process. It means that one knows only state probabilities for the output process at any instant $t \geq 0$, but in order to find the reliability function, one should know the transition intensities matrix for the MSS's output stochastic process. In other words, in order to find the reliability function $R(t)$, one should fully discover the underlying Markov process, when only the L_z-transform of this process is known. In order to solve this problem and find the underlying Markov process, this chapter introduces an inverse L_z-transform that is designated as the L_z^{-1}-transform. When the L_z^{-1}-transform is applied to a known L_z-transform of the resulting output stochastic process, it completely reveals the underlying Markov process. Based on this revealed Markov process, the reliability function of entire MSS can easily be found.

Below we introduce the L_z^{-1}-transform, suggest the method for obtaining it and demonstrate benefits of its application to the MSS reliability analysis by using some numerical examples. At this stage only Markov stochastic processes will be considered.

4.2 Inverse L_z-Transform: Definitions and Computational Procedure

4.2.1 Definitions

We consider a discrete-state continuous-time (DSCT) Markov process (Trivedi 2002) $X(t) \in \{x_1, \ldots, x_K\}$, $t \geq 0$, which has K possible states i, $(i = 1, \ldots, K)$ where the performance level associated with any state i is x_i. This Markov process is completely defined by a set of possible states $\mathbf{x} = \{x_1, \ldots, x_K\}$, the transitions intensities matrix

$\mathbf{A} = \|a_{ij}(t)\|$, $i, j = 1, \ldots, K$ and by initial states probability distribution that can be presented by the corresponding set

$$\mathbf{p}_0 = [p_{10} = \Pr\{X(0) = x_1\}, \ldots, p_{K0} = \Pr\{X(0) = x_K\}].$$

From now on for such Markov process we will use the following notation by using triplets:

$$X(t) = \{\mathbf{x}, \mathbf{A}, \mathbf{p_0}\}. \tag{4.1}$$

Remark. If functions $a_{ij}(t) = a_{ij}$ are constants, then the DSCT Markov process is said to be time-homogeneous. When $a_{ij}(t)$ are time-dependent, then the resulting Markov process is non-homogeneous.

Definition 4.1. (Lisnianski 2012a). The L_z-transform of a discrete-state continuous-time Markov process $X(t) = \{\mathbf{x}, \mathbf{A}, \mathbf{p_0}\}$ is a function $u(z, t, \mathbf{p}_0)$ defined as

$$L_z\{X(t)\} = \mathbf{u}(z, t, \mathbf{p}_0) = \sum_{i=1}^{K} p_i(t) z^{x_i}. \tag{4.2}$$

where $p_i(t)$ is a probability that the process is in state i at time instant $t \geq 0$ for any given initial states probability distribution $\mathbf{p}_0$, and z in general is a complex variable.

Furthermore, in this chapter we will occasionally omit symbol $\mathbf{p}_0$ and simply write $u(z, t)$, keeping in mind that L_z-transform will always depend on the initial probability distribution $\mathbf{p}_0$.

Example 4.1. We consider the simplest single element, which has only two states – 1 and 2 – with corresponding performance levels $x_1 = 0$ and $x_2 = x_{nom}$ respectively. It means that state 1 is a complete failure state and state 2 is a state with nominal performance. The element's failure rate is λ and the repair rate is μ.

Suppose that at time $t = 0$ the element is in state 2, so that the initial states probability distribution is the following $\mathbf{p}_0 = \{p_{10}, p_{20}\} = \{p_1(0); p_2(0)\} = \{0, 1\}$.

As shown in (Lisnianski 2012a), the L_z-transform for the Markov process $X(t)$ that describes the element's behavior is the following;

$$L_z\{X(t)\} = u(z, t, \mathbf{p}_0) = \sum_{i=1}^{2} p_i(t) z^{x_i} = \left[\frac{\lambda}{\mu+\lambda} - \frac{\lambda}{\mu+\lambda} e^{-(\lambda+\mu)t}\right] z^0 + \left[\frac{\mu}{\mu+\lambda} + \frac{\lambda}{\mu+\lambda} e^{-(\lambda+\mu)t}\right] z^{x_{nom}}$$

It was proven in (Lisnianski 2012a) that for the discrete state continuous time Markov process $X(t) = \{\mathbf{x}, \mathbf{A}, \mathbf{p_0}\}$, where transition intensities $a_{ij}(t)$ are continuous functions of time, there exists one and only one (unique) L_z-transform. In other words, each discrete-state continuous-time Markov process with continuous transition intensities under certain initial conditions has only one (unique) L_z-transform $u(z, t)$ and each L_z-transform $u(z, t)$

will have only one corresponding DSCT Markov process $X(t)$ developing from these initial conditions.

In (Frenkel *et al.* 2012a, 2012b; Lisnianski 2012b), the L_z-transform was applied to the analysis of an aging refrigerating system and an aging system under stochastic demand and the high efficiency of the method was demonstrated.

Definition 4.2. Inverse L_z-transform (L_z^{-1}-transform) of a function

$$\mathbf{u}(z, t, \mathbf{p}_0) = \sum_{i=1}^{K} p_i(t) z^{x_i}. \tag{4.3}$$

where each $p_i(t)$ is a probability that some unknown discrete-state continuous-time Markov process $X(t)$ is in state i at time instant $t \geq 0$, x_i is a performance in this state, and z is a complex variable, is such a discrete-state continuous-time (DSCT) Markov process $X(t) = \{\mathbf{x}, \mathbf{A}, \mathbf{p_0}\}$ that

$$L_z\{X(t)\} = \mathbf{u}(z, t, \mathbf{p}_0) = \sum_{i=1}^{K} p_i(t) z^{x_i}. \tag{4.4}$$

In other words, based on the given L_z-transform $L_z\{X(t)\}$ of some DSCT Markov process $X(t)$, the inverse L_z-transform $L_z^{-1}\left\{\sum_{i=1}^{K} p_i(t) z^{x_i}\right\}$ reveals the underlying Markov process $X(t)$.

As stated above, to reveal the underlying Markov process means to find for it the following features:

- a set of possible states $\mathbf{x}$;
- a transition intensities matrix $\mathbf{A}$;
- a vector of initial conditions $\mathbf{p}_0$.

So, we can write the following

$$L_z^{-1}\{L_z\{X(t)\}\} = X(t) = \{\mathbf{x}, \mathbf{A}, \mathbf{p_0}\}, \tag{4.5}$$

if all transitions intensities in matrix $\mathbf{A}$ are a continuous function of time.

Example 4.2. We consider the element from Example 4.1. In Example 4.1 was found the L_z-transform of the Markov process $X(t) = \{\mathbf{x}, \mathbf{A}, \mathbf{p_0}\}$ that describes the element's behavior. The L_z-transform is given by the following expression:

$$L_z\{X(t)\} = u(z, t, \mathbf{p}_0) = \sum_{i=1}^{2} p_i(t) z^{x_i} = \left[\frac{\lambda}{\mu+\lambda} - \frac{\lambda}{\mu+\lambda} e^{-(\lambda+\mu)t}\right] z^0$$
$$+ \left[\frac{\mu}{\mu+\lambda} + \frac{\lambda}{\mu+\lambda} e^{-(\lambda+\mu)t}\right] z^{x_{nom}}$$

The problem is to find the underlying Markov process $X(t) = \{\mathbf{x}, \mathbf{A}, \mathbf{p_0}\}$, for which the L_z-transform is given by the above expression.

Solution. From the expression for $L_z\{X(t)\}$ one can see that the process $X(t)$ has only two states with performance levels 0 and x_{nom} respectively (see powers of z). So, the set of possible states for the process $X(t)$ is determined as $\mathbf{x} = \{0, x_{nom}\}$.

The probabilities of states $p_1(t)$, $p_2(t)$ are found from $L_z\{X(t)\}$ as coefficients of the corresponding z-powers

$$p_1(t) = \frac{\lambda}{\mu+\lambda} - \frac{\lambda}{\mu+\lambda}e^{-(\lambda+\mu)t},$$

$$p_2(t) = \frac{\mu}{\mu+\lambda} + \frac{\lambda}{\mu+\lambda}e^{-(\lambda+\mu)t}.$$

It is known from (Swift and Wirkus 2006) that such probabilities are a solution to the following system of differential equations

$$\begin{cases} \dfrac{dp_1(t)}{dt} = -\mu p_1(t) + \lambda p_2(t), \\ \dfrac{dp_2(t)}{dt} = \mu p_1(t) - \lambda p_2(t). \end{cases}$$

under initial conditions $p_1(0) = p_{10} = 0$; $p_2(0) = p_{20} = 1$.

From this system of differential equations one can obtain the matrix $\mathbf{A}$ of transition intensities for the underlying Markov process $X(t)$

$$\mathbf{A} = \begin{bmatrix} -\mu & \mu \\ \lambda & -\lambda \end{bmatrix}.$$

Thus, the inverse L_z^{-1}-transform for $L_z\{X(t)\} = u(z, t, \mathbf{p}_0) = \sum_{i=1}^{2} p_i(t) z^{x_i}$, which reveals the underlying Markov process $X(t)$ was obtained

$$L_z^{-1}\left\{\sum_{i=1}^{2} p_i(t) z^{x_i}\right\} = L_z^{-1}\left\{\left[\frac{\lambda}{\mu+\lambda} - \frac{\lambda}{\mu+\lambda}e^{-(\lambda+\mu)t}\right] z^0 \right.$$

$$\left. + \left[\frac{\mu}{\mu+\lambda} + \frac{\lambda}{\mu+\lambda}e^{-(\lambda+\mu)t}\right] z^{x_{nom}} \right\} = X(t) = \{\mathbf{x}, \mathbf{A}, \mathbf{p_0}\}.$$

where

$$\mathbf{x} = \{0, x_{nom}\};\ \mathbf{A} = \begin{pmatrix} -\mu & \mu \\ \lambda & -\lambda \end{pmatrix};\ \mathbf{p}_0 = \{p_{10}, p_{20}\} = \{p_1(0); p_2(0)\} = \{0, 1\}.$$

4.2.2 *Computational Procedure*

The problem considered in Example 4.2 is very simple. In the real world, a resulting Markov process $X(t)$ may have many states and it is impossible to obtain an analytical solution. From a computational point of view, the problem is the following.

The L_z-transform of some unknown Markov process $X(t)$ is given as

$$L_z\{X(t)\} = \mathbf{u}(z, t, \mathbf{p}_0) = \sum_{i=1}^{K} p_i(t) z^{x_i}, \tag{4.6}$$

Based on this expression for the L_z-transform, one should be able to find the primary Markov process $X(t)$, or in other words, to find the set of states $\mathbf{x}$, the set of initial conditions $\mathbf{p_0}$ and the matrix A of transitional probabilities of the process.

To find the set of states $\mathbf{x}$ and the set of initial conditions $\mathbf{p_0}$ from the given L_z-transform is not a difficult problem.

From expression (4.6) one knows a number of states K of the resulting Markov process and the performance in each state i as the corresponding value x_i. So, one determines the set of states $\mathbf{x} = \{x_1, x_2, \ldots, x_K\}$ for the underlying process $X(t)$.

In reliability interpretations we usually consider the best state as an initial state. For a coherent MSS, if all the system's elements in initial instant $t = 0$ are in the best states (in states with maximal performance), the entire MSS also will be at instant $t = 0$ in the best state. Thus, without loss of generality, for the resulting Markov process one can determine state K with maximal performance x_K as an initial state. Therefore, the only problem, which remains to be solved, is finding the transition intensity matrix A for the underlying Markov process.

4.2.2.1 Determining Matrix A

The resulting stochastic process is a Markov process that has K states and we know its state probabilities $p_i(t), i = 1, \ldots, K$ for all these states that are given by expression (4.6). Generally, probabilities for these K states may be found by solving the following system of differential equations (Lisnianski *et al.* 2010) in matrix notation

$$\frac{d\mathbf{p}(t)}{dt} = \mathbf{p}(t)\mathbf{A}, \tag{4.7}$$

where $\mathbf{p}(t) = [p_1(t), p_2(t), \ldots, p_K(t)]$ is the row-vector of the state probabilities;

$$\frac{d\mathbf{p}(t)}{dt} = \left[\frac{dp_1(t)}{dt}, \frac{dp_2(t)}{dt}, \ldots, \frac{dp_K(t)}{dt}\right]$$

is the row-vector of the state probabilities' derivatives;

$\mathbf{A}$ – transition intensities matrix (that is still unknown),

$$\mathbf{A} = \begin{bmatrix} -\left(a_{12} + \ldots + a_{1K}\right) & a_{12} & \ldots & a_{1K} \\ a_{21} & -(a_{21} + \ldots + a_{2K}) & \ldots & a_{2K} \\ \ldots & \ldots & \ldots & \ldots \\ a_{K1} & a_{K2} & \ldots & -(a_{K1} + \ldots + a_{K,K-1}) \end{bmatrix}.$$

We designate the solution of system (4.7) under initial conditions

$$\mathbf{p_0} = [\Pr\{X(0) = x_1\} = 0, \Pr\{X(0) = x_2\} = 0, \ldots, \Pr\{X(0) = x_K\} = 1]$$

as

$$\mathbf{p_A}(t) = [p_{A1}(t), p_{A2}(t), \ldots, p_{AK}(t)].$$

Ideally, matrix $\mathbf{A}$ should be determined in order to provide the following equations

$$p_{A1}(t) = p_1(t), p_{A2}(t) = p_2(t), \ldots, p_{AK}(t) = p_K(t).$$

It means that the state probabilities $p_{Ai}(t), i = 1, \ldots, K$ of the resulting stochastic process, that can be found as a solution of system (4.7), and state probabilities $p_i(t)$ in known L_z-transform of this process (expression (4.6)) should be the same.

4.2.2.2 Genetic Algorithm for Determining Matrix A

A numerical solution for this problem can be obtained by using the Genetic Algorithm (GA). The GA was inspired by nature evaluation and competition phenomenon. In the GA, the optimal selection of "offsprings" of solutions from the previous population will produce a new population of solutions. The "parents" of solutions will use the crossover and mutation operators to produce their "offsprings" of solutions. However, not all the "offsprings" of solutions can survive, which is determined by their adaptation to the environment (solution fitness). The GAs are the most popular heuristic algorithms for solving different kinds of reliability optimization problems (Levitin 2005; Lisnianski *et al.* 2010).

In accordance with GA, in each genetic cycle we randomly generate all $(K-1) \cdot (K-1)$ coefficients a_{ij} in matrix **A** for system (4.7). We should generate only $(K-1)$ elements in each row, because the diagonal element a_{ii} in each row i in matrix **A** is defined as minus the sum of all other elements in this row and should not be generated.

Finally, we should find that set of coefficients that minimize the accumulated relative error – the difference between the solutions of system (4.7) (for these solutions we use the designation $p_{iA}(t), i = 1, \ldots, K$, and the probabilities $p_i(t)$, which are presented in the L_z-transform (expression (4.6)). This error will be used in GA as a measure of accuracy.

The error, which should be minimized, is the following:

$$Err = \sum_{i=1}^{N_p} \frac{|p_{1A}(t_i) - p_1(t_i)|}{p_1(t_i)} + \ldots + \sum_{i=1}^{N_p} \frac{|p_{KA}(t_i) - p_K(t_i)|}{p_K(t_i)}, \tag{4.8}$$

where N_p = number of time points t_i, for which the probability values are compared.

It means that coefficients a_{ij} should be generated in each cycle of GA and the system of Equation (4.7) should be solved. Its solution $p_{iA}(t), i = 1, \ldots, K$ should be compared with the given state probabilities $p_i(t), i = 1, \ldots, K$ from expression (4.6) and the measure of the error (accuracy) (4.8) should be calculated.

In order to determine the number of time points N_p, one should analyze functions $p_i(t)$ from the given L_z-transform (4.6). Usually around 1000 time points should be sufficient for the reliability analysis. Approximately a quarter of these points may be taken within a steady state mode and three-quarters within transient interval of functions $p_i(t)$.

The proposed GA consists of the following steps:

1. Encode the solution and generate an initial population of solutions for coefficients a_{ij} randomly, which are within the ranges of variation: $0 \le a_{ij} \le a_{ij}^{(\max)}$.
2. Decode each solution and evaluate their accuracy by using Equation (4.8).
3. Select two solutions randomly and produce pairs of new solutions for the next generations using a crossover procedure. The one-point or two-point crossover method has been used to obtain the feasibility of solutions.
4. Allow the offsprings to mutate randomly. Mutation results in slight changes in the spring structure and maintains the diversity of solutions. This procedure avoids the premature convergence to a local optimum and facilities jumps in the solution space.

5. Apply a selection procedure that compares the new offspring with the worst solutions in the population, and select the best one. The best solution joins the population and the worst one is discarded.
6. Repeat steps 2–5, until a stopping criterion is satisfied. The stopping criterion of the GA can be the pre-determined number of genetic cycles or a number of genetic cycles without improving the solution performance or both (Lisnianski and Levitin 2003). The convergence criterion in the proposed GA is set as satisfying both a minimal number of genetic cycles and a number of genetic cycles without improving the solution performance.
7. Choose the best solution from the population.

4.3 Application of Inverse L_z-Transform to MSS Reliability Analysis

In general, if we use a straightforward Markov method for the MSS, consisting of n different multi-state elements where every element j has m_j different performance levels, we will obtain a model with $M = \prod_{j=1}^{n} m_i$ states. Therefore, a system of$M$ differential equations needs to be solved. This number M can be very large even for relatively small MSSs. Therefore, by using straightforward Markov methods one will be faced with the "curse of dimensionality." State-space diagram building or model construction for such complex MSSs is also a difficult nonformalized process that may cause numerous errors. In addition, solving models with hundreds of states can challenge the available computer resources. However, by using the L_z-transform, one would need to build and solve only n separate (relatively simple) Markov models for each element where the maximum number of states is defined as $\max\{m_1, \ldots, m_n\}$. And all other computations are formalized algebraic procedures, therefore by using the L_z-transform method, one should solve n systems of $m_1, \ldots, m_n$ differential equations respectively and the total number of differential equations will be $\sum_{i=1}^{n} m_i$. It means that computational complexity decreases drastically when the L_z-transform is applied.

When the L_z-transform for the MSS output performance Markov process is known, such important reliability measures as the MSS instantaneous availability $A(t)$, $t \geq 0$, the MSS instantaneous mean expected performance $E(t)$, the MSS instantaneous performance deficiency $D(t)$, etc. can be obtained (Frenkel *et al.* 2012a, 2012b; Lisnianski 2012a, 2012b).

However, there are some reliability measures as, for example, the reliability function $R(t)$ that still cannot be found, when only the L_z-transform of the MSS output Markov process is known.

In order to find $R(t)$, the output stochastic process should be completely revealed. The L_z^{-1}-transform can reveal an underlying Markov process when only the L_z-transform of this process is known. Concerning the MSS reliability analysis, the L_z-transform for the resulting output stochastic process of the entire MSS can be found by using Ushakov's universal generating operator and the L_z-transforms of its elements (Lisnianski 2012a). Then, the L_z^{-1}-transform should be applied and the underlying output stochastic process for the entire system can be obtained. Based on that, additional important reliability measures as, for example, MSS reliability function and Mean Time To Failure can be found.

So, if the L_z-transform

$$L_z\{X(t)\} = \sum_{i=1}^{K} p_i(t) z^{x_i}, \tag{4.9}$$

of the entire MSS's output stochastic process $X(t) \in \{x_1 \ldots , x_K\}$ is known, then the L_z^{-1}-transform should be applied in order to reveal the underlying Markov process:

$$L_z^{-1}\left\{\sum_{i=1}^{K} p_i(t) z^{x_i}\right\} = X(t) = \{\mathbf{x}, \mathbf{A}, \mathbf{p_0}\}, \tag{4.10}$$

The resulting output process has K states $\mathbf{x} = \{x_1, \ldots , x_K\}$ that are arranged in the ascending order $x_1 \leq x_2 \leq \ldots \leq x_K$. So, state K is the best state with maximal performance x_K and state 1 is the worst state with minimal performance x_1 (usually $x_1 = 0$). The reliability function $R(t)$ is treated as a probability that the process $X(t)$ (4.10) that begins at $t = 0$ from the best state K will reach at time t the worst state 1. In order to find $R(t)$, one should constrain all transitions from state 1 to any other states or, in other words, in matrix $\mathbf{A}$ of the resulting process (4.11) all transition intensities a_{Kj}, $j = 1, \ldots , K - 1$ should be zeroed ($a_{Kj} = 0$, for $j = 1, \ldots , K - 1$). We designate the matrix with zeroed transition intensities $a_{1i} = 0$, $i = 2, \ldots , K$ as $\mathbf{A_0}$

$$\mathbf{A}_0 = \begin{bmatrix} 0 & 0 & \ldots & 0 \\ a_{21} & -\left(a_{21} + \ldots + a_{2K}\right) & \ldots & a_{2K} \\ \ldots & \ldots & \ldots & \ldots \\ a_{K1} & a_{K2} & \ldots & -(a_{K1} + \ldots + a_{K,K-1}) \end{bmatrix}, \tag{4.11}$$

Reliability function $R(t)$ may be found after solving the following system of differential equations in a matrix notation

$$\frac{d\mathbf{p}(t)}{dt} = \mathbf{p}(t)\mathbf{A}_0, \tag{4.12}$$

where

$\mathbf{p}(t) = [p_1(t), p_2(t), \ldots , p_K(t)]$ – row-vector of state probabilities;

$\dfrac{d\mathbf{p}(t)}{dt} = \left[\dfrac{dp_1(t)}{dt}, \dfrac{dp_2(t)}{dt}, \ldots , \dfrac{dp_K(t)}{dt}\right]$ – row-vector of state probabilities' first derivatives;

Then the reliability function can be obtained

$$R(t) = \sum_{i=2}^{K} p_i(t), \tag{4.13}$$

Based on the reliability function, we can obtain the Mean Time to Failure (MTTF) as time up to the first entrance into state 1

$$MTTF = \int_0^{\infty} R(t)dt, \tag{4.14}$$

4.4 Numerical Example

In order to illustrate the L_z^{-1}-transform application to the MSS reliability analysis, let us consider an MSS, which consists of three multi-state elements: the first and the second elements have two possible performance levels and the third one has three possible performance levels. The MSS's structure and state-transition diagrams of the elements are presented in Figure 4.1. The elements' performance levels are as follows:

$$X_1(t) = \{x_{11}, x_{12}\} = \{0, 1.5\},$$
$$X_2(t) = \{x_{21}, x_{22}\} = \{0, 2\},$$
$$X_3(t) = \{x_{31}, x_{32}, x_{33}\} = \{0, 1.8, 4\}.$$

This MSS was considered as Example 2.4 in (Lisnianski *et al.* 2010) where the MSS's reliability function $R(t)$ was obtained by using the straightforward Markov method. Here we will find the reliability function and MTTF by using the suggested inverse L_z^{-1}-transform method and compare the results.

The failure rates and repair rates are as follows:

$$a_{21}^{(1)} = 7 \text{ year}^{-1}, a_{12}^{(1)} = 100 \text{ year}^{-1}, a_{21}^{(2)} = 10 \text{ year}^{-1}, a_{12}^{(2)} = 80 \text{ year}^{-1},$$
$$a_{31}^{(3)} = 0 \text{ year}^{-1}, a_{21}^{(3)} = 7 \text{ year}^{-1}, a_{13}^{(3)} = 0 \text{ year}^{-1}, a_{12}^{(3)} = 120 \text{ year}^{-1}, a_{23}^{(3)} = 110 \text{ year}^{-1}.$$

The MSS structure function is given by the expression

$$X(t) = f\{X_1(t), X_2(t), X_3(t)\} = \min\{X_3(t), X_1(t) + X_2(t)\}.$$

The problem is to calculate the system reliability function $R(t)$ and Mean Time to Failure. The failure is treated as the system entrance into the state with performance 0.

Solution. Applying the procedure described earlier, we proceed as follows. According to the Markov method, we build the systems of differential equations for each element (using the state-transitions diagrams presented in Figure 4.1).

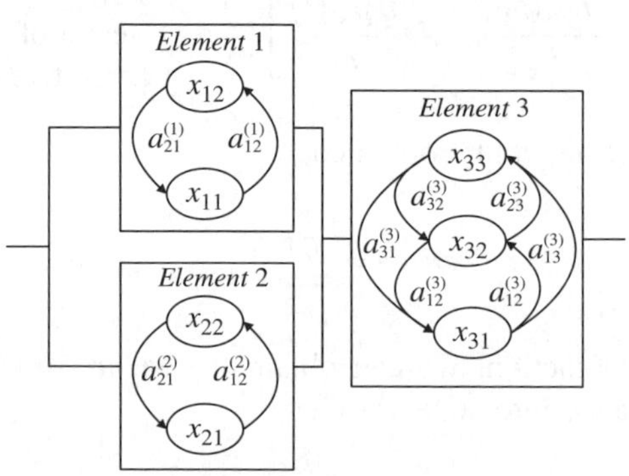

Figure 4.1 The MSS's structure and the state-transition diagram of its elements

For the first element:

$$\begin{cases} \dfrac{dp_{11}(t)}{dt} = -a_{12}^{(1)} p_{11}(t) + a_{21}^{(1)} p_{12}(t), \\ \dfrac{dp_{12}(t)}{dt} = -a_{21}^{(1)} p_{12}(t) + a_{12}^{(1)} p_{11}(t). \end{cases}$$

The initial conditions are: $\mathbf{p_{10}} = \{p_{11}(0), p_{12}(0)\} = \{0, 1\}$.

For the second element:

$$\begin{cases} \dfrac{dp_{21}(t)}{dt} = -a_{12}^{(2)} p_{21}(t) + a_{21}^{(2)} p_{12}(t), \\ \dfrac{dp_{22}(t)}{dt} = -a_{21}^{(2)} p_{22}(t) + a_{12}^{(2)} p_{21}(t). \end{cases}$$

The initial conditions are: $\mathbf{p_{20}} = \{p_{21}(0), p_{22}(0)\} = \{0, 1\}$.

For the third element:

$$\begin{cases} \dfrac{dp_{31}(t)}{dt} = -a_{12}^{(3)} p_{31}(t) + a_{21}^{(3)} p_{32}(t), \\ \dfrac{dp_{32}(t)}{dt} = a_{32}^{(3)} p_{33}(t) - (a_{21}^{(3)} + a_{23}^{(3)}) p_{32} + a_{12}^{(3)} p_{31}(t), \\ \dfrac{dp_{33}(t)}{dt} = -a_{32}^{(3)} p_{22}(t) + a_{23}^{(3)} p_{32}(t). \end{cases}$$

The initial conditions are: $\mathbf{p_{30}} = \{p_{31}(0), p_{32}(0), p_{33}(0)\} = \{0, 0, 1\}$.

By solving these three systems of ordinary differential equations we can obtain the L_z-transforms for three processes $X_1(t), X_2(t), X_3(t)$:

Process 1

$$X_1(t): \quad \mathbf{x}_1 = \{x_{11}, x_{12}\} = \{0, 1.5\}, \mathbf{p}_1(t) = \{p_{11}(t), p_{12}(t)\}, \mathbf{p_{10}}.$$

The associated L_z-transform: $L_z\{X_1(t)\} = \sum_{i=1}^{2} p_{1i}(t) z^{x_{1i}}$.

Process 2

$$X_2(t): \quad \mathbf{x}_2 = \{x_{21}, x_{22}\} = \{0, 2.0\}, \mathbf{p}_2(t) = \{p_{21}(t), p_{22}(t)\}, \mathbf{p_{20}}.$$

The associated L_z-transform:

$$L_z\{X_2(t)\} = \sum_{i=1}^{2} p_{2i}(t) z^{x_{2i}}.$$

Process 3

$$X_3(t): \quad \mathbf{x}_3 = \{x_{31}, x_{32}, x_{33}\} = \{0, 1.8, 4.0\}, \mathbf{p}_3(t) = \{p_{31}(t), p_{32}(t), p_{33}(t)\}, \mathbf{p_{30}}.$$

The associated L_z-transform: $L_z\{X_3(t)\} = \sum_{i=1}^{3} p_{3i}(t) z^{x_{3i}}$.

Now by applying Ushakov's operator Ω_f over all the L_z-transforms of the individual elements, one can obtain the Lz-transform $L_z\{X(t)\}$ associated with the output performance stochastic process $X(t)$ of the entire MSS:

$$L_z\{X(t)\} = \Omega_f\{L_z[X_1(t)], L_z[X_2(t)], L_z[X_3(t)]\},$$

where $f\{X_1(t), X_2(t), X_3(t)\} = \min\{X_1(t) + X_2(t), X_3(t)\}$ is the system structure function. Based on the known rules for a series-parallel MSS (Lisnianski *et al.* 2010), one can finally obtain the L_z-transform for the output performance Markov process

$$L_z\{X(t)\} = \sum_{i=1}^{5} p_i(t) z^{x_i},$$

where

$$\begin{array}{ll} x_1 = 0 & p_1(t) = p_{11}(t)p_{21}(t) + p_{31}(t)p_{12(t)} + p_{31}(t)p_{11}(t)p_{22}(t) \\ x_2 = 1.5 & p_2(t) = p_{12}(t)p_{21}(t)[p_{32}(t) + p_{33}(t)]; \\ x_3 = 1.8 & p_3(t) = p_{32}(t)p_{22}(t); \\ x_4 = 2.0 & p_4(t) = p_{33}(t)p_{11}(t)p_{22}(t); \\ x_5 = 3.5 & p_5(t) = p_{33}(t)p_{12}(t)p_{22}(t) \end{array}$$

Now by using the inverse L_z^{-1}-transform, the underlying Markov process can be revealed.

As one can see from the obtained L_z-transform, the underlying output process has five states

$$\mathbf{x} = \{x_1, x_2, x_3, x_4, x_5\} = \{0, 1.5, 1.8, 2.0, 3.5\}.$$

The corresponding states probabilities are as follows:

$$\mathbf{p}(t) = \{p_1(t), p_2(t), p_3(t), p_4(t), p_5(t)\}.$$

We suggest that initial state is the best state x_5.

In general, states probabilities for five-states Markov process can be obtained from the following system of ordinary differential equations:

$$\begin{cases} \dfrac{dp_{1A}(t)}{dt} = -\left(\sum\limits_{i=2}^{5} a_{1i}\right) p_{1A}(t) + \sum\limits_{i=2}^{5} a_{i1} p_{iA}(t), \\ \dfrac{dp_{2A}(t)}{dt} = -\left(\sum\limits_{i=1, i\neq 2}^{5} a_{2i}\right) p_{2A}(t) + \sum\limits_{i=1, i\neq 2}^{5} a_{i2} p_{iA}(t), \\ \dfrac{dp_{3A}(t)}{dt} = -\left(\sum\limits_{i=1, i\neq 3}^{5} a_{3i}\right) p_{3A}(t) + \sum\limits_{i=1, i\neq 3}^{5} a_{i3} p_{iA}(t), \\ \dfrac{dp_{4A}(t)}{dt} = -\left(\sum\limits_{i=1, i\neq 4}^{5} a_{4i}\right) p_{4A}(t) + \sum\limits_{i=1, i\neq 4}^{5} a_{i4} p_{iA}(t), \\ \dfrac{dp_{5A}(t)}{dt} = -\left(\sum\limits_{i=1}^{4} a_{5i}\right) p_{5A}(t) + \sum\limits_{i=1}^{4} a_{i5} p_{iA}(t). \end{cases}$$

under given initial conditions $p_{1A}(0) = p_{2A}(0) = p_{3A}(0) = p_{4A}(0) = 0$, $p_{5A}(0) = 1$.

The solution of this system $p_{1A}(t), p_{2A}(t), p_{3A}(t), p_{4A}(t), p_{5A}(t)$ should be numerically close to the probabilities $p_1(t), p_2(t), p_3(t), p_4(t), p_5(t)$ that were found above by using the L_z-transform.

In accordance with the GA procedure, transition intensities a_{ij} need be found in order to minimize (4.8) for the five-states Markov process:

$$Err = \sum_{i=1}^{N_p} \frac{|p_{1A}(t_i) - p_1(t_i)|}{p_1(t_i)} + \ldots + \sum_{i=1}^{N_p} \frac{|p_{5A}(t_i) - p_5(t_i)|}{p_5(t_i)}$$

In each GA cycle in this example, the solution $p_{1A}(t), p_{2A}(t), p_{3A}(t), p_{4A}(t), p_{5A}(t)$ should be obtained for the period of 0.15 year. During 0.15 year the transient mode for the solution $p_1(t), p_2(t), p_3(t), p_4(t), p_5(t)$ will be completely over and the process will be in a steady-state condition. For the comparison we shall take 1000 points – one point for 0.00015 year.

The population size in the GA is 100. A binary string is used to encode a solution as shown in Figure 4.2. As shown in Figure 4.2, a binary bit represents the transition intensity. The simple one-point crossover procedure has been used to generate the next feasible generations. The offspring will mutate with probability, which avoids premature convergence to a local optimum and facilities jumps in the solution space. The mutation probability is 0.005. The convergence criterion in the proposed GA is set as satisfying both the minimal number of genetic cycles (500 cycles) and a number of genetic cycles without improving the solution performance (50 cycles). The GA converges to optimal solutions by performing about 700 genetic cycles.

The transition intensities of the resulting output Markov process, which minimize (4.8), are as follows:

$$a_{12} = 95.000, a_{13} = 120.000, a_{14} = 80.000, a_{15} = 0.00,$$
$$a_{21} = 10.63, a_{23} = 96.88, a_{24} = 0.00, a_{25} = 50.0,$$
$$a_{31} = 8.0, a_{32} = 50.0, a_{34} = 191.09, a_{35} = 50.0,$$
$$a_{41} = 10.07, a_{42} = 0.00, a_{43} = 10.75, a_{45} = 387.5,$$
$$a_{51} = 0.000, a_{52} = 11.31, \ a_{53} = 12.75, a_{54} = 7.$$

So, the underlying Markov process $X(t)$ was completely revealed:

$$X(t) = \{\mathbf{x}, \mathbf{A}, \mathbf{p_0}\},$$

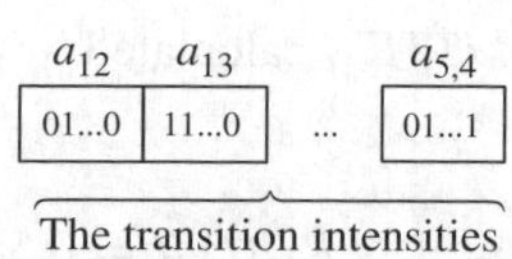

Figure 4.2 The GA chromosome structure

where $\mathbf{x} = \{x_1, x_2, x_3, x_4, x_5\} = \{0, 1.5, 1.8, 2.0, 3.5\}$;

$$\mathbf{A} = \begin{bmatrix} -295 & 95 & 120 & 80 & 0 \\ 10.63 & -157.51 & 96.88 & 0 & 50.0 \\ 8.0 & 50.0 & -299.09 & 191.09 & 50.0 \\ 10.07 & 0 & 10.75 & -408.32 & 387.5 \\ 0 & 11.31 & 12.75 & 7 & -31.06 \end{bmatrix};$$

$$\mathbf{p_0} = \{0, 0, 0, 0, 1\}.$$

In order to find the reliability function $R(t)$, all transitions from worst state 1 should be constrained. It means that all $a_{1i}, i = 2, \ldots, K$ should be zeroed. So we will have the matrix:

$$\mathbf{A}_0 = \begin{bmatrix} 0 & 0 & 0 & 0 & 0 \\ 10.63 & -157.51 & 96.88 & 0 & 50.0 \\ 8.0 & 50.0 & -299.09 & 191.09 & 50.0 \\ 10.07 & 0 & 10.75 & -408.32 & 387.5 \\ 0 & 11.31 & 12.75 & 7 & -31.06 \end{bmatrix}.$$

The reliability function may be obtained as the following (see expression (4.13))

$$R(t) = \sum_{i=2}^{5} p_i(t),$$

where functions $p_i(t)$ are obtained by solving the system of ordinary differential equations (4.12)

$$\frac{d\mathbf{p}(t)}{dt} = \mathbf{p}(t)\mathbf{A}_0$$

under initial conditions $\mathbf{p_0} = \{0, 0, 0, 0, 1\}$.

In Figure 4.3 one can see the graph of the reliability function $R_L(t)$ that was calculated by using the inverse L_z-transform method and the reliability function $R_M(t)$ that was calculated by a conventional straightforward Markov method in (Lisnianski *et al.* 2010). As one can see, the reliability curves representing those two solutions are positioned so close together that the difference between them cannot be distinguished visually. (The difference is in the 4th digit after the decimal point.)

Mean time to failure $MTTF_L$ calculated by using the L_z-transform method can be obtained by using expression (4.14)

$$MTTF_L = \int_0^\infty R_L(t)dt = 0.569 \text{ years}.$$

Note that the mean time to failure $MTTF_M$ calculated by using the straightforward Markov method is almost the same:

$$MTTF_M = \int_0^\infty R_M(t)dt = 0.568 \text{ years}.$$

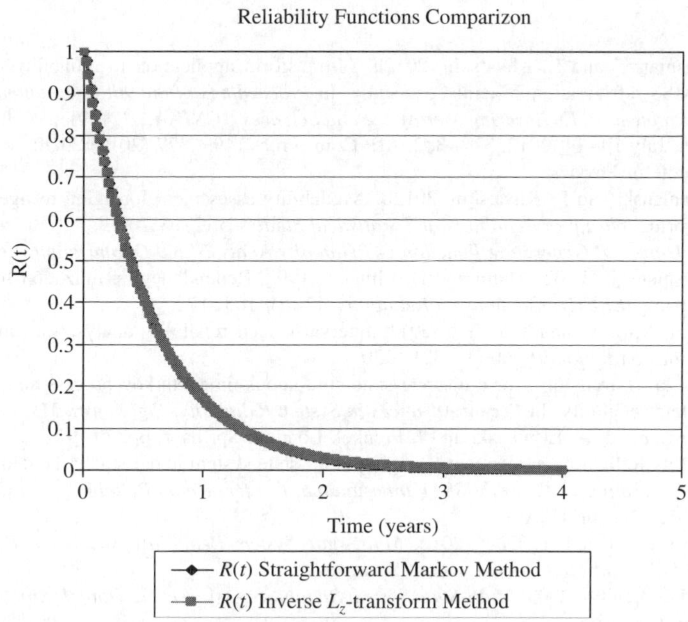

Figure 4.3 Graphs of reliability functions calculated by using the straightforward Markov method and by using the inverse L_z-transform

The error for $MTTF_L$ calculation is less than 0.2% which is a very good accuracy for reliability problems where high uncertainty in failure data is expected.

It should be noted that the resulting output Markov process that was found by using the L_z^{-1}-transform has only five states. The Markov process, which was built by using the straightforward Markov method in (Lisninaski *et al.* 2010) in order to perform the reliability analysis for this example has 12 states. Therefore, even for this simple example, the computational complexity decreases drastically with the use of the L_z^{-1}-transform.

4.5 Conclusion

This chapter has introduced and mathematically formulated the inverse L_z-transform (L_z^{-1}-transform) for the discrete-state continuous-time Markov process. The L_z^{-1}-transform application to the MSS reliability analysis expands the number of problems that can potentially be solved.

Based on the L_z-transform it is possible to use Ushakov's universal generating operator and corresponding universal generating functions techniques in order to find a reliability function of MSS and mean time to failure. Until now these characteristics could not have been found by using universal generating functions.

It was shown that the use of the L_z^{-1}-transform reduces the computational complexity compared to a traditional Markov method. The numerical example presented in this chapter has demonstrated the advantages of the proposed methodology.

References

Frenkel, I., A. Lisnianski and L. Khvatskin. 2012a. L_z-transform application to reliability measures estimation of aging MSS refrigeration system: Case study. In *Proceedings of the 9th International Conference on Mathematical Problems in Engineering, Aerospace and Sciences (ICNPAA2012)*, Vienna University of Technology, Austria, July 10–14, 2012. 377–382. AIP Conf. Proc. 1493, 377 (2012); DOI: 10.1063/1.4765516. American Institute of Physics.

Frenkel, I., A. Lisnianski and L. Khvatskin. 2012b. Availability assessment for aging refrigeration system by using L_z-transform. *Journal of Reliability and Statistical Studies* 5 (2): 33–43.

Levitin, G. 2005. *Universal Generating Function in Reliability Analysis and Optimization*. London: Springer.

Levitin, G., A. Lisnianski, H. Ben-Haim and D. Elmakis. 1998. Redundancy optimization for series–parallel multi-state systems. *IEEE Transactions on Reliability* 47 (2): 165–172.

Li, C.-Y., X. Chen, X.-S. Yi and J.-Y. Tao. 2011. Interval-valued reliability analysis of multi-state systems. *IEEE Transactions on Reliability* 60 (1): 321–330.

Lisnianski, A. 2012a. L_z-transform for a discrete-state continuous-time Markov process and its application to multi-state system reliability. In *Recent Advances in System Reliability: Signatures, Multi-state Systems and Statistical Inference*, ed. A. Lisnianski and I. Frenkel. London: Springer, pp. 79–95.

Lisnianski, A. 2012b. Reliability assessment for aging multi-state system under stochastic demand by using L_z-transform. In *Proceedings of the 18th ISSAT International Conference on Reliability and Quality in Design*. July 26–28, 2012, Boston, USA.

Lisnianski, A., I. Frenkel and Y. Ding. 2010. *Multi-state System Reliability Analysis and Optimization for Engineers and Industrial Managers*. London: Springer.

Lisnianski, A. and G. Levitin. 2003. *Multi-State System Reliability: Assessment, Optimization and Applications*. Singapore: World Scientific.

Lisnianski, A., G. Levitin, H. Ben-Haim and D. Elmakis. 1996. Power system structure optimization subject to reliability constraints. *Electric Power Systems Research* 39 (2): 145–152.

Natvig, B. 2011. *Multistate Systems Reliability: Theory with Applications*. New York: Wiley.

Swift, R.J. and S.A. Wirkus. 2006. *A Course in Ordinary Differential Equations*. New York: Chapman & Hall.

Tian, Z., M. Zuo and H. Huang. 2008. Reliability-redundancy allocation for multi-state series–parallel systems. *IEEE Transactions on Reliability* 57 (2): 303–310.

Trivedi, K. 2002. *Probability and Statistics with Reliability, Queuing and Computer Science Applications*. New York: Wiley.

Ushakov, I. 1986. A universal generating function. *Soviet Journal of Computer and System Sciences* 24: 37–49.

Ushakov, I. 1987. Optimal standby problem and a universal generating function. *Soviet Journal of Computer and System Sciences* 25: 61–73.

Yeh, W.C. 2006. The k-out-of-n acyclic multistate-node networks reliability evaluation using the universal generating function method. *Reliability Engineering and System Safety* 91 (7): 800–808.

Yeh, W.C. 2009. A convolution universal generating function method for evaluating the symbolic one-to-all target subset reliability function the acyclic multi-state information networks. *IEEE Transactions on Reliability* 58 (3): 476–484.

5

On the L_Z-Transform Application for Availability Assessment of an Aging Multi-State Water Cooling System for Medical Equipment

Ilia Frenkel,[1] Anatoly Lisnianski[2] and Lev Khvatskin[1]
[1] *Center for Reliability and Risk Management, Industrial Engineering and Management Department, SCE-Shamoon College of Engineering, Israel*
[2] *The Israel Electric Corporation Ltd, Israel*

5.1 Introduction

Today's high-tech medical equipment requires effective cooling and precise temperature control or cooling below the ambient temperature. As heat loads continue to increase, more and more equipment manufacturers are desperately looking for more efficient cooling methods to remove high-wattage density heat loads for medical imaging equipment, medical lasers, and other devices. Removing heat from these systems efficiently is necessary to reduce the thermal stress which will, in turn, lead to increased accuracy, longevity and reduced energy consumption. Greater cooling efficiencies will also allow medical equipment to be operated for longer periods without overheating, providing a greater return on investment for medical facilities and a likely increase in availability of the equipment to patients.

For MRI scanners to operate at the highest levels of efficiency, the magnet inside the scanner has to be as cool as possible, therefore liquid helium is often used as the refrigerant. The cooling system analyzed in this chapter has water-cooled chillers that are responsible for removing heat from the MRI scanner.

Applied Reliability Engineering and Risk Analysis: Probabilistic Models and Statistical Inference, First Edition.
Edited by Ilia B. Frenkel, Alex Karagrigoriou, Anatoly Lisnianski and Andre Kleyner.

The cooling system for magnetic resonance inspection (MRI) plant is based on a cold water supply and has to provide a temperature of return water from the magnet heat exchanger between 8 and 18 degrees Celsius. In the case of a temperature increase, the fluid helium of the magnet cooling system begins evaporating and after approximately two hours of operation with the return water temperature around 25 degrees, the MRI plant automatically shuts down. The damage of this kind of emergency failure causes the suspension of medical treatment for several days. Only specially trained staff members are allowed to restart a magnet and fix the fluid helium evaporation, resulting in direct financial losses.

If the water cooling system does not fix the demand conditions, the aftermath will be catastrophic: evaporation of helium is strengthened and the helium stream escapes through special tubes with the risk of destroying not only the MRI plant but also neighboring buildings. Thus, this type of system needs several levels of reserve.

In this chapter we consider a part of the reserve water cooling system for a new MRI plant in an Israeli hospital. The system consists of three cooling water chillers, a heat exchanger, two circulation pumps and an electric board.

The nominal performance of the entire system is $24 \cdot 10^4$ BTU/hr and the performance of each chiller is $12 \cdot 10^4$ BTU/hr. The system demand is not constant and depends on the demand for MRI scans, which vary widely with time and have to distribute MRI scan capacity among several medical departments. The cooling demand value varies from $10 \cdot 10^4$ to $18 \cdot 10^4$ BTU/hr.

Due to the system's nature, a fault in a single unit has only a partial effect on the entire performance: it reduces only the system's cooling capacity. Partial failure of the chillers, pumps or different elements of the electric board leads to a partial system failure (the reduction of the output cooling capacity) as well as to complete system failures. Therefore, the water cooling system can be treated as multi-state system (MSS), where components and the entire system in general have an arbitrary finite number of states corresponding to the different performance rates (Lisnianski and Levitin 2003; Lisnianski *et al.* 2010; Natvig 2011). The performance rate (cool capacity) of the system at any instant t is interpreted as a discrete-state continuous-time stochastic process $G(t)$ (Lisnianski *et al.* 2010). Water cooling systems like the ones discussed in the present work, often have aging components, and thus one could consider in general a non-homogeneous Markov model with some transition rates (intensities) being time-dependent, as the suitable reliability model. Such a model, however, is complex enough: even in relatively simple cases it has hundreds of states. So, it is rather difficult to build the model and to solve the corresponding system of differential equations by using a straightforward Markov method.

In recent years a specific approach called the universal generating function (UGF) technique has been widely applied to MSS reliability analysis (Lisnianski and Levitin 2003; Levitin 2005). The approach was primarily introduced by Ushakov (1986, 1987). The UGF technique allows one to algebraically obtain the entire MSS performance distribution through the performance distributions of its elements. However, the main restriction of this powerful technique is that theoretically it may only be applied to random variables and therefore, concerning MSS reliability, it operates only with steady-states performance distributions. Thus, the UGF technique cannot be applied when an aging system is under consideration. In order to extend the UGF technique application to dynamic MSS reliability analysis a special transform, called L_z-transform, was introduced (Lisnianski 2012) for discrete-state continuous-time Markov processes.

The L_z-transform was successfully applied to an availability analysis of real-world aging multi-state systems under constant demand (Frenkel *et al.* 2012) and its efficiency was demonstrated. In practice, there are aging multi-state systems that are functioning under variable stochastic demand. In the present chapter, the L_z-transform is applied to a real MSS aging water cooling system that is functioning under variable stochastic demand and its availability is analyzed. It is shown that the L_z-transform application drastically simplifies the availability computation for such a system compared with the straightforward Markov method.

5.2 Brief Description of the L_z-Transform Method

In this chapter the L_z-transform method is implemented for the evaluation of availability for an MSS aging water cooling system functioning under stochastic demand. The method was introduced by Lisnianski (2012) where one can find its detailed description and corresponding mathematical proofs. Briefly, the description of the method is as follows.

We consider a discrete-state continuous-time (DSCT) Markov process (Trivedi 2002) $X(t) \in \{x_1, \ldots, x_K\}$, which has K possible states i, $(i = 1, \ldots, K)$ where the performance level associated with any state i is x_i. This Markov process is completely defined by the set of possible states $\mathbf{x} = \{x_1, x_2, \ldots, x_K\}$, the transition intensities matrix depending on time $A = (a_{ij}(t))$, $i, j = 1, 2, \ldots, K$ and by the initial states probability distribution given by $\mathbf{p}_0 = [p_{10} = \Pr\{X(0) = x_1\}, \ldots, p_{K0} = \Pr\{X(0) = x_K\}]$.

According to (Lisnianski 2012), the L_z-transform of a DSCT Markov process $X(t)$ is defined by the following expression

$$L_z\{X(t)\} = \sum_{i=1}^{K} p_i(t) z^{g_i}, \tag{5.1}$$

where $p_i(t)$ is the probability that the process is in state i at time instant $t \geq 0$ for a given initial states probability distribution $\mathbf{p}_0$, g_i is the performance level of state i, and z is a complex variable.

In general, any element j in MSS can have k_j different states corresponding to different performances, represented by the set $\mathbf{g}_j = \{g_{j1}, \ldots, g_{jk_j}\}$, where g_{ji} is the performance rate of element j in the state i, $i \in \{1, 2, \ldots, k_j\}$, and $j \in \{1, \ldots, n\}$, where n is the number of elements in the MSS.

According to the L_z-transform method, at first, a Markov model of stochastic process should be built for each multi-state element in MSS. Based on this model, state probabilities $p_{ji}(t) = \Pr\{G_j(t) = g_{ji}\}$, $i \in \{1, \ldots, k_j\}$ for every MSS's element can be obtained as a solution of the corresponding system of differential equations under the given initial conditions. These probabilities define the output stochastic process $G_j(t)$ for each element j in the MSS.

Then the individual L_z-transform for each element j should be found:

$$L_z\{G_j(t)\} = \sum_{i=1}^{k_j} p_{ji}(t) z^{g_{ji}}, \; j = 1, \ldots, n \tag{5.2}$$

At the next stage, the L_z-transform of the output stochastic process for the entire MSS should be defined based on the previously determined L_z-transform for each element

j and the system structure function f, given by $G(t) = f(G_1(t), \ldots, G_n(t))$, which produces the output stochastic process of the entire MSS based on the stochastic processes of all the MSS's elements.

Lisnianski (2012) has shown that in order to find the L_z-transform of the resulting DSCT Markov process $G(t)$, which is the single-valued function $G(t) = f(G_1(t), \ldots, G_n(t))$ of n independent DSCT Markov processes $G_j(t)$, $j = 1, \ldots, n$, one can apply Ushakov's universal generating operator (Ushakov 1986, 1987) to all the individual L_z-transforms $L_z\{G_j(t)\}$ over all time points $t \geq 0$

$$L_z\{G(t)\} = \Omega_f\{L_z[G_1(t)], \ldots, L_z[G_n(t)]\}. \tag{5.3}$$

Therefore, by using Ushakov's operator Ω_f over all the L_z-transforms of individual elements, one can obtain the resulting L_z-transform $L_z\{G(t)\}$ associated with the output performance stochastic process $G(t)$ of the entire MSS. The implementation of Ushakov's operator which is well established for many different structure functions f (Lisnianski and Levitin 2003; Levitin 2005), drastically reduces the computational burden associated with such calculations.

The resulting L_z-transform $L_z\{G(t)\}$ is associated with the output performance stochastic process for the entire MSS and MSS reliability measures can be easily derived from the resulting L_z-transform.

If the L_z-transform

$$L_z\{G(t)\} = \sum_{k=1}^{K} p_k(t) z^{g_k} \tag{5.4}$$

of the entire MSS's output stochastic process $G(t) \in \{g_1, \ldots, g_K\}$ is known, then the important system's reliability measures can be easily obtained. For example, the system availability for the demand level w at instant $t \geq 0$

$$A(t) = \sum_{g \geq w} p_k(t). \tag{5.5}$$

In other words, in order to find the MSS's instantaneous availability, one should summarize all probabilities in the L_z-transform from terms where the powers of z are greater or equal to demand w.

5.3 Multi-state Model of the Water Cooling System for the MRI Equipment

5.3.1 System Description

We consider an MRI water cooling system used in Israeli hospitals. The system consists of four sub-systems:

1. three 2-level chillers;
2. a 5-level aging heat exchanger;
3. two 2-level aging pumps;
4. the electrical board.

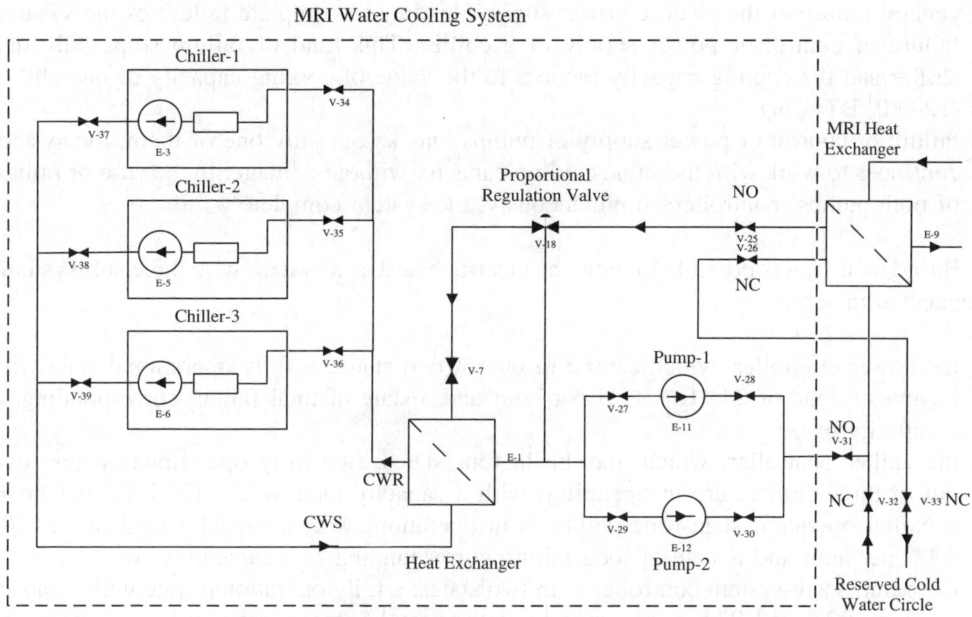

Figure 5.1 Principal scheme of the water cooling system

The principal scheme of the water cooling system is presented in Figure 5.1. The nominal performance of the whole system is $24 \cdot 10^4$ BTU/hr. The performance of each chiller is $12 \cdot 10^4$ BTU/hr. Therefore, connecting chillers as a 2-out-of-3 sub-system results in nominal performance, which is required for the functioning of the system.

To improve the temperature regulation of the MRI device, the cooling system is separated into two circles joined by a heat exchanger. The maximum cool capacity of the heat exchanger corresponds to the nominal capacity of two simultaneously working chillers. The functioning of the heat exchanger is characterized by a degradation process. For simplification we describe the heat exchanger as a device with five states of performance degradation: a fully operational state with a capacity of $24 \cdot 10^4$ BTU/hour, states of partial failure corresponding to capacities $20 \cdot 10^4$, $18 \cdot 10^4$, $16 \cdot 10^4$ of BTU per hour, and a total failure corresponding to a capacity of 0. The heat exchanger degradation process is an increasing process, so that the device possesses the aging property.

Cold water circulation in the second circle is realized by two pumps: a main and a reserve. Each pump supports the nominal cooling capacity of the system. In case of failure of one pump, the second automatically enters into operation. So only after the breakdown of both pumps will the system fail completely. Each pump can be in one of two states: a fully operational state with a capacity load of $24 \cdot 10^4$ BTU per hour, and a state of total failure corresponding to a capacity of 0. Pumps are aging devices, but wear of pumps does not influence the performance of the cooling system.

The principal role of the electrical board is the power supply and control of general parts of the cooling system. It is possible to distinguish failures of the electric board into three types:

- general failure of the electric power supply, leading to complete failure of the system;
- failure of control or power supply of a chiller. This kind of failure stops only one chiller and the cooling capacity reduces to the value of cooling capacity of one chiller ($12 \cdot 10^4$ BTU/hr).
- failure of control or power supply of pumps knocks out only one of them; the system continues to work with the same cooling capacity without damage. In the case of failure of both pumps' controllers simultaneously, the system completely fails.

Based on this, it is possible to view the electric board as a system with three sub-systems connected in series:

- the power controller, which can be in one of two states: a fully operational state with a capacity load of $24 \cdot 10^4$ BTU per hour and a state of total failure corresponding to a capacity of 0;
- the chiller controller, which may be in four states: two fully operational states (two out of three chillers are in operation) with a capacity load of $24 \cdot 10^4$ BTU per hour, a partial operation state (one chiller is in operation) with a capacity load of $12 \cdot 10^4$ BTU per hour and a state of total failure corresponding to a capacity of 0;
- the pump's sub-system controller with two states: a fully operational state with a capacity load of $24 \cdot 10^4$ BTU per hour and a state of total failure corresponding to a capacity of 0.

Working conditions may be of varying demand depending on the demand for MRI scans, which differ widely with time and have to distribute the MRI scan capacity among several medical departments. This situation is typical of hospital needs of the MRI system. Usually, for such systems, demand can be described by a two-level Markov process with a peak level and a low level. The cool demand value changes from $10 \cdot 10^4$ (low level) to the $18 \cdot 10^4$ BTU/hr (peak level).

5.3.2 *The Chiller Sub-System*

The chiller sub-system consists of three connected in parallel packaged chillers. Each chiller has no possibility of cooling regulation, so their performance can be allowed in one of two principal states: full capacity or zero capacity (the so-called two states regulation). According to the equipment type, the cooling capacity of each chiller is $12 \cdot 10^4$ BTU/hr; two chillers working simultaneously can cover a maximal cooling load needed for MRI functioning. So from a structural reliability point of view, this sub-system could be presented as a 2-out-of-3 system. For simplification, we will present the system capacity load in 10^4 BTU per hour units.

Using the state-transitions diagram (Figure 5.2) in accordance with the Markov method we can build the following system of differential equations for each chiller ($i = 1, 2, 3$):

$$\begin{cases} \dfrac{dp_{i1}^{Ch}(t)}{dt} = -\lambda^{Ch} p_{i1}^{Ch}(t) + \mu^{Ch} p_{i2}^{Ch}(t), \\ \dfrac{dp_{i2}^{Ch}(t)}{dt} = \lambda^{Ch} p_{i1}^{Ch}(t) - \mu^{Ch} p_{i2}^{Ch}(t). \end{cases} \tag{5.6}$$

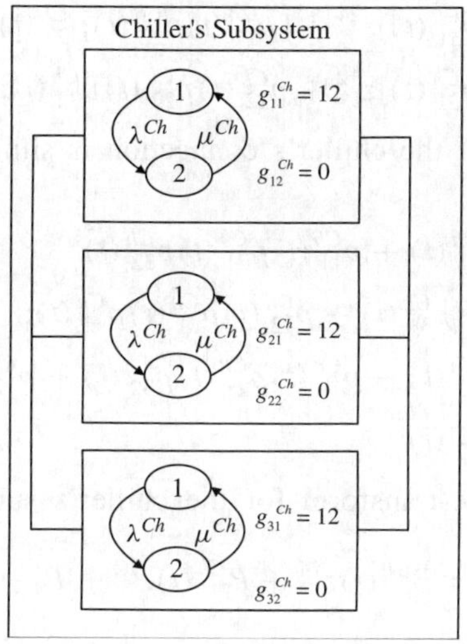

Figure 5.2 Structure and state-transition diagram of the chiller sub-system

Initial conditions are: $p_{i1}^{Ch}(0) = 1$; $p_{i2}^{Ch}(0) = 0$.

A numerical solution for probabilities $p_{i1}^{Ch}(t)$ and $p_{i2}^{Ch}(t)$ $(i = 1, 2, 3)$ can be obtained for each of these three systems of differential equations using MATLAB®. Therefore, for each chiller we can obtain the following output performance stochastic processes:

$$\begin{cases} \mathbf{g}_1^{Ch} = \{g_{11}^{Ch}, g_{12}^{Ch}\} = \{12, 0\}, \\ \mathbf{p}_1^{Ch}(t) = \{p_{11}^{Ch}(t), p_{12}^{Ch}(t)\}; \end{cases} \quad \begin{cases} \mathbf{g}_2^{Ch} = \{g_{21}^{Ch}, g_{22}^{Ch}\} = \{12, 0\}, \\ \mathbf{p}_2^{Ch}(t) = \{p_{21}^{Ch}(t), p_{22}^{Ch}(t)\}; \end{cases}$$

$$\begin{cases} \mathbf{g}_3^{Ch} = \{g_{31}^{Ch}, g_{32}^{Ch}\} = \{12, 0\}, \\ \mathbf{p}_3^{Ch}(t) = \{p_{31}^{Ch}(t), p_{32}^{Ch}(t)\}. \end{cases}$$

Having the sets $\mathbf{g}_i^{Ch}$, $\mathbf{p}_i^{Ch}(t)$, $i = 1, 2, 3$ one can define for each chiller Lz-transforms, associated with the chiller's output performance stochastic process:

$$L_z\{g_1(t)\} = p_{11}^{Ch}(t)z^{g_{11}^{Ch}} + p_{12}^{Ch}(t)z^{g_{12}^{Ch}} = p_{11}^{Ch}(t)z^{12} + p_{12}^{Ch}(t)z^0,$$

$$L_z\{g_2(t)\} = p_{21}^{Ch}(t)z^{g_{21}^{Ch}} + p_{22}^{Ch}(t)z^{g_{22}^{Ch}} = p_{21}^{Ch}(t)z^{12} + p_{22}^{Ch}(t)z^0,$$

$$L_z\{g_3(t)\} = p_{31}^{Ch}(t)z^{g_{31}^{Ch}} + p_{32}^{Ch}(t)z^{g_{32}^{Ch}} = p_{31}^{Ch}(t)z^{12} + p_{32}(t)z^0.$$

Using the composition operator Ω_{fpar} (Lisnianski and Levitin 2003; Lisnianski *et al.* 2010) for three chillers, connected in parallel, one can obtain the L_z-transform $L_z\{G^{Ch}(t)\}$ for the chiller's sub-system, where the powers of z are found as the sum of the powers of the corresponding terms:

$$L_z\{G^{Ch}(t)\} = \Omega_{fpar}(g_1^{Ch}(t), g_2^{Ch}(t), g_3^{Ch}(t))$$
$$= p_{11}^{Ch}(t)p_{21}^{Ch}(t)p_{31}^{Ch}(t)z^{36} + [p_{11}^{Ch}(t)p_{21}^{Ch}(t)p_{32}^{Ch}(t) + p_{11}^{Ch}(t)p_{22}^{Ch}(t)p_{31}^{Ch}(t)$$

$$+ p_{12}^{Ch}(t)p_{21}^{Ch}(t)p_{31}^{Ch}(t)]z^{24} + [p_{11}^{Ch}(t)p_{22}^{Ch}(t)p_{32}^{Ch}(t) + p_{12}^{Ch}(t)p_{21}^{Ch}(t)p_{32}^{Ch}(t)$$
$$+ p_{12}^{Ch}(t)p_{22}^{Ch}(t)p_{31}^{Ch}(t)]z^{12} + p_{12}^{Ch}(t)p_{22}^{Ch}(t)p_{32}^{Ch}(t)z^{0}.$$

Taking into consideration that the chiller's connection is similar to 2-out-of-3:G and using the following notations:

$$P_1^{Ch}(t) = p_{11}^{Ch}(t)p_{21}^{Ch}(t)p_{31}^{Ch}(t) + p_{11}^{Ch}(t)p_{21}^{Ch}(t)p_{32}^{Ch}(t)$$
$$+ p_{11}^{Ch}(t)p_{22}^{Ch}(t)p_{31}^{Ch}(t) + p_{12}^{Ch}(t)p_{21}^{Ch}(t)p_{31}^{Ch}(t);$$
$$P_2^{Ch}(t) = p_{11}^{Ch}(t)p_{22}^{Ch}(t)p_{32}^{Ch}(t) + p_{12}^{Ch}(t)p_{21}^{Ch}(t)p_{32}^{Ch}(t) + p_{12}^{Ch}(t)p_{22}^{Ch}(t)p_{31}^{Ch}(t);$$
$$P_3^{Ch}(t) = p_{12}^{Ch}(t)p_{22}^{Ch}(t)p_{32}^{Ch}(t);$$

one can obtain the resulting L_z-transform for the chiller's sub-system in the following form:

$$L_z\{G^{Ch}(t)\} = P_1^{Ch}(t)z^{24} + P_2^{Ch}(t)z^{12} + P_3^{Ch}(t)z^{0}. \tag{5.7}$$

5.3.3 The Heat Exchanger Sub-System

As mentioned earlier, the heat exchanger is a device with five states of performance degradation: a fully operational state with a capacity of $24 \cdot 10^4$ BTU/hour, states of partial failure corresponding to capacities $22 \cdot 10^4, 20 \cdot 10^4, 16 \cdot 10^4$ of BTU per hour, and a total failure corresponding to a capacity of 0.

According to the Markov method, we can build the state-transitions diagram (Figure 5.3) and the following system of differential equations:

$$\begin{cases} \dfrac{dp_1^{HE}(t)}{dt} = -4(\lambda^{HE} + \lambda^{HEt})p_1^{HE}(t) + \mu^{HE}p_2^{HE}(t), \\ \dfrac{dp_2^{HE}(t)}{dt} = 4\lambda^{HE}p_1^{HE}(t) - (3\lambda^{HE} + \lambda^{HEt} + 2\mu^{HE})p_2^{HE}(t) + 2\mu^{HE}p_3^{HE}(t); \\ \dfrac{dp_3^{HE}(t)}{dt} = 3\lambda^{HE}p_2^{HE}(t) - (2\lambda^{HE} + \lambda^{HEt} + 3\mu^{HE})p_3^{HE}(t) + 3\mu^{HE}p_5^{HE}(t); \\ \dfrac{dp_4^{HE}(t)}{dt} = 2\lambda^{HE}p_3^{HE}(t) - (2\lambda^{HE} + \lambda^{HEt} + 3\mu^{HE})p_4^{HE}(t) + 4\mu^{HE}p_5^{HE}(t); \\ \dfrac{dp_5^{HE}(t)}{dt} = \lambda^{HE}p_1^{HE}(t) + \lambda^{HEt}p_2^{HE}(t) + \lambda^{HEt}p_2^{HE}(t) + (\lambda^{HE} + \lambda^{HEt})p_4^{HE}(t) \\ \quad -4\mu^{HE}p_5^{HE}(t). \end{cases} \tag{5.8}$$

Initial conditions are: $p_1^{HE}(0) = 1$; $p_2^{HE}(0) = p_3^{HE}(0) = p_4^{HE}(0) = p_5^{HE}(0) = 0$.

A numerical solution for probabilities $p_i^{HE}(t), \mathrm{i} = 1, \ldots, 5$ can be obtained for each of these systems of differential equations using MATLAB®. Therefore, one obtains for the heat exchanger the following output performance stochastic processes:

$$\mathbf{g}^{HE} = \{g_1^{HE}, g_2^{HE}, g_3^{HE}, g_4^{HE}, g_5^{HE}\} = \{24, 22, 20, 16, 0\};$$
$$\mathbf{p}^{HE}(t) = \{p_1^{HE}(t), p_2^{HE}(t), p_3^{HE}(t), p_4^{HE}(t), p_5^{HE}(t)\}.$$

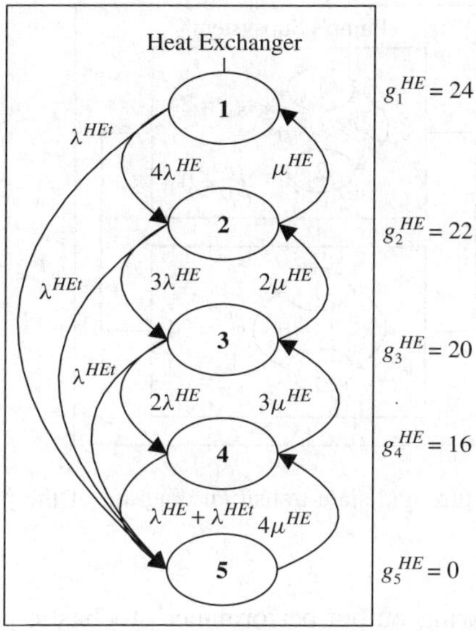

Figure 5.3 State-transition diagram of the heat exchanger

Having the sets $\mathbf{g}^{HE}$, $\mathbf{p}^{HE}(t)$ one can define the L_z-transforms associated with the heat exchanger's output performance stochastic process:

$$L_z\{G^{HE}(t)\} = p_1^{HE}(t)z^{24} + p_2^{HE}(t)z^{22} + p_3^{HE}(t)z^{20} + p_4^{HE}(t)z^{16} + p_5^{HE}(t)z^{0}. \quad (5.9)$$

5.3.4 The Pump Sub-System

As presented earlier, each pump supports full cooling capacity of the system. In case of failure of one pump, the second automatically turns on. So only after the breakdown of both pumps will the system fail completely. Each pump can be in one of two states: a fully operational state with a capacity load of $24 \cdot 10^4$ BTU per hour, and a state of total failure corresponding to a capacity of 0.

According to the Markov method, we build the state-transitions diagram (Figure 5.4) and the following system of differential equations for each pump ($i = 1, 2$):

$$\begin{cases} \dfrac{dp_{i1}^P(t)}{dt} = -\lambda^P p_{i1}^P(t) + \mu^P p_{i2}^P(t), \\ \dfrac{dp_{i2}^P(t)}{dt} = \lambda^P p_{i1}^P(t) - \mu^P p_{i2}^P(t). \end{cases} \quad (5.10)$$

The initial conditions are: $p_{i1}^P(0) = 1$; $p_{i2}^P(0) = 0$.

A numerical solution for probabilities $p_{i1}^P(t)$ and $p_{i2}^P(t)$ ($i = 1, 2$) can be obtained for each of these two systems of differential equations using MATLAB®. Therefore, for each

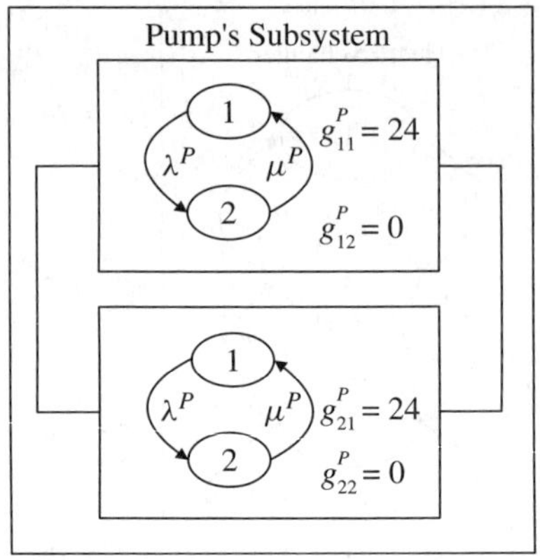

Figure 5.4 Structure and state-transition diagram of the pump's sub-system

pump we obtain the following output performance stochastic processes:

$$\begin{cases} \mathbf{g}_1^P = \{g_{11}^P, g_{12}^P\} = \{24, 0\}, \\ \mathbf{p}_1^P(t) = \{p_{11}^P(t), p_{12}^P(t)\}; \end{cases} \quad \begin{cases} \mathbf{g}_2^P = \{g_{21}^P, g_{22}^P\} = \{24, 0\}, \\ \mathbf{p}_2^P(t) = \{p_{21}^P(t), p_{22}^P(t)\}; \end{cases}$$

Having the sets $\mathbf{g}_i^P$, $\mathbf{p}_i^P(t), i = 1, 2$ one can define for each pump the L_z-transforms associated with the pump's output performance stochastic process:

$$L_z\{g_1^P(t)\} = p_{11}^P(t)z^{g_{11}^P} + p_{12}^P(t)z^{g_{12}^P} = p_{11}^P(t)z^{24} + p_{12}^P(t)z^0,$$
$$L_z\{g_2^P(t)\} = p_{21}^P(t)z^{g_{21}^P} + p_{22}^P(t)z^{g_{22}^P} = p_{21}^P(t)z^{24} + p_{22}^P(t)z^0.$$

Using the composition operator Ω_{fpar} for two pumps, connected in parallel, one obtains the L_z-transform $L_z\{G^P(t)\}$ for the pump sub-system, where the powers of z are found as maximum values of powers of the corresponding terms:

$$L_z\{G^P(t)\} = \Omega_{f_{par}}(g_1^P(t), g_2^P(t))$$
$$= [p_{11}^P(t)p_{21}^P(t) + p_{11}^P(t)p_{22}^P(t) + p_{21}^P(t)p_{12}^P(t)]z^{24} + p_{12}^P(t)p_{22}^P(t)z^0.$$

Using the following notations:

$$P_1^P(t) = p_{11}^P(t)p_{21}^P(t) + p_{11}^P(t)p_{22}^P(t) + p_{21}^P(t)p_{12}^P(t);$$
$$P_2^P(t) = p_{12}^P(t)p_{22}^P(t);$$

one can obtain the resulting L_z-transform for the pump sub-system in the following form:

$$L_z\{G^P(t)\} = P_1^P(t)z^{24} + P_2^P(t)z^0. \tag{5.11}$$

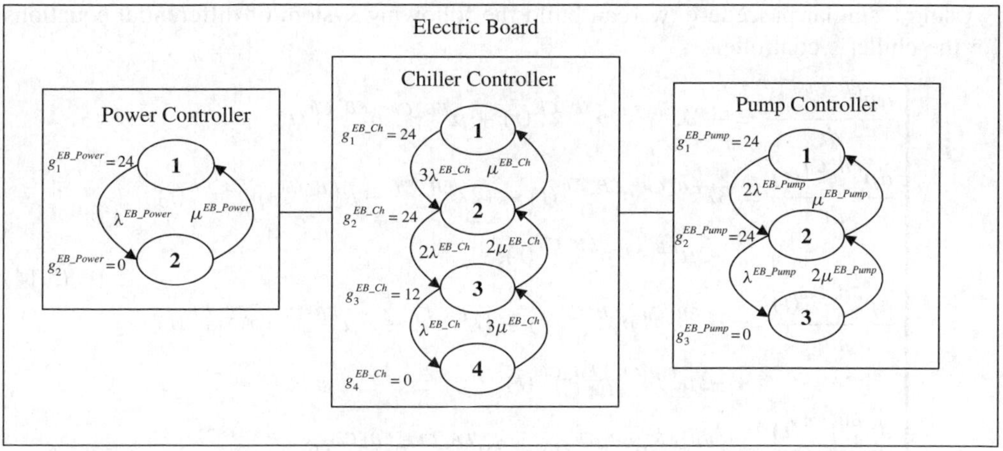

Figure 5.5 Structure and state-transition diagram of the electric board sub-system

5.3.5 The Electric Board Sub-System

The structure and state-transition diagram of the electric board sub-system is presented in Figure 5.5.

According to the Markov method, we can build the following system of differential equations for the power controller:

$$\begin{cases} \dfrac{dp_1^{EB\_Power}(t)}{dt} = -\lambda^{EB\_Power} p_1^{EB\_Power}(t) + \mu^{EB\_Power} p_2^{EB\_Power}(t), \\ \dfrac{dp_2^{EB\_Power}(t)}{dt} = \lambda^{EB\_Power} p_1^{EB\_Power}(t) - \mu^{EB\_Power} p_2^{EB\_Power}(t). \end{cases} \tag{5.12}$$

Initial conditions are: $p_1^{EB\_Power}(0) = 1$; $p_2^{EB\_Power}(0) = 0$.

A numerical solution for probabilities $p_1^{EB\_Power}(t)$ and $p_2^{EB\_Power}(t)$ can be obtained by using MATLAB®. Therefore, for the power controller, we can obtain the following output performance stochastic process:

$$\begin{cases} \mathbf{g}^{EB\_Power} = \left\{ g_1^{EB\_Power}, g_2^{EB\_Power} \right\} = \{24, 0\}, \\ \mathbf{p}^{EB\_Power}(t) = \{ p_1^{EB\_Power}(t), p_2^{EB\_Power}(t) \}. \end{cases}$$

Having the sets $\mathbf{g}^{EB\_Power}$, $\mathbf{p}^{EB\_Power}(t)$, one can define for the power controller the L_z-transform associated with the power controller's output performance stochastic process:

$$\begin{aligned} L_z\{G^{EB\_Power}(t)\} &= p_1^{EB\_Power}(t) z^{g_1^{EB\_Power}} + p_2^{EB\_Power}(t) z^{g_2^{EB\_Power}} \\ &= p_1^{EB\_Power}(t) z^{24} + p_2^{EB\_Power}(t) z^{0}. \end{aligned} \tag{5.13}$$

Using a similar procedure, we can build the following system of differential equations for the chiller's controller:

$$\begin{cases} \dfrac{dp_1^{EB\_Ch}(t)}{dt} = -3\lambda^{EB\_Ch} p_1^{EB\_Ch}(t) + \mu^{EB\_Ch} p_2^{EB\_Ch}(t), \\ \dfrac{dp_2^{EB\_Ch}(t)}{dt} = 3\lambda^{EB\_Ch} p_1^{EB\_Ch}(t) - (2\lambda^{EB\_Ch} + \mu^{EB\_Ch}) p_2^{EB\_Ch}(t) \\ \qquad +2\mu^{EB\_Ch} p_3^{EB\_Ch}(t), \\ \dfrac{dp_3^{EB\_Ch}(t)}{dt} = 2\lambda^{EB\_Ch} p_2^{EB\_Ch}(t) - (\lambda^{EB\_Ch} + 2\mu^{EB\_Ch}) p_3^{EB\_Ch}(t) \\ \qquad +3\mu^{EB\_Ch} p_4^{EB\_Ch}(t), \\ \dfrac{dp_4^{EB\_Ch}(t)}{dt} = \lambda^{EB\_Ch} p_3^{EB\_Ch}(t) - 3\mu^{EB\_Ch} p_4^{EB\_Ch}(t). \end{cases} \tag{5.14}$$

The initial conditions are: $p_1^{EB\_Ch}(0) = 1$; $p_2^{EB\_Ch}(0) = p_3^{EB\_Ch}(0) = p_4^{EB\_Ch}(0) = 0$.

A numerical solution for probabilities $p_1^{EB\_Ch}(t)$; $p_2^{EB\_Ch}(t)$; $p_3^{EB\_Ch}(t)$; $p_4^{EB\_Ch}(t)$ can be obtained using MATLAB®. Therefore, one obtains for the power controller the following output performance stochastic process:

$$\begin{cases} \mathbf{g}^{EB\_Ch} = \left\{g_1^{EB\_Ch}, g_2^{EB\_Ch}, g_3^{EB\_Ch}, g_4^{EB\_Ch}\right\} = \{24, 24, 12, 0\}, \\ \mathbf{p}^{EB\_Ch}(t) = \{p_1^{EB\_Ch}(t), p_2^{EB\_Ch}(t), p_3^{EB\_Ch}(t), p_4^{EB\_Ch}(t)\}. \end{cases}$$

Having the sets $\mathbf{g}^{EB\_Ch}$, $\mathbf{p}^{EB\_Ch}(t)$, one can define for the chiller controller the L_z-transform, associated with the power controller's output performance stochastic process:

$$\begin{aligned} L_z\{G^{EB\_Ch}(t)\} &= p_1^{EB\_Ch}(t) z^{g_1^{EB\_Ch}} + p_2^{EB\_Ch}(t) z^{g_2^{EB\_Ch}} \\ &\quad + p_3^{EB\_Ch}(t) z^{g_3^{EB\_Ch}} + p_4^{EB\_Ch}(t) z^{g_4^{EB\_Ch}} \\ &= (p_1^{EB\_Ch}(t) + p_2^{EB\_Ch}(t)) z^{24} + p_3^{EB\_Ch}(t) z^{12} + p_4^{EB\_Ch}(t) z^{0}. \end{aligned} \tag{5.15}$$

Using the same procedure, we build the following system of differential equations for pump's controller:

$$\begin{cases} \dfrac{dp_1^{EB\_Pump}(t)}{dt} = -\lambda^{EB\_Pump} p_1^{EB\_Pump}(t) + \mu^{EB\_Pump} p_2^{EB\_Pump}(t) \\ \dfrac{dp_2^{EB\_Pump}(t)}{dt} = 2\lambda^{EB\_Pump} p_1^{EB\_Pump}(t) - (\lambda^{EB\_Pump} + 2\mu^{EB\_Pump}) p_2^{EB\_Pump}(t) \\ \qquad + 2\mu^{EB\_Pump} p_3^{EB\_Pump}(t) \\ \dfrac{dp_3^{EB\_Pump}(t)}{dt} = \lambda^{EB\_Pump} p_2^{EB\_Pump}(t) - 2\mu^{EB\_Pump} p_3^{EB\_Pump}(t). \end{cases} \tag{5.16}$$

The initial conditions are: $p_1^{EB\_Pump}(0) = 1$; $p_2^{EB\_Pump}(0) = p_3^{EB\_Pump}(0) = 0$.

A numerical solution for probabilities $p_1^{EB\_Pump}(t)$; $p_2^{EB\_Pump}(t)$; $p_3^{EB\_Pump}(t)$ can be obtained using MATLAB®. Therefore, for the pump's controller we obtain the following output performance stochastic process:

$$\begin{cases} \mathbf{g}^{EB\_Pump} = \left\{g_1^{EB\_Pump}, g_2^{EB\_Pump}, g_3^{EB\_Pump}\right\} = \{24, 24, 0\}, \\ \mathbf{p}^{EB\_Pump}(t) = \{p_1^{EB\_Pump}(t), p_2^{EB\_Pump}(t), p_3^{EB\_Pump}(t)\}. \end{cases}$$

Having the sets $\mathbf{g}^{EB\_Pump}$, $\mathbf{p}^{EB\_Pump}(t)$, one can define for the chiller controller the L_z-transform associated with the power controller's output performance stochastic process:

$$\begin{aligned} L_z\{G^{EB\_Pump}(t)\} &= p_1^{EB\_Pump}(t)z^{g_1^{EB\_Pump}} + p_2^{EB\_Pump}(t)z^{g_2^{EB\_Pump}} \\ &\quad + p_3^{EB\_Pump}(t)z^{g_3^{EB\_Pump}} \\ &= (p_1^{EB\_Pump}(t) + p_2^{EB\_Pump}(t))z^{24} + p_3^{EB\_Pump}(t)z^0. \end{aligned} \tag{5.17}$$

Using the composition operator Ω_{fser} for the electric board sub-system consisting of three elements, connected in series, we obtain the L_z-transform $L_z\{G^{EB}(t)\}$ for the electric board sub-system, where the powers of z are found as minimum of powers of corresponding terms:

$$\begin{aligned} L_z\{G^{EB}(t)\} &= \Omega_{f_{ser}}(G^{EB\_Power}(t), G^{EB\_Ch}(t), G^{EB\_Pump}(t)) \\ &= \Omega_{f_{ser}}([p_1^{EB\_Power}(t)z^{24} + p_2^{EB\_Power}(t)z^0], \\ &\quad [(p_1^{EB\_Ch}(t) + p_2^{EB\_Ch}(t))z^{24} + p_3^{EB\_Ch}(t)z^{12} + p_4^{EB\_Ch}(t)z^0], \\ &\quad [(p_1^{EB\_Pump}(t) + p_2^{EB\_Pump}(t))z^{24} + p_3^{EB\_Pump}(t)z^0]) \\ &= p_1^{EB}(t)z^{24} + p_2^{EB}(t)z^{12} + p_3^{EB}(t)z^0, \end{aligned} \tag{5.18}$$

where

$$\begin{aligned} p_1^{EB}(t) &= p_1^{EB\_Power}(t) \cdot (p_1^{EB\_Ch}(t) + p_2^{EB\_Ch}(t)) \cdot (p_1^{EB\_Pump}(t) + p_2^{EB\_Pump}(t)); \\ p_2^{EB}(t) &= p_1^{EB\_Power}(t) \cdot p_3^{EB\_Ch}(t) \cdot (p_1^{EB\_Pump}(t) + p_2^{EB\_Pump}(t)); \\ p_3^{EB}(t) &= p_1^{EB\_Power}(t) \cdot \{p_4^{EB\_Ch}(t) \cdot (p_1^{EB\_Pump}(t) + p_2^{EB\_Pump}(t)) + p_3^{EB\_Pump}(t)\} \\ &\quad + p_2^{EB\_Power}(t); \end{aligned}$$

5.3.6 *Model of Stochastic Demand*

The working conditions of the MRI system may be described as a variable demand which is typical of hospital needs. Usually, for such systems, demand can be presented by daily two-level demand curves, as shown in Figure 5.6(a). In this model, the demand is represented as a continuous time Markov chain with two: $\mathbf{w} = \{w_1, w_2\}$, where w_1 is the peak level and w_2 the low level. The state-transition diagram for Markov process $W(t)$ is presented in Figure 5.6(b).

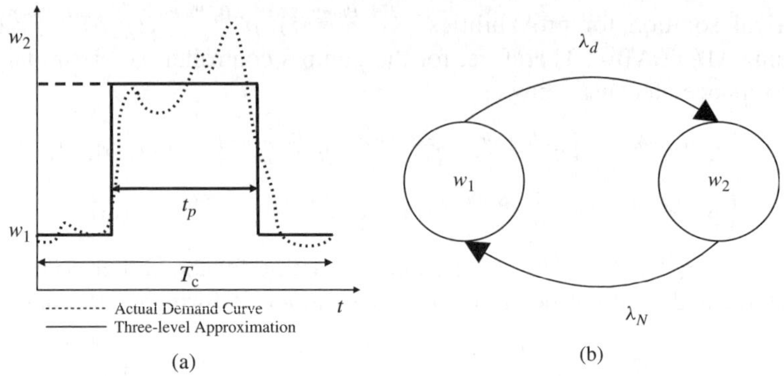

Figure 5.6 Two-level demand model: (a) approximation of actual demand curve and (b) state-transition diagram

When the cycle time T_c and the mean duration of the peak t_p are known, the transition intensities of the model can be obtained as (Lisnianski *et al.* 2010):

$$\lambda_d = \frac{1}{T_c - t_p}, \quad \lambda_N = \frac{1}{t_p},$$

where λ_d is the transition intensity from the first demand level to the second and λ_N is the transition intensity from the second demand level to the first low one.

The transition intensities for the demand are represented by the following matrix

$$\mathbf{A}_w = \begin{pmatrix} -\lambda_d & \lambda_N \\ \lambda_d & -\lambda_N \end{pmatrix}$$

where $\lambda_d = 547.5\ \text{year}^{-1}$, $\lambda_N = 2190\ \text{year}^{-1}$.

According to the Markov method, the systems of differential equations for daily stochastic demand are as follows:

$$\begin{cases} \dfrac{dp_{w1}(t)}{dt} = -\lambda_d p_{w1}(t) + \lambda_N p_{w2}(t) \\ \dfrac{dp_{w2}(t)}{dt} = \lambda_d p_{w1}(t) - \lambda_N p_{w2}(t) \end{cases} \tag{5.19}$$

The first state is the system's initial state, so $\mathbf{p}_{w0} = \{p_{w1}(0), p_{w2}(0)\} = \{1, 0\}$.

A numerical solution for probabilities $p_{w1}(t)$ and $p_{w2}(t)$ can be obtained analytically or using MATLAB®. Therefore, one obtains for demand the following output performance stochastic process:

$$\begin{cases} \mathbf{g}_w = \{g_{w1}(t), g_{w2}(t)\} = \{18, 10\}, \\ \mathbf{p}_w(t) = \{p_{w1}(t), p_{w2}(t)\}. \end{cases}$$

Using the sets $\mathbf{g}_w$, $\mathbf{p}_w(t)$, one can define for demand w, the L_z-transform associated with the demand stochastic process:

$$L_z\{G_w(t)\} = p_{w1}(t)z^{g_{w1}(t)} + p_{w2}(t)z^{g_{w2}(t)} = p_{w1}(t)z^{18} + p_{w2}(t)z^{10}. \tag{5.20}$$

5.3.7 Multi-State Model for the MRI Cooling System

All the sub-systems are connected in series. So the L_z-transform associated with the whole system is:

$$\begin{aligned} L_z\{G_s(t)\} &= \Omega_{f_{ser}}(G^{Ch}(t), G^{HE}(t), G^{P}(t), G^{EB}(t)) \\ &= \Omega_{f_{ser}}(P_1^{Ch}(t)z^{24} + P_2^{Ch}(t)z^{12} + P_3^{Ch}(t)z^{0}, p_1^{HE}(t)z^{24} \\ &+ p_2^{HE}(t)z^{22} + p_3^{HE}(t)z^{20} + p_4^{HE}(t)z^{16} + p_5^{HE}(t)z^{0}, \\ &P_1^{P}(t)z^{24} + P_2^{P}(t)z^{0}, p_1^{EB}(t)z^{24} + p_2^{EB}(t)z^{12} + p_3^{EB}(t)z^{0}) \end{aligned}$$

After simple algebra, where the powers of z are found as minimum values of the powers of the corresponding terms, the final expression of the whole system's L_z-transform is of the following form:

$$L_z\{G_s(t)\} = p_{s1}(t)z^{24} + p_{s2}(t)z^{22} + p_{s3}(t)z^{20} + p_{s4}(t)z^{16} + p_{s5}(t)z^{12} + p_{s6}(t)z^{0}, \tag{5.21}$$

where

$$\begin{aligned} p_{s1}(t) &= P_1^{Ch}(t)p_1^{HE}(t)P_1^{P}(t)p_1^{EB}(t); \\ p_{s2}(t) &= P_1^{Ch}(t)p_2^{HE}(t)P_1^{P}(t)p_1^{EB}(t); \\ p_{s3}(t) &= P_1^{Ch}(t)p_3^{HE}(t)P_1^{P}(t)p_1^{EB}(t); \\ p_{s4}(t) &= P_1^{Ch}(t)p_4^{HE}(t)P_1^{P}(t)p_1^{EB}(t); \\ p_{s5}(t) &= (1 - p_5^{HE}(t))P_1^{P}(t)(P_1^{Ch}(t)p_2^{EB}(t) + P_2^{Ch}(t)p_1^{EB}(t) + P_2^{Ch}(t)p_2^{EB}(t)); \\ p_{s6}(t) &= (P_1^{Ch}(t) + P_2^{Ch}(t))(1 - p_5^{HE}(t))(P_1^{P}(t)p_3^{EB}(t) + P_2^{P}(t)) \\ &+ (P_2^{Ch}(t)p_5^{EB}(t) + P_3^{Ch}(t))(P_1^{P}(t) + P_2^{P}(t)); \end{aligned} \tag{5.22}$$

When the resulting stochastic process $Y(t) = f\{G_S(t), G_W(t)\} = G_S(t) - G_W(t)$ drops to level zero, such an event is treated as a failure. Processes $G_S(t), G_W(t)$ are independent.

Thus, we consider here an MRI cooling system and two-level variable demand system, and the block-diagram for the MSS reliability computation is presented in Figure 5.7.

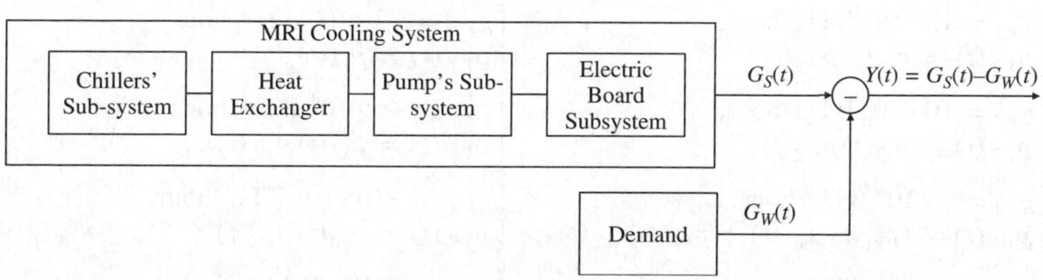

Figure 5.7 Block diagram for the MSS reliability computation

Thus, the L_z-transform, associated with the whole system with respect to the stochastic demand is calculated as follows:

$$
\begin{aligned}
L_z\{Y(t)\} &= L_z\{G_S(t) - G_W(t)\} = \Omega_{f_{\text{minus}}}(L_z\{G_S(t)\}, L_z\{G_W(t)\}) \\
&= \Omega_{f_{\text{minus}}}(p_{s1}(t)z^{24} + p_{s2}(t)z^{22} + p_{s3}(t)z^{20} + p_{s4}(t)z^{16} + p_{s5}(t)z^{12} \\
&\quad + p_{s6}(t)z^{0}, p_{w1}(t)z^{18} + p_{w2}(t)z^{10}) \\
&= \sum_{i=1}^{6}\sum_{j=1}^{2} p_{si}(t)p_{wj}(t)z^{f(g_{si},w_j)} = \sum_{i=1}^{6}\sum_{j=1}^{2} p_{si}(t)p_{wj}(t)z^{g_{si}-w_j} \\
&= p_{s1}(t)p_{w1}(t)z^{6} + p_{s2}(t)p_{w1}(t)z^{4} + p_{s3}(t)p_{w1}(t)z^{2} + p_{s4}(t)p_{w1}(t)z^{-2} \\
&\quad + p_{s5}(t)p_{w1}(t)z^{-6} + p_{s6}(t)p_{w1}(t)z^{-18} + p_{s1}(t)p_{w2}(t)z^{14} + p_{s2}(t)p_{w2}(t)z^{12} \\
&\quad + p_{s3}(t)p_{w2}(t)z^{10} + p_{s4}(t)p_{w2}(t)z^{6} + p_{s5}(t)p_{w2}(t)z^{2} + p_{s6}(t)p_{w2}(t)z^{-10} \\
&= p_{s1}(t)p_{w2}(t)z^{14} + p_{s2}(t)p_{w2}(t)z^{12} + p_{s3}(t)p_{w2}(t)z^{10} + (p_{s1}(t)p_{w1}(t) \\
&\quad + p_{s4}(t)p_{w2}(t))z^{6} + p_{s2}(t)p_{w1}(t)z^{4} + p_{s3}(t)p_{w1}(t) + p_{s5}(t)p_{w2}(t))z^{2} \\
&\quad + p_{s4}(t)p_{w1}(t)z^{-2} + p_{s5}(t)p_{w1}(t)z^{-6} + p_{s6}(t)p_{w2}(t)z^{-10} \\
&\quad + p_{s6}(t)p_{w1}(t)z^{-18}.
\end{aligned}
$$

Therefore, the final L_z-transform is given by:

$$
\begin{aligned}
L_z\{Y(t)\} = \sum_{i=1}^{10} p_{Yi}(t)z^{g_i} &= p_{Y1}(t)z^{14} + p_{Y2}(t)z^{12} + p_{Y3}(t)z^{10} + p_{Y4}(t)z^{6} + p_{Y5}(t)z^{4} \\
&\quad + p_{Y6}(t)z^{2} + p_{Y7}(t)z^{-2} + p_{Y8}(t)z^{-6} + p_{Y9}(t)z^{-10} + p_{Y10}(t)z^{-18}, \qquad (5.23)
\end{aligned}
$$

where the capacities $\mathbf{g_Y} = \{g_{Y1}, g_{Y2}, \ldots, g_{Y10}\}$ and the state probabilities $\mathbf{p_Y}(t) = \{p_{Y1}(t), p_{Y2}(t), \ldots, p_{Y10}(t)\}$ of output performance stochastic process for the entire MSS, are as follows:

$$\begin{cases} g_{Y1} = 14 \cdot 10^4 \text{ BTU/hour}, \\ p_{Y1}(t) = p_{s1}(t)p_{w2}(t), \end{cases} \qquad \begin{cases} g_{Y6} = 2 \cdot 10^4 \text{ BTU/hour}, \\ p_{Y6}(t) = (p_{s3}(t)p_{w1}(t) + p_{s5}(t)p_{w2}(t)); \end{cases}$$

$$\begin{cases} g_{Y2} = 12 \cdot 10^4 \text{ BTU/hour}, \\ p_{Y2}(t) = p_{s2}(t)p_{w2}(t); \end{cases} \qquad \begin{cases} g_{Y7} = -2 \cdot 10^4 \text{ BTU/hour}, \\ p_{Y7}(t) = p_{s4}(t)p_{w1}(t); \end{cases}$$

$$\begin{cases} g_{Y3} = 10 \cdot 10^4 \text{ BTU/hour}, \\ p_{Y3}(t) = p_{s3}(t)p_{w2}(t); \end{cases} \qquad \begin{cases} g_{Y8} = -6 \cdot 10^4 \text{ BTU/hour}, \\ p_{Y8}(t) = p_{s5}(t)p_{w1}(t); \end{cases}$$

$$\begin{cases} g_{Y4} = 6 \cdot 10^4 \text{ BTU/hour}, \\ p_{Y4}(t) = (p_{s1}(t)p_{w1}(t) + p_{s4}(t)p_{w2}(t)); \end{cases} \qquad \begin{cases} g_{Y9} = -10 \cdot 10^4 \text{ BTU/hour}, \\ p_{Y9}(t) = p_{s6}(t)p_{w2}(t); \end{cases}$$

$$\begin{cases} g_{Y5} = 4 \cdot 10^4 \text{ BTU/hour}, \\ p_{Y5}(t) = p_{s2}(t)p_{w1}(t); \end{cases} \qquad \begin{cases} g_{Y10} = -18 \cdot 10^4 \text{ BTU/hour}, \\ p_{Y10}(t) = p_{s6}(t)p_{w1}(t); \end{cases}$$

Based on the resulting L_z-transform $L_z\{Y(t)\}$ of the entire MSS, one can obtain the MSS instantaneous availability for the MRI cooling system by using the expression

$$A(t) = \sum_{g_Y \geq 0} p_{Yi}(t) = \sum_{i=1}^{6} p_{Yi}(t). \tag{5.24}$$

Additionally, using the whole system's L_z-transform (5.21)–(5.22), one obtains the MSS instantaneous availability for the MRI cooling system, working under constant demand. For peak level constant demand $w = 18$, and for low level constant demand $w = 10$, we obtain

$$A_{18}(t) = \sum_{g_s \geq 18} p_{si}(t) = \sum_{i=1}^{3} p_{si}(t),$$

$$A_{10}(t) = \sum_{g_s \geq 10} p_{si}(t) = \sum_{i=1}^{5} p_{si}(t). \tag{5.25}$$

5.4 Availability Calculation

Calculations were performed using the following failure and repair rates. The failure rate of each chiller is $\lambda^{Ch} = 3$ year^{-1}. The repair rate of each chiller is $\mu^{Ch} = 365$ year^{-1} or mean repair time is one day.

The failure and repair rates corresponding to the heat exchanger are:

$$\lambda^{HE} = 0.1 + 0.33t^2 \text{ year}^{-1}, \lambda^{HEt} = 0.2 \text{ year}^{-1}, \mu^{HE} = 200 \text{ year}^{-1}.$$

The failure rate of each pump is $\lambda^P = 3 + t$ year^{-1}. The repair rate of the pump is $\mu^P = 365$ year^{-1} or mean repair time is one day.

The failure rate of the power controller in the electric board is 1 year^{-1}. The failure rates of all the other controllers (chillers and pumps) are 2 year^{-1}. The mean repair time of all controllers is one day.

As one can see, the failure rates of the heat exchanger and the pumps are increasing functions of time, these elements possessing the aging property.

The calculated MSS instantaneous availabilities $A(t)$ for stochastic demand (solid line), $A_{18}(t)$ (dashed line) and $A_{10}(t)$ (dashed-dotted line) are presented in Figure 5.8. The curves in Figure 5.8 show that the availability of an aging system decreases with time. The required availability level is 0.99 (ASHRAE Handbook 2010) and is presented by a straight dotted line.

As one can see from the results presented in Figure 5.8, in order to provide the required availability level, replacement of aging parts of compressors and pumps should take place after 8.1 years.

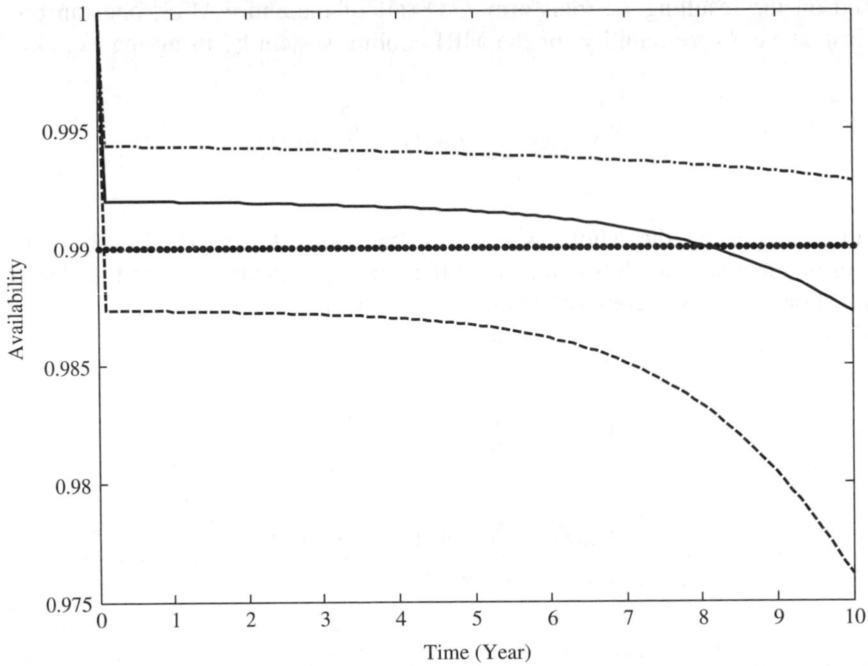

Figure 5.8 MSS instantaneous availability for variable demand A(t) (solid line), $A_{18}(t)$ (dashed line) and $A_{10}(t)$ (dash-dotted line) and A = 0.99 (dotted line) - required availability level

5.5 Conclusion

In this chapter the L_z-transform was applied to a real-world problem, that is, the availability assessment for an aging MSS using as an example a cooling system for MRI equipment under stochastic demand.

It was demonstrated that the L_z-transform method is well formalized and suitable for practical application in reliability engineering for a real-world MSSs analysis. It supports an engineering decision and also determines a system structure to provide the required availability level for complex multi-state aging systems. The proposed method provides a drastic decrease in the computational burden compared with a straightforward Markov method which in this case study would have required a model with 3840 states to be built and solved.

Acknowledgments

This work was partially supported by the Internal Funding Program of the Shamoon College of Engineering (SCE).

References

ASHRAE Handbook. 2010. *Refrigeration*. Atlanta, GA: American Society of Heating Refrigerating and Air-Conditioning Engineers.

Frenkel, I. and L. Khvatskin. 2012. Reliability decisions for supermarket refrigeration system by using combined stochastic process and universal generation function method: case study. In: *Recent Advances in System Reliability. Signatures, Multi-state Systems and Statistical Inference*, ed. A. Lisnianski and I. Frenkel. London: Springer, pp. 97–112.

Frenkel, I., L. Khvatskin and A. Lisnianski. 2010. Structure decision making based on universal generating functions for refrigeration system. *Journal of Applied Quantitative Methods* 5 (3): 397–412.

Frenkel, I., A. Lisnianski and L. Khvatskin. 2012. Availability assessment for aging refrigeration system by using L_z-transform. *Journal of Reliability and Statistical Studies* 5 (2): 33–43.

Levitin, G. 2005. *Universal Generating Function in Reliability Analysis and Optimization*. London: Springer.

Lisnianski, A. 2012. L_z-transform for a discrete-state continuous-time Markov process and its application to multi-state system reliability. In *Recent Advances in System Reliability. Signatures, Multi-state Systems and Statistical Inference*, ed. A. Lisnianski and I. Frenkel. London: Springer, pp. 79–95.

Lisnianski, A., I. Frenkel and Y. Ding. 2010. *Multi-state System Reliability Analysis and Optimization for Engineers and Industrial Managers*. London: Springer.

Lisnianski, A. and G. Levitin. 2003. *Multi-State System Reliability: Assessment, Optimization and Applications*. Singapore: World Scientific.

Natvig, B. 2011. *Multistate Systems Reliability. Theory with Applications*. New York: Wiley.

Trivedi, K. 2002. *Probability and Statistics with Reliability, Queuing and Computer Science Applications*. New York: Wiley.

Ushakov, I. 1986. A universal generating function. *Soviet Journal of Computer and System Sciences* 24: 37–49.

Ushakov, I. 1987. Optimal standby problem and a universal generating function. *Soviet Journal of Computer and System Sciences* 25: 61–73.

6

Combined Clustering and L_z-Transform Technique to Reduce the Computational Complexity of a Multi-State System Reliability Evaluation

Yi Ding
Department of Electrical Engineering, Technical University of Denmark, Denmark

6.1 Introduction

The well-known binary system reliability models are too simple to represent many complex engineering systems, which may experience more than two states. Recently multi-state system (MSS) reliability models have been comprehensively developed and studied (Lisnianski and Levitin 2003; Lisnianski *et al.* 2010; Natvig 2011), which provides a powerful tool to accurately represent engineering systems in real life. The universal generating function (UGF) method is the primary algorithm for reliability evaluation of MSS, which was first introduced by (Ushakov 1986) and further developed by (Lisnianski and Levitin 2003; Levitin 2005; Ding and Lisnianski 2008). The UGF method is usually used to determine the output performance distribution of MSS. However, the main restriction on this method is that it may only be applied to random variables and, thus, it can operate with only steady-state performance distribution. To evaluate the non steady-state (dynamic) reliability of MSS, an L_z-transform technique has been developed (Lisnianski 2012a) and applied to solve some real-world problems (Frenkel *et al.* 2012; Lisnianski 2012b).

Applied Reliability Engineering and Risk Analysis: Probabilistic Models and Statistical Inference, First Edition.
Edited by Ilia B. Frenkel, Alex Karagrigoriou, Anatoly Lisnianski and Andre Kleyner.

In spite of the fact that combining Ushakov's universal generating operator and L_z-transform technique drastically decreases the computational efforts, the dynamic reliability assessment of real-world problems is still a difficult problem and decreasing the number of MSS's states is still desirable.

To tackle this important problem, some advanced research has been conducted. A method for reducing the system states by clustering states with similar performance characteristics has been developed (Ding *et al.* 2010), which has been used for steady-state reliability assessment of MSS.

It seems that the idea of combining the clustering method and the L_z-transform technique is promising to approximate the dynamic MSS reliability and decrease the computational complexity. In this chapter, a framework is proposed, which can provide accurate time-varying reliability indices for real-life large-scale engineering systems, for example, electric power systems. In Section 6.2, the L_z-transform for dynamic reliability evaluation for MSS is briefly introduced. The clustering composition operator is developed to reduce the computational dimension for the L_z-transform in Section 6.3. The corresponding computational procedures are discussed in Section 6.4. A numerical example is provided in Section 6.5.

6.2 The L_z-Transform for Dynamic Reliability Evaluation for MSS

The performance level $G(t)$ of each multi-state component is a discrete-state continuous-time Markov process, which takes values from the set $\mathbf{g} = \{g_1, \ldots, g_K\}$. It is also assumed that the set is ordered so that $g_{j+1} \geq g_j$ for any j. The general state-space diagram of the multi-state component considering minor and major failures and repairs (Lisnianski and Levitin 2003) is presented in Figure 6.1, which graphically represents a discrete-state continuous-time (DSCT) Markov process (Trivedi 2002).

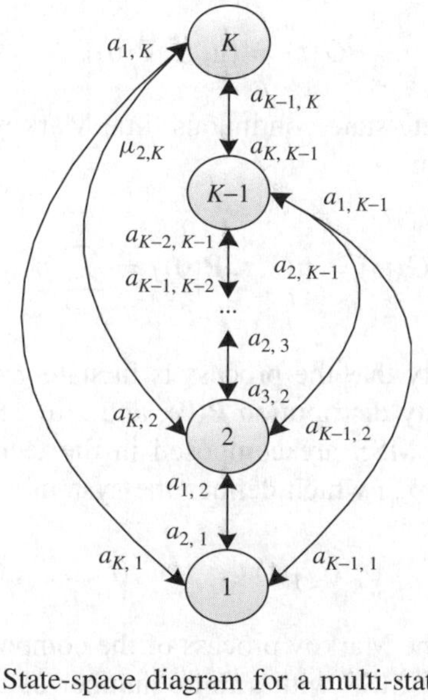

Figure 6.1 State-space diagram for a multi-state component

As shown in Figure 6.1, failures cause the component transition from the state l to the state j, $(l > j)$ with corresponding transition intensity a_{lj}. Repairs cause the component transition from the state j to the state $l\,(j < l)$ with corresponding transition intensity a_{jl}. The following system of differential equations (Kolmogorov equations) can be written for the state probabilities of the component:

$$\begin{cases} \dfrac{dp_K(t)}{dt} = \sum\limits_{l=1}^{K-1} a_{l,K}p_l(t) - p_K(t)\sum\limits_{l=1}^{K-1} a_{K,l} \\ \dfrac{dp_j(t)}{dt} = \sum\limits_{l=j+1}^{K} a_{l,j}p_l(t) + \sum\limits_{l=1}^{j-1} a_{l,j}p_l(t) - \\ \qquad p_j(t)\left(\sum\limits_{l=1}^{j-1} a_{j,l} + \sum\limits_{l=j+1}^{K} a_{j,l}\right), \text{ for } 1 < j < K \\ \dfrac{dp_1(t)}{dt} = \sum\limits_{l=2}^{K} a_{l,1}p_l(t) - p_1(t)\sum\limits_{l=2}^{K} a_{1,l} \end{cases} \tag{6.1}$$

with the initial conditions $p_K(0) = 1$, $p_{K-1}(0) = \ldots = p_1(0) = 0$..

Solving this system of differential equations one can obtain the state probabilities $p_j(t)$, $j = 1, \ldots, K$, which define probabilities that at time instant $t > 0$ the component will be in state j. Therefore, the Markov process representing the multi-state component can be defined by the set of performance levels $\{g_1, \ldots, g_K\}$, transitions intensities matrix $\mathbf{A} = (a_{lj}), l, j = 1, \ldots, K$ and by the initial conditions of states' probability:

$$\mathbf{P}(0) = \{p_K(0) = \Pr(G(0) = g_K), \ldots, p_1(0) = \Pr(G(0) = g_1)\}.$$

The Markov process of the multi-state component can be represented using the triplet:

$$G(t) = \{\mathbf{g}, \mathbf{A}, \mathbf{P}(0)\} \tag{6.2}$$

The L_z-transform of a discrete-state continuous-time Markov process can be represented as (Lisnianski 2012a, 2012b):

$$L_Z\{G(t)\} = u(z, t, \mathbf{P}(0)) = \sum_{j=1}^{K} p_j(t)z^{g_j} \tag{6.3}$$

where $p_j(t)$ is the probability that the process is in state j at time instant $t \geq 0$ for any given initial states' probability distribution $\mathbf{P}(0)$, and z in general, is a complex variable.

All m components in the MSS are composed in the technical structure according to the given structure function ψ_m, which defines the system's stochastic performance level $G_{sys}(t)$:

$$G_{sys}(t) = \psi_m\{G_1(t), \ldots, G_i(t), \ldots, G_m(t)\}, \tag{6.4}$$

where $G_i(t)$, $1 \leq i \leq m$, is the Markov process of the component i, $G_{sys}(t)$ represents the Markov process of the entire MSS with a finite number of different performance levels.

We designate:

K_m = number of performance levels for the MSS (state number)

$p_j^{sys}(t)$ = probability that stochastic performance of the MSS will be at performance level j

g_j^{sys}, $j = 1, \ldots, K_m$ at time instant $t > 0$.

The L_z-transform of the Markov process of the entire MSS can be represented as

$$L_Z\{G_{sys}(t)\} = \sum_{j=1}^{K_m} p_j^{sys}(t) \cdot z^{g_j^{sys}} \tag{6.5}$$

By using Ushakov's composition operator Ω_{ψ_m} over the L_z-transform of the individual components, the system L_z-transform can be obtained as:

$$L_Z\{G_{sys}(t)\} = \sum_{j=1}^{K_m} p_j^{sys}(t) \cdot z^{g_j^{sys}} = \Omega_{\psi_m}\{L_Z\{G_1(t)\}, \ldots, L_Z\{G_m(t)\}\} \tag{6.6}$$

Taking into account Equation (6.3) and using the general definition of the composition operator (Lisnianski and Levitin 2003; Levitin 2005, Lisnianski 2012a, 2012b), one can obtain the following expression:

$$\begin{aligned} L_Z\{G_{sys}(t)\} &= \Omega_{\psi_m}\{L_Z\{G_1(t)\}, \ldots, L_Z\{G_m(t)\}\} \\ &= \Omega_{\psi_m}\left\{\sum_{j_1=1}^{k_1} p_{1j_1}(t)\, z^{g_{1,j_1}}, \ldots, \sum_{j_m=1}^{k_m} p_{m,j_m}(t) z^{g_{m,j_m}}\right\} \\ &= \sum_{j_1=1}^{k_1}\sum_{j_2=1}^{k_2}\cdots\sum_{j_m=1}^{k_m}\left(\prod_{i=1}^{m} p_{i,j_i}(t) z^{\psi_m(g_{1,j_1},\ldots,g_{m,j_m})}\right). \end{aligned} \tag{6.7}$$

The composition operators for various types of MSS (parallel, series, series-parallel or bridge configurations) have been well developed (Lisnianski and Levitin 2003; Levitin 2005). Based on these composition operators, the system L_z-transform can be obtained by using simple algebraic operations.

After obtaining the system L_z-transform, the system's reliability measures can easily be evaluated. The entire MSS availability $A(t)$ at time instant $t > 0$ can be evaluated as:

$$A(t) = \sum_{j=1}^{K_m} p_j^{sys}(t) \cdot \mathbf{1}(g_j^{sys} \geq D(t)) \tag{6.8}$$

where $\mathbf{1}(True) \equiv 1$, $\mathbf{1}(False) \equiv 0$, $D(t)$ is the system demand at time instant $t \geq 0$.

6.3 Clustering Composition Operator in the L_z-Transform

Generally one can reduce the computational resources by collecting the identical items in the L_z-transform, which is used in the UGF techniques (Lisnianski and Levitin 2003;

Levitin 2005). However, it depends on how many identical items exist in the intermediate calculation of UGF techniques. Collecting the identical items may even make the calculation slower because collecting identical items itself also takes time (Li and Zuo 2008).

Though the performance levels of components or the system are unlikely to be identical, there does exist a similarity among them. For example, suppose the capacities of a generating unit are 100 MW, 98 MW, 60 MW, 59 MW, 58 MW and 0 MW corresponding to different states. Obviously the capacities can be grouped into three clusters: (100 MW, 98 MW), (60 MW, 59 MW, 58 MW) and (0 MW), respectively. If we use clustered states to replace the original states, we may greatly reduce the computational burden.

There are many clustering methods such as k-mean, mixture of Gaussians and fuzzy-c-mean (FCM), which can be used to partition data into different subsets. The FCM is used to determine the characteristics of clustered states (Ding *et al.* 2010). For more technical details of FCM, one can refer to Bezdek *et al.* (1984).

The objective of FCM algorithm is to minimize the specified function F (Bezdek *et al.* 1984):

$$\min F = \sum_{j=1}^{K}\sum_{s=1}^{S}(U_{sj})^m \|g_j - c_s\|^2,\ 1 \le m < \infty \tag{6.9}$$

where K is the number of original states, S is the number of specified clustered states, g_j represents the performance level of the jth original state, c_s signifies the performance level of the center of the sth clustered state, m is a specified real number greater than 1, U_{sj} is the membership grade, which represents the weighting factor between g_j and c_s, and $\|g_j - c_s\|$ is any norm expressing the similarity between g_j and c_s.

The objective function F represents the distance from any given g_i to a cluster center c_s weighted by the U_{sj}. During the minimization process of F, the K original states can be partitioned into S clustered states (Ding *et al.* 2010).

The clustering composition operator φ_C is developed to obtain the L_z-transform of the clustered states:

$$\varphi_C(L_Z\{G(t)\}) = \varphi_C\left(\sum_{j=1}^{K} p_j(t)\, z^{g_j}\right) = \sum_{s=1}^{S} p_s(t)\cdot z^{c_s} \tag{6.10}$$

In Equation (6.10), the cluster center c_s represents the performance level of the clustered state s, and $p_s(t)$ is the probability of the clustered state s at time instant $t \ge 0$, which can be calculated by the following equation:

$$p_s(t) = \sum_{j=1}^{K} p_j(t)\cdot \mathbf{1}(U_{sj} == \max(U_{s1}, \ldots, U_{sK})) \tag{6.11}$$

where $\mathbf{1}(True) \equiv 1,\ \ \mathbf{1}(False) \equiv 0$.

The calculation of $p_s(t)$ is based on apportioning the probabilities of the original states into the specified clustered states. From Equation (6.11), it can be seen that the probability of an original state will be apportioned to the closest clustered state.

6.4 Computational Procedures

The following procedure is used to build the proposed framework of combining the clustering method and the L_z-transform technique to approximate the dynamic MSS reliability:

1. Suppose a MSS consists of n^*sub-systems. Each sub-system i^* consists of n_{i^*} components with identical or similar characteristics.
2. Develop the L_z-transform of the sub-system i^* using the technique introduced in Section 6.2.
3. Suppose the sub-system i^* has S possible clustered states. Determine the performance level and the probability of the sub-system i^* $(1 \leq i^* \leq n^*)$ being in the clustered state s. The techniques for determining the performance level and the probability of the state s are discussed in Section 6.3.
4. Determine the L_z-transform of the MSS and evaluate the system reliabilities.

6.5 Numerical Example

An application of the proposed method is to evaluate the time-varying availability of a complex electric power generation system, which consists of various multi-state generating units and has different demand levels (Ding *et al.* 2009).

Suppose the generation system has five generating units: two identical coal-generating units, two gas-generating units, and one oil-generating unit. The reliability data of generating units come from the Israeli power system as shown in Tables 6.1–6.3, respectively (Ding *et al.* 2006). The coal-generating unit, the gas-generating unit and the oil-generating unit have 10 states, 10 states and 11 states, respectively. The total generating capacity is 1404 MW. The generation system can have up to 110,000 states.

The two coal-generating units are combined into subsystem 1 and the two gas-generating units are combined into subsystem 2. For subsystem 1 and subsystem 2, the original states are clustered into 20 clustered states and 10 clustered states, respectively.

Suppose the system demand is 1000 MW. The system instant availabilities evaluated by the exact L_z-transform technique and the proposed framework are shown in Figure 6.2,

Table 6.1 Reliability parameters of the coal unit

State no.	Average capacity (MW)	MTTR (hr)	MTTF (hr)
1	360	0	–
2	328	6.9	311
3	303	6.7	1464
4	282	6.0	2221
5	255	7.8	1389
6	233	3.9	3187
7	204	6.4	1120
8	181	7.7	1022
9	124	4.3	1219
10	0	93.5	749

Table 6.2 Reliability parameters of the gas unit

State no.	Average capacity (MW)	MTTR (hr)	MTTF (hr)
1	228	0	–
2	223	92	228
3	215	15	13
4	194	18	35
5	174	5	102
6	153	3	16
7	136	6	64
8	115	13	15
9	96	6	103
10	0	64	111

Table 6.3 Reliability parameters of the oil unit

State no.	Average capacity (MW)	MTTR (hr)	MTTF (hr)
1	228	0	–
2	224	34	29
3	212	61	32
4	198	54	24
5	173	18	48
6	153	10	46
7	133	15	75
8	112	14	63
9	96	10	52
10	68	31	64
11	0	119	68

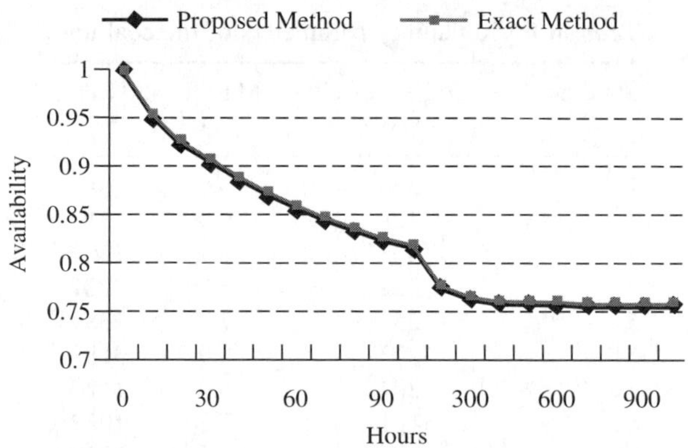

Figure 6.2 Instant system availabilities evaluated by exact method and proposed method

respectively. It can be seen from Figure 6.2 that the error between values obtained by the exact L_z-transform technique and the proposed framework is very small, which ranges from 0 to 0.69%.

The computational efficiency is also compared. The computer programs for two approaches were developed in MATLAB 7.8, and were implemented on a laptop with a 2.67 GHZ CPU. The CPU time of the exact L_z-transform technique and the proposed framework for the proposed example is 97.937 seconds and 0.281 seconds, respectively. Obviously the computational efficiency of the proposed framework is much higher than the exact L_z-transform technique.

6.6 Conclusion

In this chapter a framework of combining the clustering method and the L_z-transform technique for dynamic reliability assessment of multi-state system has been presented. The proposed framework can reduce the computational dimension for a complex system and therefore simplify computational complexity. From the presented numerical example, we can observe that the proposed framework can provide accurate instant availability for electric power generating system while the computational time is greatly reduced.

References

Bezdek, J., R. Ehrlich and W. Full. 1984. FCM: The fuzzy-c-means clustering algorithm. *Computers and Geosciences* 10 (2–3): 191–203.

Ding, Y. and A. Lisnianski. 2008. Fuzzy universal generating functions for multi-state system reliability assessment. *Fuzzy Sets and Systems* 159 (3): 307–324.

Ding, Y., A. Lisnianski and I. Frenkel. 2006. Fuzzy based state reduction technique for multi-state system reliability assessment. In *Proceedings of the International Conference on Statistical Methods for Biomedical and Technical Systems*, ed. F. Vonta. BIOSTAT2006, May 29–31, Limassol, Cyprus. Cyprus: Kantzilaris Publ., pp. 269–274.

Ding, Y., A. Lisnianski, P. Wang, L. Goe and L. P. Chiang. 2009. Dynamic reliability assessment for bilateral contract electricity providers in the restructured power systems. *Electric Power System Research* 79 (10): 1424–1430.

Ding, Y., M. Zuo, A. Lisnianski and W. Li. 2010. A framework for reliability approximation of multi-state weighted *k*-out-of-*n* systems. *IEEE Transactions on Reliability* 59 (2): 297–308.

Frenkel I., A. Lisnianski and L. Khvatskin. 2012. Availability assessment for aging refrigeration system by using L_z-transform. *Journal of Reliability and Statistical Studies* 5 (2): 33–43.

Levitin, G. 2005. *Universal Generating Function in Reliability Analysis and Optimization*. London: Springer.

Li, W. and M. Zuo. 2008. Reliability evaluation of multi-state weighted *k-out-of-n* systems. *Reliability Engineering and System Safety* 93 (1): 160–167.

Lisnianski, A. 2012a. L_z-transform for a discrete-state continuous-time Markov process and its application to multi-state system reliability. In *Recent Advances in System Reliability. Signatures, Multi-state Systems and Statistical Inference*, ed. A. Lisnianski and I. Frenkel. London: Springer, pp. 79–95.

Lisnianski, A. 2012b. Reliability assessment for aging multi-state system under stochastic demand by using L_z-transform. In *Proceedings of the 18th ISSAT International Conference on Reliability and Quality in Design*. July 26–28, 2012, Boston, USA.

Lisnianski, A., I. Frenkel and Y. Ding. 2010. *Multi-state System Analysis and Optimization for Engineers and Industrial Managers*. London: Springer.

Lisnianski, A and G. Levitin. 2003. *Multi-State System Reliability: Assessment, Optimization and Applications*. Singapore: World Scientific.

Natvig, B. 2011. *Multi-state Reliability Theory with Applications*. New York: John Wiley & Sons, Inc.

Trivedi, K. 2002. *Probability and Statistics with Reliability, Queuing and Computer Science Applications*. New York: John Wiley & Sons, Inc.

Ushakov, I. 1986. A universal generating function. *Soviet Journal of Computer and System Sciences* 24: 37–49.

7

Sliding Window Systems with Gaps

Gregory Levitin
The Israel Electric Corporation Ltd, Israel

7.1 Introduction

A linear consecutive k-out-of-r-from-n:F system consists of n ordered elements and fails if it contains a group of r consecutive elements in which at least k elements fail. This system was formally introduced in (Griffith 1986), but had been mentioned previously in (Saperstein 1972, 1973; Naus 1974; Nelson 1978; Tong 1985) in connection with tests for non-random clustering, quality control and inspection procedures, service systems, and radar problems. Different types of consecutive k-out-of-r-from-n:F systems were studied in (Papastavridis and Sfakianakis 1991; Sfakianakis *et al.* 1992; Cai 1994; Psillakis 1995; Habib and Szántai 2000; Dutuit and Rauzy 2001; Levitin 2004), where exact and approximate algorithms have been suggested for evaluating system reliability.

The sliding window system (SWS) introduced in (Levitin 2003) is the multi-state generalization of the k-out-of-r-from-n:F system. This system consists of n linearly ordered multi-state elements (MEs). Each ME j can have H_j different states: from complete failure up to perfect functioning. A performance rate is associated with each state. The SWS fails if the sum of the performance rates of any r consecutive MEs is lower than the demand W. Further, we will refer to the group of r consecutive elements with cumulative performance lower than W as failed window and to the group of r consecutive elements with a cumulative performance greater than or equal to W as working window. The SWS fails if it contains at least one failed window.

Applied Reliability Engineering and Risk Analysis: Probabilistic Models and Statistical Inference, First Edition.
Edited by Ilia B. Frenkel, Alex Karagrigoriou, Anatoly Lisnianski and Andre Kleyner.

In (Levitin and Ben Haim 2011; Levitin and Dai 2012; Xiang and Levitin 2012), three new models that generalize the SWS to the case of allowed failed windows have been suggested:

- k-out-of-e SWS (k/eSWS) that fails if contains at least k failed windows;
- m-consecutive SWS (mCSWS) that fails if contains at least m consecutive overlapping failed windows;
- m-gap SWS (mGSWS) that fails if any pair of failed windows is separated by less than m working windows.

The introduction of the generalized SWS models is motivated by the following examples. Consider a heating system that should provide a certain temperature along the line with moving parts (Figure 7.1). The temperature at each point of the line is determined by a cumulative effect of r closest heaters. Each heater consists of several electrical heating elements. The heating effect of each heater depends on the availability of its heating elements and, therefore, can vary discretely (if the heaters are different, the number of different levels of heat radiation and the intensity of the radiation at each level are specific to each heater). In order to provide the temperature, which is not less than some specified value at each point on the line, any r adjacent heaters should be in states where the sum of their radiation intensity is greater than an allowed minimum W. If the heating intensity drops below the desired level at some part of the line, a cold spot is created. The existence of the cold spots can cause a deterioration of the moving parts.

The k/eSWS model corresponds to the case when the parts cannot be heated up to the desired temperature if there exist at least k cold spots along the line.

The mCSWS model corresponds to the case when the parts have enough time to cool down and deteriorate if the heating intensity drops below the desired level along the considerable part of the line (corresponding to m inter-heater spans).

The mGSWS model corresponds to the case when moving along any single cold spot the parts have not enough time to cool down and deteriorate because of thermal inertia. However, in the case of repeated temperature drops, if the gap between the cold spots is too short, the parts have not enough time to warm up, which causes their deterioration.

A variety of other systems also fit the model, for example, radar systems that should provide certain radiation intensity or combat systems that should provide certain fire density along a defense line without considerable number of gaps, etc. The model can be also applied to system integration testing where groups of r interconnected modules

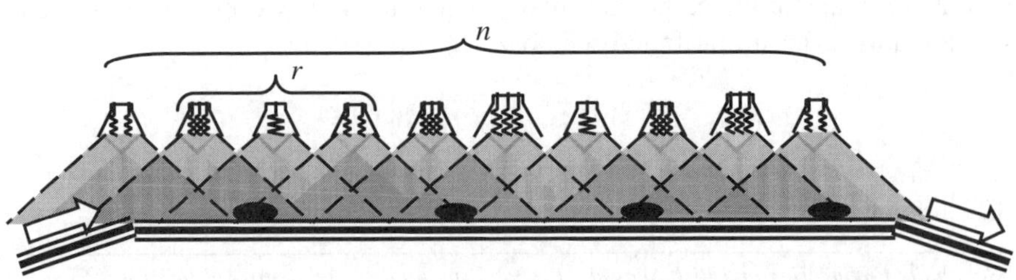

Figure 7.1 Heating system represented by the SWS model

are tested consecutively (Lazzarini Lemos *et al.* 2009; Pakala *et al.* 2011) and to wireless sensor networks with data fusion applied to multi-target tracking where the correct tracking depends on the availability of groups of r sensors supplying the data to a fusion algorithm (Wang and Wang 2007; Gage and Murphy 2010).

7.2 The Models

We assume that all n ordered MEs in the system are mutually independent. Each ME j is characterized by its random performance rate G_j and can be in one of H_j different states. Each state $h \in \{1, 2, \ldots, H_j\}$ of ME j is characterized by its probability $p_{j,h}$ and performance rate $g_{j,h} : p_{j,h} = \Pr(G_j = g_{j,h})$ and $\sum_{h=1}^{H_j} p_{j,h} = 1$. The pair of vectors $(p_{j,1}, \ldots, p_{j,H_j})$ and $(g_{j,1}, \ldots, g_{j,H_j})$ represents the pmf of the random variable G_j.

The group of r consecutive elements numbered from f to $f + r - 1$ fails if

$$\sum_{j=f}^{f+r-1} G_j < W. \tag{7.1}$$

7.2.1 The k/eSWS Model

The total number of groups of r consecutive elements in the system consisting of n elements is $e = n - r + 1$. The first group of r consecutive elements consists of elements $1, 2, \ldots, r$, the last, $n - r + 1$-th group, consists of elements $n - r + 1, n - r + 2, \ldots, n$. The entire k/eSWS does not fail if the condition

$$1\left(\sum_{j=f}^{f+r-1} G_j < W\right) = 1. \tag{7.2}$$

holds for less than k groups. Therefore, the k/eSWS reliability can be defined as

$$R_{k/e} = \Pr\left\{\sum_{f=1}^{m-r+1} 1\left(\sum_{j=f}^{f+r-1} G_j < W\right) < k\right\}. \tag{7.3}$$

7.2.2 The mCSWS Model

Consider the set of m overlapping consecutive groups of r consecutive elements (windows) starting from element s. The total number of elements in such a set is $m + r - 1$. The first window consists of elements $s, s + 1, \ldots, s + r - 1$, the last, m-th window consists of elements $s + m - 1, s + m, \ldots, s + m + r - 2$. If each one of m consecutive windows fails

$$\prod_{f=s}^{s+m-1} 1\left(\sum_{j=f}^{f+r-1} G_j < W\right) = 1. \tag{7.4}$$

If the left side of expression (7.4) equals 0, the failure condition for m consecutive groups of r consecutive elements starting from element s does not hold.

In the system consisting of n linearly ordered elements the sets of m overlapping consecutive windows can start from elements $1, \ldots, n-r-m+2$. The entire system does not fail if condition (7.4) does not hold for any set of m overlapping consecutive windows. Therefore, the system reliability can be defined as

$$R_{mC} = \Pr\left\{ \sum_{s=1}^{n-r-m+2} \left[\prod_{f=s}^{s+m-1} 1\left(\sum_{j=f}^{f+r-1} G_j < W \right) \right] = 0 \right\}. \tag{7.5}$$

7.2.3 The mGSWS Model

The condition that any pair of failed windows is separated by at least m overlapping working windows is equivalent to the condition that any set of $m+1$ overlapping consecutive windows can contain no more than one failed window. Consider the set of $m+1$ overlapping consecutive windows starting from element s. The total number of elements in such set is $m+r$. The first window consists of elements $s, s+1, \ldots, s+r-1$, the last, $m+1$-th window consists of elements $s+m, s+m+1, \ldots, s+m+r-1$. If no more than one window among $m+1$ consecutive windows fails

$$\sum_{f=s}^{s+m} 1\left(\sum_{j=f}^{f+r-1} G_j < W \right) \leq 1. \tag{7.6}$$

In the system consisting of n linearly ordered elements, the sets of $m+1$ overlapping consecutive windows can start from elements $1, \ldots, n-r-m+1$. The entire system does not fail if condition (7.6) holds for any set of $m+1$ overlapping consecutive windows. Therefore, the system reliability can be defined as follows:

$$R_{mG} = \Pr\left\{ \prod_{s=1}^{n-r-m+1} 1\left(\sum_{f=s}^{s+m} 1\left(\sum_{j=f}^{f+r-1} G_j < W \right) \leq 1 \right) = 1 \right\}. \tag{7.7}$$

The reliability evaluation problem is to obtain R defined in Equations (7.3), (7.5) or (7.7) given the pmf of random performance for all system elements: $(p_{j1}, \ldots, p_{jH_j})$ and $(g_{j1}, \ldots, g_{jH_j})$ for $j = 1, \ldots, n$.

7.2.4 Interrelations among Different Models

The working state condition for mGSWS is much stronger than for mCSWS. Indeed, if any pair of failed windows is separated by at least $m > 0$ working windows, the system contains no consecutive failed windows. The failure conditions for k/eSWS and mGSWS do not interdepend directly. The system can contain at least k failed windows, but each pair of these windows can be separated by at least m working windows. On the contrary, the system can contain less than k failed windows, but some of these windows can be separated by less than m working windows. However, the minimal gap between two failed windows in the system with the total of $e = n - r + 1$ windows when k among them are failed is $\lfloor (e-k)/(k-1) \rfloor$. Thus, if $m > (e-k)/(k-1)$, the failure of k/eSWS constitutes the failure of mGSWS.

The failure conditions for k/eSWS and mCSWS also do not interdepend directly. The system can contain at least k failed windows, but less than m consecutive failed windows. On the contrary, when $m < k$, the system can contain less than k failed windows, but at least m consecutive failed windows. However, if SWS with e windows contains at least $k > (e + 1)(m - 1)/m$ failed windows, it cannot contain less than m consecutive failed windows. Thus, for $k > (e + 1)(m - 1)/m$, the failure of k/eSWS constitutes the failure of mCSWS. If $m \geq k$, the failure of mCSWS constitutes the failure of k/eSWS.

7.3 Reliability Evaluation Technique

The procedure for a system reliability evaluation is based on the universal z-transform, also called the u-function or universal moment generating function technique, introduced in (Ushakov 1986), and proved to be very effective for the reliability evaluation of different types of multi-state systems (Levitin 2005).

7.3.1 Determining u-functions for Individual Elements and their Groups

The u-function of a discrete random variable X is defined as a polynomial

$$u(z) = \sum_{\omega=1}^{\Omega} q_k z^{x_\omega}, \tag{7.8}$$

where the variable X has Ω possible values x_ω, and q_ω is the probability that X is equal to x_ω.

In our case, the u-function can define an ME performance rate distribution, that is, it represents all the possible states of the ME j by relating the probabilities of each state p_{jh} to the performance rate g_{jh} of the ME in the form

$$u_j(z) = \sum_{h=1}^{H_j} p_{j,h} z^{g_{j,h}}. \tag{7.9}$$

To represent the performance distribution of a group consisting of r consecutive MEs, one has to modify the u-function by replacing the random value X with the random vector $\boldsymbol{Y} = \{Y(1), \ldots, Y(r)\}$ consisting of random performance values corresponding to all the MEs belonging to the group.

The vector element $Y(j)$ is equal to the performance rate of the j-th one out of r consecutive MEs. When the f-th group of MEs is considered (the group of MEs numbered from f to $f + r - 1$), the element $Y_f(j)$ of the random vector $\boldsymbol{Y}_f$ represents the performance of the $f + j - 1$-th ME of the system.

The random variable $Y(j)$ representing G_{f+j-1} takes values $g_{f+j-1,1}, \ldots, g_{f+j-1,H_{f+j-1}}$ corresponding to the performance rates of the $f + j - 1$-th ME in different states. Each combination of states of MEs belonging to the group constitutes a state of the group. The total number of different states of the f-th group Ω_f is equal to the number of possible combinations of the states of the individual MEs belonging to the group. As all the MEs are s-independent, and each ME j has H_j states, the total number of states of the f-th group of r MEs is $\Omega_f = \prod_{j=1}^{r} H_{f+j-1}$. The performance values of the elements of the group in any state k are represented by the realization $\boldsymbol{y}_{f,k}$ of the random vector $\boldsymbol{Y}_f$. The probability of any state of the group is equal to the product of the probabilities of the corresponding states of the individual MEs.

For example, the vector-u-function for a group of two MEs s and t ($r = 2$) takes the form

$$U(z) = \sum_{a=1}^{H_s}\sum_{b=1}^{H_t} q_{a,b} z^{\{g_a, g_b\}}, \tag{7.10}$$

where $q_{a,b}$ is a probability of an event in which ME s is in state a, and ME t is in state b. For independent MEs, $q_{a,b} = p_{s,a}p_{t,b}$. Therefore, the vector-u-function of the group can be obtained by applying over u-functions of individual MEs of the operator

$$U(z) = u_s(z) \otimes u_t(z) = \sum_{a=1}^{H_s} p_{s,a} z^{g_{s,a}} \otimes \sum_{b=1}^{H_t} p_{t,b} z^{g_{t,b}} = \sum_{a=1}^{H_s}\sum_{a=1}^{H_t} p_{s,a}p_{t,b} z^{\{g_{s,a}, g_{t,b}\}}. \tag{7.11}$$

Applying the similar operator over u-functions of r consecutive MEs starting from f-th ME, one obtains the u-function corresponding to the group containing these MEs:

$$U_f(z) = \otimes(u_f(z), u_{f+1}(z), \ldots, u_{f+r-1}(z))$$
$$= \sum_{h_1=1}^{H_f}\sum_{h_2=1}^{H_{f+1}} \cdots \sum_{h_r=1}^{H_{f+r-1}} p_{f,h_1} \times p_{f+1,h_2} \cdots \times p_{f+r-1,h_r} z^{\{g_{f,h_1}, g_{f+1,h_2}, \ldots, g_{f+r-1,h_r}\}}. \tag{7.12}$$

Simplifying this representation, one obtains

$$U_f(z) = \sum_{\omega=1}^{\Omega_f} Q_{f,\omega} z^{\boldsymbol{y}_{f,\omega}}. \tag{7.13}$$

The obtained u-function defines all the possible states of the group of r MEs.

Having the vector $\boldsymbol{y}_{f,\omega}$ representing the performance rates of MEs belonging to the group in any state k, one can obtain the cumulative performance of the group in this state as

$$\sigma(\boldsymbol{y}_{f,w}) = \sum_{j=1}^{r} y_{f,\omega}(j). \tag{7.14}$$

7.3.2 *Determining u-functions for all the Groups of r Consecutive Elements*

The considered system contains exactly $n - r + 1$ groups of r consecutive elements, and each element can belong to no more than r such groups. To obtain the vector-u-functions corresponding to all the groups of r consecutive elements, the following procedure is introduced.

First, define the vector-u-function $U_{1-r}(z)$ as

$$U_{1-r}(z) = z^{\boldsymbol{y}_0}, \tag{7.15}$$

where the vector $\boldsymbol{y}_0$ consists of r zeros.

Then, define the operator $\underset{\leftarrow}{\otimes}$ over vector-u-function $U(z)$, and u-function of individual ME $u(z)$, as

$$U_f(z)\underset{\leftarrow}{\otimes}u_j(z)=\sum_{\omega=1}^{\Omega_f}Q_{f,k}z^{\mathbf{y}_{f,\omega}}\underset{\leftarrow}{\otimes}\sum_{h=1}^{H_j}p_{j,h}z^{g_{j,h}}=\sum_{\omega=1}^{\Omega_f}\sum_{h=1}^{H_j}Q_{f,\omega}p_{j,h}z^{\varphi(\mathbf{y}_{f,\omega},g_{j,h})}, \tag{7.16}$$

where operator φ over arbitrary vector $\mathbf{y}$ and value g shifts all the vector elements one position to the left: $y(s-1)=y(s)$ for $1<s\le r$. It also adds the value g to the right position: $y(r)=g$ (the first element of vector $\mathbf{y}$ disappears after applying the operator). For example, $\varphi(\{31201\},2)=\{12012\}$. The operator φ removes the performance value of the first (leftmost) ME of the group, and adds the performance value of the next (not considered yet) ME to the group preserving the order of MEs belonging to the group. Therefore, applying the operator $\underset{\leftarrow}{\otimes}$ over the vector-u-function representing the performance distribution of the group of r MEs starting from the f-th ME, one obtains the vector-u-function representing the performance distribution of the group of r MEs starting from the $f+1$-th ME.

Then, use the operator $\underset{\leftarrow}{\otimes}$ in sequence as follows:

$$U_{j+1-r}(z)=U_{j-r}(z)\underset{\leftarrow}{\otimes}u_j(z) \tag{7.17}$$

For $j=1,\ldots,n$, one obtains vector-u-functions for all the possible groups of r consecutive MEs: $U_1(z),\ldots,U_{n-r+1}(z)$. The vector-u-function for the first group $U_1(z)$ is obtained after applying the operator $\underset{\leftarrow}{\otimes}$ r times. In the vector-u-function $U_f(z)$ (for $f>0$), the value $y(s)$ of vector $\mathbf{y}$ corresponds to the performance rate of ME $f-1+s$.

7.3.3 Detecting the System Failure

To detect the combinations of working and failed groups corresponding to the system failure we modify the vector-u-function (7.13) by incorporating an integer counter c_ω:

$$U_f(z)=\sum_{\omega=1}^{\Omega^*_f}Q_{f,\omega}z^{c_\omega,\mathbf{y}_{f,\omega}}, \tag{7.18}$$

where Ω^*_f is the total number of distinguished combinations of c_ω and $\mathbf{y}_{f,\omega}$. Initially, the value of 0 is assigned to the counter c_ω, and the operations (7.15) and (7.16) are modified as follows.

$$U_{1-r}(z)=z^{0,\mathbf{y}_0}, \tag{7.19}$$

$$U_f(z)\underset{\leftarrow}{\otimes}u_j(z)=\sum_{\omega=1}^{\Omega^*_f}Q_{f,\omega}z^{c_\omega,\mathbf{y}_{f,\omega}}\underset{\leftarrow}{\otimes}\sum_{h=1}^{H_j}p_{j,h}z^{g_{j,h}}$$
$$=\sum_{\omega=1}^{\Omega^*_f}\sum_{h=1}^{H_j}Q_{f,\omega}p_{j,h}z^{\rho(c_\omega,\sigma(\mathbf{y}_{f,\omega})),\varphi(\mathbf{y}_{f,\omega},g_{j,h})} \tag{7.20}$$

where $\rho(c_\omega, \sigma(\mathbf{y}_{f,\omega}))$ is the counter updating rule, which is based on the current value of the counter and the cumulative performance of the group of MEs in state ω represented by the vector $\mathbf{y}_{f,\omega}$. The updating rule must allow one to detect a system failure using the equality $c_\omega = c^*$. The updating rules $\rho(c_\omega, \sigma(\mathbf{y}_{f,\omega}))$ and the values of c^* for the three types of considered systems are introduced in the following sub-section.

The terms $Q_\omega z^{c_\omega, \mathbf{y}_\omega}$ of the vector-u-function $U_h(z)$ with $c_\omega = c^*$ correspond to the total system failure. The following operator δ collects the probabilities Q_ω corresponding to such terms, and obtains the overall probability that elements represented by the vector-u-function $U_h(z)$ cause the system failure.

$$\delta(U_h(z)) = \sum_{\omega=1}^{\Omega^*_h} Q_\omega 1(c_\omega = c^*) \tag{7.21}$$

7.3.4 Updating the Counter

7.3.4.1 k/eSWS

Using the updating rule

$$\rho(c_\omega, \sigma(\mathbf{y})) = c_\omega + 1(\sigma(\mathbf{y}) < W). \tag{7.22}$$

one lets the counter represent the total number of failed groups in the system. The counter c_ω takes the value $c^* = k$ when the total number of failed groups in the system reaches k.

7.3.4.2 mCSWS

Using the updating rule

$$\rho(c_\omega, \sigma(\mathbf{y})) = \begin{cases} c_\omega + 1 & \text{if } \sigma(\mathbf{y}) < W \\ 0 & \text{if } \sigma(\mathbf{y}) \geq W \end{cases}. \tag{7.23}$$

one lets the counter represent the total number of consecutive failed groups as the counter is zeroed by any group which is in the working state. The counter c_ω takes the value $c^* = m$ when the number of consecutive failed groups reaches m.

7.3.4.3 mGSWS

It can be seen that with the updating rule

$$\rho(c_\omega, \sigma(\mathbf{y})) = \begin{cases} m + 1\left(c_\omega > 0\right) & \text{if } \sigma(\mathbf{y}) < W \\ \max\{0, c_\omega - 1\} & \text{if } \sigma(\mathbf{y}) \geq W. \end{cases} \tag{7.24}$$

counter c_ω takes the value of $c^* = m + 1$ only if the number of working groups between two failed groups is less than m. Consider, for example, a sequence of values $\sigma(\mathbf{y}_\omega)$ 543452. According to (7.24), for $m = 3$, and $W = 4$, the counter consecutively takes values 0,0,3,2,1,4. The value of 4 indicates that the gap between groups with $\sigma(\mathbf{y}_\omega) < W$ in the sequence is less than 3. On the contrary, the sequence 434543 produces the sequence of the counter values 0,3,2,1,0,3 that contains no values of $c^* = m + 1$, indicating that the sequence of values $\sigma(\mathbf{y}_\omega)$ contains no values less than 4 separated by less than 3 values greater than or equal to 4.

7.3.5 Recursive Determination of System Failure Probability

Let E_i be the sub-SWS consisting of first i consecutive elements from the set $\{1, \ldots, n\}$ and Λ_i be the event when the sub-SWS E_i fails. Note that for $i < r + 1$, the event Λ_i cannot occur as the SWS E_i contains no more than one window. The system failure can be considered as a combination of the mutually exclusive failure events $\Lambda_{r+1} \cup (\Lambda_{r+2} \cap \overline{\Lambda}_{r+1}) \cup \ldots \cup (\Lambda_n \cap \overline{\Lambda}_{n-1})$, where $(\Lambda_i \cap \overline{\Lambda}_{i-1})$ means that SWS E_{i+1} fails, whereas SWS E_i does not fail. The entire system failure probability can be obtained as

$$F = \Pr\{\Lambda_{r+1}\} + \Pr\{\Lambda_{r+2} \cap \overline{\Lambda}_{r+1}\} + \ldots + \Pr\{\Lambda_n \cap \overline{\Lambda}_{n-1}\}. \tag{7.25}$$

The probability $\Pr\{\Lambda_{r+1}\}$ can be obtained as $\Pr\{\Lambda_{r+1}\} = \delta(U_2(z))$. To obtain the probability $\Pr\{\Lambda_{r+2} \cap \overline{\Lambda}_{r+1}\}$, one has to perform the following three-step procedure:

(1) Remove all the terms with $c_\omega = c^*$ from $U_2(z)$ (excluding the case when the cases when the failure of E_{r+2} is caused by the failure of SWS E_{r+1}); (2) apply the operator $\underleftarrow{\otimes}$ over the truncated vector-u-function $U_2(z)$ and $u_{r+2}(z)$ to get $U_3(z)$; and (3) apply the operator δ over $U_3(z)$.

Recursively, to obtain the probability $\Pr\{\Lambda_i \cap \overline{\Lambda}_{i-1}\}$, one removes all the terms with $c_\omega = c^*$ from $U_{i-r}(z)$ (excluding the case when the failure of SWS E_i is caused by the failure of SWS E_{i-1}), obtains $U_{i-r+1}(z) = U_{i-r}(z) \underleftarrow{\otimes} u_i(z)$, and finally applies $\delta(U_{i-r}(z))$.

7.3.6 Computational Complexity Reduction

Consider a vector-u-function $U_f(z)$. For each combination of values $y_f(2), \ldots, y_f(r)$, it contains H_f terms corresponding to different values of $y_f(1)$ (realizations of random performance G_f). After applying the operator $\otimes$, $y_f(1)$ disappears from the vector y_f being replaced with $y_f(2)$. This produces H_f terms with the same vector y_{f+1} in the vector-u-function $U_{f+1}(z)$. The coefficients of these terms corresponding to vector $\mathbf{y}_{f+1}$ are equal to the probability that the MEs of the $f + 1$-th group have states represented by $\mathbf{y}_{f+1}$, while element f is in one of its H_f states. If the terms with identical vector $\mathbf{y}_{f+1}$ have the same values of the counter, by summing the coefficients of the terms (collecting like terms in $U_{f+1}(z)$), one can obtain a single term for vector $\mathbf{y}_{f+1}$ with a coefficient equal to the overall probability that the $f + 1$-th group of elements has a state represented by $\mathbf{y}_{f+1}$. Therefore, the number of different terms in each vector-u-function $U_{f+1}(z)$ can be considerably reduced.

7.3.7 Algorithm for System Reliability Evaluation

The following algorithm finds the reliability of SWS with allowed gaps based on (7.25) and the considerations earlier.

1. Initialization:

$$F = 0;\ U_{1-r}(z) = z^{0,\mathbf{y}_0} \text{ } (\mathbf{y}_0 \text{ consists of } r \text{ zeros}).$$

2. For $j = 1, \ldots, r-1$:
   Obtain $U_{j-r+1}(z) = U_{j-r}(z) \underset{\leftarrow}{\otimes} u_j(z)$ and zero counters in each term of $U_{j-r+1}(z)$.
3. Main loop: repeat the following for $j = r, \ldots, n$.
   (a) Obtain $U_{j-r+1}(z) = U_{j-r}(z) \underset{\leftarrow}{\otimes} u_j(z)$, and collect the like terms in the obtained vector-u-function.
   (b) Add the value $\delta(U_{j-r+1}(z))$ corresponding to $\Pr\{\Lambda_j \cap \overline{\Lambda}_{j-1}\}$ in (7.25) to F;
   (c) Remove all of the terms with a counter value equal to c^* from $U_{j-r+1}(z)$.
4. Obtain the system reliability as $R = 1 - F$.

In step 2 of the algorithm, the counters must be zeroed because, for $j = 1, \ldots, r-1$, the obtained u-functions do not represent full groups of r consecutive elements (windows). The first full group consisting of elements numbered from 1 to r corresponds to $j = r$.

Analytical and numerical examples of SWS reliability evaluation can be found in (Levitin and Ben Haim 2011; Levitin and Dai 2012; Xiang and Levitin 2012).

7.4 Conclusion

The chapter presents three newly introduced types of multi-state systems that generalize the linear multi-state sliding window system to the case of allowed gaps (groups of consecutive elements with insufficient cumulative performance). The considered models differ in the constraints that are imposed on the total number of gaps, on the number of consecutive gaps and on the length of the inter-gap spans. Possible applications of the new models are in manufacturing, radar and military systems, system integration testing, and multi-target tracking.

The interrelations among the three models are discussed and it is shown that no model can be reduced to any other one. The recursive algorithm for system reliability evaluation is suggested. This algorithm is based on the universal generating function technique. The basic steps of the algorithm are the same for all the models. By changing the counter updating rul, one can adjust the algorithm to any specific model.

Future research can be focused on the optimal element sequencing in the considered systems and on the reliability evaluation of the sliding window systems with combined gap constraints.

References

Cai, J. 1994. Reliability of a large consecutive-*k*-out-of-*r*-from-*n*:*F* system with unequal component-reliability. *IEEE Transactions on Reliability* 43 (1): 107–111.

Dutuit, Y. and A. Rauzy. 2001. New insights into the assessment of *k*-out-of-*n* and related systems. *Reliability Engineering and System Safety* 72 (3): 303–314.

Gage, J. and R.R. Murphy. 2010. Sensing assessment in unknown environments: A survey. *IEEE Transactions on Systems, Man and Cybernetics* 40 (1): 1–12.

Griffith, W. 1986. On consecutive *k*-out-of-*n* failure systems and their generalizations. In *Reliability and Quality Control*, ed. A. P. Basu. New York: Elsevier, pp. 157–165.

Habib, A. and T. Szántai. 2000. New bounds on the reliability of the consecutive k-out-of-r-from-n: F system. *Reliability Engineering and System Safety* 68 (2): 97–104.

Lazzarini Lemos, O., I. Franchin and P. Masiero. 2009. Integration testing of Object-Oriented and Aspect-Oriented programs: a structural pairwise approach for Java. *Science of Computer Programming* 74 (10): 861–878.

Levitin, G. 2003. Linear multi-state sliding window systems. *IEEE Transactions on Reliability* 52 (2): 263–269.

Levitin, G. 2004. Consecutive k-out-of-r-from-n system with multiple failure criteria. *IEEE Transactions on Reliability* 53 (3): 394–400.

Levitin, G. 2005. *Universal Generating Function in Reliability Analysis and Optimization*. London: Springer.

Levitin, G. and H. Ben Haim. 2011. Consecutive sliding window systems. *Reliability Engineering and System Safety* 96 (10): 1367–1374.

Levitin, G. and Y. Dai. 2012. k-out-of-n sliding window systems. *IEEE Transactions on Systems, Man, and Cybernetics* 42 (3): 707–714.

Naus, J. 1974. Probabilities for a generalized birthday problem. *Journal of the American Statistical Association* 69 (347): 810–815.

Nelson, J. 1978. Minimal-order models for false-alarm calculations on sliding windows. *IEEE Transactions on Aerospace Electronic Systems* AES-14 (2): 351–363.

Pakala, H. G. M, K. V. S. V. N. Raju and I. Khan. 2011. Integration testing of multiple embedded processing components. In *Advanced Computing. Communications in Computer and Information Science*, 133, Part 2, 200–209. Proceedings of the First International Conference on Computer Science and Information Technology, CCSIT 2011, Bangalore, India, January 2–4, 2011, Part III, edited by N. Meghanathan, B.K. Kaushik and D. Nagamalai. Heidelberg: Springer.

Papastavridis, S. and M. Sfakianakis. 1991. Optimal-arrangement and importance of the components in a consecutive-k-out-of-r-from-n: F system. *IEEE Transactions on Reliability* 40 (3): 277–279.

Psillakis, Z. 1995. A simulation algorithm for computing failure probability of a consecutive-k-out-of-r-from-n: F system. *IEEE Transactions on Reliability* 44 (3): 523–531.

Saperstein, B. 1972. The generalized birthday problem. *Journal of the American Statistical Association* 67 (338): 425–428.

Saperstein, B. 1973. On the occurrence of n successes within N Bernoulli trails. *Technometrics* 15 (4): 809–881.

Sfakianakis, M., S. Kounias and A. Hillaris. 1992. Reliability of consecutive-k-out-of-r-from-n:F systems. *IEEE Transactions on Reliability* 41 (3): 442–447.

Tong, Y. 1985. A rearrangement inequality for the longest run with an application in network reliability. *Journal of Applied Probability* 22: 386–393.

Ushakov, I. 1986. Universal generating function. *Soviet Journal of Computer and System Sciences* 24 (5): 118–129.

Wang, X. and S. Wang. 2007. Collaborative signal processing for target tracking in distributed wireless sensor networks. *Journal of Parallel and Distributed Computing* 67 (5): 501–515.

Xiang, Y. and G. Levitin. 2012. Linear m-gap sliding window systems. *IEEE Transactions on Reliability* 61 (2): 560–568.

8

Development of Reliability Measures Motivated by Fuzzy Sets for Systems with Multi- or Infinite-States

Zhaojun (Steven) Li[1] and Kailash C. Kapur[2]
[1]*Electro-Motive Diesel: A Caterpillar Company, USA*
[2]*University of Washington, USA*

8.1 Introduction

The Institute of Electrical and Electronic Engineers (IEEE) states that reliability is the ability of a system or component to perform its required functions under stated conditions for a specified period of time (Weininger *et al.* 2010). The word ability is very complex and means many things. One element of ability is related to probability, which has been used in the literature to quantify reliability and also is easy for mathematical analysis. Moreover, the majority of reliability research effort assumes that the system or component performance is either satisfactory or unsatisfactory, that is, success or failure in a bi-valued logic sense. However, for many degradable systems, the system performance deteriorates continuously and a single failure threshold value is not sufficient to differentiate between system success and failure. A degradable system can experience a continuous performance transition or deterioration over its design life. Thus, from the customers' point of view, various performance levels correspond to different degrees of success. In other words, the definition for success is imprecise and cannot be determined in terms of one single threshold value. By allowing a partial truth value for a statement, for example, the statement that a system is successful or performing its function to some

Applied Reliability Engineering and Risk Analysis: Probabilistic Models and Statistical Inference, First Edition.
Edited by Ilia B. Frenkel, Alex Karagrigoriou, Anatoly Lisnianski and Andre Kleyner.

extent, fuzzy set theory and fuzzy logic can well capture such continuous performance degradation and quantify the imprecision involved in customers' perception of system success and failure definition.

The above-mentioned imprecision is very common in decision-making since human beings use a lot of linguistic variables (Zadeh 1975) such as success, failure, age, safety, and beauty which take imprecise natural language or sentences rather than precise numerical numbers as their values. For example, the linguistic variable of age could take very young, young, not young, not old, old, very old as its values and each linguistic value is essentially a fuzzy set, that is, "a class of objects with a continuum of grades of membership" (Zadeh 1965). Specifically, the objects of a fuzzy set belong to the set to a degree of membership. The fuzzy set concept is dramatically different from the concept of the traditional crisp sets whose elements/objects either belong to the set or not belong to the set. Fuzzy sets are consistent with multi-valued logic thinking and allow partial truth or degrees of membership, whereas crisp sets are consistent with the bi-valued logic thinking, that is, either true or false. The membership function of a fuzzy set can mathematically quantify the imprecise linguistic variable such as success and failure and it forms the foundation for the proposed fuzzy reliability model where the success events are considered as fuzzy sets rather than crisp sets.

To define reliability, the meaning that a system performs or functions or succeeds needs to be well understood and defined from the viewpoint of the customer. Since the voices and needs from the customers are usually very generic, the voice of customers needs to be translated to some technological or engineering requirements using substitute characteristics such that the VOC can be designed into the systems and products (Kapur 1998). Some examples of the substitute characteristics are the strength of a structure, the resistance of a resistor, the current or voltage in an electrical system, shrinkage, wear or some other deterioration of a part, the level of flow of a fluid to meet some demand, and the amount of power generated relative to the demand/maximum capacity. The system performance is closely related to the values of these substitute characteristic variables and they are also well understood by the users and the system designers. Since the word success or failure is typically related to the values of these substitute characteristics, the membership function of the fuzzy success event can be defined based on the substitute characteristic variable and used to quantify the imprecision in the success definition. On the other hand, the substitute characteristic variable can be precisely measured and historical or empirical data can be used to fit the substitute characteristic variables with appropriate probability models. With the imprecision of the linguistic word of success quantified using a membership function and the stochastic behavior of the substitute characteristic variable modeled using probability models, reliability can be mathematically defined and evaluated in the fuzzified success event.

The membership functions of the fuzzy success events and the probability models for the substitute characteristic variable are sufficient to evaluate the static fuzzy reliability. For dynamic fuzzy reliability evaluation, two issues have to be addressed. First, since there exists no single failure threshold value distinguishing between success and failure, the time to fuzzy failure involves not only randomness but also fuzziness and the concept of fuzzy random variable is introduced to characterize and develop the dynamic fuzzy reliability measures. Second, the physics of failure for the substitute characteristics needs to be well understood and modeled by appropriate stochastic degradation processes.

Stochastic degradation processes and fuzzy random variables as well as related properties are investigated in this chapter.

8.2 Models for Components and Systems Using Fuzzy Sets

In the literature, fuzzy reliability has been extensively investigated; however, the existing fuzzy reliability models and methods so far deal with the uncertainty in estimating the component reliability/probability, failure rate, transition rate, and failure time under either the traditional binary state or multi-state reliability methods. The universal set over which the membership function of the fuzzy number/fuzzy variable is defined is not well motivated and interpreted in a meaningful way (Li and Kapur 2010, 2011, 2012). In addition, the research so far has not emphasized how to define and evaluate fuzzy reliability based on defining success or failure using fuzzy sets and their membership functions from customers' perspectives. The concept of using the substitute characteristic to develop the membership function for a fuzzy set has not been explored properly in the literature.

8.2.1 *Binary Reliability and Multi-State Reliability Model*

For binary reliability, the state space for both components and the system is restricted in the set $S = \{0, 1\}$, that is, failure or success. A state indicator variable X can be defined over the substitute variable y. Suppose the substitute variable is a "larger the better" type, then $X = 1$, if $Y \geq y_0$; $X = 0$, if $Y < y_0$, where y_0 is a threshold value for the substitute variable which differentiates between success ($X = 1$) and failure ($X = 0$) (Figure 8.1(a)). Suppose the stochastic behavior of the substitute random variable Y can be modeled using empirical data and probability models such as the cumulative distribution function $F(y)$ or the probability density function $f(y)$ can be developed. The reliability R for the binary model can be evaluated as,

$$R = P(X = 1) = E[X] = P(Y \geq y_0) = \int_{Y \geq y_0} dF(y). \tag{8.1}$$

It is noted that the characteristic function or indicator function for a crisp set can be also written as $\mu_S(y)$. In this way, the binary reliability in Equation (8.1) can be written as $R = E[\mu_S(y)]$, where $\mu_S(y) = 1$,if $Y \geq y_0$; $\mu_S(y) = 0$,if $Y < y_0$.

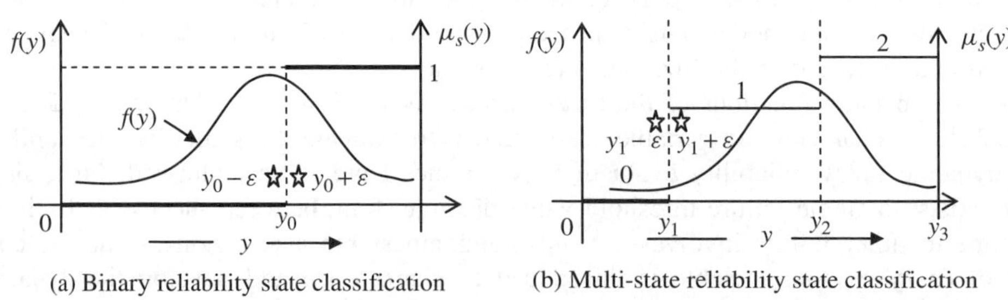

Figure 8.1 Reliability models under (a) binary and (b) multi-state assumption

Instead of using one single threshold value to differentiate success from failure, multi-state reliability model uses multiple threshold values to classify the states for a component or system. For instance, the multi-state approach (Figure 8.1(b)) classifies the state of the component as 2 if $y \in (y_2, y_3]$,state 1 if $y \in (y_1, y_2]$, and state 0 if $y \in [0, y_1]$. In general, the states of the component may be classified as $i = 0, 1, 2, \ldots \ldots M$. This multi-state classification is an extension of the binary state reliability modeling, and it approximates the continuous performance levels by simplifying and discretizing y with finite number of states.

8.2.2 *Definition of Fuzzy Reliability*

The binary and multi-state assumptions for reliability models imply that the values in the neighborhood of a threshold value y_0 or y_1, for example, $y_0 \pm \varepsilon$ or $y_1 \pm \varepsilon$, respectively represent very different states, where ε is a small positive quantity (Figure 8.1). However, there is little difference between $y_0 - \varepsilon$ and $y_0 + \varepsilon$ or $y_1 - \varepsilon$ and $y_1 + \varepsilon$ from the customer's point of view. Thus, there are no sharp boundary values to define a state using the traditional set concept and we propose the fuzzy set concept to characterize the continuous state transition process. Figure 8.2 shows the fuzzy state representation using the membership function $\mu_S(y)$ over the substitute variable Y, and where $f(y)$ is the probability density function of the substitute variable Y.

In Figure 8.2, each value of the substitute variable corresponds to a degree of success, for example, for the value of y_0, the degree of success is α. It is important to note that the membership function of a fuzzy set is an extension of the characteristic function of a crisp set and it is the foundation for the fuzzy reliability definition and models in this chapter.

With the success or failure events treated as fuzzy sets, the continuous substitute characteristic values y can exhibit different degrees of success or failure. As a natural extension of the traditional binary state reliability evaluation given by Equation (8.1), the fuzzy reliability of a component can be evaluated as the probability of the fuzzy success event (Zadeh 1965). Thus, we define fuzzy reliability as:

$$R = P[\textit{fuzzy success}] = \int \mu_S(y) dF(y) = E[\mu_S(Y)], \tag{8.2}$$

where $\mu_S(y)$ is the membership function for the fuzzy success event and $F(y)$ is the cumulative distribution function for the substitute characteristic random variable. The fuzzy reliability definition in Equation (8.2) degenerates to the classic binary reliability

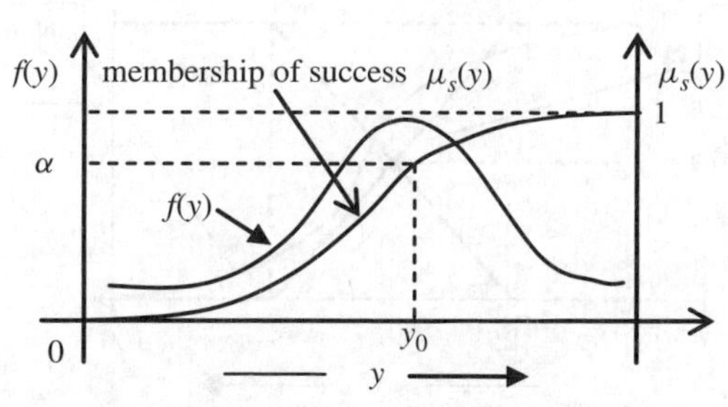

Figure 8.2 Reliability models under fuzzy state representation

model when the membership function of a fuzzy success event is replaced with the characteristic function of a crisp success event. This fuzzy reliability definition is also an analog to the system's performance measure based on the state expectation definition given by $E[X] = \sum_{i=0}^{M} i * \Pr[X = i]$ which has been traditionally used for both binary and multi-state reliability modeling (Brunelle and Kapur 1999). It should be emphasized that in Equation (8.2), reliability is an expectation of the membership function for the fuzzy success state and it is not a probability in the traditional sense. This concept of expectation is also consistent with the reliability measure for multi-state systems.

8.2.3 Fuzzy Unreliability: A Different Perspective

Fuzzy reliability can also be defined in a negative way. In other words, we can define the fuzzy unreliability in terms of the fuzzy failure event,

$$\mathrm{R} = P[\textit{fuzzy failure}] = \int \mu_F(y)dF(y) = E[\mu_F(Y)], \tag{8.3}$$

where $\mu_F(y)$ is the membership function for the fuzzy failure event. If the membership function for a fuzzy failure event is assumed to be the standard complement of the membership function for a fuzzy success event, that is, $\mu_F(y) = 1 - \mu_S(y)$, then the fuzzy reliability and fuzzy unreliability of a component/system will sum to one. However, the standard complement relationship between fuzzy success and fuzzy failure events is based on the bi-valued logic and is not necessary for our approach.

The fuzzy reliability definition in Equation (8.2) does not impose any restriction on the relationship between the fuzzy success and fuzzy failure membership functions. It is often true that $\mu_F(y) \neq 1 - \mu_S(y)$ in fuzzy set theory. $\mu_F(y) \neq 1 - \mu_S(y)$ Figure 8.3 shows the differences between the standard complement fuzzy set F and the nonstandard complement fuzzy set F' of the fuzzy set S.

8.2.4 Evolution from Binary State to Multi-State and to Fuzzy State Reliability Modeling

The previous comparisons for state representation under binary, multi-state, and fuzzy state assumptions show that the fuzzy state representation captures more information in terms

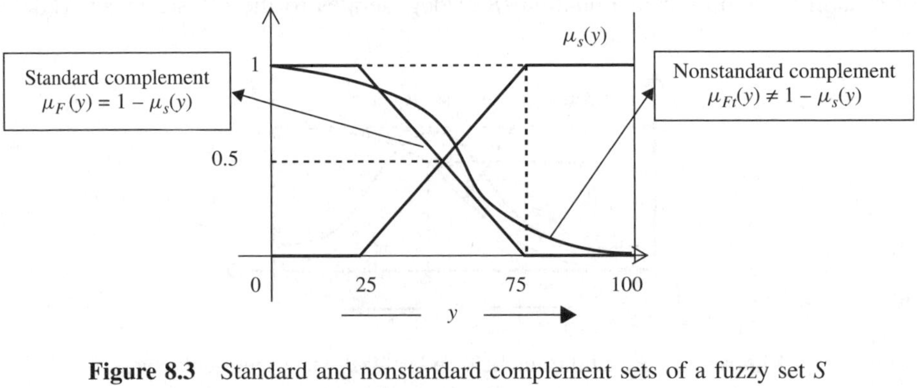

Figure 8.3 Standard and nonstandard complement sets of a fuzzy set S

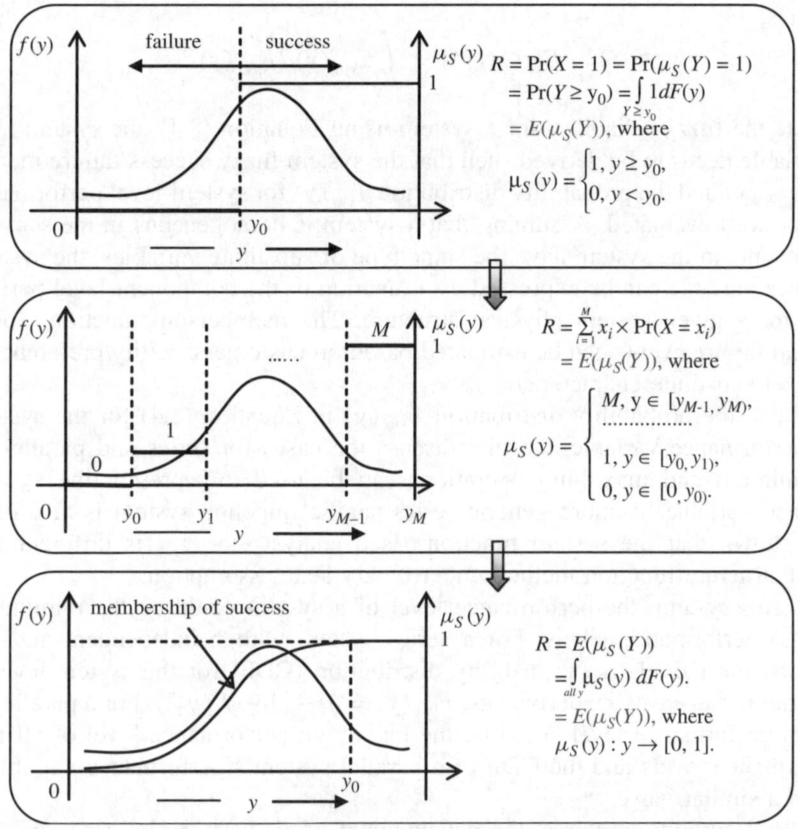

Figure 8.4 Reliability modeling from binary state to multi-state to fuzzy state

of the system's continuous performance from the perspective of the customer and system designers. The binary and multi-state representations are a simplification and discretization of the continuous performance levels. In addition, the fuzzy state representation using membership functions integrates the customer's preferences and judgments for system performance and can provide more meaningful and informed information for decision making. Figure 8.4 shows the evolving processes from binary state and multi-state to fuzzy state representation when the state indicator variable X in traditional binary state and multi-state reliability models is respectively written as $\mu_S(y) = 0$ or 1, and $\mu_S(y) = 0$, 1, ... M, for different y values.

8.3 Fuzzy Reliability for Systems with Continuous or Infinite States

To evaluate fuzzy system reliability, the linguistic words of system success and failure events are defined as fuzzy sets. The probability distribution and the membership function for system-level substitute/performance variable need to be derived from the component-level substitute characteristics. In this thought process, the system fuzzy reliability can be evaluated as,

$$R = E[\mu_{sys}(y)] = \int \mu_{sys}(y) dF_{sys}(y). \tag{8.4}$$

To evaluate the fuzzy reliability for a system using Equation (8.4), the system-level substitute variable needs to be derived such that the system fuzzy success/failure membership function $\mu_{sys}(y)$ and the probability distribution $f_{sys}(y)$ for system-level performance variable can be well evaluated. Assuming that a system is homogeneous in the sense that all the components in the system have the same type of substitute variables, the system-level performance variable can be expressed as a function of the component-level performance variables for a given system physical structure. The membership functions for system success and failure events can be estimated based on customers' utility/preference for the system-level substitute characteristic y.

To evaluate the probability distribution $F_{sys}(y)$ in Equation (8.4) for the system-level random performance variable, we first discuss the cases for series and parallel systems where minimum and maximum operations can be used to represent the system-level performance variable. A more generic series-parallel pipeline system is also discussed, and it is shown that the system function-based analysis for y_{sys} is different from the traditional structure function method under binary state assumption.

For a series system, the performance level of a system could be the minimum of all individuals' performance levels. For a series system with n independent and identical components, the cumulative probability distribution (CDF) for the system-level performance variable can easily be derived as, $F_{sys}(y) = 1 - (1 - F(y)^n)$. For a parallel system, the system performance level could be the maximum performance level of all individuals' performance levels and the CDF of a parallel system's performance variable can be derived in a similar way.

For general system structures, the development of $F_{sys}(y)$ can become very complex and we propose to evaluate the system-level performance variable through system functional analysis.

8.4 Dynamic Fuzzy Reliability

The fuzzy reliability defined in Equation (8.2) can easily be extended for dynamic fuzzy reliability evaluation by considering time-dependent probability distributions for the performance characteristic variable. It is reasonable to assume that the success membership function defined over the performance characteristic Y does not change over time since customers' perception of fuzzy success based on Y is relatively stable. Thus, dynamic fuzzy reliability is defined as,

$$R(t) = P[\mathit{fuzzy\ success}] = \int_y \mu_S(y) dF(y_t(t)) = E[\mu_S(Y)]. \tag{8.5}$$

where $F(y(t))$ is the cumulative probability distribution function for the performance characteristic variable at a given time t, and $E[\mu_S(Y)]$ is a function of time t. From Equation (8.5), it is noted that the probability distributions over time for the performance characteristic variable and the fuzzy success membership function $\mu_S(y)$ suffice to evaluate the dynamic fuzzy reliability. To evaluate this dynamic fuzzy reliability, the performance degradation processes need to be well understood and modeled such that the probability distributions for the performance variable Y over time can be well quantified and evaluated using probabilistic models.

The dynamic behavior for fuzzy reliability can also be investigated by evaluating the time to fuzzy failure which is a fuzzy random variable. In the following section, the concept of fuzzy random variable is introduced to model the time to fuzzy failure when the failure threshold is imprecise and no single threshold value can differentiate between success and failure.

8.4.1 Time to Fuzzy Failure Modeled by Fuzzy Random Variable

In the traditional dynamic reliability model, for a given failure threshold value y_0, time to failure is a random variable $X(\omega)$ due to random variation in the degradation processes (Figure 8.5). When the system state is considered as a fuzzy event, it exhibits different degrees of failure from [0, 1] over the performance range of performance characteristic values from y_I to y_M, where y_I is the initial/ideal state of the system and y_M is the minimum performance value for a decreasing degradation path model with a "larger the better" performance characteristic. For a given degree of system failure α, the observed time to fuzzy failure is a random interval $[\tilde{X}_\alpha^-(\omega), \tilde{X}_\alpha^+(\omega)]$. Thus, time to fuzzy failure is not only random but also fuzzy, that is, a fuzzy random variable.

The concept of fuzzy random variable $\tilde{X}$ is used to model the time to fuzzy failure, and it can be formally defined over a probability space as follows (Kwakernaak 1978; Liu *et al.* 1997): Let $(\Omega, \mathcal{F}, \mathcal{P})$ be a probability space. A function $\tilde{X} : \Omega \to \mathcal{F}_0(R)$ is called a fuzzy random variable on $(\Omega, \mathcal{F}, \mathcal{P})$, if for any $\alpha \in (0, 1]$ and $\omega \in \Omega$,

$$\tilde{X}_\alpha(\omega) = \{x : x \in R, \tilde{X}(\omega)(x) \geq \alpha\} = [\tilde{X}_\alpha^-, \tilde{X}_\alpha^+] \tag{8.6}$$

is a random interval, i.e., $\tilde{X}_\alpha^-$ and $\tilde{X}_\alpha^+$ are two random variables and measurable with respect to the σ-algebra $\mathcal{F}$, where $\mathcal{F}_0(R)$ is a collection of all bounded and closed fuzzy random variables on $\mathbb{R}$. The fuzzy random variable degenerates to a random variable when $\alpha = 1$ since $\tilde{X}_\alpha^- = \tilde{X}_\alpha^+$.

The expectation of a fuzzy random variable $E[\tilde{X}]$ is a fuzzy number (Liu *et al.* 1997). For a given value α, the expectation of a fuzzy random variable can be expressed as

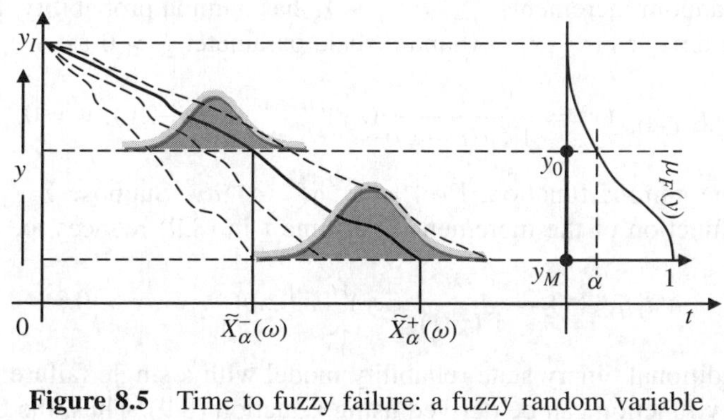

Figure 8.5 Time to fuzzy failure: a fuzzy random variable

$E[\tilde{X}_\alpha] = (E[\tilde{X}_\alpha^-], E[\tilde{X}_\alpha^+])$. The membership function for the expectation of fuzzy random variables, for example, the membership function for the expectation of time to fuzzy failure, can be evaluated using the Resolution Identity (Wu 2001),

$$\begin{aligned}\mu_{E[\tilde{X}_\alpha]}(x) &= sup_{0<\alpha\leq 1}\alpha * 1_{(E[\tilde{X}_\alpha])},\\ &= max\{\alpha : 0 < \alpha \leq 1, x \in (E[\tilde{X}_\alpha])\},\\ &= max\{\alpha : 0 < \alpha \leq 1, x \in (E[\tilde{X}_\alpha^-], E[\tilde{X}_\alpha^+])\},\end{aligned} \tag{8.7}$$

where x is a value in the support of the expectation of the fuzzy random variable and $1_{(E[\tilde{X}_\alpha])}$ is the indicator function. Given the underlying degradation mechanism, the membership function of $E[\tilde{X}_\alpha]$ in Equation (8.7) can be evaluated by solving the following optimization problem for each fixed x_0 over all $\alpha \in (0, 1]$,

$$\begin{aligned} &Max \quad \alpha\\ s.t. \quad &E[\tilde{X}_\alpha^-] \leq x_0 \leq E[\tilde{X}_\alpha^+]\\ &0 < \alpha \leq 1.\end{aligned} \tag{8.8}$$

In Equation (8.8), the expectations of $E[\tilde{X}_\alpha^-]$ and $E[\tilde{X}_\alpha^+]$ can be evaluated based on some stochastic degradation processes. The stochastic performance degradation processes and the methods for evaluating $E[\tilde{X}_\alpha^-]$ and $E[\tilde{X}_\alpha^+]$ are investigated in the following sections.

8.4.2 Stochastic Performance Degradation Model

The stochastic performance degradation process can be modeled in different ways such as physics-based models, data-driven models and sometimes a combination of the physics of failure with data-driven method. Most of the time, the performance characteristic such as the amount of degradation can be well measured even though the underlying failure mechanism cannot be well modeled. The gamma stochastic process can capture both the unit-by-unit uncertainty for degradation and the temporal uncertainty over time (Noortwijk 2009), thus we use the gamma process to model the stochastic performance degradation for the performance variable over time.

In mathematical terms, a stochastic process Y_t is a gamma process such that for all $0 \leq s < t$, the random increment, $Y_{t-s} = Y_t - Y_s$ has gamma probability density function with a shape parameter $r(t-s) > 0$ and a scale parameter $\lambda > 0$ as,

$$f_{r(t-s),\lambda}(y) = \frac{\lambda}{\Gamma(\gamma(t-s))}(\lambda y)^{r(t-s)-1}exp(-\lambda y), \; y > 0, \tag{8.9}$$

where $\Gamma(.)$ is the gamma function, $\Gamma(r) = \int_0^\infty y^{r-1}e^{-y}dy$. Suppose $Y_{s=0} = 0$, the probability density function of the increment Y_t at time t in (8.9) reduces to,

$$f_{r(t),\lambda}(y) = \frac{\lambda}{\Gamma(\gamma(t))}(\lambda y)^{r(t)-1}exp(-\lambda y), \; y > 0. \tag{8.10}$$

Under the traditional binary state reliability model with a single failure threshold value y_0, the reliability function can be derived using Equation (8.9). Thus, the expectations for

$E[\tilde{X}_\alpha^-]$ and $E[\tilde{X}_\alpha^+]$ in Expression (8.8) can be further evaluated. The following section elaborates on how the membership function for the expectation of time to fuzzy failure can be evaluated. In addition, the probability density function in Equation (8.9) can also be used for the dynamic fuzzy reliability evaluation in Equation (8.5).

8.4.3 Membership Function Evaluation for the Expectation of Time to Fuzzy Failure

To evaluate the membership function for the expectation of time to fuzzy failure, the gamma stochastic process is used to model the continuous performance degradation process. For the decreasing path degradation process (Figure 8.5) with an initial performance level of y_I, the associated cumulative failure time distribution function $F(t)$ for any given performance level y_0 can be derived as,

$$\begin{aligned} F(t) &= P(T \le t) = P(Y_t \ge y_I - y_0), \\ &= \int_{y_I-y_0}^{\infty} f_{r(t),\lambda}(y)dy = \int_{y_I-y_0}^{\infty} \frac{\lambda}{\Gamma(r(t))}(\lambda y)^{r(t)-1} exp(-\lambda y)dy, \\ &= \int_{y_I-y_0}^{\infty} \frac{1}{\Gamma(r(t))}(\lambda y)^{r(t)-1} exp(-\lambda y)d(\lambda y) \text{ by change of variable, } z = \lambda y, \\ &= \int_{\lambda(y_I-y_0)}^{\infty} \frac{1}{\Gamma(r(t))} z^{r(t)-1} exp(-z)dz = \frac{\Gamma(r(t), \lambda(y_I - y_0))}{\Gamma(r(t))}, \end{aligned} \tag{8.11}$$

where $\Gamma(.,.)$ is the upper incomplete gamma function given by $\Gamma(r, y) = \int_y^{\infty} u^{r-1} e^{-u}du, y > 0, r > 0$.

For any degree of failure α, the performance characteristic value $y = \mu^{-1}(\alpha)$ can be calculated using the membership function for the fuzzy failure event. Hence, we can evaluate $E[\tilde{X}_\alpha^-]$ as,

$$\begin{aligned} E[\tilde{X}_\alpha^-] &= \int_0^{\infty} (1 - F(t))dt \\ &= \int_0^{\infty} \left(1 - \frac{\Gamma\left(r\left(t\right), \lambda(y_I - \mu^{-1}(\alpha))\right)}{\Gamma(r(t))}\right) dt, \end{aligned} \tag{8.12}$$

and

$$E[\tilde{X}_\alpha^+] = \int_0^{\infty} \left(1 - \frac{\Gamma\left(r\left(t\right), \lambda(y_I - \mu^{-1}(1))\right)}{\Gamma(r(t))}\right) dt. \tag{8.13}$$

The minimum value t_{min} and maximum value t_{max} for the expectation of time to fuzzy failure can be computed by setting $\alpha = 0$ and 1, respectively. Thus,

$$t_{min} = \int_0^{\infty} \left(1 - \frac{\Gamma\left(r\left(t\right), \lambda(y_I - \mu^{-1}(0))\right)}{\Gamma(r(t))}\right) dt = \int_0^{\infty} \left(1 - \frac{\Gamma\left(r\left(t\right), 0\right)}{\Gamma(r(t))}\right) dt = 0, \tag{8.14}$$

and,

$$t_{max} = \int_0^{\infty} \left(1 - \frac{\Gamma\left(r\,(t)\,,\lambda(y_I - \mu^{-1}(1))\right)}{\Gamma(r(t))}\right) dt = \int_0^{\infty} \left(1 - \frac{\Gamma\left(r\,(t)\,,\lambda(y_I - y_T)\right)}{\Gamma(r(t))}\right) dt. \tag{8.15}$$

For any fixed $t_0 \in [t_{min}, t_{max}]$, the membership value can be evaluated by solving the optimization problem in Expression (8.8),

$$\begin{aligned} &Max \quad \alpha \\ \text{s.t.} \quad & E[\tilde{X}_\alpha^-] = \int_0^{\infty} \left(1 - \frac{\Gamma\left(r\,(t)\,,\lambda(y_I - \mu^{-1}(\alpha))\right)}{\Gamma(r(t))}\right) dt \le t_0, \\ & t_0 \le E[\tilde{X}_\alpha^+] = \int_0^{\infty} \left(1 - \frac{\Gamma\left(r\,(t)\,,\lambda(y_I - \mu^{-1}(1))\right)}{\Gamma(r(t))}\right) dt, \\ & 0 < \alpha \le 1. \end{aligned} \tag{8.16}$$

For each value t_0 in $[t_{min}, t_{max}]$, the corresponding membership value in Expression (8.16) can be solved using MATLAB®. By choosing various values for $t_0 \in [t_{min}, t_{max}]$, the membership function for the expectation of time to fuzzy failure can be evaluated.

8.4.4 Performance Measures for Dynamic Fuzzy Reliability

This section first introduces the methods for comparing fuzzy numbers, for example, the expectation of time to fuzzy failure. Then, we propose two customer-centric reliability performance metrics, that is, the normalized integrated expected state value (NIESV) and the utility/disutility-based reliability performance measures for dynamic fuzzy reliability performance evaluation. These measures can capture customers' cumulative experience with the system performance behavior over a mission time and are thus more comprehensive than the traditional measures such as the mean time to failure (MTTF) and the reliability function $R(t)$.

8.4.4.1 Methods for Comparing Fuzzy Numbers

Different system/product designs may result in different membership functions for the expectation of time to fuzzy failure. In other words, the membership functions of the fuzzy numbers resulting from various system designs need to be compared for performance evaluation. The cardinality, degree of inclusion, energy measure, and entropy measure are some metrics for fuzziness measures (Pedrycz and Gomid 2007). The cardinality of a fuzzy set A defined over a universal set of X can be expressed as $Card(A) = \sum_{x \in X} \mu_A(x)$ or $\int_X \mu_A(x)dx$ for discrete and continuous membership functions, respectively. The degree of inclusion of a fuzzy number A in B, denoted as $||A \subset B||$, is defined as follows,

$$||A \subset B|| = \frac{1}{Card(X)} \int_X (A \subset B)dx \tag{8.17}$$

where $A \subset B = 1$, if $\mu_A(x) \leq \mu_B(x)$; otherwise $A \subset B = 1 - \mu_A(x) + \mu_B(x)$. The degree of inclusion indicates the extent that one fuzzy number is contained in the other. For example, if $\mu_A(x) \leq \mu_B(x)$ for all $x \in X$, the degree of inclusion of A in B is equal to 1, i.e. $||A \subset B|| = 1$. Both the cardinality and the degree of inclusion of fuzzy numbers are utilized later to compare the membership functions for the expectations of time to fuzzy failure and hence for comparing different system designs.

8.4.4.2 Reliability Performance Measure Using the NIESV Metric

For dynamic reliability performance evaluation, the reliability metric $R(t)$ alone may not be sufficient to differentiate and compare different system designs. For instance, in Figure 8.6, $R_B(T^*) < R_A(T^*)$, but it is hard to say that Design B is worse than Design A over some target life T^* since Design B sustains a higher reliability level for most of the time in $[0, T^*]$ and thus the customer will experience higher cumulative value over the period $[0, T^*]$.

The reliability performance measure called the normalized integrated expected state value (NIESV) for the system over some target time duration T^* can be used for system performance evaluation (Liu, 2007). For the fuzzy reliability model, the NIESV can be expressed as follows.

$$\begin{aligned}\text{NIESV}(T^*) &= \frac{1}{T^*}\int_0^{T^*} R(t)dt \\ &= \frac{1}{T^*}\int_0^{T^*} E[\mu_S(y)]dt \\ &= \frac{1}{T^*}\int_0^{T^*}\int_{y_I}^{y_M} \mu_S(y)f(y(t))dy\,dt. \end{aligned} \tag{8.18}$$

The metric NIESV(T^*) essentially represents the ratio of the area for the shaded part in Figure 8.6 to the ideal or the maximum value of the area which is $1 \times T^*$. A higher value of the NIESV(T^*) indicates that the system stays longer at high performance levels over the time period $[0, T^*]$. Thus, NIESV(T^*) is an effective metric in evaluating system

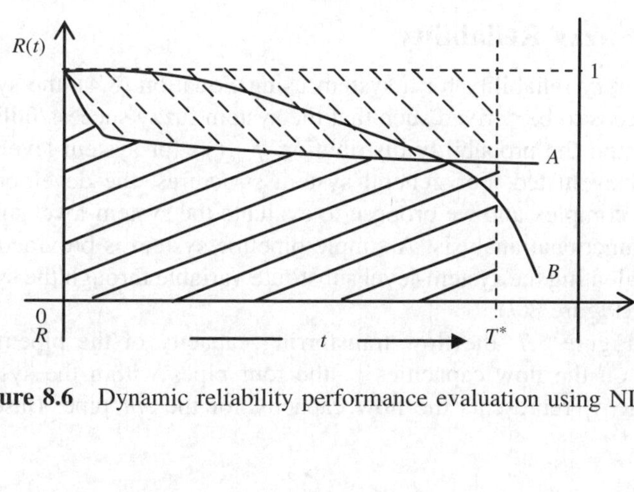

Figure 8.6 Dynamic reliability performance evaluation using NIESV

designs, especially when other reliability metrics such as $R(t)$ and MTTF are not able to differentiate different system designs.

8.4.4.3 Reliability Performance Measure Based on the Disutility Function of Customers

Another method for dynamic fuzzy reliability evaluation is to introduce the utility or disutility function for the membership values for the expectation of time to fuzzy failure. It is reasonable to assume that customers' utility would be lower if the membership value for the expectation of time to fuzzy failure deviates more from zero. The difference between the membership value and zero is denoted by $d(t)$ which is equal to the membership value $\mu(t)$, where t is one possible value for the expectation of time to fuzzy failure. The system that gives the customer the smaller expected loss $E[d(t)]$, should be considered as more reliable and a better system. A common utility function that considers the effect of variance is the exponential utility function of the form $U(x) = 1 - exp\left(-\frac{x}{r}\right)$, where r is a scale parameter (Nicholson 1998). We use the similar model for the disutility function with bigger x indicating greater disutility.

$$DU(d(t)) = 1 - exp\left(-\frac{d(t)}{r}\right) = 1 - exp\left(-\frac{\mu(t)}{r}\right) \tag{8.19}$$

Instead of using the disutility of a single membership value for the expectation of time to fuzzy failure, the accumulated disutility for the membership values of all possible expectation values of time to fuzzy failure is proposed for the evaluation of system performance, which can be expressed as,

$$\int_0^t DU(d(\tau))d\tau = \int_0^t \left[1 - exp\left(-\frac{\mu(\tau)}{r}\right)\right] d\tau \tag{8.20}$$

This cumulative disutility measure for any given value t for the life of the system is used for evaluating system's dynamic reliability performance in the following numerical examples.

8.5 System Fuzzy Reliability

To evaluate the fuzzy reliability for a system using Equation (8.4), the system-level substitute variable needs to be derived such that the system fuzzy success/failure membership function $\mu_{sys}(y)$ and the probability distribution $f_{sys}(y)$ for system-level substitute variable can be well evaluated. For general system structures, the development of $F_{sys}(y)$ can become very complex and we propose to evaluate the system-level substitute variable through system functional analysis. A simple pipeline system is provided to demonstrate the method of evaluating the system-level substitute variable through the system functional analysis method (Figure 8.7).

As shown in Figure 8.7, the flow transferring capacity of the pipeline system from A to B depends on the flow capacities of the four pipes within the system. Suppose a substitute variable y_i represents the flow capacity for the ith pipe. Based on the given

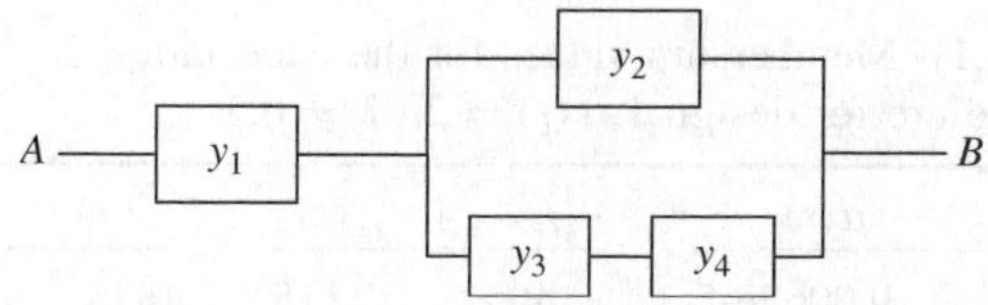

Figure 8.7 A pipeline system with flow capacity represented by y_i

system structure, the system substitute variable y_{sys} can be expressed as,

$$y_{sys} = min\{y_1, (y_2 + min\{y_3, y_4\})\}. \tag{8.21}$$

It should be noted that Equation (8.21) is derived based on the system functional analysis and is different from the one derived using the traditional binary state structure function, which would be $y_{sys} = min\{y_1, max(y_2, min\{y_3, y_4\})\}$. Further research is needed to develop the substitute characteristic for system performance where the subsystems or components have different types of substitute characteristics for their reliability or performance evaluation.

8.6 Examples and Applications

This section shows examples to evaluate the time to fuzzy failure for either a component or system.

8.6.1 Reliability Performance Evaluation Based on Time to Fuzzy Failure

The application of the dynamic behavior of the time to fuzzy failure is demonstrated through evaluating two designs where the performance characteristic is larger the better with performance deteriorating from $y_I = 10$ to $y_M = 2$. The first design has gamma degradation process characterized by a linear shape parameter $r(t) = 2t$, and a scale parameter $\lambda = 0.5$. Each performance level of $y \in (2, 10)$ exhibits a fuzzy failure membership value α, which is given by $\mu_F(y) = \frac{10-y}{8}$, a monotonic fuzzy failure membership function defined over the performance characteristic variable y. The upper limit of time to fuzzy failure is $t_{\max} = \int_0^\infty \left(1 - \frac{\gamma(2t,4)}{\gamma(2t)}\right) dt = 1.2362$. By solving the nonlinear optimization problem of Expression (8.8) for various values of $t_0 \in [0, 1.2362]$, the membership function of the expectation of time to fuzzy failure is given in Table 8.1.

The second system degrades over time with a different shape parameter $r(t) = 2.7t$, and a scale parameter $\lambda = 0.6$. The membership function for the expectation of time to fuzzy failure of this system is given in Table 8.2.

For comparison, the membership functions of the expectation of time to fuzzy failure of the two gamma degradation processes are plotted in Figure 8.8. It is observed that the membership values for the expectation of time to fuzzy failure for the degradation process with $r(t) = 2.7t, \lambda = 0.6$ (design 2) are smaller than for the degradation process with $r(t) = 2t, \lambda = 0.5$ (design 1) over [0, 0.4], and vice versa over the interval [0.4, 1).

Table 8.1 Membership values for the expectation of time to fuzzy failure for design 1, $r(t) = 2t$, $\lambda = 0.5$

t	$\mu(t)$	t	$\mu(t)$	t	$\mu(t)$
0.05	9.0063e-5	0.40	0.1115	0.75	0.4237
0.10	0.0034	0.45	0.1406	0.80	0.5048
0.15	0.0125	0.50	0.1733	0.85	0.6078
0.20	0.0262	0.55	0.2103	0.90	0.7492
0.25	0.0431	0.60	0.2525	0.95	0.9803
0.30	0.0630	0.65	0.3009	1.00	1.0000
0.35	0.0860	0.70	0.3571	1.24	1.0000

Table 8.2 Membership values for the expectation of time to fuzzy failure for design 2, $r(t) = 2.7t$, $\lambda = 0.6$

t	$\mu(t)$	t	$\mu(t)$	t	$\mu(t)$
0.05	5.1621e-5	0.40	0.1142	0.75	0.5770
0.10	0.0025	0.45	0.1485	0.80	0.7736
0.15	0.0103	0.50	0.1887	0.82	0.9058
0.20	0.0229	0.55	0.2364	0.85	1.0000
0.25	0.0397	0.60	0.2937	0.90	1.0000
0.30	0.0603	0.65	0.3641	0.95	1.0000
0.35	0.0851	0.70	0.4539	0.965	1.0000

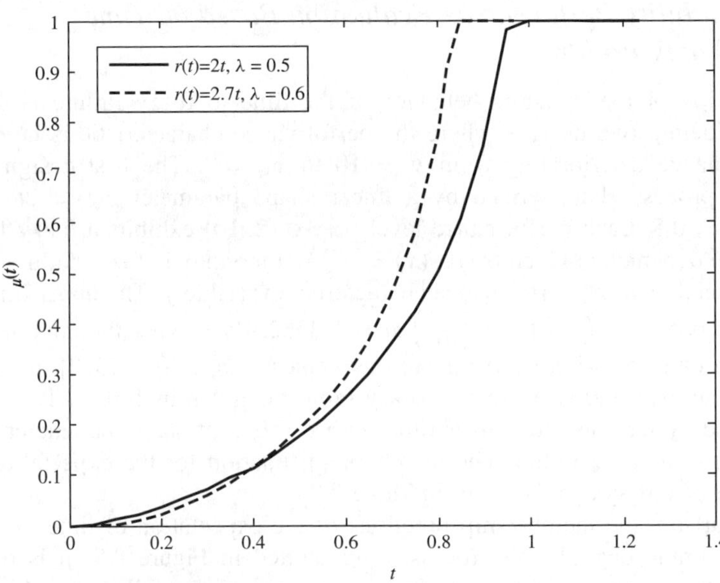

Figure 8.8 Membership functions of the expectation of time to fuzzy failure

The performance of the two system designs are evaluated using the reliability measures developed in Section 8.4.4, that is, the methods for fuzzy number comparison and the disutility-based metric. Since the cardinality of fuzzy number measures the magnitude of fuzzy numbers, the cardinality value of a fuzzy number is an indication of the customer's perception to the fuzzy event. In our case, larger cardinality value implies worse system performance/design since the fuzzy number is the expectation of time to fuzzy *failure*. The cardinality values for design 1 and design 2 using the equation given in section 8.5 are 0.4885 and 0.5547, respectively. So design 1 is better than design 2 in terms of the cardinality measure.

The degree of inclusion for the two systems is evaluated using Equation (8.17), and $||A \subset B|| = 0.959$, which is close to one. The high degree of inclusion value indicates that most membership values for design 1 are lower than that for design 2. So we conclude that design 1 is better than design 2 in terms of the degree of inclusion.

System performance evaluation by incorporating customer's utility/disutility is more customer-centric since the customers' and designer's preferences for system behavior are considered. The accumulated disutility of the two systems is evaluated based on Equation (8.20) with a scale parameter $r = 2$. The cumulative disutility values of design 1 and design 2 are 0.1502 and 0.1741, respectively. Based on the disutility values, we also conclude that design 1 is better. The dynamic fuzzy reliability of the two systems is evaluated from various aspects using three metrics, and the same conclusion is reached for this example.

8.6.2 *Example for System Fuzzy Reliability Modeling*

In this section, we present an example for system fuzzy reliabiity evaluation. The failure of components within the system is defined as fuzzy sets based on the performance variable of flow rate.

8.6.2.1 Time to Fuzzy Failure Analysis for the Pipeline System

For the proposed pipeline system in Section 8.5, the substitute variable for the system can be characterized using the flow capacity for each pipeline. Suppose the flow capacity for the pipelines follow Weibull distributions with the following probability density functions:

$$f_i(y) = \frac{\beta}{\alpha_i(t)}\left(\frac{y}{\alpha_i(t)}\right)^{\beta-1} exp\left[-\left(\frac{y}{\alpha_i(t)}\right)^{\beta}\right], y \geq 0, i = 1, 2, 3, 4, \tag{8.22}$$

where $\alpha_i(t) > 0, \beta > 0$. $\alpha_i(t)$ is the characteristic flow capacity value for the Weibull distribution of the *ith* pipeline which decreases over time and is used to characterize the performance degradation process. β is the shape parameter. The mean flow capacity level $m(t)$ for the substitute variable defined in Equation (8.22) at any given time t can be expressed as follows,

$$m_i(t) = \alpha_i(t)\Gamma\left(1 + \frac{1}{\beta}\right), \tag{8.23}$$

where $\Gamma(x)$ is the gamma function: $\Gamma(x) = \int_0^{\infty} y^x e^{-y} dy$, $\alpha_i(t)$ can be any monotonically non-increasing function characterizing the performance degradation process.

From Equation (8.22), it is easy to see that $f_i(y) > 0$ for $y > 0$, so the CDF for Y is strictly increasing in y.

Through system function analysis, the system flow capacity can be expressed as a function of the capacities of the components as in Equation (8.21), that is, $y_{sys} = min\{y_1, (y_2 + min\{y_3, y_4\})\}$. It is well known that if the CDFs of two random variables are strictly increasing functions, the CDF of the minimum and sum of the two random variables are also increasing functions.

Since the system-level substitute variable, $Y_{sys} = \min\{Y_1, (Y_2 + min\{Y_3, Y_4\})\}$, is a function under the operators of minimum and summation, the CDF for y_{sys} is also a strictly increasing function for $y_{sys} \in (0, \infty)$ (Li 2011), and we can evaluate the membership function for the time to fuzzy failure.

The probability distribution for the system-level substitute variable which characterizes the system performance degradation over time is derived for the pipeline system. The mean time to fuzzy failure for the system for each level y, that is, the function $t_E = f(y)$ can be evaluated as follows:

$$
\begin{aligned}
t_E &= E[\tilde{T}]_y = \int_{t=0}^{\infty} [1 - F_{sys}(y, t)]dt \\
&= \int_{t=0}^{\infty} exp\left[-\left(\frac{y}{\alpha_1(t)}\right)^{\beta}\right] \int_y^{\infty}\int_0^{w} \left\{\frac{\beta}{\alpha_2(t)}\left(\frac{w-z}{\alpha_2(t)}\right)^{\beta-1} exp\left[-\left(\frac{w-z}{\alpha_2(t)}\right)^{\beta}\right]\right\} \\
&\left\{\left(\frac{1}{(\alpha_3(t))^{\beta}} + \frac{1}{(\alpha_4(t))^{\beta}}\right)\beta z^{\beta-1} exp\left[-\left(\frac{1}{(\alpha_3(t))^{\beta}} + \frac{1}{(\alpha_4(t))^{\beta}}\right)z^{\beta}\right]\right\} dzdwdt.
\end{aligned}
\tag{8.24}
$$

It is noted that t_E is a function of y. By change of variable, $z = wu$, the varying limits integral in Equation (8.24) can be converted to definite triple integral as follows,

$$
\begin{aligned}
t_E &= E[\tilde{T}]_y \\
&= \int_{t=0}^{\infty} exp\left[-\left(\frac{y}{\alpha_1(t)}\right)^{\beta}\right] \int_y^{\infty}\int_0^{1} \left\{\frac{\beta}{\alpha_2(t)}\left(\frac{w(1-u)}{\alpha_2(t)}\right)^{\beta-1} exp\left[-\left(\frac{w(1-u)}{\alpha_2(t)}\right)^{\beta}\right]\right\} \\
&\left\{\left(\frac{1}{(\alpha_3(t))^{\beta}} + \frac{1}{(\alpha_4(t))^{\beta}}\right)\beta(wu)^{\beta-1} exp\left[-\left(\frac{1}{(\alpha_3(t))^{\beta}} + \frac{1}{(\alpha_4(t))^{\beta}}\right)(wu)^{\beta}\right] wdudwdt.\right.
\end{aligned}
\tag{8.25}
$$

The definite triple integral in Equation (8.25) can be numerically evaluated for each value of y in a programming application such as MATLAB®. There exists a degree of failure for each y in the space domain, that is, $\mu_{F_{sys}}(y)$, and the degree of failure for the corresponding mean time to failure $E[\tilde{T}]_y$ can be evaluated. In this way, the fuzziness with respect to failure time in the time domain can be evaluated from the fuzziness with respect to the substitute variable in the space domain.

8.6.2.2 Decision Analysis Using Fuzzy Reliability Performance Information

This section investigates how the system fuzzy reliability information, that is, the membership function of the expectation of time to fuzzy failure can be utilized to assess different system designs in decision-making. We propose a utility/disutility-based reliability performance measure for dynamic fuzzy reliability performance evaluation. A utility/disutility function for the membership values for the expectation of time to fuzzy failure is introduced and the measure of accumulated disutility over a given time period is used for decision support.

It is reasonable to assume that customers' utility would be lower if the membership value for the expectation of time to fuzzy failure deviates more from zero. The difference between the membership value and zero is denoted by $d(t)$ which is equal to the membership value $\mu(t)$, where t is one possible value for the expectation of time to fuzzy failure. The system that gives the customer the smaller expected loss $E[d(t)]$, should be considered as more reliable and a better system. A common utility function that considers the effect of variance is the exponential utility function of the form $U(x) = 1 - exp\left(-\frac{x}{r}\right)$, where r is a scale parameter (Nicholson 1998). We use the similar model for the disutility function with bigger x indicating greater disutility,

$$DU(d(t)) = 1 - exp\left(-\frac{d(t)}{r}\right) = 1 - exp\left(-\frac{\mu(t)}{r}\right). \tag{8.26}$$

Instead of using the disutility of a single membership value for the expectation of time to fuzzy failure, the accumulated disutility for the membership values of all possible expectation values of time to fuzzy failure is proposed for the evaluation of system performance, which can be expressed as,

$$\int_0^t DU(d(\tau))d\tau = \int_0^t \left[1 - exp\left(-\frac{\mu(\tau)}{r}\right)\right] d\tau \tag{8.27}$$

This cumulative disutility measure for any given life time t of the system can be used to evaluate a system's dynamic reliability performance.

8.6.3 Numerical Results

In this section, we show an illustrative example to demonstrate how the proposed fuzzy reliability model can capture more realistic and meaningful information for systems reliability performance evaluation. The system fuzzy reliability for the pipeline system given in Figure 8.4 is evaluated. The substitute variable indicating the system performance is the flow capacity of each pipeline. Suppose the flow capacities for the pipelines follow Weibull distributions with the probability density functions given by Equation (6.1). For computational simplicity, $\alpha_i(t)$ is assumed to degrade over time according to a linear form as $\alpha_i(t) = a_i - b_i t$, where a_i and b_i are positive constant reflecting the degradation behavior. Specifically, suppose the characteristic performance for the four pipelines deteriorates with the following parameters: $\alpha_1(t) = 10 - 0.02t$, $\alpha_2(t) = 5 - 0.01t$, $\alpha_3(t) = 5 - 0.01t$, and $\alpha_4(t) = 6 - 0.02t$. A common shape parameter $\beta = 2$ is assumed for all the pipelines within the system. Through communications with the customers and system designer, the pipeline system has an initial flow capacity of 12 barrels per second. When the flow capacity falls below 2 barrels per second, the system is assumed to have a 100% degree of failure. If the flow capacity stays above 10 barrels per second, the system has zero degree of failure. With this information, the S-shaped membership function for system fuzzy failure can be evaluated in terms of the system performance variable.

The fuzziness with respect to time to fuzzy failure is evaluated for the pipeline system. To evaluate the membership function for time to fuzzy failure, the stochastic degradation process for the system substitute variable y_{sys} is derived based on the Weibull distribution-based flow capacity for each pipeline as described in Section 8.4. By applying the generalized extension principle and the propositions in Section IV, the membership values for the expectation of time to fuzzy failure are evaluated (Table 8.3) and plotted (Figure 8.9). It is noted that $y_I - y_0$ indicates the amount of degradation, and a bigger value corresponds to a higher degree of failure.

Table 8.3 Membership values for the expectation of time to fuzzy failure for the pipeline system

t_E	$\mu_{E[\tilde{T}]}(t_E)$	$y_I - y_0$	y_0	t_E	$\mu_{E[\tilde{T}]}(t_E)$	$y_I - y_0$	y_0
5.7330	0.008	2.5	9.5	47.0084	0.617	6.5	5.5
8.0420	0.031	3.0	9.0	55.2907	0.719	7.0	5.0
11.0495	0.070	3.5	8.5	63.5794	0.805	7.5	4.5
14.8637	0.125	4.0	8.0	71.5148	0.875	8.0	4.0
19.5672	0.195	4.5	7.5	78.7542	0.930	8.5	3.5
25.1982	0.281	5.0	7.0	85.0210	0.969	9.0	3.0
31.7321	0.383	5.5	6.5	90.1424	0.992	9.5	2.5
39.0654	0.500	6.0	6.0	94.0697	1.000	10	2.0

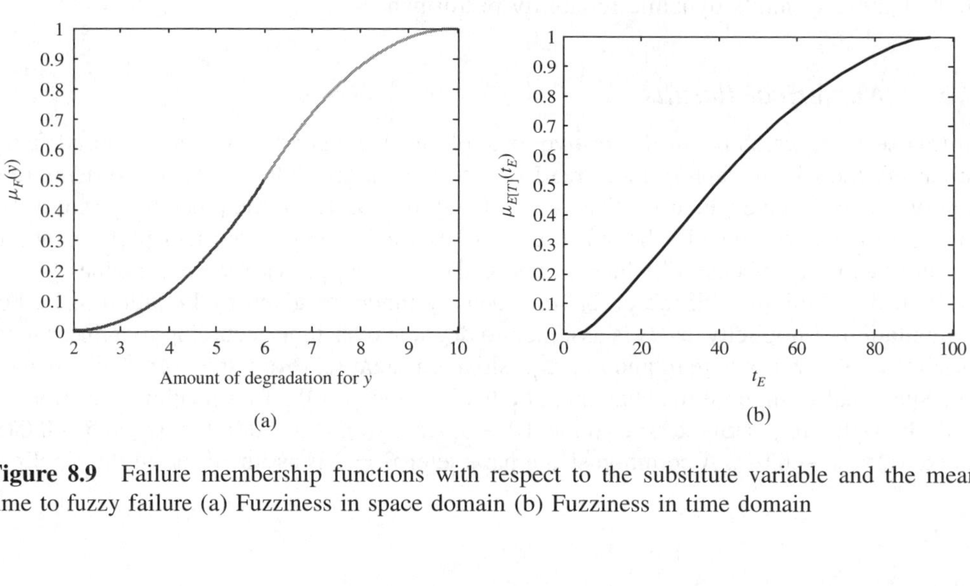

Figure 8.9 Failure membership functions with respect to the substitute variable and the mean time to fuzzy failure (a) Fuzziness in space domain (b) Fuzziness in time domain

The membership function for the expectation of time to fuzzy failure shown in Table 8.3 is consistent with the membership function of fuzzy failure event defined over the substitute variable. For instance, a higher degree of fuzzy failure corresponds to a higher level of degradation amount $y_I - y_0$ or a lower level of the substitute variable Y_0, which is associated with a higher expectation for the time to fuzzy failure. It is noted that under the traditional crisp failure definition, only one point in Figure 6.2(b) can be evaluated for a given failure threshold value rather than the whole picture for expected time to fuzzy failure with various degrees.

Using Equation (8.20) the cumulative disutility value over a time period of 94.07 is calculated, which is 23.1432. This metric can be used to choose a more reliable system among multiple systems with distinct stochastic performance degradations.

8.7 Conclusion

In this chapter, we propose a new perspective to model the reliability of a component or system using the theory and methods for fuzzy sets. We also give a comprehensive literature search for traditional fuzzy reliability. We propose to use a substitute variable which is well understood by the customer to fuzzify the states (success or failure) of a component or system. We show that the proposed measures based on fuzzy thinking are more meaningful and customer-centric for reliability-related decision-making processes. The approach and methods presented are a generalization for the traditional binary and multi-state reliability and can also be used as an alternative for modeling the continuous state behavior for systems that degrade over time. We illustrated how fuzzy modeling of the states for components and a system can capture more realistically the reliability performance information for decision-making processes by the system designer and the customer. The fuzzy success event is defined over the levels of the system performance or substitute characteristics and we can use it to develop the membership function for the fuzzy success or failure event. The fuzzy reliability modeling incorporates the customer's understanding and utility through the fuzzy success/failure membership function, thus the fuzzy reliability can capture more information and is more realistic and meaningful than the traditional reliability models. The perspective of regarding success and failure as fuzzy state instead of fuzzifying some numerical values in the traditional binary/multi-state models has not been well investigated in the literature.

One of the future research efforts is to investigate further the development for time to fuzzy failure. Under the fuzzy state assumption, the traditional failure in terms of a single failure threshold value has no meaning and degrees of failure have to be considered. How to efficiently evaluate the membership function for the expectation of time to fuzzy failure under any stochastic performance degradation process needs to be further investigated. Other related reliability measures such as the variance of time to fuzzy failure and mean residual life also need to be investigated. In addition, maintenance optimization and system health management, for example, system health diagnostics and prognostics under fuzzy success and failure consideration are some other potential research areas.

The membership function evaluation for the fuzzy success/failure events is a critical part of the proposed fuzzy reliability model. Utility theory and measurement theory can be used as tools for integrating utility and decision preferences into the evaluation of the membership function and very little research has been done in this area.

As a complementary theory to the probability measure (Zadeh 1995; Singpurwalla and Booken 2004), a possibility measure can also be applied to characterize the uncertainty behavior for the substitute characteristic variable, and corresponding fuzzy reliability models can be developed. Such reliability models will extend the probability-based reliability definition to a more general framework under possibility theory assumption. Fuzzy reliability under possibility theory will be another area for the future research.

References

Brunelle, R. D. and K. C. Kapur. 1999. Review and classification of reliability measures for multistate and continuum models. *IIE Transactions* 31 (12): 1171–1180.

Kapur, K. C. 1998. An integrated customer-focused approach for quality and reliability, *International Journal of Reliability, Quality and Safety Engineering* 5 (2): 101–113.

Kwakernaak, H. 1978. Fuzzy random variables – I. Definitions and theorems, *Information Science* 15 (1), 1–29.

Li, Z. 2011. Customer-centric reliability measures based on fuzzy set theory, PhD dissertation, University of Washington. (Publication No. 3501570.)

Li, Z. and K. C. Kapur. 2010. Models and customer-centric system performance measures using fuzzy reliability. In *Proceedings of the 2010 IEEE International Conference on Systems, Man and Cybernetics (SMC)*, Istanbul, 819–825.

Liu, Y. and K. C. Kapur. 2007. Customer's cumulative experience for reliability of nonrepairable aging multistate system. *Quality Technology and Quantitative Management* 4 (2): 69–78.

Li, Z. and K. C. Kapur. 2011. Models and measures for fuzzy reliability and its relationship to multi-state reliability. *International Journal of Performability Engineering* 7 (3): 240–251.

Li, Z. and K. C. Kapur. 2012. Continuous-state reliability measures based on fuzzy sets. *IIE Transactions* 44 (11): 1033–1044.

Liu, Y., Z., Qiao and G. Wang. 1997. Fuzzy random reliability of structures based on fuzzy random variables. *Fuzzy Sets and Systems* 86 (3): 345–355.

Nicholson, W. 1998. *Microeconomics Theory: Basic Principles and Extensions*. New York: The Dryden Press, Harcourt Brace College Publishers.

Noortwijk, J. M. V. 2009. A survey of the application of gamma processes in maintenance. *Reliability Engineering and System Safety* 94 (1): 2–21.

Pedrycz, W. and F. Gomide. 2007. *Fuzzy Systems Engineering: Toward Human-Centric Computing*. Hoboken, NJ: John Wiley & Sons.

Singpurwalla, N. D. and J. M. Booken. 2004. Membership functions and probability measures of fuzzy sets. *Journal of the American Statistical Association* 99 (467): 867–877.

Weininger, S., K. C. Kapur and M. Pecht. 2010. Exploring medical device reliability and its relationship to safety and effectiveness. *IEEE Transactions on Components and Packaging Technologies* 33 (1): 240–245.

Wu, H. C. 2001. Evaluate fuzzy Riemann integrals using the Monte Carlo method. *Journal of Mathematical Analysis and Applications* 264 (2): 324–343.

Zadeh, L. A. 1965. Fuzzy sets. *Information and Control* 8 (3): 338–353.

Zadeh, L. A. 1975. The concept of linguistic variable and its application to approximate reasoning – I. *Information Science* 8 (3): 199–249.

Zadeh, L. A. 1995. Discussion: Probability theory and fuzzy logic are complementary rather than competitive. *Technometrics* 37 (3): 271–272.

9

Imperatives for Performability Design in the Twenty-First Century

Krishna B. Misra
International Journal of Performability Engineering, India

9.1 Introduction

In 1987, the United Nations released the Brundtland Report (A/RES/42/187 1987), which defined sustainable development as "development which meets the needs of the present without compromising the ability of future generations to meet their own needs." Inherently the concept of sustainable development is intertwined with the concept of carrying capacity (Postel 1994). Theoretically, the long-term result of environmental degradation is the inability to sustain human life. Such degradation on a global scale could imply extinction for humanity.

It is indeed true that over-exploitation of resources, on the one hand, and their wastage, on the other, have led humans to surpass the carrying capacity of Earth. Realizing the gravity of the situation, more than 1700 scientists, including 102 Noble Laureates collectively signed a Warning to Humanity in 1992, which reads as follows:

> No more than one or few decades remain before the chance to avert the threats we confront, will be lost and the prospects for humanity immeasurably diminished … A new ethics is required – a new attitude towards discharging responsibility for caring for ourselves and for Earth … this ethics must motivate a great movement, convincing reluctant leaders, reluctant governments and reluctant people themselves to effect the needed changes.

Applied Reliability Engineering and Risk Analysis: Probabilistic Models and Statistical Inference, First Edition.
Edited by Ilia B. Frenkel, Alex Karagrigoriou, Anatoly Lisnianski and Andre Kleyner.

Economists and environmentalist agree that the protection of the Earth's environment and achieving better living standards for humans are closely interlinked (Earth Summit+5 1997) and are in fact mutually reinforcing. Slowing down the increase in population (which is nearly 6.5 billion) at present (The State of the World Population 1999), especially in the face of rising per capita demand for natural resources, can take pressure off the environment and buy time to improve the living standards on a sustainable basis.

In fact, global prosperity considerably depends on our ability to use the Earth's resources, wisely, more efficiently, distributing them more equitably, and in reducing their wastages while reducing their overall consumption. Unless we can accelerate this process, serious social tensions are likely to arise between peoples, resulting in wars between the nations over the increased competition to usurp the remaining scarce and ever costlier resources of the Earth. We have seen this occur in the case of the ever-increasing cost of oil, and the Gulf War is an example.

The United Nations (Earth Summit+5 1997) took an initiative in this direction by declaring a *Decade of Education for Sustainable Development*, starting in January of 2005. In the US also, a non-partisan multi-sector group called the *U.S. Partnership for the Decade of Education for Sustainable Development* has been constituted. Any organization and individuals or groups of youth, business, religion communities, etc. can join and share resources and success stories to create a sustainable future.

9.2 Strategies for Sustainable Development

In this chapter, an attempt has been made to present a holistic view of the problem to effectively implement a sustainable development programme, which should not be considered in isolation of economic and performance implications. Only then will we be able to realize the benefits of sustainable development for prosperity in the twenty-first century.

9.2.1 The Internalization of Hidden Costs

First of all, the classical economic theories that have often treated nature as a bottomless well of resources and infinite sink for waste must be discarded. Environmental costs, that is, the cost of preventing or controlling pollution and ecological disruption, must be internalized. In fact, it is our inability to deal with the economic nature of environmental pollution that has been largely responsible for the destruction of the Earth's ecological systems at a faster rate. In fact, many hidden environmental costs incurred for resource exploitation need to be passed on to the consumer or to the user.

To preserve the Earth's environment (Misra 1996, 2008) for future generations, the internalization of hidden costs of environment preservation will have to be accounted for, sooner or later, in order to be able to produce sustainable products in the long run. It is therefore logical to add these hidden costs to the cost of acquiring a product or a system. According to some surveys, consumers are willing to pay up to 10% more to have an environmentally preferred product. It is also known that consumers in Europe are more willing to pay a premium than the consumers in the United States, but the definition of what attributes would be attached to an environmentally preferred product or system is still being discussed and is still at the initial stage.

9.2.2 Mitigation Policies

The economic implications of sustainable development include costs and overall welfare. The costs of various mitigation policies have to be considered. For example, mitigation options in the energy sector may be classified into those that improve energy efficiency and those that reduce the use of carbon-intensive fuels. Energy efficiency improvement reduces reliance on the energy supply; which is likely to improve a nation's energy security. Switching to low carbon energy supply sources is the other mitigation category in the energy sector which reduces air pollutio, with significant benefits of less greenhouse gases (GHG). For example, ethanol production from sugar waste has created a new industry and generated employment opportunities and tax revenue for the Government of Brazil. New industries create new jobs and income.

In the buildings sector, energy efficiency options are characterized as integrated and include efficient designs and siting, including passive solar technologies, and designs and urban planning to limit the heat island effect. In developing countries, efficient cooking stoves that use clean biomass fuels are an important option.

In the transport sector, the energy efficiency measures may be categorized into those that are vehicle-specific and those that address transportation planning. Vehicle-specific programmes focus on improvement of the technology and vehicle operations. In the industrial sector, energy efficiency options may be classified as those aimed at mass-produced products and systems, and those that are process-specific. The potential for cost-effective measures is significant in this sector. Efficiency measures in both categories would have a positive impact on the environment. To the extent the measures improve productivity; they would increase economic output and hence add to the government tax revenue. However, mitigation costs are just one part of the broader economic impacts of sustainable development. Other impacts include growth and distribution of income, employment and availability of jobs, government fiscal budgets, and competitiveness of the economy or sector within a globalizing market.

9.2.3 Dematerialization

The dematerialization of a product means using less material to deliver the same level of functionality to a user. A material can be anything from an unprocessed raw material to a finished product.

The United Nations Environment Programme (UNEP) defines dematerialization as:

> the reduction of total material and energy throughput of any product and service, and thus the limitation of its environmental impact. This includes reduction of raw materials at the production stage, of energy and material inputs at the use stage, and of waste at the disposal stage.

Dematerialization considers natural resources as well as wastes involved in the product's life-cycle. It literally means using less materials. It entails actions at every stage of the production and consumption chain:

- resource savings in material extraction;
- improved eco-design of products;

- technological innovations in the production process;
- environmentally friendly consumption;
- recycling of waste, etc.

The important economic consideration that is part of the dematerialization strategy for developing sustainable products and systems is to utilize the obsolete products at the end of their life for recycling or reuse. If obsolete materials are not recycled, raw materials have to be processed afresh to make new products. This represents a colossal loss of resources in the energy used at every stage of material realization, and the transport and environmental damage caused by these processes is huge. In 1998, it was estimated that 6 million tonnes of electrical equipment waste, arising in Europe, entailed a loss of resources to the order of:

- 2.4 million tons of ferrous metal
- 1.2 million tons of plastic
- 652,000 tons of copper
- 336,000 tons of aluminium
- 336,000 tons of glass

as well as the loss of heavy metals such as lead, mercury, etc. The production of all these raw materials and the goods made from them would have caused enormous environmental damage through mining, transport and energy use. In fact, recycling 1 kg of aluminum saves 8 kg of bauxite, 4 kg of chemical products and 14 Kilowatts of electricity. Therefore consideration of end-of-life treatment is going to become an integral part of product design soon.

9.2.3.1 Remanufacturing

From both an environmental as well as an economic consideration, the end-of-life treatment of products and systems is now becoming the liability of the manufacturers and distributors, eventually. The WEEE Directive of the European Union is the first step in that direction, at least in the electrical and electronic sector. The WEEE Directive (2002/96/EC), as passed by European Union, is aimed at preventing waste electrical and electronic equipment from ending up in landfill, and is promoting the level of recycling and reuse in the electrical and electronic sector. This directive requires all manufacturers and importers of electric and electronic equipment to meet the cost of collection, treatment and recovery of their waste electrical and electronic equipment at the end of their useful life.

Sustainability design of products requires the manufacturers to reclaim responsibility for their products at the end of the product's life. This is as an alternative to landfill or incineration and includes: maintenance, recycling for scrap material, and remanufacturing.

While maintenance extends the product life through individual upkeep or repair upon specific failures, remanufacturing is a production batch process involving disassembly, cleaning, refurbishment and replacement of worn-out parts, in defective or obsolete products. Scrap-material recycling involves separating a product into its constituent materials and reprocessing the material. Remanufacturing involves recycling at a parts level as opposed to at the scrap-material level. It is actually in effect the recycling of materials while preserving value-added components. Remanufacturing saves about 40–60% of the cost of manufacturing a completely new product and requires only 20% of energy.

Several major companies are resorting to remanufacturing. Xerox is an example in this case. IBM also established a facility in Endicott, New York, as a reutilization and remanufacturing center. UNISYS and Hewlett Packard also use this strategy. It must, however, be stated that remanufacturing is not suitable for all types of products, it is appropriate only for those products that are technologically mature and where a large fraction of the product can be used after refurbishment.

It may be mentioned here that a designer must account for the various costs associated with recycling and remanufacturing including: the first cost, the recycling cost, and the cost of failure during disassembly and reassembly. The first cost is the cost of manufacturing and the first assembly. The recycling cost includes the cost of extracting material or the cost of separating parts of different materials. Both maintenance and remanufacturing involve disassembly and reassembly and part reuse and failures can occur during these phases. Therefore, the consequences of the above failures are weighted by their probabilities of occurrence. For example, rivets and welds usually get destroyed during the disassembly. The other part of the cost includes the cost of a part getting damaged during assembly or disassembly. The other part of the cost includes the cost of damage done to a part when a fastener is extracted. Maintenance costs are the costs associated with disassembly or assembly whereas the remanufacturing cost is the total cost under all the mentioned heads.

9.2.3.2 Recycling and Reuse

The waste generated even in the electrical and electronic sector is not small. For example, in a small country like Ireland, an estimated 35,000–82,000 tons of electrical and electronic equipment waste was generated in 2001. This amounted to 9–18 kg per person. Each year, more than 100 million computers are sold and over 1 million computers are disposed of in landfill sites. The rest are recycled for parts or material. Ecomicro, a recycling company in Bordeaux, France, is reported to resort to recycling of components out of 1500 tons of obsolete or unusable computers annually. In fact, the market for refurbished computers has increased by 500% since 1996 but less than 20% of all discarded computers are recycled. Currently, 40% of all ink cartridge, which is 1.5 million printer cartridges, are recycled annually.

Today the electronic waste recycling business is a large and rapidly increasing business in the developed world. Electronic waste processing systems have matured in recent years, following increased regulatory, public, and commercial scrutiny, and a commensurate increase in entrepreneurial interest. Part of this development has involved greater diversion of electronic waste from energy-intensive down-cycling processes (e.g. conventional recycling), where equipment is returned to a raw material form. This diversion is achieved through reuse and refurbishing. The environmental and social benefits of reuse include decreased demand for new products and virgin raw materials (with their own environmental issues); larger quantities of pure water and electricity for associated manufacturing; less packaging per unit; the availability of technology to wider sections of society due to the greater affordability of products; and reduced use of landfills. Audio-visual components, televisions, VCRs, stereo equipment, mobile phones, other handheld devices, and computer components contain valuable elements and substances suitable for reclamation, such as lead, copper, and gold. Almost all electronic products contain lead and tin (as solder) and copper (as wire and printed circuit board tracks), though the use of lead-free solder is now becoming quite common.

Reuse is an alternative option to recycling because it extends the lifespan of a device. Devices still need eventual recycling, but by allowing others to purchase used electronics, recycling can be postponed and value gained from device use.

9.2.4 Minimization of Energy Requirement

According to the ICCA (International Council of Chemical Associations) report, energy-saving products installed in homes in the United States prevented nearly 283 million tons of CO_2 emissions in 2010 – equivalent to the greenhouse gas emissions of 50 million passenger vehicles. Studies also show (Parry *et al.* 2007) that if this trend continues, more than 7 billion tons of emissions can be avoided by 2050 in the US alone – equivalent to the CO_2 emissions of more than 1.2 billion passenger vehicles.

Eco-labeling is a promising market-based approach to improving the environmental performance of products through consumer choice. While eco-labeling itself is not new, eco-labeling to promote energy efficiency or sustainability is a more recent phenomenon. Five energy-labeling programs in the US are evaluated: Green Seal, Scientific Certification Systems, Energy Guide, Energy Star, and Green-e. Of these, the first four certify energy-efficient appliances while the last one certifies renewable electricity. Energy Guide and Energy Star are government-run programs, the rest are privately administered.

9.3 Reappraisal of the Performance of Products and Systems

The ultimate value of any product system or service must be judged by its performance, either expected or specified. Without being more specific, we can say that what applies to the performance of a product also applies to a system or services Therefore, in order to define the desired performance of a product, it is important to consider the following aspects:

- definition or objective of a product
- criteria of acceptable performance
- definition of failure malfunctioning
- expected time of operation
- operating conditions
- maintenance conditions
- tests and sampling procedures.

How well a product meets its technical performance requirements (Misra 1992, 1993) depends on various characteristics, such as quality, reliability, availability, safety and efficiency. A product having these attributes is usually expected to perform well over its lifetime, incurring minimum lifecycle costs, which include design and development, manufacturing, and maintenance costs. Generally, these costs are also inter-dependent on the attributes of performance. For example, a highly reliable product will have lower maintenance costs. But to produce a reliable product, we may have to incur an increased cost of design and manufacturing. The poor deficient performance attributes not only affect the lifecycle costs but also have effects in terms of environmental consequences. Degraded performance attributes reflect more on the material and energy demand and wastes, and cause more environmental pollution when reckoned over a given period of time. Obviously, a product with poor quality, reliability, maintainability, availability, safety or efficiency will incur more lifecycle costs and would be uneconomical to use.

No one can ever dispute the need for a product to be reliable. We are all aware of the problems of less reliability in household products such as TV sets and automobiles. Moreover, general utility services such as railways, airlines, water supply, electricity and the military organizations can hardly ignore the costs of unreliability. Product manufacturers often suffer high costs of failure under warranty. Arguments begin when we try to quantify reliability values, or try to assign financial or other benefit values to levels of reliability. In fact, market competition, the compulsion to reduce product costs, the pressure of schedules and deadlines, the cost of failures, rapid development of new materials, methods and complex systems, and safety considerations all emphasize greatly the need to develop better products.

The simplest, purely producer-oriented view of reliability is that in which a product is assessed against some specification or a set of performance attributes, and when it has passed these, then it is delivered to the customer. In other words, the only concern of a manufacturer appears to be the *product quality* and the customer is happy to accept the product as long as it is often supported with a warranty. At best, the customer may also have some protection in law, so that he may claim redress for failures occurring within a reasonable time, usually the warranty period. However, this approach provides no guarantee of performance over a period of time, particularly outside the warranty period. Even within a warranty period, the customer usually has no grounds for further action if the product fails once, twice or several times, provided that the manufacturer repairs the product as promised each time. If it fails often, the manufacturer will suffer high warranty costs, and the customers will suffer inconvenience. But outside the warranty period, it is only the customer who is left to suffer the consequences of failures. Of course, the manufacturer may also probably incur a loss of reputation and possibly future business.

Therefore, we have the requirement of a time-based concept of quality. The inspectors' concept is not time-dependent nor does it ensure that the product will be able to function satisfactorily under the actual conditions of environment of use. The quality tests either pass a product or fail it. On the other hand, reliability is usually concerned with failures in the time domain of the product's use. This distinction marks the difference between traditional quality control and reliability engineering. Moreover, whether failures occur or not and the times of their occurrence, can never be forecast accurately. Therefore, reliability implies an aspect of engineering uncertainty, which is often reflected in its probabilistic definition. Thus, the usual engineering definition of reliability runs as: *the probability that a product will perform the intended function without failure under stated conditions of use over a stated period of time.*

9.4 Dependability and Environmental Risk are Interdependent

The ever-increasing complexity of modern products and systems increases the possibility of malfunction of any of their constituent subsystems and therefore necessitates very high reliability of their components and subsystems, to ensure human safety and the protection of the surrounding environment from damage in case of an accident. It is basically a failure of some component or equipment or control in a large system that leads to major catastrophic consequences for the environment and humans. Especially the high-risk systems such as nuclear power plants and very many chemical plants require operational safety of the highest order. Any failure in such a plant may be financially or economically disastrous and environmentally damaging. For example, the methyl isocyanate gas leakage in the Bhopal Union Carbide plant in India in 1984 was the worst industrial catastrophe (Sriramachari 2004), which resulted in the immediate death of 2259 people and some 8000 in the first weeks of disaster and another 8000 who have died from gas-related diseases.

Very recently, the oil spill from a British Petroleum (BP) well in the Gulf of Mexico that flowed for three months in 2010, starting from April 20, 2010, was another environmental disaster. Finally, the wellhead was capped on July 15, 2010, after releasing about 4.9 million barrels of crude oil. It is the largest accidental marine oil spill in the history of the petroleum industry. The spill has caused extensive damage to marine and wildlife habitats as well as the Gulf's fishing and tourism industries. Skimmer ships, floating containment booms, anchored barriers, and sand-filled barricades along shorelines were used in an attempt to protect hundreds of miles of beaches, wetlands and estuaries from the spreading oil. It is estimated that this disaster could result in a cost of more than US$3 billion to BP in environmental clean-up and compensation. The eventual loss to BP may turn out to be of the order of several billion dollars. According to a report published on 27 July, 2010, BP posted a loss of US$17 billion in the second quarter in charges related to the Deepwater Horizon rig explosion and oil clean-up.

Even on the basis of economic considerations, a designer is left with no option but to look for the high reliability of systems as the cost of downtime works out to a crushing sum. For example, the power replacement cost when a moderate-sized nuclear plant is shut, may run to over US$80000. A loss of several billion dollars was reported besides the loss of human lives in the total failure of the U.S. *Challenger* mission.

Therefore, it is imperative that we must have the systems' reliability and safety of the highest order if we intend to avoid financial losses and environmental damages on account of failures or accidents.

9.5 Performability: An Appropriate Measure of Performance

The true performance of a product or a system or a service should not only be based on its technical performance which is judged by dependability, which is the composite reflection of various performance attributes such as quality, reliability, maintainability and safety but should also take into consideration how the processes were made during its production or manufacturing so that the environmental consequences of its production are accounted for. If a product leaves a bad environmental footprint, it is not a sustainable product. Therefore a product must be sustainable as well as dependable. In the process of an earnest attempt to define an appropriate and simple performance measure, reflecting the new concept, which includes dependability and sustainability criteria, several terms from the literature were assessed by the author but none was found more appropriate than the term *performability*, which incidentally was introduced by John Meyer in 1980 in the context of evaluation of highly reliable aircraft control computers for use by NASA, mainly to reflect a combined attribute for reliability and availability (Meyer 1980). The author feels there is no reason why the same term cannot be used to connote a much wider meaning than originally intended by Meyer.

Performability can be used to mean a combined attribute of many performance measures such as quality, reliability, maintainability or availability, safety and finally sustainability because all these measures are interrelated. The aim of this chapter is only to show that interrelationship among various performance measures and to propose and use a composite attribute like *performability* to reflect a much wider performance attribute. We would also like to standardize various terms that are in vogue and have not found rational connotation, convincingly. For example, the term *dependability* has been widely used in the reliability literature and is often used confusingly in various contexts. It has also been used to include more than one attribute related to performance. Therefore, it was considered logical and appropriate to extend the meaning of performability to include attributes like dependability and sustainability and to interrelate them with the hope of achieving some standardization. At the same time, it was felt that the term *survivability* (which is often confusingly taken to mean reliability sometimes) should only mean to reflect probability of a product surviving, which means it should reflect attributes like quality and reliability and maintainability) but should not include safety. However, if we combine survivability with an attribute like safety, the system or product should become dependable.

Therefore, with the foregoing explanation, it is also considered appropriate to widen the definition of the term *performability* to include sustainability in the context of the changed scenario of the twenty-first century in order to reflect a holistic view of designing, producing and using products, systems or services, which will satisfy all the performance requirements of a customer to the best possible extent. This would mean that products and systems are not only dependable (implying survivability and safety) but are also sustainable. The picture that emerges of the interrelationship of various prevalent terms with the explanation provided above is shown in Figure 9.1.

9.5.1 Performability Engineering

Now we may even be in a position to define *Performability engineering* in simple terms as the entire engineering effort that goes into improving the performance of a system that not only ensures its high quality, reliability, maintainability and safety but is also sustainable.

Implicit in this definition is not only the high performance of a system but also its minimum lifecycle costs. We cannot separate environmental problems from the economics of clean production and clean technologies. Likewise, improved performance should necessarily imply less environmental pollution, less material and energy requirements, waste minimization, and finally conservation and efficient utilization of available resources,

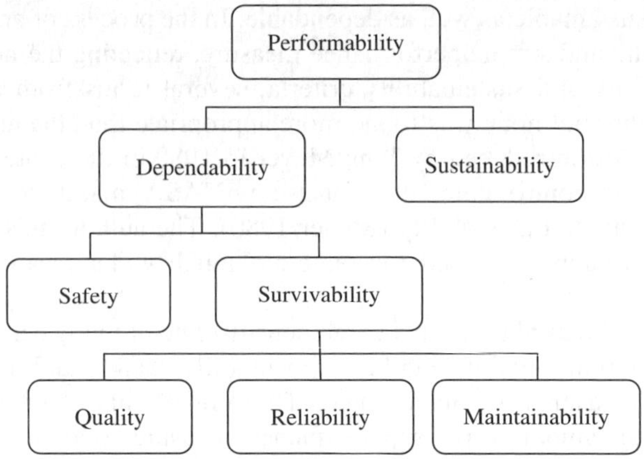

Figure 9.1 Interrelationship of various attributes

which in turn would result into minimum lifecycle costs. These problems are best tackled at the design stage of a system.

Therefore, an aggregate attribute such as *performability* would reflect a designer's entire effort in achieving not only a *dependable* product but also a *sustainable* product. In fact, it can be called an improving 3-S, namely, *survivability, safety* and *sustainability*. It may be emphasized here that the usual definition of dependability ignores the environmental consequences that accompany products, systems and services during manufacturing It is evident that in order to produce truly optimal designs economically, consideration of sustainability should not be overlooked. These attributes are very much influenced by the design, raw materials, fabrication, techniques and manufacturing processes. These attributes are interrelated and reflect the level or grade of the product so designed and utilized which is expressed through dependability.

The lifecycle activities of a product or system are shown in Figure 9.2. Performability takes a holistic view of various activities and processes and takes stock of what is being produced and what is being wasted. We conserve and economize on materials, energy, and avoid wastes to optimize a product or system's design over its entire lifecycle.

In fact, performability engineering not only aims at producing products, systems and services that are dependable but involves developing economically viable and safe processes (clean production and clean technologies) that would entail minimal environmental pollution, require minimum quantities of raw material and energy and yield safe products of acceptable quality and reliability that can be disposed of at the end of their life without causing any adverse effects on the environment. The WEEE Directive of the EU is a step in that direction.

This would also necessitate the efficient use of natural resources and the use of non-waste technologies which would ensure that all raw materials and energy are used in the most rational and integrated way to curb all kinds of wastages while maximizing the performance. Obviously, less material and energy consumption – either through dematerialization, reuse or recycle or through proper treatment (clean-up technology) – would lead to less environmental degradation. Similarly, a better design would result in prolonging

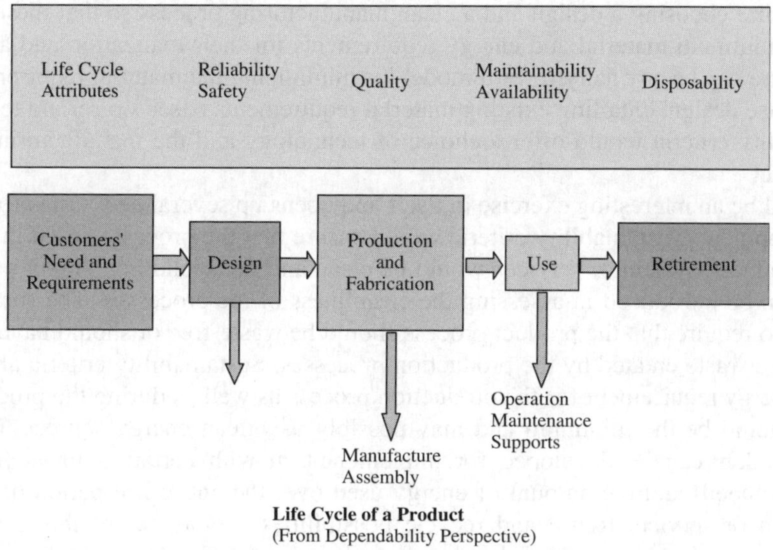

Figure 9.2 Lifecycle activities

the lifespan of a product and hence would ensure less adverse effects on the environment over a given period of time. In other words, we must integrate the entire lifecycle of activities of survivability with that of environmental lifecycle considerations to improve the product or system performance within the technological barriers, with the minimum cost.

At every stage of the lifecycle of a product, be it extraction of material, manufacturing, use or disposal, energy and materials are required, as inputs and emissions (gaseous, solid effluents or residues) are always associated which influence the environmental health of our planet.

Unless we consider all these factors, we cannot call our design of products, systems and services truly optimal from an engineering point of view. This would necessitate bringing synergetic consideration of all constituent areas of performability.

Sensing the possibility of using performability for the design of systems and products in future, the author started an *International Journal of Performability Engineering* (Misra 2013) in India in 2005 and has ensured the publication of this journal for the past 8 years, emphasizing the necessity of sustainability in the design of products and systems. Reliability or dependability is just a small part of this consideration.

9.6 Towards Dependable and Sustainable Designs

We can now formulate the proposition that truly optimal systems, products, and services would be the ones that not only have the attributes of quality, reliability, maintainability and safety but also sustainability. This means that they should have high performability over the entire lifecycle periods. Much needs to be done to formulate a methodology to design such high performability systems, products or services. The sustainability criteria would require choosing a design and a clean manufacturing process so that these products must use minimum material and energy requirements for their realization and also in use phase. There can be a penalty/reward model for minimizing the material requirement from the base case design, entailing existing material requirements based on certain technology. Sustainability criteria would offer a choice of technology and the manufacturing process of a product.

This will be an interesting exercise in itself and opens up several new vistas for research and development. Sustainability criteria would ensure that the processes used in manufacture of products, system or services would be clean and non-polluting. Again a reward/or penalty can be introduced in assessing the cleanliness of the processes. The sustainability criteria also require that the product process should be waste-free or should have means of utilizing the waste created by the production processes. Sustainability criteria also require that the energy requirement for the production process as well as during the product maintenance should be the minimum and may possibly use clean energy sources. Reward/or penalty models can be developed for implementation with certain sources (renewable sources included) and the amount of energy used over the entire life period of the product, system or service. Reuse and recycle possibilities should be suitably rewarded in the design model. Alternative technologies such as nanotechnology and bioengineering also widen the possibilities of such realization of design and offer several new design possibilities for products and systems.

The concept presented in this chapter opens up new research areas for the design of high performability systems, products and services. This chapter only discusses the concept of such a possibility and what may eventually happen in the future.

9.7 Conclusion

The concept of performability opens up entirely new possibilities of designing, producing, maintaining and disposing off the products, system and services in the future, which is going to make these systems more environmentally friendly as well as economic over their entire lifecycle while excelling in dependability. This concept is far wider than just designing for reliability which has been used for nearly a century for performance. However, there is much to be done in this area. It is hoped that this chapter will help generate interest and ideas in that direction.

References

A/RES/42/187. 1987. *Report of the World Commission on Environment and Development*. New York: United Nations, General Assembly Resolution 42/187. 11 December 1987.

Earth Summit+5. 1997. *Backgrounder*. New York: United Nations Department of Public Information (DPI).

Meyer, J.F. 1980. On evaluating the performability of degradable computing systems. *IEEE Transaction on Computers* C29 (8): 720–731.

Misra, K. B. 1992. *Reliability Analysis and Prediction: A Methodological Treatment*. Amsterdam: Elsevier.

Misra, K. B. (ed.) 1993. *New Trends in System Reliability Evaluation*. Amsterdam: Elsevier.

Misra, K. B. (ed.) 1996. *Clean Production: Environmental and Economic Perspectives*. Berlin: Springer.

Misra, K. B. (ed.) 2008. *Handbook of Performability Engineering*. London: Springer.

Misra, K. B. 2013. Sustainable designs of products and systems. *International Journal of Performability Engineering* 9 (2): 175–190.

Parry M.L., O.F. Canziani, J.P. Palutikof, P.J. van der Linden and C.E. Hanson (eds) 2007. *Climate Change 2007: Impacts, Adaptation and Vulnerability*. Contribution of Working Group II to the Fourth Assessment Report of the Intergovernmental Panel on Climate Change. Cambridge: Cambridge University Press.

Postel, S. 1994. Carrying capacity: Earth's bottom line. In *State of the World*. 3–21. New York: W W Norton.

Sriramachari, S. 2004. The Bhopal gas tragedy: an environmental disaster. *Current Science* 86 (7): 905–920.

The State of the World Population. 1999. *Six Billion: A Time for Choices*, ed. A. Marshall. New York: United Nations Population Fund (UNFPA).

Part Two

Networks and Large-Scale Systems

10

Network Reliability Calculations Based on Structural Invariants

Ilya B. Gertsbakh[1] and Yoseph Shpungin[2]
[1]*Department of Mathematics, Ben-Gurion University of the Negev, Israel*
[2]*Software Engineering Department, Shamoon College of Engineering, Israel*

10.1 First Invariant: D-Spectrum, Signature

Let us consider a binary systems with binary components, where each component $i, i = 1, \ldots, m$ can be in two states: *up* (operational) or *down* (failure). The component's state is denoted by a binary variable $x_i = 1(0)$ where 1 indicates the *up* state and 0 – the down state. The system's state depends on $\mathbf{x} = (x_1, x_2, \ldots, x_m)$, and is determined by the structure function

$$\phi(\mathbf{x}) \to \{UP, DOWN\},$$

where *UP* and *DOWN* denote system states, operational and non-operational, respectively. It will be assumed that the system is *monotone* (Barlow and Proschan 1975; Gertsbakh and Shpungin 2011a, 2012), which means that $\phi(\mathbf{1}) \to UP$, $\phi(\mathbf{0}) \to DOWN$ and component failure either causes no change in its state or causes the transition from *UP* to *DOWN*.

It is desirable to generate a visual image of a monotone binary system. For this purpose we will consider a special class of systems called networks. Network $\mathcal{N}$ is a triple $\mathcal{N} = (V, E, T)$, where E is a set of edges, $|E| = m$, V is a set of nodes or vertices, $|V| = n$, and T is a set of special nodes called terminals, $T \subseteq V$, $|T| = k$.

Typically, the network states are defined in terms of so-called *T-terminal connectivity*. If T has only two nodes s, t, the network is *UP* if and only if there is a path of operational edges connecting s and t. If $T = V$, the network is *UP* if and only if there is a tree of operational edges which provides a path between any pair of network nodes. Another important type of network states is related to the number of so-called *clusters*. A cluster is a set of connected nodes which contains at least one terminal. Suppose that $|T| = 4$.

Applied Reliability Engineering and Risk Analysis: Probabilistic Models and Statistical Inference, First Edition.
Edited by Ilia B. Frenkel, Alex Karagrigoriou, Anatoly Lisnianski and Andre Kleyner.

Then, for example, we define that $\mathcal{N}$ is in the *UP* state if and only if the number of isolated clusters does not exceed 2.

Definition 10.1. anchor of the permutation (Gertsbakh and Shpungin 2009, 2011b) Let $e_1, e_2, \ldots, e_m$ be the components of the system subject to failure. Denote by $\pi = (e_{i_1}, e_{i_2}, \ldots, e_{i_m})$ a permutation of system components. Suppose that initially they all are *up*. Then, let's begin turning them *down* by moving along π from left to right. Fix the first component e_{i_r} when the system state becomes *DOWN*. The ordinal number r of this component in the permutation is called the *anchor* of π and denoted by $r(\pi)$.#

Consider the set of all $m!$ permutations and assign the probability of $1/m!$ to each. Define the probability of the event $A(i) = \{r(\pi) = i\}$ as

$$f_i = P(A(i)) = \frac{\#\text{ of permutations with } r(\pi) = i}{m!}. \tag{10.1}$$

The set $\{f_1, f_2, \ldots, f_m\} = \mathbf{f}$ is a proper discrete density. Let us denote by Y the integer-valued random variable (r.v.) having this density.

Definition 10.2. D-spectrum.
The cumulative distribution function (CDF) $F(x) = \sum_{i=1}^{x} f_i$ of the r.v. Y is called the Destruction spectrum (*D-spectrum*) of the system.

Balakrishnan *et al.* (2011) and Samaniego (1985, 2010) consider binary coherent systems consisting of m components whose lifetimes are independent identically distributed (i.i.d.) random variables $X_1, X_2, \ldots, X_m$ with common CDF $H(t) = P(X \leq t)$. Below is the definition of signature (Samaniego, 2010):

Definition 10.3. Signature.
Assume that the lifetimes of systems n components are i.i.d. according to the continuous distribution $H(t)$. The signature $\mathbf{s}$ of the system is an m-dimensional probability vector whose i-th element s_i is equal to the probability that the i-th component failure causes the system to fail. In brief, $s_i = P(T = X_{(i:m)})$, where T is the failure time of the system and $X_{(i:m)}$ is the i-th order statistic of the m component failure times. #

The signature was first introduced by F.J. Samaniego in his paper (Samaniego, 1985). The most important fact discovered in this paper is the following formula for the CDF of system lifetime T:

$$P(T \leq t) = \sum_{i=1}^{m} s_i \cdot P(X_{(i:m)} \leq t). \tag{10.2}$$

An equivalent of the signature was suggested independently several years later in (Elperin *et al.* 1992) and termed the system ID (Internal Distribution).

Several interesting facts follow from Definition 10.3 and (10.2). The i.i.d. assumption implies that system components fail in a random order, and all the possible $m!$ such orders are equally probable. From here it follows that $\mathbf{s} = \mathbf{f}$. The next surprise comes if we substitute the explicit expression for the CDF of the order statistics, see (David 1981) into (10.2):

$$P(X_{(r:m)} \leq t) = \sum_{i=r}^{m} \frac{m!}{i!(m-i)!}[H(t)]^i[1 - H(t)]^{(m-i)} \tag{10.3}$$

Now substitute (10.3) into (10.2) and collect all the terms having $[H(t)]^i[1 - H(t)]^{(m-i)}$ as a multiple. Denote $S_x = s_1 + \ldots s_x$ and $H(t) = q$. Then (10.2) takes the following form:

$$P(T \leq t) = \sum_{x=1}^{m} S_x \cdot m! \cdot q^x (1-q)^{(m-x)}/(x!(m-x)!). \tag{10.4}$$

$P(T \leq t)$ is the probability that the system is *DOWN* at a fixed time t. $H(t)$ is the probability that component with CDF $H(\cdot)$ is *down* at a fixed time t. Denote $H(t) = q$. We see therefore that the probability that the system is *DOWN* at the time t depends only on the cumulative signature and q. Imagine that for each system component we carry out an independent binary lottery and declare it to be *up* with probability $p = 1 - q$ and *down* with probability q. Then, remembering that $\mathbf{s} = \mathbf{f}$ and $S_x = F(x)$ for $x = 1, \ldots, m$, we obtain the following formula for system *DOWN* probability $Q(q)$:

$$P(DOWN) = Q(q) = \sum_{x=1}^{m} C(x) \cdot q^x (1-q)^{(m-x)}, \tag{10.5}$$

where

$$C(x) = F(x)\frac{m!}{x!(m-x)!} \tag{10.6}$$

Now, let us forget for a moment about D-spectra and signatures and return to the above lottery. Let us write the expression for the system *DOWN* probability $Q(q)$. In other words, we have to find the probability of all $\mathbf{x}$ such that $\phi(\mathbf{x}) \rightarrow DOWN$. All these sets can be classified according to the number j of failed components in the system. Note that if $\mathbf{x}$ has j *down* components, then the probability associated with this state equals $q^j(1-q)^{(m-j)}$. Let $D(j)$ be the number of system states which have exactly j *down* components (and therefore $(m-j)$ *up* components). Thus, the probability that the system is *DOWN* equals

$$Q(q) = \sum_{j=1}^{m} D(j) q^j (1-q)^{(m-j)}. \tag{10.7}$$

Now, comparing (10.7) and (10.5) we come to an important conclusion that $C(y)$ equals the number of system failure sets having exactly y *down* components. Therefore, system D-spectrum $F(x)$ depends entirely on system structure function $\phi(\mathbf{x})$ and *does not* depend on stochastic mechanism governing component failures. A multidimensional parameter (a distribution or a set) which depends only on the system structure function will be called a structural invariant.

Let us consider an example illustrating the properties of D-spectrum.

Example 10.1. Bridge system. Let us consider a bridge-type system (Figure 10.1) where nodes s and t are terminals. The system is considered to have failed if there is

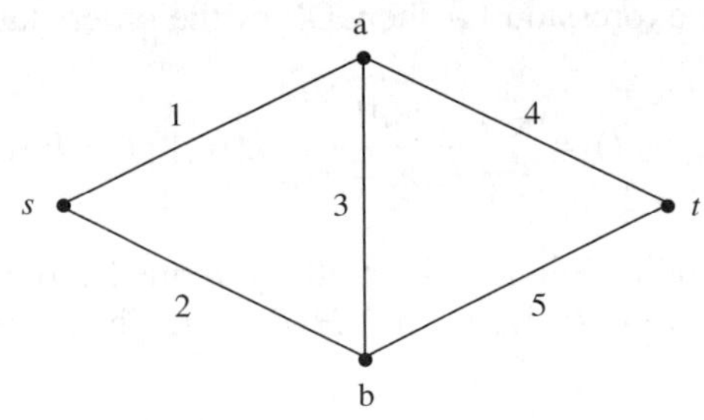

Figure 10.1 Bridge network

no connection between them. Components subject to failures are edges numbered from 1 to 5. There are 5!=120 permutations of edge numbers. The bridge never fails if only one edge is *down*. Therefore $f_1 = 0$. It is easy to find out that there are 24 permutations in which the bridge fails after two edges are *down*. Therefore $f_2 = 1/5$. There are exactly 24 permutations with the anchor is in the fourth position. Thus $f_4 = 1/5$. After any four edges are *down*, the bridge always fails, $f_5 = 0$. Therefore $f_3 = 3/5$, and the D-spectrum is

$$F(1) = 0, F(2) = 1/5; F(3) = 4/5, F(4) = F(5) = 1.$$

According to (10.6), $D(2) = 2, D(3) = 8, D(4) = 4, D(5) = 1$. Now let us check $D(3)$. There are $5!/(2!3!) = 10$ subsets of size 3. Only two of them, (1,3,4) and (2,3,5), do not cause bridge failure if all edges in these sets fail. The remaining 8 sets are the failure sets of size 3. Suppose each edge fails with probability q independently of other edges. By (10.5), the bridge is *DOWN* with probability

$$Q(q) = 2q^2(1-q)^3 + 8q^3(1-q)^2 + 4q^4(1-q) + 5q^5.$$

Suppose now that we are interested in estimating bridge-type road system reliability in case of an earthquake over a period of 5 years. According to the seismic forecast, there is a probability of 0.5 that there is will be no earthquakes during this period, and with probability 0.4 there will be a single earthquake of minor type causing edge destruction probability $q_1 = 0.1$, and the probability of 0.1 that there will be a strong earthquake, as a result of which each edge fails with probability 0.5. Then the probability of bridge survival during the 10-year period equals

$$R = 1 - [0.5 \cdot Q(0) + 0.4 \cdot Q(0.1) + 0.1 \cdot Q(0.5)] = 0.932037.$$

Note. Suppose that the probability q of the component being down is not known exactly (as it is usually the case in the real life) and lies in the interval $q \in [q_{\min}, q_{\max}]$. Since system reliability is a monotone function of its component reliability (Barlow and Proschan, 1975), we have the following bounds:

$$Q(q_{\min}) \leq Q(q) \leq Q(q_{\max}).$$

where $Q(q)$ is given by (10.5).

10.2 Second Invariant: Importance Spectrum. Birnbaum Importance Measure (BIM)

Suppose that a monotone system consists of independent components and component i fails with probability q_i. Then system reliability is a function of $p_i = 1 - q_i$, $i = 1, \ldots, m$:

$$P(UP; \mathbf{p}) = 1 - P(DOWN; \mathbf{q}) = \Psi(p_1, p_2, \ldots, p_m). \tag{10.8}$$

The Birnbaum Importance Measure (BIM) of component j (see Barlow and Proschan, 1975; Birnbaum, 1969) equals

$$BIM_j = \frac{\partial \Psi(p_1, p_2, \ldots, p_m)}{\partial p_j}. \tag{10.9}$$

Using pivotal decomposition (Barlow and Proschan, 1975) it is easy to represent BIM_j in the following form:

$$BIM_j = Q(p_1, p_2, \ldots, 0_j, \ldots, p_m) - Q(p_1, p_2, \ldots, 1_j, \ldots, p_m), \tag{10.10}$$

where the first term is the *DOWN* probability of the original system in which component j is permanently *down*, and the second term is the *DOWN* probability of the original system in which component j is replaced by a fully reliable one. The component BIMs is crucial for approximate calculation of system reliability $\Psi(p_1, p_2, \cdots, p_m)$ and for optimal system design problems (see, for example, Gertsbakh and Shpungin 2011b, 2012). The use of BIM is limited by the necessity to know the analytic form of the reliability function $\Psi(\cdot)$. This problem, however, can be resolved if all $p_i \equiv p = 1 - q$ for all i, by means of the so-called *Importance spectrum*.

It follows from (10.6) that

$$F(x) = \frac{D(x)}{m!/(x!(m-x)!)},$$

where $D(x)$ is the number of failure sets with exactly x components *down*. Obviously,

$$D(x) = D(x; 0_j) + D(x; 1_j), \tag{10.11}$$

where the first and the second terms are the numbers of failure sets with x *down* components and with component j *down* or *up*, respectively.

Definition 10.4. The collection of

$$\left\{ F(x; 0_j) = \frac{D(x; 0_j)}{m!/(x!(m-x)!)} \right\}, x = 1, \ldots, m, j = 1, 2, \ldots, m,$$

is called the *Importance spectrum*.

Obviously, the Importance spectrum, like the D-spectrum, is a structural invariant of the system.

Let us present the following theorem proved in (Gertsbakh and Shpungin 2009, 2011b):

Theorem 10.1. If the system consists of identical and independent components with failure probability q, then

$$BIM_j = m!\left[\sum_{x=1}^{m} F(x;0_j)q^{x-1}(1-q)^{(m-x)}/(x!(m-x)!) \right.$$
$$\left. -\sum_{x=1}^{m}(F(x)-F(x;0_j))q^{x}(1-q)^{(m-x-1)}/(x!(m-x)!)\right]. \tag{10.12}$$

The outline of the proof is provided below. If $p_j \equiv p = 1-q$, then the first and the second sum in (10.12) equals the first and the second term in (10.10), respectively. The probabilistic measure of the failure sets with x components being *down*, given that component j is *down*, equals $D(x;0_j)q^{(x-1)}(1-q)^{(m-x)}$.

The following statement follows from this theorem:

Corollary. (Gertsbakh and Shpungin 2009, 2011b)
If

$$F(x;0_i) \geq F(x;0_j) \text{for all } x = 1, \ldots, m,$$

then

$$BIM_i \geq BIM_j \tag{10.13}$$

for all $q \in [0, 1]$.

Theorem 10.1 and the Corollary have obvious applications to the optimal choice of the components which have to be reinforced in order to provide the maximal reliability increase.

Now let us show how Theorem 10.1 can be used to approximate system reliability in the case of nonequal component reliabilities $\mathbf{p} = (p_1, p_2, \ldots, p_m)$. Suppose that the difference between the maximal component reliability $p_{\max}$ and the minimal component reliability $p_{\min}$ is not large, say, $\Delta = (p_{\max} - p_{\min}) \leq 0.2$. Write the Taylor expansion of the reliability function taking p_0 in the "middle" between $p_{\max}$ and $p_{\min}$:

$$\Psi(\mathbf{p}) \approx \Psi(p_0, \ldots, p_0) + \sum_{i=1}^{m} BIM_i(p_0 - p_i). \tag{10.14}$$

In (10.14), all terms are known if we know the first two invariants.

Remark. Typically, the contribution of higher order derivatives to the approximation (10.14) is of magnitude 0.001 − 0.002. If we want to have a more accurate approximation we need to know the second-order partial derivatives $\frac{\partial^2\Psi}{\partial p_i \partial p_j}$. As shown in (Gertsbakh and Shpungin 2012), they also can be obtained using specially constructed structural invariants.

10.3 Example: Reliability of a Road Network

Let us consider a road network having 22 nodes, four of which are terminals and 37 edges, see Figure 10.2. Components subject to failure are the roads represented by network edges. The edges may fail as a result of an earthquake. Network failure is defined as network disintegration into four isolated clusters, that is, the network is *DOWN* if all four terminals are isolated from each other.

Table 10.1 presents the edge cumulative spectra for this network, obtained by means of Monte Carlo simulation of 10^5 edge permutations. The spectra $F(x)$, $G(x)$ and $H(x)$ are related to the disintegration of the network into 2, 3 and 4 clusters, respectively.

Suppose that we can reinforce 4 edges, that is, replace them by more reliable ones. We will omit here the intermediate results of the calculations. The final results are the following. The edges 1, 7, 9, 18 are the most critical with respect to the above defined network *DOWN* state. The edges marked in bold in Figure 10.2 are the best candidates for the reinforcement.

Table 10.2 presents the gain $\Delta = BIM_1 \cdot 0.1$ in network reliability if q_1 is reduced by 0.1 (by replacing edge 1), and also the "true" value δ of this gain obtained by a crude Monte Carlo simulation.

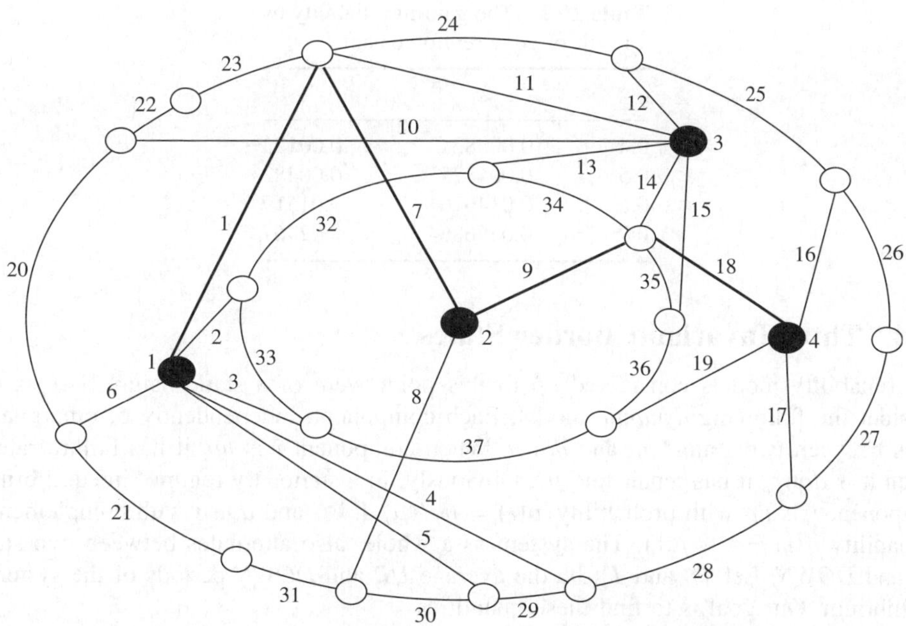

Figure 10.2 Road network with 22 nodes, 4 terminals and 37 edges

Table 10.1 The first, second and third network cumulative edge spectra

x	$F(x)$	$G(x)$	$H(x)$	x	$F(x)$	$G(x)$	$H(x)$
1	0	0	0	19	.65879	.17631	.01316
2	0	0	0	20	.75415	.27167	.03022
3	.00015	0	0	21	.83288	.37965	.05919
4	.00064	0	0	22	.89231	.49292	.10311
5	.00158	0	0	23	.93435	.60471	.16298
6	.00365	0	0	24	.96236	.70634	.23766
7	.00707	0	0	25	.98000	.79030	.32496
8	.01247	0	0	26	.98994	.85661	.41878
9	.02026	0	0	27	.99514	.90869	.51534
10	.03300	.00004	0	28	.99777	.94370	.61008
11	.05143	.00015	0	29	.99933	.96800	.69840
12	.07952	.00063	0	30	.99974	.98309	.77652
13	.12035	.00167	0	31	.99995	.99211	84404
14	.17556	.00447	.00002	32	1	.99637	.90041
15	.24650	.01198	.00010	33	1	.99860	.94151
16	.33689	.02730	.00049	34	1	.99968	.97203
17	.44133	.05611	.00164	35	1	1	.99122
18	.55098	.10548	.00485	36	1	1	1

Table 10.2 The gain in reliability as a result of edge reinforcement

q	Δ	δ
0.4	0.0008243	0.00123
0.5	0.0048753	0.00482
0.6	0.0146164	0.01519
0.7	0.0256864	0.02467

10.4 Third Invariant: Border States

The reliability models considered up to this point were of a static nature. Let us now consider the following *dynamic* model. Each component, independently of others, alternates between two states: *up* and *down*. When component i is *up*, it has failure rate λ_i. When it is *down*, it has repair rate μ_i. Obviously, in a stationary regime (in equilibrium), component i is *up* with probability $p(i) = \mu_i/(\mu_i + \lambda_i)$ and *down* with complementary probability $q(i) = 1 - p(i)$. The system as a whole, also alternates between two states, *UP* and *DOWN*. Let T_U and T_D be the average *UP* and *DOWN* periods of the system in equilibrium. Our goal is to find these quantities.

From the availability theory (see Barlow and Proschan 1975), it is known that the stationary system reliability $R = \Psi(p(1), p(2), \ldots, p(m))$ can be expressed via T_U and T_D as follows:

$$R = \Psi(p(1), p(2), \ldots, p(m)) = \frac{T_U}{T_U + T_D} \tag{10.15}$$

The book (Gertsbakh and Shpungin 2009) presents the relationship needed to find out T_U and T_D. The quantity

$$\rho = \frac{1}{T_U + T_D} \tag{10.16}$$

is called system *DOWN* $\to$ *UP transition rate*. This rate can be found by using so-called system *border states*.

Definition 10.5. Border state.
The system border state is a *DOWN* state whose Manhattan distance from *UP* equals 1.

In words: a border state becomes an *UP* state by turning a *single* component from *down* into *up*. The border states also are system structural invariants.

Let us define for each border state σ the set A_σ of all *down* components which, being turned from *down* to *up*, bring the system to *UP*. We will call A_σ the *activating set* for the border state σ. Denote by B the set of all border states and by $P(\sigma)$ the static probability of border state σ.

Theorem 10.2. (Gnedenko *et al.* 1969; Gertsbakh and Shpungin 2009)

$$\rho = \sum_{\sigma \in B} P(\sigma) \cdot \mu(\sigma), \tag{10.17}$$

where

$$\mu(\sigma) = \sum_{j \in A(\sigma)} \mu_j.\# \tag{10.18}$$

The following example demonstrates the use of this theorem.

Example 10.2. Let us consider a small network shown in Figure 10.3.
All network nodes are terminals. The components subject to failure are edges. Edge j has failure rate $\lambda_j = j/10$, $j = 1, 2, 3, 4$, and repair rate $\mu_j = 1$. The network is *UP* if all terminals are connected to each other. Obviously, the network is *UP* if any three edges

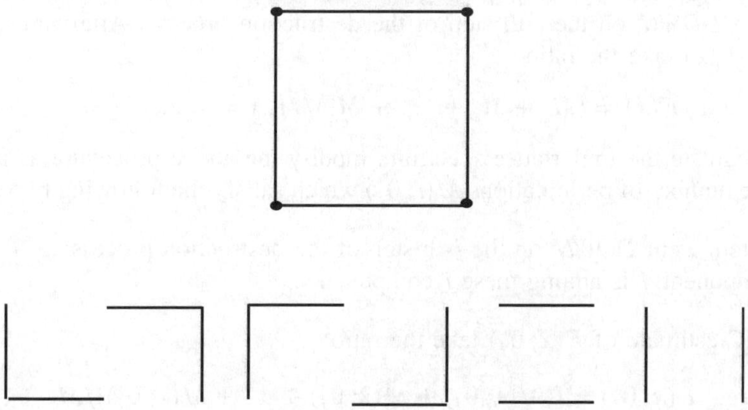

Figure 10.3 Network with four terminals and the set of its border states

or all four edges are *up*. Figure 10.3 shows all six border states. Each such state turns the network into *UP* if any one of two *down* edges becomes *up*. Since all repair rates are the same, $\mu(\sigma) = 2$ for any σ. Put

$$p(j) = \frac{\mu_j}{\mu(j) + \lambda_j} = 10/(10 + j), j = 1, 2, 3, 4, q(j) = 1 - p(j) = j/(10 + j).$$

The network static reliability R equals

$$R = \prod_{j=1}^{4} p(j) + \sum_{i=1}^{4} q(i) \prod_{j \neq i} p(j).$$

By Theorem 10.2,

$$\rho = 2 \cdot [p(1)p(2)q(3)q(4) + p(1)p(3)q(2)q(4) + p(1)p(4)q(2)q(3) \\ + p(2)p(3)q(1)q(4) + p(2)p(4)q(1)q(3) + p(3)p(4)q(1)q(2)].$$

After substituting the values of $p(j)$ and $q(j)$, we obtain $R = 0.8325$ and $\rho = 0.2914$, which gives, using (10.15) and (10.16), $T_U = 2.86$ and $T_D = 0.58$.

10.5 Monte Carlo to Approximate the Invariants

Precise calculation of the structural invariants is, as a rule, NP-hard, and is therefore available via direct enumeration only for small networks, say, when $m < 10$. There were several attempts in the literature to tabulate the signatures for all systems having a given number of components. For example, (Navarro and Rubio 2010) present a list of all coherent systems with $m = 5$ components. We believe that for $m \geq 10$ the only feasible approach is to use Monte Carlo (MC) simulation to approximate the D-spectra and the Importance Spectra.

All MC algorithms to approximate the D-spectra are in fact quite straightforward. To estimate $F(x)$, simulate M random permutations $\pi = (i_1, i_2, \ldots, i_n)$ of component numbers and imitate a sequential destruction of components by moving along a permutation from left to right and by remembering the number M_i of such permutations that the system went *DOWN* on the i-th step of the destruction process. Afterwards, as an MC estimate of $F(x)$ take the ratio

$$\hat{F}(x) = (M_1 + M_2 + \ldots + M_x)/M, x = 1, 2, \ldots, m.$$

To approximate the Importance spectrum, modify the above procedure as follows. Let us count the number of permutations $M(i; 0_j)$ which satisfy the following two conditions:

(a) the system went *DOWN* on the i-th step of the destruction process;
(b) the component j is among these i components.

As an MC estimate of $F(x; 0_j)$ take the ratio

$$\hat{F}(x; 0_j) = [(M(1; 0_j) + M(2; 0_j) + \cdots + M(x; 0_j)]/M, \\ x = 1, 2, \ldots, m, j = 1, \ldots, m.$$

It is important to stress that the number of Monte Carlo runs and therefore the computer time for the above MC is not a critical issue since we are dealing with system *invariants*. For any given system (network) they are calculated "once and for all". The book (Gertsbakh and Shpungin 2009) presents approximate D-spectra for a series of typical networks, for all-terminal connectivity *UP* criterion, namely, for several complete graphs, cubic networks H_4, H_5, H_6, and several small butterfly networks. Our experience shows that for a small-to-medium sized network with 20–40 nodes and 30–60 edges, $M = 10^5 – 10^6$ provides quite an accurate approximation with relative error 0.5% to system *DOWN* probability (calculated by (10.5)) and the component BIMs (calculated by (10.12)).

The most time- consuming part of implementing the above MC procedures is finding the "critical" moment in the sequential destruction when the network changes its state from *UP* to *DOWN*. It seems that when checking the terminal connectivity it is necessary to construct a spanning tree on each step of the destruction process. It turns out, however, that there exist an ingenious algorithm discovered by M. Lomonosov that constructs for each permutation π a *single* spanning tree, see (Gertsbakh and Shpungin 2009).

It is important to note that in finding the permutation anchor it is always possible to achieve a substantial acceleration of the MC algorithm by using the well-known *bisection* method. It works in the following way. Simulate permutation π. Assume that the first $Int[m/2]$ components in π are *down*. Check the system state. If it is *DOWN*, the anchor is among the first $Int[m/2]$ components. Repeat the search by dividing the left tail into two approximately equal parts and proceed. Otherwise, repeat the similar operation on the right tail of the permutation. On average, it will demand $O(\log_2[m])$ checks. It provides a substantial acceleration compared to the straightforward approach of checking system step on each of m steps of the destruction process.

Now let us discuss the MC procedures for estimating the transition rate ρ If a network has $m < 15$ components subject to failure, it is a feasible solution to enumerate all border states and then use formula (10.16). For $m = 12$, the number of such states does not exceed 4096. The straightforward enumeration becomes impractical for larger values of m, and another technique is needed here. An extremely efficient method of computing ρ is based on an ingenious graph-theoretic construction known as an *evolution process* based on *Lomonosov's "turnip"* (Elperin *et al.* 1991). A detailed description can be found in (Gertsbakh and Shpungin 2009, Chapter 9). Let us briefly describe the idea of MC estimation of ρ in this evolution process when the components subject to failure are network edges.

First, the static model with given $p_1, p_2, \ldots, p_m$ values is transformed into a *dynamic* model where edge i is replaced by an element having birth rate $\lambda_i = -\ln(1 - p_i)$ and therefore an exponentially distributed birth time. When we take a snapshot of the whole network at time $t_0 = 1$, we see edge i born (i.e. being *up*) with probability p_i. Now consider a sequence of network states in an *evolution* process which starts from a "root" state with no edges at all and proceeds with the sequential birth of the first edge which has the smallest birth time, then with the birth of an edge which has the second smallest birth time, and so on, until the network enters the *UP* state. To illustrate the evolution process, let us turn to the network shown in Figure 10.3. It starts from "no edge" state. Suppose edge 1 is born first, edge 4 is born next, and edge 3 is born afterwards. Then the sequence of states of this evolution process is the following *trajectory* ω:

$$\omega : (0, 0, 0, 0) \Rightarrow (1, 0, 0, 0) \Rightarrow (1, 0, 0, 1) \Rightarrow (1, 0, 1, 1) = UP$$

Note that the state directly preceding the *UP* state is one of six border states shown in Figure 10.3.

Now, denote by Ω the set of all trajectories $\{\omega\}$ and rewrite (10.17) as

$$\rho = \sum_{\omega \in \Omega} \Pr(\omega) P(\sigma|\omega)\mu(\sigma|\omega)$$

Having this representation, the obvious unbiased MC estimate of ρ based on M simulated trajectories is

$$\hat{\rho} = \frac{\sum_{i=1}^{M} P(\sigma|\omega_i)\mu(\sigma|\omega_i)}{M}.$$

10.6 Conclusion

We considered the most important and useful structural invariants of a binary coherent system: the D-spectrum, the Importance Spectrum, and the Border States. They all depend only on the system structure function and the failure definition of the system. They do not depend on stochastic mechanism which governs the component behavior. The most efficient use of these invariants takes place when the components fail independently and are identical, for example, in the case of a random attack on network nodes or edges, or in the case that system components are independent and have identically distributed lifetimes. The first of the above three invariants was discovered by J.F. Samaniego in his pioneering work in 1985 and termed by the author as signature. As a rule, finding out the invariants is a NP-hard problem, and a good practical solution is to use Monte Carlo procedures for their numerical calculation. For relatively large systems, the respective calculations may demand a considerable amount of CPU time, say, several hours. This, however, is not a serious obstacle because the invariant, once computed, may serve as a reliability estimation for many versions of the component lifetime data.

There are of course many other structural parameters which are invariants and may be used in reliability estimation. Let us mention two of them: the number of minimal cut sets and the multidimensional D-spectrum. The former is used in reliability approximation for highly reliable systems consisting of identical components in the so-called Burtin-Pittel approximation (Gertsbakh and Shpungin 2009). The multidimensional D-spectrum is a natural extension of the D-spectrum for the case when the system has more than two states, and gradually deteriorates from its *UP* state to its *DOWN* state via several intermediate states $DOWN-1, DOWN-2, \ldots, DOWN-k$. More information on this invariant can be found in (Gertsbakh and Shpungin 2011b).

References

Balakrishnan, N., J. Navarro and F.J. Samaniego. 2011. Signature representation and preservation results for engineered systems and applications to statistical inference. In: *Recent Advances in Reliability: Signatures, Multi-State Systems and Statistical Inference*, ed. A. Lisnianski and I. Frenkel. London: Springer, pp. 1–22.

Barlow, R.E. and F. Proschan. 1975. *Statistical Theory of Reliability and Life Testing*. New York: Holt, Rinehart and Winston.

Birnbaum, Z.W. 1969. On the importance of different components in a multicomponent system. In *Multivariate Analysis*, vol. II, ed. P.R. Krishnaiah. New York: Academic Press, pp. 581–592.

David, H.A. 1981. *Order Statistics*. New York: Wiley.

Elperin, T., I.B. Gertsbakh and M. Lomonosov. 1991. Estimation of network reliability using graph evolution models. *IEEE Transactions on Reliability* 40 (5): 572–581.

Elperin, T., I.B. Gertsbakh and M. Lomonosov. 1992. An evolution model for Monte Carlo estimation of equilibrium network renewal parameters. *Probability in Engineering and Informational Sciences* 6 (4): 457–469.

Gertsbakh, I.B. and Y. Shpungin. 2009. *Models of Network Reliability: Analysis, Combinatorics, and Monte Carlo*. Boca Raton, FL: CRC Press.

Gertsbakh, I.B. and Y. Shpungin. 2011a. Using D-spectra in network Monte Carlo: estimation of system reliability and component importance. In: *Recent Advances in Reliability: Signatures, Multi-State Systems and Statistical Inference*, ed. A. Lisnianski and I. Frenkel. London: Springer, pp. 23–32.

Gertsbakh, I.B. and Y. Shpungin. 2011b. *Network Reliability and Resilience*. Heidelberg: Springer.

Gertsbakh, I.B. and Y. Shpungin. 2012. Combinatorial approach to computing importance indices in coherent systems. *Probability in Engineering and Informational Sciences* 26 (1): 82–89.

Gnedenko, B.V., Yu.K. Belyaev and A.D. Solovyev. 1969. *Mathematical Methods of Reliability Theory*. New York: Academic Press.

Navarro, J. and R. Rubio. 2010. Computation of signatures of coherent systems with five components. *Communications in Statistics: Simulation and Computation* 39 (1): 68–84.

Samaniego, F.J. 1985. On closure of the IFR under formation of coherent systems. *IEEE Transactions on Reliability* 34 (1): 69–72.

Samaniego, F.J. 2010. *System Signatures and Their Applications in Engineering Reliability*. New York: Springer.

11

Performance and Availability Evaluation of IMS-Based Core Networks

Kishor S. Trivedi,[1] Fabio Postiglione[2] and Xiaoyan Yin[1]
[1]*Department of Electrical and Computer Engineering, Duke University, USA*
[2]*Department of Electrical and Computer Engineering, University of Salerno, Italy*

11.1 Introduction

The Next Generation Network (NGN) (ITU 2009) aims to integrate wired, wireless and INTERNET technologies to support all kinds of services such as real-time audio/video communication and data transports. The IP Multimedia Subsystem (IMS) (3GPP 2008) is a standardized NGN architecture for delivering multimedia services via a Session Initiation Protocol (SIP) infrastructure. Such a system is expected to provide a high level of end-to-end Quality of Service (QoS) (Hui and Yeung 2003) to customers in the presence of failures occurring in the network elements. Therefore, performance and availability assessments are essential for system deployment.

The availability and performance of IMS-based core networks have been evaluated in (Mendiratta and Pant 2007; Pant *et al.* 2008; Guida *et al.* 2008). However, these cited papers have not investigated system availability in face of changes in performance demand, system configuration and the resulting cost. In a recent paper (Guida *et al.* 2010a), an analytic model based on the Universal Generating Function (UGF) technique (Ushakov 1986; Lisnianski and Levitin 2003) is proposed to evaluate the steady-state availability with one performance demand level, and a redundancy optimization problem (Ushakov 1987; Tian *et al.* 2009) is solved to maintain high availability so as to minimize the cost.

Applied Reliability Engineering and Risk Analysis: Probabilistic Models and Statistical Inference, First Edition.
Edited by Ilia B. Frenkel, Alex Karagrigoriou, Anatoly Lisnianski and Andre Kleyner.

In this work, we propose three alternative analytic models to that proposed in (Guida *et al.* 2010a) for the independent software recovery scheme. In addition, three other models are developed to capture the software recovery with dependencies. The proposed six analytic models are all hierarchical models (Trivedi 2001; Zang *et al.* 2003; Kim *et al.* 2009), with the same top-level Reliability Block Diagram (RBD) model. These models are compared with respect to the accuracy, execution time and the influences of different input parameters. Steady-state availability is evaluated in face of changes in system performance demand. In addition, for each level of performance demand, a redundancy optimization problem is solved to find the best system configuration while maintaining the high availability and minimizing the cost.

11.2 IMS-Based Core Network Description

IMS is standardized by the Third Generation Partnership Project (3GPP) for the next generation mobile network. The IMS network consists of the access network and the core network. Access networks enable User Equipments (UEs) to connect to an IMS core network based on different radio technologies. The IMS core network is a collection of functions to control signaling transmissions, such as calls and other kinds of sessions between UEs. In the IMS core network, the signaling flows are managed by a set of Call Session Control Functions (CSCFs) (i.e. signaling servers) that can be classified into three types: (1) a Proxy CSCF (P-CSCF) server is a SIP proxy that is the first contact point to the IMS network for the UE, which forwards the SIP signaling to the subscriber's Home Network (HN); (2) a Serving CSCF (S-CSCF) is located in the HN and manages multimedia sessions; and (3) an Interrogating CSCF (I-CSCF), also located in the HN, is responsible for selecting appropriate S-CSCF for all requests entering the HN. The Home Subscriber Server (HSS) is a user database that contains subscription-related information to support the above signaling servers. The HSS is located in the HN and typically is contacted by the I-CSCF to identify the S-CSCF to be assigned to users to route call set-up requests.

Figure 11.1 presents a typical call set-up network scenario for two UEs. The *home networks* represent the network domains of telecom operators that users are subscribed to, whereas the *visited networks* represent other service providers' networks that roaming users can attach to. The signaling flow is established between two UEs through the interactions among servers including three kinds of CSCF and the HSS.

To satisfy the QoS requirement of the telecommunication operators' subscribers, a performance demand (i.e. the number of sessions supported per second) can be specified for the signaling network. Hence, we can define the system availability as the probability that the service network can meet the performance demand. Since all servers in the service network can be modeled as a series system, the system performance demand is met as long as all servers can meet the performance demand (i.e. the number of sessions forwarded by each server per time unit is not less than the demand). Therefore, we can use RBD to model such a series system in the top level. Because servers are exposed to various failures and repairs, each server can be modeled in the lower level using different techniques. Therefore, the overall IMS-based signaling core network can be captured by a hierarchical system availability model (Trivedi 2001; Zang *et al.* 2003; Kim *et al.* 2009).

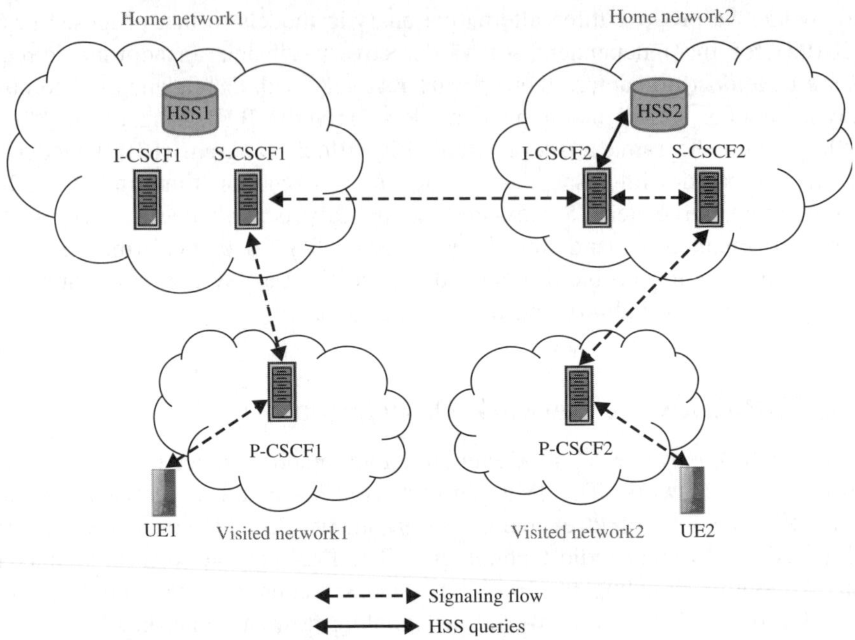

Figure 11.1 Call set-up scenario

The system availability is defined as the probability that the IMS core network satisfies the performance demand. The performance metric studied in this work is the number of call set-up sessions that an IMS core network can handle.

In order to maintain high system availability, redundancy is necessary in most practical cases. Hence, the parallel redundancy is proposed as shown in Figure 11.2 for the CSCF and HSS servers. Each server contains a set of independent parallel nodes. Each node is composed of a service logic software part and a core part (Pant *et al.* 2008) as shown in Figure 11.3. The service logic software part consists of concurrent software instances

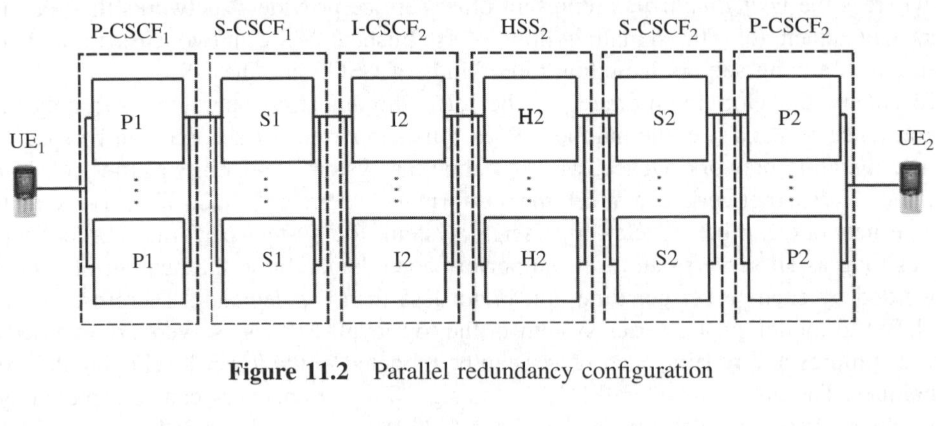

Figure 11.2 Parallel redundancy configuration

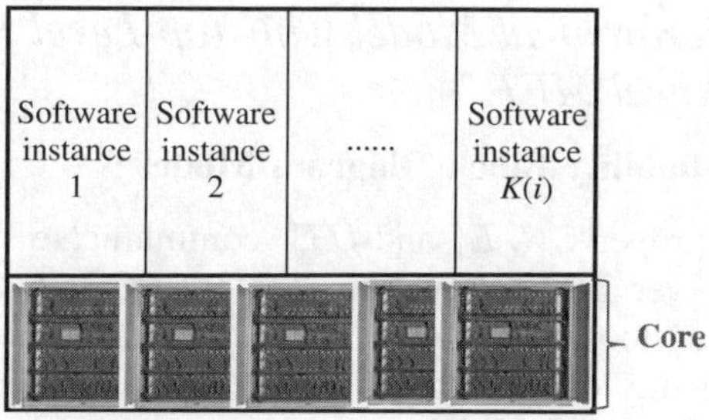

Figure 11.3 A node in server i

to handle a certain number of call set-up sessions. The core part consists of hardware and the operating system. For the analytic model of the next section, the following parameters are used:

i: server, $i \in$ {P-CSCF$_1$, S-CSCF$_1$, I-CSCF$_2$, HSS$_2$, S-CSCF$_2$, P-CSCF$_2$};
$h(i)$: number of parallel nodes in server i; **h** is the corresponding vector;
$K(i)$: number of software instances per node in server i; **K** is the corresponding vector;
$c(i)$: performance capacity of each software instance in server i; **c** is the corresponding vector;
$Cost(i)$: cost per node of server i; **Cost** is the corresponding vector;
λ_s: failure rate of a software instance; μ_s: recovery rate of a software instance;
λ_c: failure rate of a core part; μ_c: repair rate of a core part;
W: performance demand.

In this work, we assume that the nodes of the same server are identical. However, this assumption can easily be generalized to the case of different parallel nodes within a server. Furthermore, the nodes are independent of each other in terms of both failure and recovery behavior. The recovery dependencies regarding software instances within a node will be taken into account. We construct analytic models for both independent software recovery mode and possible software recovery mode with dependencies.

11.3 Analytic Models for Independent Software Recovery

We assume that, for a single node, the core repair and each software instance recovery within a node are all independent of each other. Three alternative two-level hierarchical models will be proposed to analyze the system steady-state availability. The top-level models for these hierarchical models are all RBD, whereas the lower level has three alternatives: Multi-state Fault Tree (MFT) (Smith *et al.* 2008; Trivedi *et al.* 2008), (binary) Fault Tree (FT), and Stochastic Reward Net (SRN) model (Ciardo *et al.* 1989), respectively. The lower-level model is used to capture the behavior of a single server along the communication path. For the MFT model, a component (i.e. event) represents a node in a server and has multiple states. By contrast, in the (binary) FT model, a component represents either a core or a software instance in a node and has two states only.

11.3.1 Model 1: Hierarchical Model with Top-Level RBD and Lower-Level MFT

11.3.1.1 Top-Level Reliability Block Diagram Model

For the system to work properly (UE_1 and UE_2 communicate successfully), the performance level for each server in the signaling path P-CSCF$_1$, S-CSCF$_1$, I-CSCF$_2$, HSS$_2$, S-CSCF$_2$, P-CSCF$_2$ needs to satisfy the demand W sessions/sec. Assuming that each server behaves independently of the others, the system availability is the product of the availability for each server in steady state:

$$A_{sys} = A_{P-CSCF_1} \cdot A_{S-CSCF_1} \cdot A_{I-CSCF_2} \cdot A_{HSS_2} \cdot A_{S-CSCF_2} \cdot A_{P-CSCF_2} \tag{11.1}$$

11.3.1.2 Lower-Level Multi-State Fault Tree Model

The availability of each server in the signaling path is the probability that the performance level C of the server satisfies the performance demand W (i.e. $C \geq W$). The performance level C of a server is the total capacity of all the nodes in this server.

Recall that a node in server i, consists of a core and $K(i)$ software instances. The failure and recovery behaviors of the core and software instances will influence the current performance level of this node. Hence, each node in server i can be modeled as a multi-state component, with each state corresponding to a performance level (i.e. capacity) of this node. Since nodes in a given server are identical, the probability and the corresponding capacity for node j in server i being in state k are given by:

$$P(i, j, k) = P(i, k) = \begin{cases} A_{core} \cdot \left[\binom{K(i)}{k} A_{inst}{}^{k} (1 - A_{inst})^{K(i)-k} \right] & k = 0, 1, 2, \ldots, K(i) \\ 1 - A_{core} & k = K(i) + 1 \end{cases} \tag{11.2}$$

$$C(i, j, k) = C(i, k) = \begin{cases} k \cdot c(i) & k = 0, 1, 2, \ldots, K(i) \\ 0 & k = K(i) + 1 \end{cases} \tag{11.3}$$

where:

i denotes a server {P-CSCF$_1$, S-CSCF$_1$, I-CSCF$_2$, HSS$_2$, S-CSCF$_2$, P-CSCF$_2$};
j denotes the index of a node in a server $\{1, 2, \ldots, h(i)\}$;
k denotes the states of a node;
$k = 0, 1, 2, \ldots, K(i)$ represents the state in which the core and k software instances are working properly;
$k = K(i) + 1$ represents the state in which the core has failed;
$A_{core} = \mu_c/(\lambda_c + \mu_c)$ is the steady-state availability for a single core;
$A_{inst} = \mu_s/(\lambda_s + \mu_s)$ is the steady-state availability for a single software instance.

Hence, the summations of the performance levels in all the nodes within server i will be the performance level of this server:

$$C(i, \mathbf{k}(i)) = C(i, 1, k_1) + C(i, 2, k_2) + \cdots + C(i, h(i), k_{h(i)}) \tag{11.4}$$

where $\mathbf{k}(i) = [k_1, k_2, \ldots, k_{h(i)}]$ is the vector of the node states of server i.

Due to the independence across nodes, the product of the corresponding probabilities for each node's state gives the probability that the server i has that specific performance level $C(i, \mathbf{k}(i))$:

$$P(i, \mathbf{k}(i)) = P(i, 1, k_1) \cdot P(i, 2, k_2) \cdot \cdots \cdot P(i, h(i), k_{h(i)}) \tag{11.5}$$

When the performance level $C(i, \mathbf{k}(i))$ of the server satisfies the performance demand W, the corresponding probability will contribute to the availability of the server. Therefore, the steady-state availability of server i is given by:

$$A_i = \sum_{k_1=0}^{K(i)+1} \sum_{k_2=0}^{K(i)+1} \cdots \sum_{k_{h(i)}=0}^{K(i)+1} Sat(i, \mathbf{k}(i)) \cdot P(i, \mathbf{k}(i)), \tag{11.6}$$

where

$$Sat(i, \mathbf{k}(i)) = \begin{cases} 1 & if \;\; C(i, \mathbf{k}(i)) \geq W \\ 0 & \text{otherwise} \end{cases}. \tag{11.7}$$

Hence, the overall system steady-state availability can easily be obtained using (11.1).

11.3.2 Model 2: Hierarchical Model with Top-Level RBD and Lower-Level FT

11.3.2.1 Top-Level RBD Model

The top-level model is the same as that in Section 11.3.1.1 and the system steady-state availability is computed again by (11.1).

11.3.2.2 Lower-Level FT Model

In addition to the multi-state fault tree model approach, a single fault tree can also be used to model the availability of an individual server. As previously described, the availability of server i is the probability that the performance level of the server satisfies the performance demand W. Since server i contains $h(i)$ nodes, and each node contains $K(i)$ software instances, the total number of software instances in the server is $h(i) \cdot K(i)$. Server i is considered available if at least $\lceil W/c(i) \rceil$ out of $h(i) \cdot K(i)$ software instances work properly. Based on such an analysis, the fault tree model is constructed and shown in Figure 11.4.

In the fault tree model, an inverse triangle represents a repeated event (i.e. the event with the same name is the same physical component), whereas the circle represents a basic event. The repeated event C_j represents the core in node j, whereas $S_{j,k}$ represents the software instance k in node j. A software instance contributes to the system's performance only when it and the corresponding core in the same node are all working properly. Hence, they are connected using an *Or* gate. The k/n gate in the fault tree represents that if at least k out of n inputs events fails, the system will fail (i.e. server i is unavailable). Hence:

$$n = h(i) \cdot K(i); \qquad k = n - \left\lceil \frac{W}{c(i)} \right\rceil + 1 \tag{11.8}$$

Note that in Equation (11.8), n represents the total number of input events of the k/n gate.

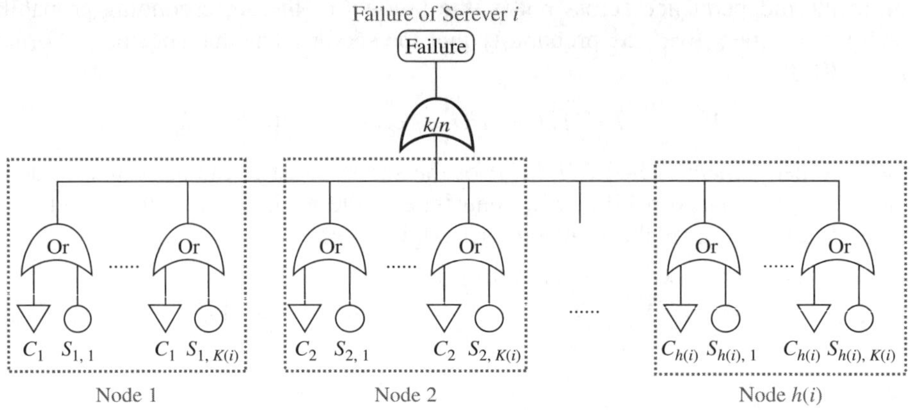

Figure 11.4 Fault tree model for server i

By solving the fault tree model in Figure 11.4, the server i's availability A_i is given by:

$$A_i = \sum_{m=\left\lceil \frac{W}{c(i)K(i)} \right\rceil}^{h(i)} \sum_{l=\left\lceil \frac{W}{c(i)} \right\rceil}^{m \cdot K(i)} \left[\binom{h\,(i)}{m} A_{core}^{m} (1 - A_{core})^{h(i)-m} \cdot \binom{m \cdot K\,(i)}{l} A_{inst}^{l} (1 - A_{inst})^{m \cdot K(i) - l} \right] \tag{11.9}$$

Next, the overall system steady-state availability can be obtained using Equation (11.1).

11.3.3 *Model 3: Hierarchical Model with Top-Level RBD and Lower-Level SRN*

11.3.3.1 Top-Level RBD Model

The top-level model is still the same and the system steady-state availability is given again by Equation (11.1). The individual server's availability will be derived via the lower-level model below.

11.3.3.2 Lower-Level SRN Model

In this section, the SRN model is presented to compute the availability of an individual server. Due to the space limitation, only the SRN model for node j in server i is illustrated in Figure 11.5. The overall SRN for server i can be easily constructed by varying j through their range, where $j \in [1, h(i)]$.

Figure 11.5(a) represents the failure and recovery behavior of all $K(i)$ software instances in node j server i. The number of tokens in place SU_{ij} is the number of operational software instances and the number of tokens in place SD_{ij} represents the number of failed software instances. The initial number of tokens in place SU_{ij} is $K(i)$.

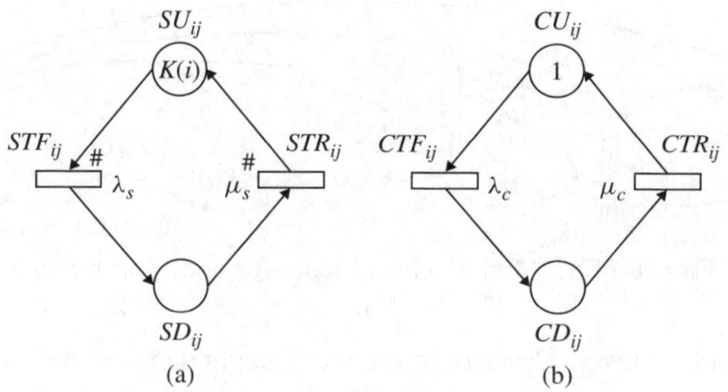

Figure 11.5 SRN model for node j in server i

The firing of transition STF_{ij} represents the failure of one of the software instances. The firing rate of the transition STF_{ij} is marking-dependent: Rate(STF_{ij}) $= \lambda_s \cdot \#SU_{ij}$, where λ_s is failure rate of a single software instance and $\#SU_{ij}$ is the number of tokens in place SU_{ij}. Similarly, transition STR_{ij} represents the recovery of a failed software instance and is marking-dependent on the number of tokens in place SD_{ij}. Figure 11.5(b) shows the failure and repair behavior of the core in node j server i. Therefore, by constructing the overall SRN model for server i, we can obtain the steady-state availability of server i by defining its *reward function* as:

$$r_i = \begin{cases} 1 \text{ if } \sum_{j=1}^{h(i)} [\#(CU_{ij}) \cdot \#(SU_{ij})] \geq \left\lceil \frac{W}{c(i)} \right\rceil \\ 0 \text{ otherwise} \end{cases} \tag{11.10}$$

11.4 Analytic Models for Recovery with Dependencies

In Section 11.3, three different analytic models were presented to analyze the system steady-state availability when core repair and each software instance recovery within a node are all independent. However, in practice, possible dependencies may arise. We consider the following two dependencies for the recovery mode within the same node: software instances within a node share the recovery process, that is, only one software instances within a node can be under recovery at one time; after a failed core is repaired, the failed software instances within the same node will be simultaneously recovered due to the reboot of the node. Since the fault tree model cannot capture the dependency between different components, we extend the other two analytic models based on MFT and SRN in Section 11.3 to incorporate these dependencies.

11.4.1 Model 4: Hierarchical Model with Top-Level RBD, Middle-Level MFT and Lower-Level CTMC

Similar to the independent recovery case, each node in the server i is modeled as a multi-state component with each state corresponding to a performance level (i.e. capacity) of

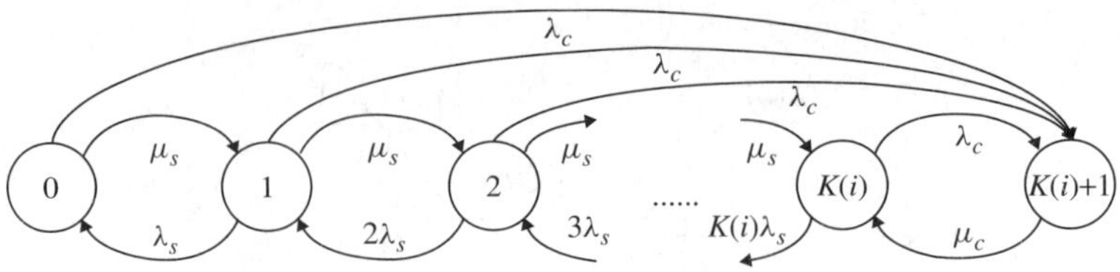

Figure 11.6 Markov chain model for node j in server i

this node in the middle level. Due to the recovery dependencies, the probability for node j in server i being in state k cannot be easily derived as in Section 11.3.1.2. Therefore, the Continuous-Time Markov Chain (CTMC) model in Figure 11.6 is used in the lower level to compute such probabilities.

In the CTMC model, the state $k(k \in [0, K(i)+1])$ represents the same state definition for a multi-state component (i.e. a node) in Section 11.3.1.2:

$k = 0, 1, 2, \ldots, K(i)$ presents the case that the core and k software instances are working properly;
$k = K(i) + 1$ presents the case that the core has failed.

As shown in Figure 11.6, the phenomena that software instances have the shared recovery with parameter μ_s and the case that all failed software instances will be recovered simultaneously when a failed core is repaired are incorporated into the model. Since each node j in server i has the same CTMC model, by solving the above CTMC, the state k's steady-state probability $\pi_{i,k}$ can be obtained, which is the probability for a multi-state component for node j in server i being in state k. Therefore, the probability and the associated capacity for a multi-state node (the same for every node in the server i) are given by:

$$P(i, j, k) = P(i, k) = \pi_{i,k} \quad k = 0, 1, 2, \ldots, (K(i) + 1) \tag{11.11}$$

$$C(i, j, k) = C(i, k) = \begin{cases} k \cdot c(i) & k = 0, 1, 2, \ldots, K(i) \\ 0 & k = K(i) + 1 \end{cases} \tag{11.12}$$

By using the same Equations (11.4)–(11.7) and (11.1), the overall system steady-state availability can be computed. An alternative approach, based on the Universal Generating Function (UGF), is proposed for such a model.

11.4.2 Model 5: Alternative Approach for Model 4 based on UGF

A convenient approach to evaluate system availability can be based on the UGF methodology, originally introduced by Ushakov (1986). From Equation (11.4), the discrete random variable describing its stationary performance levels is

$$C = \min_i C(i, \mathbf{k}(i)), \tag{11.13}$$

whose values are in the set

$$\{C_1, \ldots, C_K\}, \tag{11.14}$$

where $K = \prod_i [K(i) + 2]^{h(i)}$ is the number of different states of the overall signaling system. The associated (steady-state) probability set is

$$\{P_1, \ldots, P_K\}. \tag{11.15}$$

The UGF of the stationary output performance C is the polynomial-shape u-function

$$u_{sys}(z) = \sum_{k=1}^{K} P_k z^{C_k}. \tag{11.16}$$

By exploiting the fundamental operators for u-functions of series-parallel multi-state systems, it is possible to compute $u_{sys}(z)$, that is, all P_k and C_k in (11.16), on the basis of the steady-state behavior of all elements composing the system. Since the IMS core network has the parallel redundancy configuration represented in Figure 11.2, it can be modeled as a series-parallel system with *flow dispersion* (Lisnianski and Levitin 2003). Consequently, for each server i, the output performance of the $h(i)$ parallel nodes of the server is given by the sum of the $h(i)$ capacities of those nodes, and the overall u-function can be written as (Guida *et al.* 2010b):

$$u_i(z) = \sum_{k_1=0}^{K(i)+1} \sum_{k_2=0}^{K(i)+1} \cdots \sum_{k_{h(i)}=0}^{K(i)+1} \left(\prod_{j=1}^{h(i)} P(i, j, k_j) z^{\sum_{j=1}^{h(i)} C(i, j, k_j)} \right) \tag{11.17}$$

where $C(i, j, k_j)$ is the capacity (11.12), and $P(i, j, k_j)$ is the corresponding steady-state probability (11.11), derived by solving the CTMC in Figure 11.6.

On the other hand, the overall output performance of servers connected in series is simply given by the minimum of individual servers' capacities. In order to compute the u-function for m series components, the following σ operator is adopted:

$$u_s(z) = \sigma(u_1(z), u_2(z), \ldots u_m(z)). \tag{11.18}$$

For a pair of generic components connected in series, the operator (11.18) is

$$\sigma(u_1(z), u_2(z)) = \sigma\left(\sum_{i=1}^{I} \varepsilon_i z^{e_i}, \sum_{j=1}^{J} \varphi_j z^{f_j}\right) = \sum_{i=1}^{I} \sum_{j=1}^{J} \varepsilon_i \varphi_j z^{\min\{e_i, f_j\}},$$

where the parameters e_i and f_j are the performance levels of the two components, while ε_i and φ_j are the corresponding steady-state probabilities, $\forall i = 1, .., I$ and $\forall j = 1, .., J$.

Finally, the UGF of the IMS-based core network in (11.16) can be obtained by substituting server i u-function (11.17) for each i into (11.18), viz.

$$u_{sys}(z) = \sum_{k=1}^{K} P_k z^{C_k} = \sigma\left(u_{\text{P-CSCF}_1}(z), u_{\text{S-CSCF}_1}(z), u_{\text{I-CSCF}_2}(z), u_{\text{HSS}_2}(z), u_{\text{S-CSCF}_2}(z), u_{\text{P-CSCF}_2}(z)\right). \tag{11.19}$$

The earlier polynomial-shape function provides immediately all the system performance levels (the exponents of z) and their corresponding steady-state probabilities (the corresponding coefficients), that can be used to compute the steady-state system availability as follows

$$A_{sys} = \sum_{C_k \geq W} P_k \tag{11.20}$$

11.4.3 Model 6: Hierarchical Model with Top-Level RBD and Lower-Level SRN

The partial SRN model for node j in server i is illustrated in Figure 11.7. The overall SRN for server i can easily be constructed by varying j through its range, where $j \in [1, h(i)]$.

Compared to Figure 11.5 for the independent recovery case, two modifications are made in Figure 11.7 to capture the two recovery dependencies. First, the marking-dependence for transition STR_{ij} is removed to represent the shared recovery among software instances within a node. Second, an immediate transition t_{ij} is added to capture the scenario that when a failed core is repaired, all failed software instances are recovered simultaneously. If there is a token in place CU_{ij} (i.e. the core is working properly), the immediate transition t_{ij} is disabled. Otherwise, the immediate transition will be enabled. This ensures that all failed software instances are simultaneously recovered the moment the failed core is repaired.

By using the same reward function (11.10), the steady-state availability for server i can be computed. The overall system steady-state availability can be calculated based on (11.1).

11.5 Redundancy Optimization

The analytic models in Section 11.3 and Section 11.4 can be used to compute the overall system steady-state availability, given the input parameters including the parallel redundancy configurations $h(i)$ for each server. In practice, we usually want to minimize the cost while maintaining high availability to satisfy the requirement. Hence, we adopt

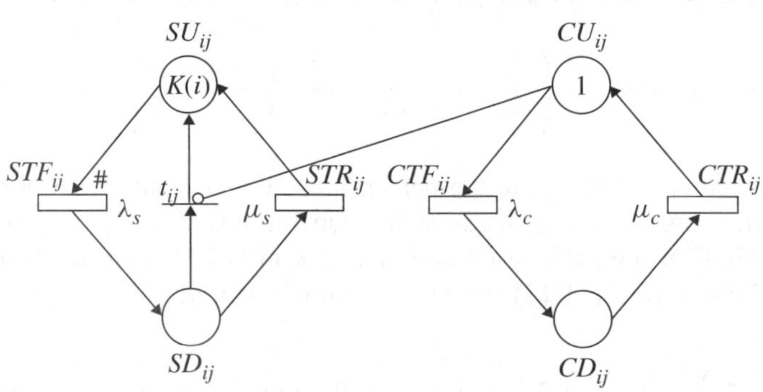

Figure 11.7 SRN model for node j in server i

the optimization problem presented in (Guida *et al.* 2010a) to find the optimal redundancy configuration $\mathbf{h}_{opt}$ providing the minimal cost and satisfying the steady-state availability requirement A_0. Under the assumption of identical costs *Cost*(i) of the nodes of each server i, the cost of the whole system is:

$$Cost_{sys}(\mathbf{h}) = \sum_i Cost(i) \cdot h(i). \tag{11.21}$$

Given the set $H_{sys} = \{\mathbf{h} : A_{sys}(W, \mathbf{h}) \geq A_0\}$ of the configurations satisfying the steady-state availability requirement, the redundancy optimization problem can be written as:

$$\mathbf{h}_{opt} = \arg \min_{\mathbf{h} \in H_{sys}} Cost_{sys}(\mathbf{h}). \tag{11.22}$$

The problem (11.22) is solved by an exhaustive search. However, many quicker heuristic techniques are available, such as Genetic Algorithms (Lisnianski and Levitin 2003).

11.6 Numerical Results

In this section, we compare the proposed models with respect to the accuracy, execution time and recovery scheme. Next, based on the numerical results for model 1 as shown in Table 11.1, we further evaluate the system steady-state availability and cost according to the variation of the performance demand and redundancy configuration.

11.6.1 Model Comparison

11.6.1.1 Accuracy and Execution Time Comparison

Given a software recovery scheme, different analytic models are proposed. Therefore, the numerical results of these models need to be compared to cross-validate their correctness and verify the accuracy.

The same input parameters will be used for these six models for accuracy and execution time comparison. The definitions for the input parameters are at the end of Section 11.2 and the values are the same as those in (Guida *et al.* 2010a), in line with system engineers' experience in real operation scenarios. Their list follows:

$\lambda_s = 7.716 \times 10^{-7}\text{sec}^{-1}$, i.e., 2 software instance failures per month;
$\mu_s = 1.667 \times 10^{-3}\text{sec}^{-1}$, i.e., a mean recovery time of 10 minutes;
$\lambda_c = 6.342 \times 10^{-8}\text{sec}^{-1}$, i.e., 2 core failures per year;
$\mu_c = 3.472 \times 10^{-5}\text{sec}^{-1}$, i.e., a mean repair time of 8 hours;
$\mathbf{h} = [3, 3, 3, 2, 3, 3]$; $\mathbf{K} = [3, 3, 3, 2, 3, 3]$; $\mathbf{c} = [1400, 1400, 1400, 2500, 1400, 1400]$;
$W = 4000$ sessions/sec.

The numerical implementations of these analytic models are based on SHARPE (Kim *et al.* 2009) and Mathematica® software packages. As shown in Table 11.1, the three analytic models for the independent recovery have exactly the same numerical results and this is also valid for the three analytic models for the dependency case. This cross-validates the correctness of our proposed models. Moreover, the recovery scheme is observed to have little influence on the steady-state availability for the given input parameters.

Table 11.1 Accuracy and execution time comparison

Recovery scheme	Model name	System steady-state availability	Execution time (second)
	1	0.999993209	0.025
Independent	2	0.999993209	0.009
	3	0.999993209	0.052
	4	0.999993208	0.026
Dependent	5	0.999993208	0.031
	6	0.999993208	0.021

The execution times for those models, running on a computer based on an Intel®Core™2 Duo CPU L9600 @ 2.13 GHz, are all really small, which shows that the proposed hierarchical models are very efficient.

11.6.1.2 Independent vs. Dependent Recovery

As mentioned earlier, the recovery scheme has little influence on the system availability for the given input parameters in the previous section. In this section, we evaluate the steady-state availability by changing the input parameters λ_s, μ_s, λ_c, μ_c and **K** for both recovery schemes. Since different models for the same recovery scheme lead to exactly the same numerical results, we only compare model 3 and model 6 here. In Figures 11.8–11.12, only one parameter is varied, keeping the others the same as those in Section 11.6.1.1. The varied input parameters used in Figures 11.8–11.12 are chosen as follows:

λ_s: $[5 \times 10^{-7}, 3 \times 10^{-4}]$, i.e. mean failure time from about 23 days to about 1 hour;
μ_s: $[2 \times 10^{-4}, 2 \times 10^{-3}]$, i.e. mean recovery time from about 1.4 hours to about 8 minutes;
λ_c: $[6 \times 10^{-8}, 4 \times 10^{-7}]$, i.e. mean failure time from about 192 days to about 29 days;
μ_c: $[1 \times 10^{-5}, 3 \times 10^{-4}]$, i.e. mean repair time from about 27 hours to about 1 hour.
k: [1, 10], where $\mathbf{K} = [k, k, k, k, k, k]$.

Figures 11.8–11.12 show that the independent software recovery scheme is only beneficial when software instance failure rate λ_s is very high (i.e. mean failure time is less than 3 hours), and has no significant improvement than dependent scheme when the other parameters including μ_s, λ_c, μ_c and k vary. In other words, the analytic models for the independent software recovery scheme have good approximations for the dependent software recovery scheme when software instance failure rate is relatively small (mean failure time is more than 3 hours), as typically happens in telecom operators networks.

From Figure 11.9, we observe that the system availability quickly increases when μ_s increases within the range $[2 \times 10^{-4}, 1 \times 10^{-3}]$ (i.e. the mean software recovery time varies from about 1.4 hours to about 16 minutes), while it becomes relatively stable when the mean software recovery time is less than 16 minutes. A similar trend is also observed

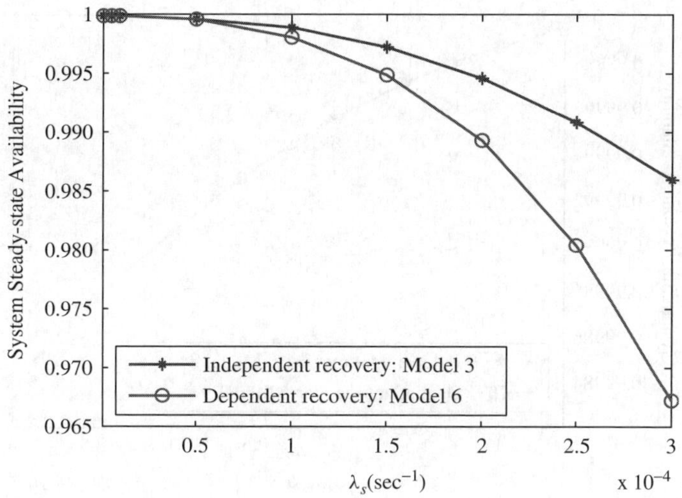

Figure 11.8 Influence of software instance failure rate

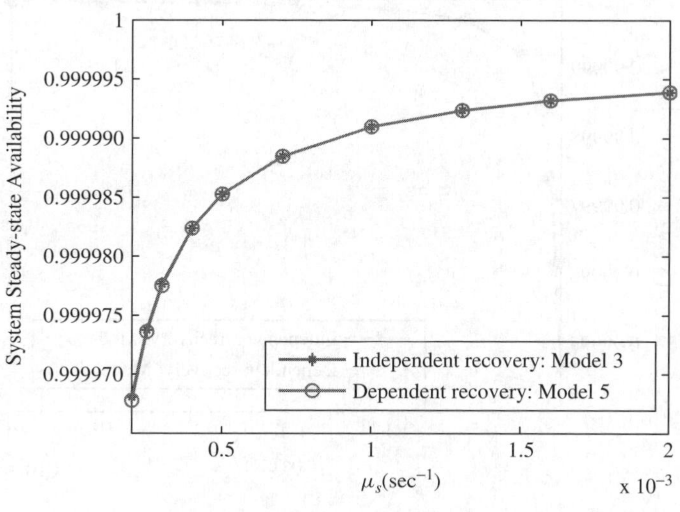

Figure 11.9 Influence of software instance recovery rate

in Figure 11.11. The system steady-state availability remains stable when μ_c is larger than $1 \times 10^{-4}\text{sec}^{-1}$ (i.e. the mean core repair time is less than 3 hours).

Figure 11.12 shows that the system steady-state availability increases sharply when the number of software instances in a node increases from 1 to 2, and remains stable when the number of software instances in a node is larger than 2. Note that such phenomena are observed for a given redundancy configuration **h** and performance demand W presented

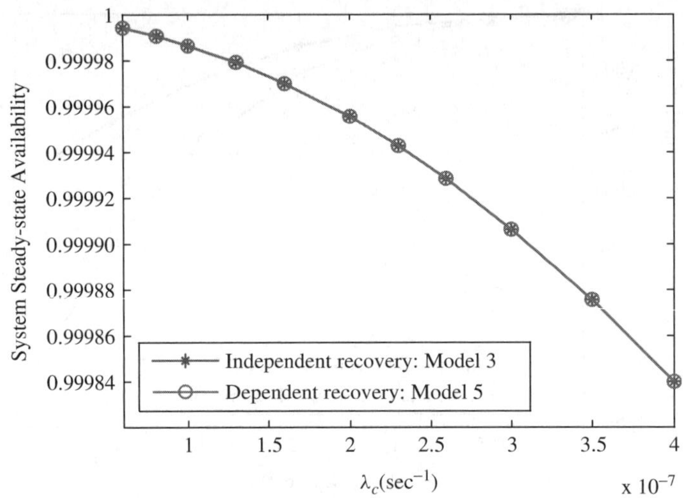

Figure 11.10 Influence of core failure rate

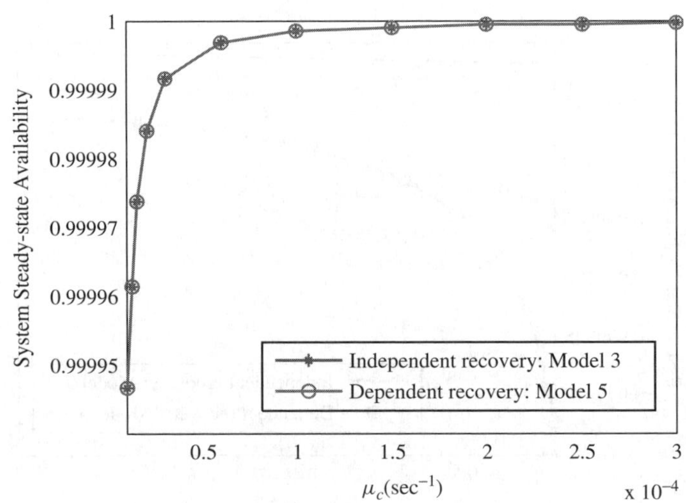

Figure 11.11 Influence of core repair rate

in Section 11.6.1.1. We will evaluate the influence of these two parameters on the system steady-state availability next.

11.6.2 Influences of Performance Demand and Redundancy Configuration

In this section, we only use model 1 for the evaluation without loss of generality. This is because the input parameters are used as in Section 11.6.1.1 except the values for **h** and W.

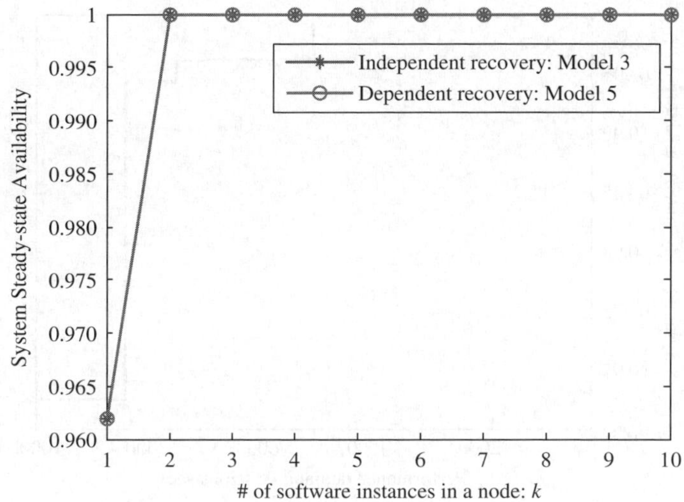

Figure 11.12 Influence of the number of software instances in a node

For any given **h** and W and specified other parameters, the independent and dependent recovery schemes are little different with respect to system steady-state availability as shown in Section 11.6.1.1. Therefore, model 1 is chosen for its simplicity of model construction.

11.6.2.1 Steady-state Availability vs. Performance Demand

In this section, the system steady-state availability is provided according to the change in the required performance demand. The system steady-state availability is shown in Figure 11.13 vs. performance demand for a fixed redundancy configuration $\mathbf{h} = [3, 3, 3, 2, 3, 3]$. As shown in Figure 11.13, the system steady-state availability decreases as the performance demand W increases. This behavior can be explained by noting that the support of the function (11.7) shrinks as W increases, thus reducing the number of the (positive) terms contributing to the summation (11.6). Furthermore, since the performance level in each state for a node is discrete, the overall system steady-state availability changes by jumps as a function of the performance demand.

11.6.2.2 Steady-State Availability vs. Redundancy Configuration

From the previous section, we know that given a specific system configuration, the system steady-state availability decreases with the increase of the performance demand. Hence, if the performance demand is too high, the obtained system steady-state availability may not satisfy the user's service requirement. Therefore, redundancy configuration needs to be changed in this case.

In order to evaluate the relationship between the system redundancy configuration and the steady-state availability, we fix the performance demand and utilize the redundancy

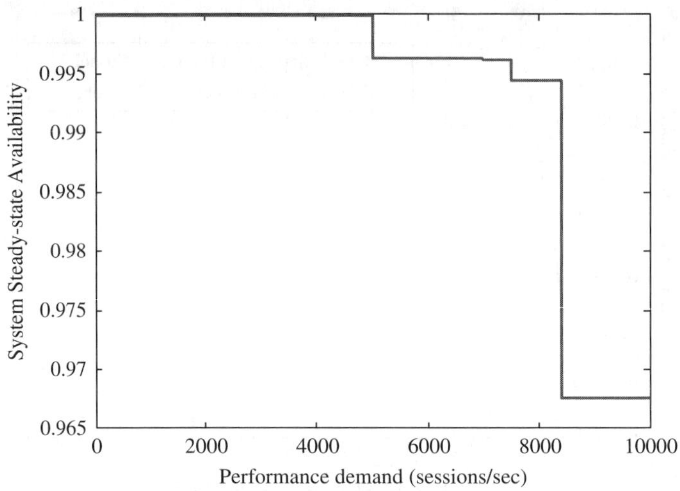

Figure 11.13 Steady-state availability vs. performance demand

optimization approach in Section 11.5 to find the optimal redundancy configuration that minimizes the cost while maintaining high availability.

Additional parameters other than those presented in Section 11.6.1.1 are as follows:

- **Cost** $= [0.9, 0.9, 0.9, 1, 0.9, 0.9]$; $A_0 = 0.99999$.

We assume that the maximum redundancy configuration involves five nodes for each server. By using an exhaustive search approach to solve the optimization problem (11.22), we obtain the best system redundancy configuration $\mathbf{h}_{opt} = [3, 3, 3, 2, 3, 3]$, which minimizes the cost and satisfies the availability requirement. Table 11.2 shows a part of the results for better comprehension, where $\mathbf{h}_{opt}$ is reported in boldface.

11.6.2.3 Steady-State Availability vs. Performance Demand and Redundancy Configuration

In the previous section, it is shown that given a performance demand, the optimal redundancy configuration can be computed to minimize the cost and maintain high availability. In this section, we evaluate the optimized cost and corresponding steady-state availability with respect to the change in performance demand, as shown in Table 11.3. As the performance demand increases, the optimized cost also increases, whereas the steady-state availability does not necessarily increase.

We remark that, given W and A_0, the optimal redundancy configuration is not always unique, because all the servers but HSS have one and the same cost (0.9). For instance, if $W = 2000$ is considered, 10 solutions (equivalent with respect to steady-state availability and cost) are available: they correspond to all the combinations of the servers with a cost of 0.9 where 2 out of 5 have 2 parallel nodes and the other 3 have 3 parallel nodes, and HSS_2 has 2 nodes. One of the said redundancy configurations has been reported in Table 11.3. For the other values of performance demand considered in Table 11.3, the

Table 11.2 Steady-state availability and cost vs. redundancy configuration

Redundancy configuration **h**	Steady-state availability	Cost
[5, 5, 5, 5, 5, 5]	0.9999999999	27.5
[5, 5, 5, 4, 4, 4]	0.9999999998	24.7
[4, 4, 4, 4, 4, 4]	0.9999999997	22.0
[3, 3, 3, 4, 4, 4]	0.9999999403	19.3
[3, 3, 3, 3, 3, 3]	0.9999998854	16.5
[3, 3, 3, 2, 3, 3]	**0.9999932086**	**15.5**
[3, 3, 3, 2, 2, 2]	0.9999765004	13.7

Table 11.3 Steady-state availability vs. performance demand and redundancy configuration

Performance demand W	Redundancy configuration **h**	Steady-state availability	Cost
2000	[2,2,3,2,3,3]	0.999990003	13.7
4000	[3,3,3,2,3,3]	0.999993208	15.5
6000	[4,4,4,4,4,4]	0.999999854	22.0
8000	[4,4,4,4,4,4]	0.999999542	22.0
10000	[5,5,5,4,5,5]	0.999999635	26.5

redundancy configuration is the only one with the (minimum) cost reported in the last column, while maintaining high availability.

11.7 Conclusion

In this work, six hierarchical analytic models are constructed to evaluate the steady-state availability of a signaling IMS-based core network, three models for the independent software recovery scheme and three models for the recovery scheme with dependencies. An optimization problem is solved to provide an optimal redundancy configuration for servers in the call set-up signaling network, while minimizing the cost and maintaining high availability. The accuracy of these models is compared to cross-validate each other. The execution time is also illustrated to show the efficiency of these proposed models. In addition, the models for the independent and dependent software recovery schemes are compared with respect to different input parameters. Moreover, the influences of the performance demand and redundancy configuration on the system steady-state availability are evaluated to provide the optimal configurations.

References

3GPP. 2008. Tech. spec. TS 23.228 version 8.5.0, IP Multimedia Subsystem (IMS); Stage 2 (Release 8). http://www.3gpp.org/ftp/Specs/html-info/23228.htm.

Ciardo, C., J. Muppala and K. S. Trivedi. 1989. SPNP: Stochastic Petri Net Package. *In Proceedings of Third*

International Workshop on Petri Nets and Performance Models, PNPM'89. 142–151. December 11–13, 1989. Kyoto.

Guida, M., M. Longo and F. Postiglione. 2008. Reliability and survivability methodologies for next generation networks. In *Proceedings of the 6th International Conference on Advances in Mobile Computing and Multimedia*, MoMM/BWCCA'08. 326–331. November 24–26, 2008. Linz, Austria, ACM.

Guida, M., M. Longo and F. Postiglione. 2010a. Performance evaluation of IMS-based core networks in presence of failures. In *Proceedings of the Global Telecommunications Conference, GLOBECOM 2010*. 1–5. December 6–10, 2010. Miami, FL, IEEE.

Guida, M., M. Longo and F. Postiglione. 2010b. Reliability analysis of next generation mobile networks. In *Reliability, Risk and Safety. Theory and Applications*, ed. R. Briš, C. Guedes Soares and S. Martorell. 1999–2006. Leiden, The Netherlands: CRC Press/Balkema-Taylor & Francis Group.

Hui, S. Y. and K. H. Yeung. 2003. Challenges in the migration to 4G mobile systems. *IEEE Communication Magazine* 41 (12): 54–59.

ITU-T. 2009. ITU-T Next Generation Networks Global Standards Initiative. Available at: http://www.itu.int/ITU-T/ngn/.

Kim, D. S., F. Machida and K. S. Trivedi. 2009. Availability modeling and analysis of a virtualized system. In: *Proceedings of the 15th IEEE Pacific Rim International Symposium on Dependable Computing*, PRDC'09. 365–371. November 16–18, 2009. Shanghai, China.

Lisnianski, A. and G. Levitin. 2003. *Multi-state system reliability*. Singapore: World Scientific.

Mendiratta V. B. and H. Pant. 2007. Reliability of IMS architecture. In *Proceedings of the Australasian Telecommunication Networks and Applications Conference*, ATNAC 2007. 1–6. December 2–5, 2007. Christchurch, New Zealand.

Pant, H., C. K. Chu, S. H. Richman, A. Jrad and G. P. O'Reilly. 2008. Reliability of next-generation networks with a focus on IMS architecture. *Bell Labs Technical Journal* 12 (4): 109–126.

Smith, W. E., K. S. Trivedi, L. A. Tomek and J. Ackaret. 2008. Availability analysis of blade server systems. *IBM System Journal* 47 (4): 621–640.

Tian, Z., G. Levitin and M. J. Zuo. 2009. A joint reliability–redundancy optimization approach for multi-state series-parallel systems. *Reliability Engineering and System Safety* 94 (10): 1568–1576.

Trivedi, K. S. 2001. *Probability and Statistics with Reliability, Queuing and Computer Science Applications*. New York: John Wiley & Sons.

Trivedi, K. S., D. Wang, D. J. Hunt, A. Rindos, W. E. Smith and B. Vashaw. 2008. Availability modeling of SIP protocol on IBM© WebSphere©. In *Proceedings of the 14th IEEE Pacific Rim International Symposium on Dependable Computing*, PRDC'08. 323–330. December 15–17, 2008. Taipei, Republic of China (Taiwan).

Ushakov, I. A. 1986. A universal generating function. *Soviet Journal of Computing System Science* 24 (5): 37–49.

Ushakov, I. A. 1987. Optimal standby problems and a universal generating function. *Soviet Journal of Computing System Science* 25 (4): 79–82.

Zang, X., D. Wang, H. Sun, and K. S. Trivedi. 2003. A BDD-based algorithm for analysis of multistate systems with multistate components. *IEEE Transactions on Computers* 52 (12): 1608–1618.

12

Reliability and Probability of First Occurred Failure for Discrete-Time Semi-Markov Systems

Stylianos Georgiadis, Nikolaos Limnios and Irene Votsi
Laboratoire de Mathématiques Appliquées de Compiègne, Université de Technologie de Compiègne, Centre de Recherches de Royallieu, France

12.1 Introduction

Semi-Markov chains constitute a generalization of Markov chains and renewal chains. For a Markov chain, the sojourn time in each state is geometrically distributed, whereas for a discrete-time semi-Markov system (DTSMS), the sojourn time distribution can be any distribution on $\mathbb{N}$. An introduction to homogeneous semi-Markov chains is given by Howard (2007) as well as by Mode and Pickens (1998) and Mode and Sleeman (2000). For non-homogeneous discrete-time semi-Markov systems, see Vassiliou and Papadopoulou (1992) and Vassiliou and Papadopoulou (1994), whereas for ergodic theory of semi-Markov chains, see Anselone (1960). Moreover, for semi-Markov replacement chains, see Gerontidis (1994). Barbu and Limnios (2006) study an empirical estimator of the discrete-time semi-Markov kernel and its asymptotic properties, with application to reliability. Chryssaphinou *et al.* (2011) establish a discrete-time reliability system with multiple components under semi-Markov hypothesis. An overview in the theory on semi-Markov chains oriented toward applications in modeling and estimation is presented in Barbu and Limnios (2008).

In this work, the investigation of the rate of occurrence of failures (ROCOF) is addressed for the first time for semi-Markov chains. The ROCOF may be interpreted as the expected number of transitions of a semi-Markov chain to a subset of its state space at a specific

Applied Reliability Engineering and Risk Analysis: Probabilistic Models and Statistical Inference, First Edition.
Edited by Ilia B. Frenkel, Alex Karagrigoriou, Anatoly Lisnianski and Andre Kleyner.

time. Firstly, a simple formula for evaluating the ROCOF is derived. As a consequence of this result, a statistical estimator of this function is proposed. The continuous-time version of the ROCOF is calculated in a wide range of scientific fields including reliability (Yeh 1997, Ouhbi and Limnios 2002, Limnios 2012) and seismology (Votsi *et al.* 2012).

Afterwards, we examine the steady-state (or asymptotic) availability by two different aspects. At first, we consider the pointwise availability and we take its limit as the time tends to infinity. Alternatively, the steady-state availability is written as the sum of the stationary distribution of the semi-Markov chain in the working states of the system (see Barbu and Limnios 2008). Two different proposed nonparametric estimators of the stationary distribution are presented (see Georgiadis and Limnios 2014) and the estimator of the availability is expressed in terms of them. As a sequence, we derive the empirical estimation of the steady-state availability through two different estimators of the stationary distribution.

Finally, we introduce the measure of the probability of the first occurred failure. Consider a sequence of disjoint subsets of the failure states of a DTSMS. Given the initial state of the system, the probability of first occurred failure expresses the probability of entry for the first time into a failure subset before any other one. An empirical estimator for the probability of first occurred failure is then proposed. The asymptotic property of the strong consistency is presented for the estimators of ROCOF, steady-state availability and probability of first occurred failure.

In Section 12.2, the necessary preliminaries of a semi-Markov model are introduced and, in the next one, the definitions of the reliability measures studied in this chapter are given. Section 12.4 depicts their empirical estimation and their strong consistency. Finally, we apply these results to a numerical example in Section 12.5.

12.2 Discrete-Time Semi-Markov Model

Let us first of all introduce some preliminaries from the theory of semi-Markov chains absolutely necessary for our purposes. Let $\mathbb{N}$ be the set of nonnegative integers and $\mathbb{N}^* = \mathbb{N}\backslash\{0\}$. Consider a stochastic system with finite state space E. We suppose that the evolution in time of the system is described by the following chains:

1. The chain $\boldsymbol{J} := (J_n)_{n\in\mathbb{N}}$ with state space E, where J_n is the system state at the n-th jump time.
2. The chain $\boldsymbol{S} := (S_n)_{n\in\mathbb{N}}$ with state space $\mathbb{N}$, where S_n is the n-th jump time. We suppose that $S_0 = 0$ and $0 < S_1 < S_2 < \ldots < S_n < S_{n+1} < \ldots$.
3. The chain $\boldsymbol{X} := (X_n)_{n\in\mathbb{N}^*}$ with state space $\mathbb{N}$, with $X_n := S_n - S_{n-1}$. Thus, for all $n \in \mathbb{N}^*$, X_n is the sojourn time in state J_{n-1}, before the n-th jump.

The chain $(\boldsymbol{J}, \boldsymbol{S}) := (J_n, S_n)_{n\in\mathbb{N}}$ is said to be a Markov renewal chain (MRC) with state space $E \times \mathbb{N}$, if for all $n \in \mathbb{N}$, $j \in E$ and $k \in \mathbb{N}$, it satisfies almost surely (a.s.)

$$\mathbb{P}(J_{n+1} = j, S_{n+1} - S_n = k | J_0, \ldots, J_n; S_0, \ldots, S_n) = \mathbb{P}(J_{n+1} = j, S_{n+1} - S_n = k | J_n).$$

If the above equation is independent of n, then $(\boldsymbol{J}, \boldsymbol{S})$ is said to be homogeneous.

The process $\boldsymbol{J}$ is the embedded Markov chain (EMC) of the MRC $(\boldsymbol{J}, \boldsymbol{S})$ with initial distribution $\alpha := (\alpha_i;\ i \in E)$, where $\alpha_i := \mathbb{P}(J_0 = i)$, and stationary distribution $\boldsymbol{\nu} = (\nu_i;\ i \in E)$. The transition matrix $\boldsymbol{P} := (P(i, j);\ i, j \in E)$ of $\boldsymbol{J}$ is given by

$$P(i, j) := \mathbb{P}(J_{n+1} = j | J_n = i), \quad n \in \mathbb{N}. \tag{12.1}$$

Moreover, for all $k \in \mathbb{N}$, we define the discrete-time counting process of the number of jumps in $[1, k] \subset \mathbb{N}$ by $N(k) := \max\{n \in \mathbb{N} : S_n \leq k\}$. The semi-Markov chain (SMC) $\boldsymbol{Z} := (Z_k)_{k\in\mathbb{N}}$ is defined as $Z_k = J_{N(k)}$, $k \in \mathbb{N}$, and it gives the system state at time k. We have also $J_n = Z_{S_n}$, $n \in \mathbb{N}$. Note that the initial distribution of the SMC $\boldsymbol{Z}$ coincides with that of $\boldsymbol{J}$.

The evolution of the DTSMS is governed by the semi-Markov kernel $\boldsymbol{q}(k) := (q_{ij}(k);\ i, j \in E)$, $k \in \mathbb{N}$, defined by

$$q_{ij}(k) := \mathbb{P}(J_{n+1} = j, X_{n+1} = k | J_n = i), \quad n \in \mathbb{N}. \tag{12.2}$$

For all $k \in \mathbb{N}^*$, the entries of the semi-Markov kernel $\boldsymbol{q}(k)$ are written as

$$q_{ij}(k) = p_{ij} f_{ij}(k), \quad i, j \in E,$$

where $f_{ij}(k) := \mathbb{P}(X_{n+1} = k | J_n = i, J_{n+1} = j)$ $i, j \in E$, $k \in \mathbb{N}$, is the conditional distribution of the sojourn time in state i given that the next visited state is j. In this case the sojourn times are attached to transitions and when a sojourn time in a state i expires, we can determine the next visited state j by using the probability of the EMC as well as the duration of this time.

Let us denote by $H_i(k)$, $k \in \mathbb{N}$, the sojourn time cumulative distribution function in any state $i \in E$, i.e.

$$H_i(k) := \mathbb{P}(X_{n+1} \leq k | J_n = i) = \sum_{j\in E}\sum_{l=0}^{k} q_{ij}(l),$$

and by $\overline{H}_i(k)$, $k \in \mathbb{N}$, the survival function in any state i, i.e.

$$\overline{H}_i(k) := \mathbb{P}(X_{n+1} > k | J_n = i) = 1 - H_i(k) = 1 - \sum_{j\in E}\sum_{l=0}^{k} q_{ij}(l).$$

Also, consider the matrices $\boldsymbol{H}(k) := diag(H_i(k);\ i \in E)$ and $\overline{\boldsymbol{H}}(k) := diag(\overline{H}_i(k);\ i \in E)$, $k \in \mathbb{N}$.

The transition function $\boldsymbol{P}(k) := (P_{ij}(k);\ i, j \in E)$, $k \in \mathbb{N}$, of the SMC $\boldsymbol{Z}$ is defined by

$$P_{ij}(k) := \mathbb{P}(Z_k = j | Z_0 = i).$$

Let $\psi(k) := (\psi_{ij}(k);\ i, j \in E)$ be the Markov renewal function given by

$$\psi_{ij}(k) := \sum_{n=0}^{k} q_{ij}^{(n)}(k),$$

where $q_{ij}^{(n)}(k) := \mathbb{P}(J_n = j, S_n = k | J_0 = i)$ is the n-fold discrete-time convolution of $q_{ij}(k)$ (see Barbu *et al.* 2004). Then, the transition function $\boldsymbol{P}(k)$ of the SMC $\boldsymbol{Z}$ can be written as

$$\boldsymbol{P}(k) = \psi * (\boldsymbol{I} - \boldsymbol{H})(k).$$

Also, the stationary distribution $\pi := (\pi_i; i \in E)$, $k \in \mathbb{N}$, of the SMC $\boldsymbol{Z}$ is defined, when it exists, by

$$\pi_i := \lim_{k\to\infty} P_{li}(k), \quad l \in E.$$

In addition, we denote by $\boldsymbol{m} := (m_i; i \in E)^\top$ the mean sojourn times of the SMC $\boldsymbol{Z}$ in any state i defined by

$$m_i := \mathbb{E}[S_1 | J_0 = i] = \sum_{n\in\mathbb{N}} \overline{H}_i(n).$$

We assume that the MRC $(\boldsymbol{J}, \boldsymbol{S})$ is irreducible and aperiodic, with finite mean sojourn times.

Then, the stationary distribution of the SMC is expressed in terms of the stationary distribution of the EMC:

$$\pi_i = \frac{\nu_i m_i}{\sum_{k\in E} \nu_k m_k}.$$

Let $\boldsymbol{U} := (U_k)_{k\in\mathbb{N}}$ be the sequence of the backward recurrence times for the SMC $\boldsymbol{Z}$ defined as follows:

$$U_k := \begin{cases} k, & \text{if } k < S_1, \\ k - S_{N(k)}, & \text{if } k \geq S_1, \end{cases}$$

where, since $S_0 = 0$, we have that $U_0 = 0$. We note that, for all $k \in \mathbb{N}$, $U_k \leq k$. The stochastic process $(\boldsymbol{Z}, \boldsymbol{U}) := (Z_k, U_k)_{k\in\mathbb{N}}$ is a Markov chain with values in $E \times \mathbb{N}$ (Limnios and Oprişan 2001). A study of this process is given by Chryssaphinou *et al.* (2008). It is worth noting that the Markov chain $(\boldsymbol{Z}, \boldsymbol{U})$ is time-homogeneous.

We denote by $\tilde{\alpha}$ the initial distribution of the Markov chain $(\boldsymbol{Z}, \boldsymbol{U})$, whereas its transition matrix $\tilde{\boldsymbol{P}} := (\tilde{P}((i, t_1), (j, t_2));\ (i, t_1), (j, t_2) \in E \times \mathbb{N})$ can be written as:

$$\begin{aligned} \tilde{P}((i, t_1), (j, t_2)) &:= \mathbb{P}(Z_{k+1} = j, U_{k+1} = t_2 | Z_k = i, U_k = t_1) \\ &= \begin{cases} q_{ij}(t_1 + 1)/\overline{H}_i(t_1), & \text{if } i \neq j, t_2 = 0, \\ \overline{H}_i(t_1 + 1)/\overline{H}_i(t_1), & \text{if } i = j, t_2 - t_1 = 1, \\ 0, & \text{otherwise,} \end{cases} \end{aligned} \tag{12.3}$$

for every $k \in \mathbb{N}$ such that $\mathbb{P}(Z_k = i, U_k = t_1) > 0$. Let us further denote by $(\tilde{\alpha}\tilde{\boldsymbol{P}}^{k-1})(i, m)$ the (i, m) element of the vector $\tilde{\alpha}\tilde{\boldsymbol{P}}^{k-1}$.

After introducing the basic notation and giving a brief description of the model under study, we proceed to the calculation of the ROCOF along with the steady-state availability and the probability of first occurred failure for the semi-Markov case.

12.3 Reliability and Probability of First Occurred Failure

Let us consider that the possible states of the semi-Markov chain $\boldsymbol{Z}$ belong to the set $E = \{1, \ldots, s\}$, $s \in \mathbb{N}^*$. Let $U := \{1, \ldots, r\}$ and $D := \{r + 1, \ldots, s\}$ be the subsets of the state space E of the working states and the down states of the system, respectively, with $U \cup D = E$ and $U \cap D = \emptyset$. We present now the reliability measures under consideration in this work.

12.3.1 Rate of Occurrence of Failures

We further consider an arbitrary subset B of the state space E, with $B \neq \emptyset$ and $B \neq E$. Initially, we are concentrating on the study of an important parameter in semi-Markov chains, the ROCOF. Before providing a formula for the evaluation of the ROCOF, let us clarify its meaning. For all $k \in \mathbb{N}$, we denote by $N_B(k)$ the counting process, up to time k, of the transitions of the SMC $\mathbf{Z}$ from B^{C} to B, namely

$$N_B(k) := \sum_{l=1}^{k} \mathbf{1}_{\{Z_{l-1} \in B^{\mathrm{C}}, Z_l \in B\}}.$$

The ROCOF is interpreted as the expected number of transitions of the SMC to the set B at time k, i.e.

$$\tilde{r}_B(k) := \mathbb{E}[N_B(k) - N_B(k-1)].$$

The following proposition gives a simple formula of the ROCOF for semi-Markov chains.

Proposition 12.1. *The ROCOF of the SMC $\mathbf{Z}$ at time k is given by*

$$\tilde{r}_B(k) = \sum_{i \in B^{\mathrm{C}}} \sum_{j \in B} \sum_{m=0}^{k-1} [(\tilde{a}\tilde{P}^{k-1})(i,m)] \tilde{P}((i,m),(j,0)). \tag{12.4}$$

12.3.2 Steady-State Availability

Another interesting measure concerning the asymptotic reliability theory is the steady-state availability of a system. First, let us define the pointwise availability A of a system at time $k \in \mathbb{N}$ as the probability that the system is operational at time k, independently of the fact that the system has failed or not in $[0, k-1]$, i.e.

$$A(k) := \mathbb{P}(Z_k \in U).$$

Proposition 12.2. *Given a stochastic system described by a SMC $\mathbf{Z}$, the pointwise availability is given by*

$$A(k) = \boldsymbol{\alpha} \boldsymbol{P}(k) \mathbf{1}_{s,r}, \tag{12.5}$$

where $\mathbf{1}_{s,r}$ is a s-column vector whose the r first elements are 1's and the last $s-r$ ones are 0's.

The steady-state availability A_∞ of a system is defined as the limit of the pointwise availability, when the limit exists, as the time tends to infinity, i.e.

$$A_\infty := \lim_{k \to \infty} A(k).$$

Proposition 12.3. *For a semi-Markov system, the steady-state availability is given by*

$$A_\infty = \sum_{i \in U} \pi_i. \tag{12.6}$$

12.3.3 Probability of First Occurred Failure

The measure of the probability of first occurred failure is introduced here. We consider the subset U of the working states, as defined in the previous subsection. We further consider the partition of the subset D to k subsets of failure states $(C_k)_{k\in\mathbb{N}^*}$ with $D = \cup_k C_k$ and $C_k \cap C_l = \emptyset$, $k \neq l$.

Denote by $T_{C_k} := \min\{n \in \mathbb{N} : Z_n \in C_k\}$ the first hitting time of the subset C_k and $\rho_{C_k} := (\rho_{C_k}(i);\ i \in U)^\top$ the column-vector of probabilities defined below. If $T_{C_j} < \infty$ for a $j \in \mathbb{N}^*$, then $T_{C_k} = +\infty$ for any $k \neq j$. Under this notation, for any $i \in U$, we define the probability of first occurred failure and get

$$\begin{aligned}\rho_{C_k}(i) &:= \mathbb{P}(T_{C_k} < \infty | J_0 = i)\\ &= \mathbb{P}_i(T_{C_k} < \infty)\\ &= \mathbb{P}_i(J_1 \in C_k, T_{C_k} < \infty) + \mathbb{P}_i(J_1 \in U, T_{C_k} < \infty)\\ &= \mathbb{P}_i(J_1 \in C_k) + \sum_{j\in U} P(i,j)\rho_{C_k}(j).\end{aligned}$$

Let $\boldsymbol{P}_{00}$ be the restriction on the subset $U \times U$ of the transition matrix $\boldsymbol{P}$ of the EMC $\boldsymbol{J}$ and $\boldsymbol{P}_{0k}$ the column-vector defined as $\boldsymbol{P}_{0k} := (P(i, C_k);\ i \in U)^\top$ with $P(i, C_k) := \sum_{j\in C_k} P(i,j)$. Then, we have

$$\rho_{C_k} = \boldsymbol{P}_{0k} + \boldsymbol{P}_{00}\rho_{C_k}.$$

Consequently, we obtain the following proposition:

Proposition 12.4. *For a stochastic system governed by the modified process $\mathbf{Z}'$ and given a failure class C_k, if the matrix $\boldsymbol{I} - \mathbf{P}_{00}$ is nonsingular, the probability of first occurred failure is given by*

$$\boldsymbol{\rho}_{C_k} = (\boldsymbol{I} - \boldsymbol{P}_{00})^{-1}\boldsymbol{P}_{0k}. \tag{12.7}$$

Note that

$$\sum_k \boldsymbol{\rho}_{C_k} = \mathbf{1},$$

where $\mathbf{1}$ is a r-dimensional column-vector with all entries equal to 1.

12.4 Nonparametric Estimation of Reliability Measures

In this section, we follow the next observational procedure concerning the statistical inference of stochastic processes: a single realization of the process is observed over the fixed time interval $[0, M]$, $M \in \mathbb{N}^*$. The asymptotic property of strong consistency is obtained as the censoring time M tends to infinity (practically, when it becomes large). We consider an observation $\mathcal{H}_M$ of the ergodic MRC $(\boldsymbol{J}, \boldsymbol{S})$, censored at a fixed arbitrary time $M \in \mathbb{N}^*$, defined by

$$\mathcal{H}_M := \begin{cases}\{J_0, X_1, J_1, \ldots, X_{N(M)}, J_{N(M)}, U_M\}, & \text{if } N(M) > 0,\\ \{J_0, U_M = M\}, & \text{if } N(M) = 0,\end{cases}$$

where $N(M)$ is the discrete-time counting process of the number of jumps in $[1, M]$ and $U_M := M - S_{N(M)}$ is the censored sojourn time in the last visited state $J_{N(M)}$. Let us consider the following set: $T_M := \{0, \ldots, M\}$.

Additionally, for all states $i, j \in E$, let us define the counting processes:

1. $N_i(M) := \sum_{n=1}^{N(M)} \mathbf{1}_{\{J_{n-1}=i\}}$ is the number of visits to state i of the EMC, up to time M.
2. $N_{ij}(M) := \sum_{n=1}^{N(M)} \mathbf{1}_{\{J_{n-1}=i, J_n=j\}}$ is the number of transitions of the EMC from i to j, up to time M.

The proposed empirical estimator $\hat{\boldsymbol{q}}(k, M) := (\hat{q}_{ij}(k, M);\ i, j \in E)$, $k \in T_M$, $M \in \mathbb{N}^*$, of the semi-Markov kernel (12.2) is defined by the following equation

$$\hat{q}_{ij}(k, M) := \frac{1}{N_i(M)} \sum_{n=1}^{N(M)} \mathbf{1}_{\{J_{n-1}=i, J_n=j, X_n=k\}}.$$

For further study of the asymptotic properties of the proposed empirical estimator, see Barbu and Limnios (2006). Once the estimator of the semi-Markov kernel is obtained, any measure concerning the SMC can be estimated, after having been expressed as a function of the semi-Markov kernel.

12.4.1 Estimation of ROCOF

The estimators $\hat{H}_i(k, M)$ and $\hat{\overline{H}}_i(k, M)$, $k \in T_M$, $M \in \mathbb{N}^*$, $i \in E$, for the sojourn time cumulative distribution functions $H_i(k, M)$ and the survival function $\overline{H}_i(k)$, respectively, are given by

$$\hat{H}_i(k, M) := \sum_{j \in E} \sum_{l=0}^{k} \hat{q}_{ij}(l, M) \quad \text{and} \quad \hat{\overline{H}}_i(k, M) := 1 - \sum_{j \in E} \sum_{l=0}^{k} \hat{q}_{ij}(l, M).$$

Consider also the estimators $\hat{\boldsymbol{H}}(k, M) := diag(\hat{H}_i(k, M);\ i \in E)$ and $\hat{\overline{\boldsymbol{H}}}(k, M) := diag(\hat{\overline{H}}_i(k, M);\ i \in E)$.

The empirical estimator $\hat{\tilde{\boldsymbol{P}}}_M := (\hat{\tilde{P}}((i, t_1), (j, t_2));\ (i, t_1), (j, t_2) \in E \times \mathbb{N})$ of the transition matrix (12.3) of the Markov chain $(\boldsymbol{Z}, \boldsymbol{U})$ has entries defined by

$$\hat{\tilde{P}}_M((i, t_1), (j, t_2)) = \begin{cases} \hat{q}_{ij}(t_1 + 1, M)/\hat{\overline{H}}_i(t_1, M), & \text{if } i \neq j, t_2 = 0, \\ \hat{\overline{H}}_i(t_1 + 1, M)/\hat{\overline{H}}_i(t_1, M), & \text{if } i = j, t_2 - t_1 = 1, \\ 0, & \text{otherwise.} \end{cases}$$

On the basis of Proposition 12.1, we propose the following estimator $\hat{\tilde{r}}_B(k, M)$, $k \in T_M$, for the ROCOF (12.4) of the DTSMS,

$$\hat{\tilde{r}}_B(k, M) = \sum_{i \in B^{\complement}} \sum_{j \in B} \sum_{m=0}^{k-1} [(\hat{\tilde{\boldsymbol{\alpha}}} \hat{\tilde{\boldsymbol{P}}}_M^{k-1})(i, m)] \hat{\tilde{P}}_M((i, m), (j, 0)), \tag{12.8}$$

where $(\hat{\boldsymbol{\alpha}}\hat{\tilde{\boldsymbol{P}}}_M^{k-1})(i,m)$ is the (i,m)-th element of the vector $\hat{\boldsymbol{\alpha}}\hat{\tilde{\boldsymbol{P}}}_M^{k-1}$. The following proposition gives the uniform strong consistency of the ROCOF estimator.

Proposition 12.5. *For any arbitrary $k \in \mathbb{N}$ fixed, the estimator (12.8) of the ROCOF (12.4) at instant k is strongly consistent in the sense that*

$$\hat{\tilde{r}}_B(k,M) \xrightarrow{a.s.} \tilde{r}_B(k), \quad \textit{as } M \to \infty.$$

Proof. We have $\hat{\tilde{P}}_M \xrightarrow{a.s.} \tilde{P}$, as $M \to \infty$, and since we deal with finite sums, we can write directly:

$$\begin{aligned}
\hat{\tilde{r}}_B(k,M) &= \sum_{i\in B^{\complement}}\sum_{j\in B}\sum_{m=0}^{k-1}[(\hat{\tilde{a}}\hat{\tilde{P}}_M^{k-1})(i,m)]\hat{\tilde{P}}_M((i,m),(j,0)) \\
&\underset{M\to\infty}{\overset{a.s.}{\to}} \sum_{i\in B^{\complement}}\sum_{j\in B}\sum_{m=0}^{k-1}[(\tilde{a}\tilde{P}^{k-1})(i,m)]\tilde{P}((i,m),(j,0)) \\
&= \tilde{r}_B(k).
\end{aligned}$$

□

12.4.2 Estimation of the Steady-State Availability

Concerning the stationary distribution π of the SMC, we consider two empirical estimators $\hat{\pi}(M) := (\hat{\pi}_i(M);\, i \in E)$ and $\tilde{\pi}(M) := (\tilde{\pi}_i(M);\, i \in E)$, defined as follows:

$$\hat{\pi}_i(M) := \frac{1}{M}\sum_{k=1}^{M}\mathbf{1}_{\{Z_{k-1}=i\}}, \tag{12.9}$$

$$\tilde{\pi}_i(M) := \frac{\hat{\nu}_i(M)\hat{m}_i(M)}{\sum_{k\in E}\hat{\nu}_k(M)\hat{m}_k(M)}, \tag{12.10}$$

where $\hat{\nu}_i(M)$ and $\hat{m}_i(M)$ are the empirical estimators of the stationary distribution ν_i of the EMC $\boldsymbol{J}$ and the mean sojourn time m_i, respectively, given by

$$\hat{\nu}_i(M) := \frac{N_i(M)}{N(M)} \quad \text{and} \quad \hat{m}_i(M) := \frac{1}{N_i(M)}\sum_{r=1}^{N_i(M)} X_{ir},$$

where X_{ir} is the r-th sojourn time is state i.

For any censoring time $M \in \mathbb{N}^*$, we consider the estimators $\hat{\psi}(k,M) := (\hat{\psi}_{ij}(k,M);\, i,j \in E)$ and $\hat{\boldsymbol{P}}(k,M) := (\hat{P}_{ij}(k,M),\, i,j \in E)$, $k \in T_M$, of the Markov renewal function $\psi(k)$ and the transition function $\boldsymbol{P}(k)$ of the SMC $\boldsymbol{Z}$, respectively, taking the form

$$\hat{\psi}(k,M) := \sum_{n=0}^{k}\hat{\boldsymbol{q}}^{(n)}(k,M) \quad \text{and} \quad \hat{\boldsymbol{P}}(k,M) = \hat{\psi} * (\boldsymbol{I} - \hat{\boldsymbol{H}})(k,M).$$

Then, the empirical estimator $\hat{A}(k, M)$ of the pointwise availability $A(k)$, $k \in T_M$, can be written as

$$\hat{A}(k, M) = \hat{\boldsymbol{\alpha}}\hat{\boldsymbol{P}}(k, M)\mathbf{1}_{s,r}. \tag{12.11}$$

Remark. Formally, the estimator of the initial distribution α is defined as $\hat{\alpha} := \delta_{Z_0}$. In our case, as one trajectory is taken into account, the estimation of the initial distributions $\boldsymbol{\alpha}$ and $\tilde{\boldsymbol{\alpha}}$ is trivial.

From Equation (12.6), we may obtain two different estimators for the steady-state availability:

$$\tilde{A}_\infty(M) = \sum_{i \in U} \tilde{\pi}_i(M), \tag{12.12}$$

$$\hat{A}_\infty(M) = \sum_{i \in U} \hat{\pi}_i(M), \tag{12.13}$$

where $\hat{A}(k, M)$, $\tilde{\pi}_i(M)$ and $\hat{\pi}_i(M)$ are given in (12.11), (12.9) and (12.10), respectively.

Proposition 12.6. *For any fixed $k \in \mathbb{N}$, the estimator (12.11) of the pointwise availability $A(k)$, and the estimators (12.12) and (12.13) of the steady-state availability A_∞, are strongly consistent, i.e.*

$$\hat{A}(k, M) \xrightarrow{a.s.} A(k),$$

$$\tilde{A}_\infty \xrightarrow{a.s.} A_\infty,$$

$$\hat{A}_\infty \xrightarrow{a.s.} A_\infty,$$

as $M \to \infty$.

Proof. The first result is obtained in Barbu and Limnios (2008). The estimators (12.9) and (12.10) of the stationary distribution of the SMC $\boldsymbol{Z}$ are strongly consistent (see Georgiadis and Limnios 2014) and, therefore, the consistency of $\tilde{A}_\infty$ and $\hat{A}_\infty$ is direct. □

12.4.3 Estimation of the Probability of First Occurred Failure

The empirical estimator $\hat{\boldsymbol{P}}_M := (\hat{P}_M(i, j);\ i, j \in E)$, $M \in \mathbb{N}^*$, of the transition matrix (12.1) of the EMC $\boldsymbol{J}$ is given by

$$\hat{P}_M(i, j) := \frac{N_{ij}(M)}{N_i(M)}.$$

Based on the estimation $\hat{\boldsymbol{P}}$ of the transition matrix (12.1), we obtain the estimators $\hat{\boldsymbol{P}}_{00}(M)$ and $\hat{\boldsymbol{P}}_{0k}(M)$ of $\boldsymbol{P}_{00}$ and $\boldsymbol{P}_{0k}$, respectively. For any failure subset C_k, the estimator

$\hat{\rho}_{C_k}(M) := (\hat{\rho}_{C_k}(i, M);\ i \in U)^\top$ of the probability of first occurred failure (12.7) takes the form

$$\hat{\boldsymbol{\rho}}_{C_k}(M) = (\boldsymbol{I} - \hat{\boldsymbol{P}}_{00}(M))^{-1}\hat{\boldsymbol{P}}_{0k}(M). \tag{12.14}$$

Proposition 12.7. *For any $M \in \mathbb{N}^*$, the estimator (12.14) of the probability of first occurred failure (12.7) of the failure subset C_k, is strongly consistent, i.e.*

$$\hat{\boldsymbol{\rho}}_{C_k}(M) \xrightarrow{a.s.} \rho_{C_k}, \quad as \quad M \to \infty.$$

Proof. We can write the difference

$$\begin{aligned}\hat{\boldsymbol{\rho}}_{C_k}(M) - \rho_{C_k} &= (\boldsymbol{I} - \hat{\boldsymbol{P}}_{00}(M))^{-1}\hat{\boldsymbol{P}}_{0k}(M) - (\boldsymbol{I} - \boldsymbol{P}_{00})^{-1}\boldsymbol{P}_{0k}\\ &= (\boldsymbol{I} - \hat{\boldsymbol{P}}_{00}(M))^{-1}(\hat{\boldsymbol{P}}_{0k} - \boldsymbol{P}_{0k}) + [(\boldsymbol{I} - \hat{\boldsymbol{P}}_{00}(M))^{-1} - (\boldsymbol{I} - \boldsymbol{P}_{00})^{-1}]\boldsymbol{P}_{0k}.\end{aligned}$$

Since $(\boldsymbol{I} - \boldsymbol{P}_{00})^{-1} = \sum_{n=0}^{\infty} \boldsymbol{P}_{00}^n$, $\hat{\boldsymbol{P}}_{00}(M) \xrightarrow{a.s.} \boldsymbol{P}_{00}$ and $\hat{\boldsymbol{P}}_{0k}(M) \xrightarrow{a.s.} \boldsymbol{P}_{0k}$, as $M \to \infty$, the desired result is obtained. □

12.5 Numerical Application

In this section, we apply the previous results to a four-state semi-Markov system, as described in Figure 12.1.

The state space of the system $E = \{1, 2, 3, 4\}$ is partitioned into the up-state set $U = \{1, 2\}$ and the down-state set $D = \{3, 4\}$. Since B is arbitrarily chosen, for the sake of simplicity, we set $B \equiv D$ and $B^{\complement} \equiv U$. The initial law $\boldsymbol{\alpha}$ and the transition matrix $\boldsymbol{P}$ of the EMC are given by

$$\boldsymbol{\alpha} = \begin{pmatrix}1 & 0 & 0 & 0\end{pmatrix} \quad \text{and} \quad \boldsymbol{P} = \begin{pmatrix} 0 & 0.8 & 0.2 & 0\\ 0.9 & 0 & 0 & 0.1\\ 1 & 0 & 0 & 0\\ 0 & 1 & 0 & 0\end{pmatrix}.$$

We suppose that the conditional sojourn-time distributions $f_{13}(\cdot)$ and $f_{24}(\cdot)$ are geometric ones defined by

$$f(k) := \begin{cases} p(1-p)^{k-1}, & \text{if } k \geq 1,\\ 0, & \text{if } k = 0,\end{cases}$$

with $p = 0.7$ and $p = 0.6$, respectively. The distributions $f_{12}(\cdot)$, $f_{21}(\cdot)$, $f_{31}(\cdot)$ and $f_{42}(\cdot)$ are discrete-time Weibull ones with

$$f(k) := \begin{cases} q^{(k-1)^b} - q^{k^b}, & \text{if } k \geq 1,\\ 0, & \text{if } k = 0,\end{cases}$$

where $(q, b) = (0.8, 1.6)$ for the transition $1 \to 2$, $(q, b) = (0.7, 1.6)$ for the transition $2 \to 1$, $(q, b) = (0.4, 0.7)$ for the transition $3 \to 1$ and $(q, b) = (0.3, 0.7)$ for the transition $4 \to 2$.

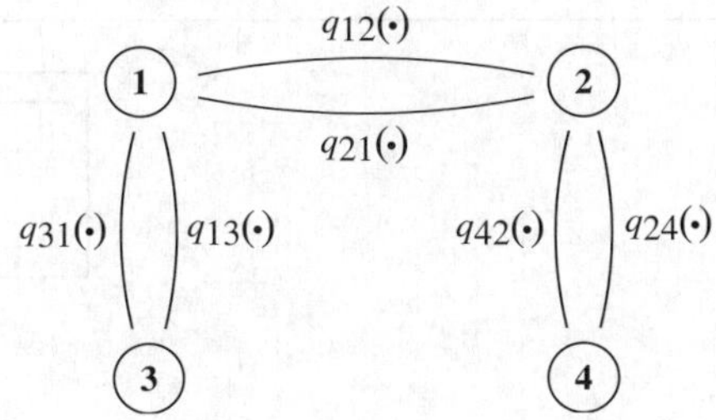

Figure 12.1 Four-state DTSMS

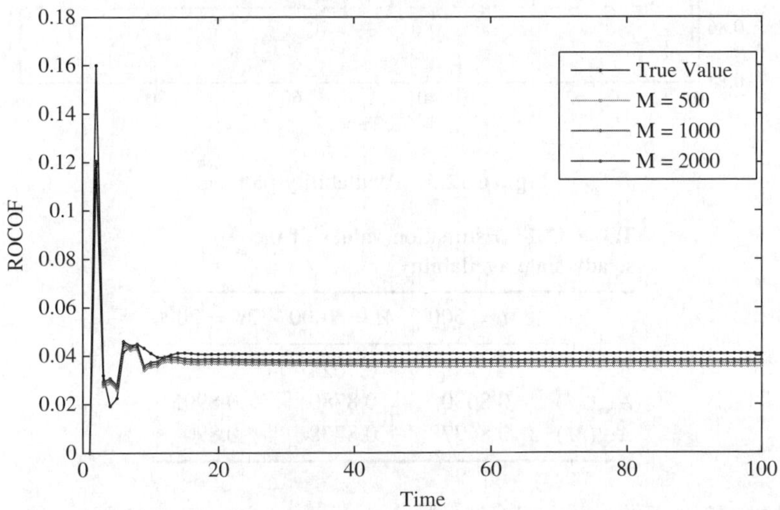

Figure 12.2 ROCOF plot

A single trajectory of the SMC up to three censoring times $M = 500$, $M = 1000$ and $M = 2000$ is obtained by means of a Monte Carlo method. In Figures 12.2 and 12.3, we present the plots of the theoretical and estimated values for the ROCOF and the availability of the system for the first 100 time units. In Table 12.1, the estimation of the steady-state availability is depicted.

Remark. Note that, since $M \to \infty$ and $k \to \infty$, the estimation of the pointwise availability converges to the steady-state availability. Thus, we may derive an alternative estimation for the steady-state availability through the availability function.

To investigate the estimation of the probability of first occurred failure, we consider two failure subsets $C_1 = \{3\}$ and $C_2 = \{4\}$ with $C_1 \cup C_2 = D$. The probability of first occurred failure and its estimation, given the subsets C_1 and C_2, is presented in Table 12.2.

The consistency of the estimators of the ROCOF, the steady-state availability and the probability of first occurred failure seem to be verified by the results of the preceding plots

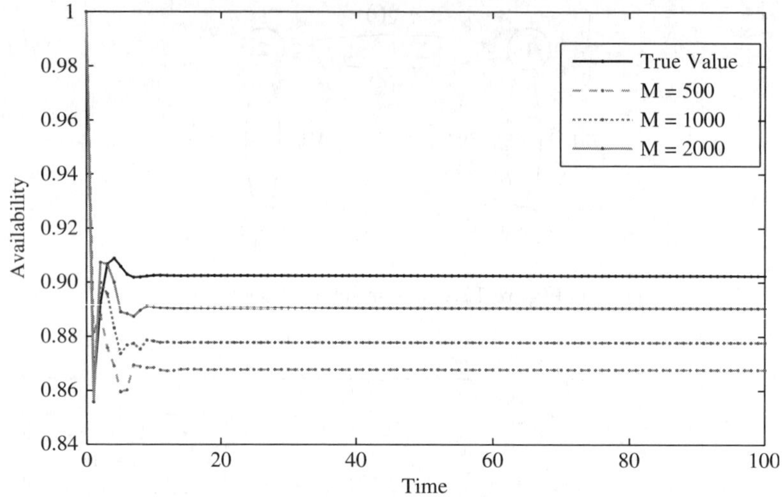

Figure 12.3 Availability plot

Table 12.1 Estimation values of the steady-state availability

	$M = 500$	$M = 1000$	$M = 2000$
A_∞		0.9024	
$\tilde{A}_\infty(M)$	0.8680	0.8780	0.8905
$\hat{A}_\infty(M)$	0.8677	0.8778	0.8904

Table 12.2 Estimation values of the probability of first occurred failure

	$M = 500$	$M = 1000$	$M = 2000$
ρ_{C_1}		$(0.7143, 0.6429)^\top$	
ρ_{C_2}		$(0.2857, 0.3571)^\top$	
$\hat{\rho}_{C_1}(M)$	$(0.5899, 0.5047)^\top$	$(0.6329, 0.5471)^\top$	$(0.6719, 0.5937)^\top$
$\hat{\rho}_{C_2}(M)$	$(0.4101, 0.4953)^\top$	$(0.3671, 0.4529)^\top$	$(0.3281, 0.4063)^\top$

and tables and, the estimated values approach the true ones as the length of the trajectory becomes larger. Also, we see that two proposed estimators of A_∞ tend to coincide, with a slightly better estimator to be $\tilde{A}_\infty(M)$.

12.6 Conclusion

In this chapter, the problem of evaluating and estimating the ROCOF and the probability of the first occurred failure is addressed for the first time for discrete-time semi-Markov systems, along with the steady-state availability. These three important reliability measures are evaluated and estimated statistically and the main result given here is the strong consistency of the proposed estimators. In addition to reliability theory, the aforementioned measures can find straightforward applications in biostatistics, seismology and several other fields.

References

Anselone, P. 1960. Ergodic theory for discrete semi-Markov chains. *Duke Mathematical Journal* 27 (1): 33–40.

Barbu, V.S., M. Boussemart, and N. Limnios. 2004. Discrete-time semi-Markov processes for reliability and survival analysis. *Communications in Statistics-Theory and Methods* 33 (11): 2833–2868.

Barbu, V.S. and N. Limnios. 2006. Empirical estimation for discrete-time semi-Markov processes with applications in reliability. *Journal of Nonparametric Statistics* 18 (7–8) 483–498.

Barbu, V.S. and N. Limnios. 2008. *Semi-Markov Chains and Hidden Semi-Markov Models toward Applications: Their Use in Reliability and DNA Analysis*. New York: Springer.

Chryssaphinou, O., M. Karaliopoulou and N. Limnios. 2008. On discrete-time semi-Markov chains and applications in words occurrences. *Communications in Statistics; Theory and Methods* 37 (8): 1306–1322.

Chryssaphinou, O., N. Limnios and S. Malefaki. 2011. Multi-state reliability systems under discrete time semi-Markovian hypothesis. *IEEE Transactions on Reliability* 60 (1): 80–87.

Georgiadis, S. and N. Limnios. 2014. Nonparametric estimation of the stationary distribution of a discrete-time semi-Markov process. *Communications in Statistics; Theory and Methods*, to appear.

Gerontidis, I. 1994. Semi-Markov replacement chains. *Advances in Applied Probability* 26 (3): 728–755.

Howard, R. 2007. *Dynamic Probabilistic Systems*, Volume II: Semi-Markov and Decision Processes. New York: Dover Publications.

Limnios, N. 2012. Reliability measures of semi-Markov systems with general state space. *Methodology and Computing in Applied Probability* 14 (4): 895–917.

Limnios, N. and G. Oprişan. 2001. *Semi-Markov Processes and Reliability*. Boston: Birkhäuser.

Mode, C. and G. Pickens. 1998. Computational methods for renewal theory and semi-Markov processes with illustrative examples. *The American Statistician* 42 (2): 143–152.

Mode, C. and C. Sleeman. 2000. *Stochastic Processes in Epidemiology: HIV/AIDS, Other Infectious Diseases and Computers*. Singapore: World Scientific.

Ouhbi, B. and N. Limnios. 2002. The rate of occurrence of failures for semi-Markov processes and estimation. *Statistics and Probability Letters* 59 (3): 245–255.

Vassiliou, P.-C. and A. Papadopoulou. 1992. Non-homogeneous semi-Markov systems and maintainability of the state sizes. *Journal of Applied Probability* 29 (3): 519–534.

Vassiliou, P.-C. and A. Papadopoulou. 1994. Asymptotic behavior of non homogeneous semi-Markov systems. *Linear Algebra and its Applications* 210: 153–198.

Votsi, I., N. Limnios, G. Tsaklidis and E. Papadimitriou. 2012. Estimation of the expected number of earthquake occurrences based on semi-Markov models. *Methodology and Computing in Applied Probability* 14 (3): 685–703.

Yeh, L. 1997. The rate of occurrence of failures. *Journal of Applied Probability* 34 (1): 234–247.

13

Single-Source Epidemic Process in a System of Two Interconnected Networks

Ilya B. Gertsbakh[1] and Yoseph Shpungin[2]
[1]*Department of Mathematics, Ben-Gurion University of the Negev, Israel*
[2]*Software Engineering Department, Shamoon College of Engineering, Israel*

13.1 Introduction

Networks attract ever growing interest in the modern research literature. Since network type systems presently appear in every sphere of our life (social networks, communication networks, Internet, etc.), their reliability, survivability and resilience studies are becoming an extremely important issue. Particular interest is attracted by the so-called cascading failure phenomena in networks, that is, massive network failures triggered by failure of a small part of the network components, see, for example, Gutfraint (2011) and many references presented there on this issue.

The majority of the works on network reliability deal with probabilistic failure description and modeling in a single network, see, for example, Gertsbakh and Shpungin (2011a, 2011b, 2011c, 2011d), Samaniego (2010), Newman (2010), Zhou *et al.* (2012). At the same time, most modern networks do not exist and function in an isolated mode. As noted in Gertsbakh and Shpungin (2012), due to technological progress, modern systems are becoming more and more coupled together. While in the past many networks would provide their functionality independently, modern systems depend on one another to provide proper functionality. For example, the financial network of banks and related financial institutions is interconnected with the business network of industrial enterprises. A failure (e.g. bankruptcy) of a particular bank in one country may cause serious problems of financing and/or industrial activity of several enterprises operating in another

Applied Reliability Engineering and Risk Analysis: Probabilistic Models and Statistical Inference, First Edition.
Edited by Ilia B. Frenkel, Alex Karagrigoriou, Anatoly Lisnianski and Andre Kleyner.

part of the world. One of the works devoted to the failures in interdependent networks is an important paper (Gertsbakh and Shpungin 2012), which discusses as an example a communication network and the power supply network. In this case, the nodes of the communication network, which control the operation of power supply stations, depend on the power supplied by the power station, while the power station operation depends on information generated and processed in the communication network.

A series of recently published works (Buldyrev *et al.* 2010; Dickison *et al.* 2012; Gao *et al.* 2012; Gertsbakh and Shpungin 2012; Li *et al.* 2012) describes various modes of network interaction, including a system of two networks and a collection of several networks united into one superstructure with given topology.

Typically, the works on network interaction and cascading failure phenomena in these networks deal with very large networks, formally with networks whose number of nodes n tends to infinity. The topology of these networks is defined only partially, and typically the only information given is their node degree distribution $\{p_k\}, k = 1, 2, 3, \ldots$. The main object of the research for these networks is finding the conditions under which the catastrophic failure (or epidemic process) spreads over an *essential* part of the network having size $O(n), n \to \infty$ (i.e. over the so-called giant component).

The present work continues the line of research initiated in Gertsbakh and Shpungin (2012) in which we considered a pair of interacting *finite* networks, each of which had a defined topology. Here we concentrate on the situation when the cascading failure process is initiated by a single failed node in one of the two interacting networks. This failure is "transmitte" to a randomly chosen node in another network, then backwards, and so on, until it exhausts itself creating some random number of failed nodes. Depending on network topologies and their failure definition, this process may lead to the failure of the interacting networks. Eventually, we find out the distribution of failed nodes in each of the two networks and the corresponding network failure probabilities.

The exposition is as follows. In Section 13.2 we describe the process of failure development and derive the distribution of the number of failed nodes for our networks. In Section 13.3 we describe how to compute the failure probabilities for each network. The crucial role here is played by so-called cumulative D-spectrum or signature, a well-developed technique widely used for calculating network reliability, see Gertsbakh and Shpungin 2011b, 2011c; Samaniego 2010. Section 13.4 presents an example of two networks with $n = 30$ and $m = 32$ nodes, respectively, which fail if their maximal connected component becomes smaller than some critical value L.

13.2 Failure Process and the Distribution of the Number of Failed Nodes

We consider two networks, A and B, having n and m nodes, respectively, $m \geq n$. Nodes of network A will be numbered $1_a, 2_a, \ldots, n_a$ etc, and the nodes of network B in a similar way: $1_b, 2_b, \ldots, m_b$.

In further exposition we will use the terminology of an epidemic process.

Initially, at time $t = 0$ node 1_a gets infected. It delivers the infection "forward" to a randomly chosen node in network B, suppose to the node 1_b. Then 1_b delivers the infection back to a randomly chosen node 2_a in A. This is the first round of the whole process. We assume that each round takes one time unit.

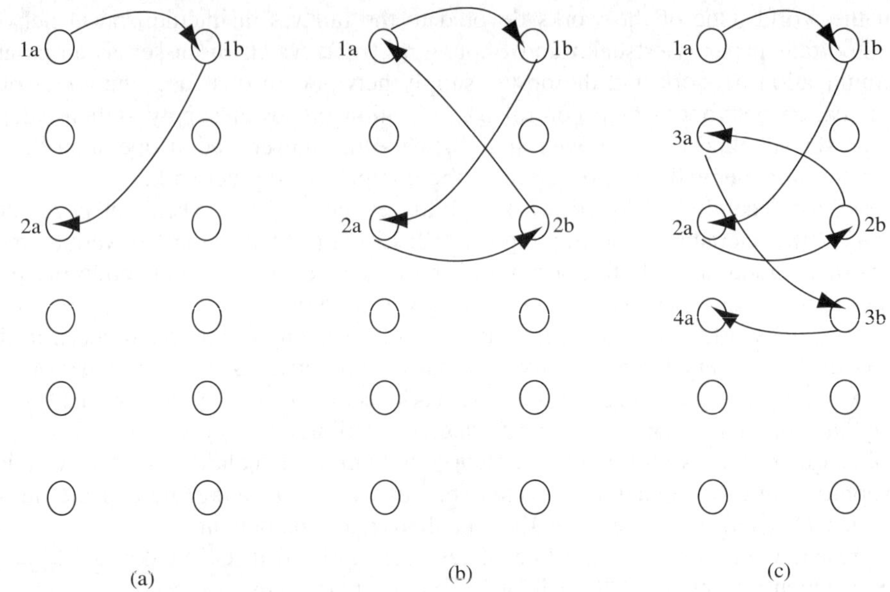

Figure 13.1 a – First round; b – second round; c – third round

The node 2_a may coincide with 1_a (with probability $1/n$). Then the process of infection delivery will stop, or equivalently, it will circulate between 1_a and 1_b without involving other nodes.

With probability $(1 - 1/n)$, node 1_b will deliver the infection to another, second node 2_a. Thus the first round ends with infecting one or two nodes in A. In the first case, the further infection spread stops, in the second, it continues. See Figure 13.1a.

The second round is the delivery of the infection from 2_a to some randomly chosen node in B, 2_b, and from 2_b to 3_a. Suppose that this round takes also one unit of time. It may happen that $2_b = 1_b$. Then the further spread of the infection will be stopped since it will begin to circulate between 1_a and 1_b. Another option is that the infection from 2_b will be delivered back to 1_a. In this case the infection spread stops with two infected nodes in A, since the further process continues to circulate $1_a \to 1_b \to 2_a \to 2_b \to 1_a \to \ldots$. Another possibility is that the randomly chosen node in A is **not** 1_a and **not** 2_a but some other node, 3_a. Thus the second round ends with infecting 2 or 3 nodes in A, see Figure 13.1b, bold lines.

Example 13.1. Let us consider the case of $n = m = 4$ and find out the probability of having the following trajectory

$$\tau = 1_a \to 1_b \to 2_a \to 2_b \to 3_a \to 3_b \to 4_a.$$

If this trajectory takes place, the total number of infected nodes in A will be 4, see Figure 13.1c.

Elementary calculations show that

$$P(\tau) = P(1_a \to 1_b \to 2_a) \cdot P(2_a \to 2_b \to 3_a) P(3_a \to 3_b \to 4_a)$$
$$= \frac{3}{4} \cdot \left(\frac{3}{4} \cdot \frac{2}{4}\right)\left(\frac{2}{4} \cdot \frac{1}{4}\right). \tag{13.1}$$

Indeed, the transition $1_a \to 2_a$ happens if and only if the backward infection from 1_b is delivered **not** to 1_a, i.e. with probability 3/4. The transition from 2_a to 3_a happens if and only if the forward infection from 2_a hits **not** $2_b = 1_b$ and gets back to A to any node other than 1_a and 2_a, which gives the probability (3/4)(2/4) = 6/16. Similarly, the transition from 3_a to 4_a takes place if and only if the forward infection from 3_a hits one of the two so far noninfected nodes in B and the backward infection hits the only remaining noninfected node in A. The corresponding probability is (2/4)(1/4) = 2/16.

Let us denote by X the total random number of infected nodes in A.

Obviously,

$$P(X = 1) = 1/4,\ P(X = 2) = (3/4) \cdot (10/16);\ P(X = 3) = (3/4) \cdot (6/16) \cdot (14/16).$$
$$P(X = 4) = (3/4) \cdot (6/16) \cdot (2/16).$$

and it is easy to verify that these probabilities sum up to one.

Lemma 13.1

(i) Suppose that after k rounds there are k infected nodes in A, $1_a, 2_a, \ldots, k_a$. Then the number of infected nodes in B is $k - 1$. Denote them $1_b, \ldots, (k-1)_b$.

(ii) Consider one round of delivering infection from node k_a to a node in B and backward to A.

The probability that the result will be $k + 1$ infected nodes in A, equals

$$\frac{m - (k-1)}{m} \cdot \frac{n-k}{n}.$$

Indeed, to create a new infected node in A it is necessary to hit one node outside $k - 1$ already infected in B and then return back to hit one of the noninfected nodes in A.

Claim 13.1-a. Suppose that the infection process lasts n rounds.

Then for $r = 2, \ldots, n$,

$$P(X = r) = \prod_{i=0}^{i=r-2} \left(\frac{m-i}{m} \cdot \frac{n-(i+1)}{n}\right)\left[1 - \frac{m-(r-1)}{m} \cdot \frac{n-r}{n}\right] \tag{13.2}$$

Proof. By Lemma 13.1, the first term of the product in (13.2) is the probability of the trajectory $1_a \to 1_b \to 2_a$. The next term of the product is the probability of trajectory $2_a \to 2_b \to 3_a$, and so on. The last term of the product is the probability of the trajectory $(r-1)_a \to (r-1)_b \to r_a$. Finally, the last term in (13.2) in square brackets is the probability of an event D that the infection from r_a does **not** spread further. The complement to

D means that the infection from r_a spreads further, that is, hits one **new** (noninfected) node in B and returns to a noninfected node in A, that is, $P(\overline{D}) = (m - (r - 1))(n - r)/nm$.

Finally, $P(X = 1) = 1/n$.

We present without proof an analogous claim about the distribution of Y, the random number of infected nodes in network B. (Note, that the infectionalways starts in a randomly chosen node of A.)

Claim 13.1-b. For $r = 3, 4, \ldots, n$,

$$P(Y = r) = \frac{n-1}{n} \cdot \prod_{i=1}^{i=r-2} \left(\frac{m-i}{m} \cdot \frac{n-i-1}{n} \right) \cdot \frac{m-r+1}{m} \cdot \left(\frac{r}{n} + \frac{(n-r) \cdot r}{n \cdot m} \right). \tag{13.3}$$

For $r = 2$,

$$P(Y = 2) = \frac{n-1}{n} \cdot \frac{m-r+1}{m} \cdot \left(\frac{r}{n} + \frac{(n-r) \cdot r}{n \cdot m} \right), \tag{13.4}$$

and

$$P(Y = 1) = \frac{1}{n} + \frac{n-1}{n} \cdot \frac{1}{m}. \tag{13.5}$$

13.3 Network Failure Probabilities

Note that network A has its own topology which *remains unchanged* during the infection spread from the initially infected node 1_a. After the infection process ends, all infected nodes in network A are declared to be failed, that is, all edges incident to the infected nodes are removed and the infected nodes become isolated, that is, all edges incident to the infected node are erased. This may cause network A failure. Our goal is to find the probability that the network A will fail as the result of its nodes' failure. Similarly, we are interesting in the failure of network B for the case that the infected nodes of B become isolated.

Denote by $F_A(r), r = 1, 2, \ldots, n$ the cumulative D-spectrum of network A. As we know, $F_A(r)$, by definition, is the probability that A fails if the number of randomly failed nodes equals r, see Appendix.

Claim 13.2. The probability G_A that the network A fails as the result of the infection process equals

$$G_A = E[F_A(X)] = \sum_{r=1}^{n} P(X = r) F_A(r), \tag{13.6}$$

where X is the number of the infected (removed) nodes in A.

The proof is obvious.

Similarly, the probability that network B fails equals

$$G_B = E[F_B(Y)] = \sum_{r=1}^{m} P(Y = r) F_B(r), \tag{13.7}$$

where $F_B(r)$ is the cumulative spectrum of network B.

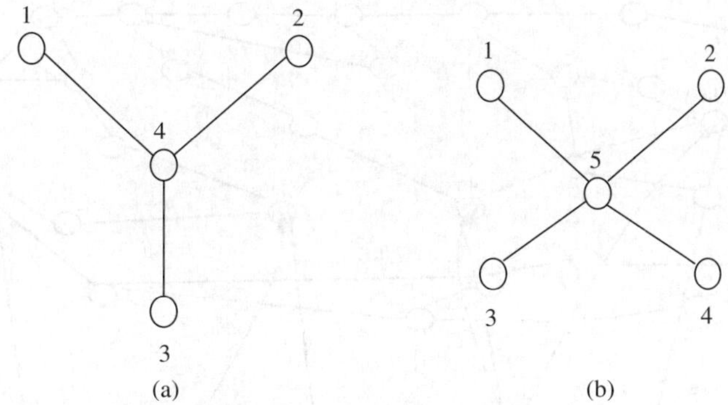

Figure 13.2 Star-type networks A (left) and B (right)

Example 13.2. two small star-type networks.

Suppose network A has $n = 4$ nodes and network B has $m = 5$ nodes, see Figure 13.2. The source of infection is one randomly chosen node in A.

Network failure, by definition, is isolation of all its nodes. So, for example, A fails if node 4 fails, or nodes 1,2,3 fail. It is easy to check that the spectrum (signature) of network A is as follows:

$$f_1 = f_2 = f_3 = f_4 = 0.25.$$

Therefore, the cumulative D-spectrum

$$F_A(1) = 0.25,\ F_A(2) = 0.5,\ F_A(3) = 0.75,\ F_A(4) = 1.$$

Similarly, for network B we find that

$$F_B(1) = 0.2,\ F_B(2) = 0.4,\ F_B(3) = 0.6,\ F_B(4) = 0.8,\ F_B(5) = 1.$$

Using formulas given in Claim 13.1-a and Claim 13.1-b, we find that

$$P(X = 1) = 0.25,\ P(X = 2) = 0.45,\ F(X = 3) = 0.255,\ F(X = 4) = 0.045,$$

and

$$P(Y = 1) = 0.4,\ P(Y = 2) = 0.42,\ P(Y = 3) = 0.162,\ P(Y = 4) = 0.018,$$

$P(Y = 5) = 0$. Finally, using formulas (13.6) and (13.7) we find out that the network A fails with probability $G_A = 0.52375$ and network B -with probability $G_B = 0.3596$.

Thus, the epidemic process originated in a single node of A causes with substantial probabilities the failure of both networks A and B.

13.4 Example

Let us consider a system of two interacting networks shown in Figure 13.3. Network A (the upper network) has $n = 30$ nodes and 40 edges. Network B has $m = 32$ nodes an 49 edges.

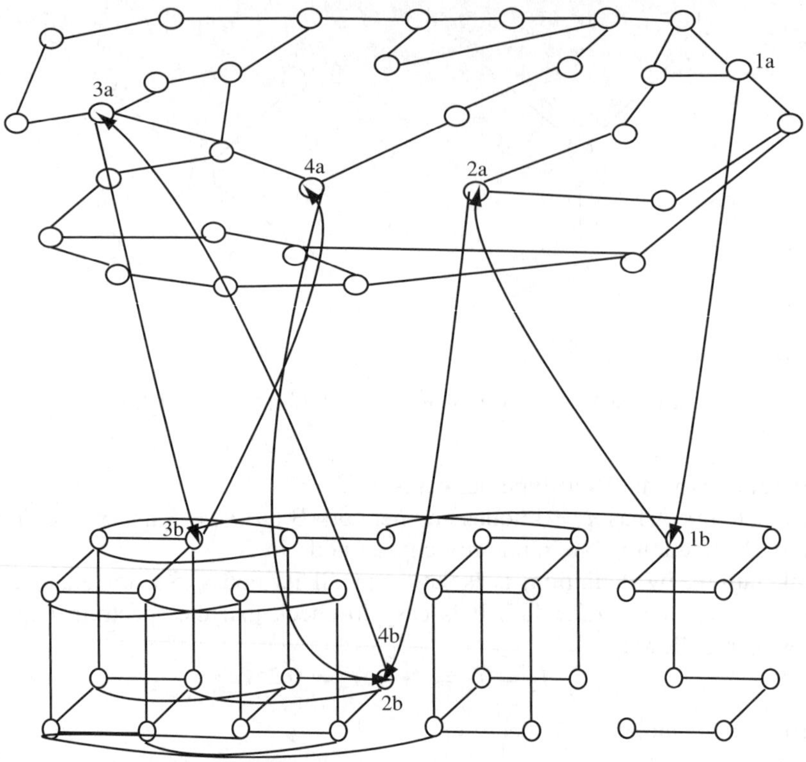

Figure 13.3 Networks A and B. Epidemic process starts at node 1_a and leads to failure of 4 nodes in A and 3 nodes in B

In both networks, the elements subject to failure are the nodes. Node failure means that all edges incident to this node are erased, and the failed node becomes isolated. By definition, network A fails if its largest connected component has $L \leq 20$ nodes.

In some sense it is an analogue of having a "giant component" which contains about not more than approximately two-thirds of the initial network. Failure of network B is defined similarly. The epidemic process starts at a randomly chosen node 1_a of A, goes to a random node of B, returns to A, and so on. Figure 13.3 shows the case when the epidemic contaminated 4 nodes in A and 3 nodes in B.

The cumulative D-spectra of both networks are given in Table 13.1. It is seen that with probability about 0.5, both networks fail when about 7 nodes fail. The details about network D-spectra (signatures) definition, calculations and references are given in the Appendix.

The next step is computing the probabilities $P(X = r)$ and $P(Y = s)$ that the epidemic process infects r and s nodes in networks A and B, respectively. The corresponding results obtained by using formulas given in Claims 13.1-a and 13.1-b are presented in Table 13.2.

Table 13.1 The cumulative node D-spectra of networks A and B

x, y	$F_A(x)$	$F_B(y)$
1	0	0
2	0.00231	0
3	0.01357	0.00802
4	0.05980	0.07092
5	0.16432	0.19832
6	0.32568	0.34607
7	0.52362	0.49860
8	0.72808	0.65359
9	0.89348	0.79708
10	0.97927	0.90741
11	1	0.97065
12	1	0.99533
13	1	1
–	1	1
–	1	1
30	1	1
31	–	1
32	–	1

Now it remains only to apply formulas (13.6) and (13.7) of Claim 13.3 to compute the failure probabilities for both networks. Calculations give the following results:

$$G_A = \mathbf{0.290}, \ G_B = \mathbf{0.239}.$$

One might say that these are quite substantial probabilities of network failure caused by a single-source epidemic. We can assume also that a two-node source epidemic processes developing independently on these networks would result in both network failure with probabilities close to 0.5.

13.5 Conclusion

In this chapter we concentrated on a two-network interaction model, in which the interaction was caused by a single epidemic source in one of the networks and the infection spread in a random forward-backward way. This simple model can be generalized in several main directions. One direction is introducing a different interaction model between two networks. For example, it might be a deterministic system of connections between the two networks in which a node i_a, $i = 1, 2, \ldots, k$, of network A is permanently connected to node j_b, $j = 1, 2, \ldots, k$ of network B. Another quite natural extension is considering a random mapping from A to B in which each infected node in one network delivers infection to several nodes in the other network. Another direction is considering a star-type structure of several networks in which one network represents the center and the rest its periphery. Each of the small peripheral networks contributes to damaging a relatively

Table 13.2 The probabilities $P(X = r)$ and $P(Y = s)$

r, s	$P(X = r)$	$P(Y = s)$
1	0.0333333	0.0635417
2	0.0926389	0.117057
3	0.136567	0.151077
4	0.158247	0.161512
5	0.15687	0.15046
6	0.137262	0.124723
7	0.1075	0.0928929
8	0.0758424	0.0624313
9	0.0483264	0.037914
10	0.0278195	0.0207951
11	0.0144499	0.0102826
12	0.00675607	0.0457079
13	0.00283389	0.00181975
14	0.00106195	0.0064591
15	0.0003537	0.000203289
16	0.00104095	0.0000563729
17	0.000026879	0.000013672
18	0.00000604	0.0000028751
19	0.00000117	0.000000519
20	0.0000002	0.00000079
21	0	0
22	0	0
–	0	0
–	0	0
30	0	0
31	–	0
32	–	0

small number of nodes in the center, but their overall impact may lead to fatal damage of the center. We plan to consider the earlier described models in our further research.

Appendix D: Spectrum (Signature)

In this section we introduce the principal topological invariant of our network, the so-called destruction spectrum (D-spectrum). We consider a binary network having binary components. Denote the network elements subject to failure (nodes or edges) as $e_1, e_2, \ldots, e_n$.

Definition 13.1. Let $\pi = \{e_{i_1}, e_{i_2}, e_{i_3}, \ldots, e_{i_n}\}$ be a permutation of network elements. Suppose initially that they all are *up*. Start turning them from *up* to *down* by moving along π from left to right. Fix the first element e_{i_r} when the network state becomes *DOWN*. The ordinal number of this element in the permutation is called the *anchor* of π and denoted $r(\pi)$.

Assume that all permutations π are equally probable and each particular permutation has probability of $1/n!$ Denote by f_i the probability that $r(\pi) = i$, $i = 1, \dots, n$.

Definition 13.2. The discrete density function $\{f_1, f_2, \dots, f_n\}$ is called network destruction spectrum (D-spectrum). ("D" stands for "destruction".)

Remark. It is important to stress that the D-spectrum is a purely topological characteristic of the network which depends only on its structure and network failure definition. The probabilities $\{f_i\}$ are induced by a uniform measure on the set of all $n!$ permutations. D-spectrum is completely separated from any information regarding the real stochastic mechanism which governs network element failure appearance (see Gertsbakh and Shpungin 2011a, 2011b).

Remark. The D-spectrum numerically coincides with the so-called *signature*, first discovered by Samaniego (1985). His definition of signature is the following (Samaniego 2010):

Assume that the lifetimes of system's n components are i.i.d. according to the continuous distribution $H(t)$. The signature s of the system is an n-dimensional probability vector whose i-th element i_s is equal to the probability that the i-th component failure causes the system to fail. In brief, $s_i = P(T) = X_{(i:n)}$, where T is the failure time of the system and $X_{(i:n)}$ is the i-th order statistic of the n component failure times.

The D-spectrum as system topological invariant was introduced six years later in the work (Elperin *et al.* 1991) under the name Internal Distribution (ID).

Definition 13.3. The cumulative distribution $F(r)$, $r = 1, 2, \dots, n$ of the anchor position is called *cumulative D-spectrum* or cumulative signature. Formally,

$$F(r) = \sum_{i=1}^{r} f_i.$$

Probabilistically, $F(r)$ is the probability that the network is *DOWN* if r of its *randomly chosen* components are *down*.

The role of cumulative D-spectrum in reliability calculations stems from the following important property. Let $C(r)$ be the number of system failure sets of size r, that is, the number of such failure sets which have r components *down*, and $n - r$ components *up*.

Then

$$C(r) = F(r) \cdot \frac{n!}{r!(n-r)!}.$$

References

Buldyrev, S. V., R. Parshani, G. Paul, H. E. Stanley and S. Havlin. 2010. Catastrophic cascade of failures in interdependent networks. *Nature* 464: 1025–1028.

Dickison, M., S. Havlin and H.E. Stanley. 2012. Epidemics on interconnected networks. *Physical Review* E 85, 066109.

Elperin, T., I.B. Gertsbakh and M. Lomonosov. 1991. Estimation of network reliability using graph evolution models. *IEEE Transactions on Reliability* 40 (5): 572–581.

Gao, J., S.V. Buldyrev, S. Havlin and H.E. Stanley. 2012. Robustness of a network formed by n interdependent networks with a one-to-one correspondence of dependent nodes. *Physical Review* E 85, 066134.

Gertsbakh, I.B. and Y. Shpungin. 2011a. Multidimensional spectra of multistate systems with binary components In *Recent advances in reliability: signatures, multi-state systems and statistical inference*, ed. Lisniansky, A. and I. Frenkel. Springer, pp. 49–61. London: Springer.

Gertsbakh, I.B. and Y. Shpungin. 2011b. *Network Reliability and Resilience*. Heidelberg: Springer.

Gertsbakh, I.B. and Y. Shpungin. 2011c. Stochastic models of network survivability. *Quality Technology and Quantitative Management* 9 (1): 45–58.

Gertsbakh, I.B. and Y. Shpungin. 2011d. Probabilistic description of network behavior under random attack on its nodes. In *Risk Analysis. Proceedings of the 4th International Conference on Risk Analysis, ICRA-4*, ed. A. Karagrigiriou. May 26–29, 2011. Limassol, Cyprus, pp. 105–112.

Gertsbakh, I.B. and Y. Shpungin. 2012. Failure development in a system of two connected networks. *Transport and Telecommunication* 13 (4): 255–260.

Gutfraint, A. 2011. Optimizing network topology for cascade resilience. In *Handbook of Optimization in Complex Networks*, ed. M.T. Thai and P.M. Pardalos. Springer, pp. 35–39. New York: Springer.

Li, W., A. Bashan, S. V. Buldyrev, H.E. Stanley and S. Havlin. 2012. Cascading failures in interdependent lattice networks: the critical role of the length of dependency links. *Physical Review Letters* 108 (22): 228702.

Newman, M.E.J. 2010. *Networks: An Introduction*. New York: Oxford University Press.

Samaniego, F.J. 1985. On closure of the IFR under formation of coherent systems. *IEEE Transactions on Reliability* R34 (1): 69–72.

Samaniego, F. J. 2010. *System Signatures and Their Applications in Engineering Reliability*. New York: Springer.

Zhou, Q., L. Gao, R. Liu and S. Cui. 2012. *Network Robustness Under Large-Scale Attacks*. New York: Springer.

Part Three

Maintenance Models

14

Comparisons of Periodic and Random Replacement Policies

Xufeng Zhao[1] and Toshio Nakagawa[2]
[1]*School of Economics and Management, Nanjing University of Technology, China*
[2]*Department of Business Administration, Aichi Institute of Technology, Japan*

14.1 Introduction

For a large and complex system which consists of many kinds of units, we should make minimal repairs at its failures and make planned replacement or preventive maintenance at periodic times. This is called periodic replacement or maintenance with minimal repairs at failures which was summarized in (Barlow and Proschan 1965; Nakagawa 2005; Tadj *et al.* 2011). However, when the system is successively executing jobs or computer procedures in offices and industries, it would be impossible or impractical to maintain or replace it in a strict periodic fashion, because sudden suspension of the jobs may create losses of production in different degrees if there is no sufficient preparation in advance (Nakagawa 2005).

By considering the factors of working times in operations, the reliability quantities of the random age replacement policy were obtained (Barlow and Proschan 1965). Several schedules of jobs that have random processing times were summarized (Pinedo 2012). The properties of replacement policies between two successive failed units, where the unit is replaced at random times, were investigated (Stadje 2003). Under the assumptions of random failure and maintenance, replacement and inspection with planned and random policies, were considered, and their comparisons were made (Nakagawa *et al.* 2011). Combining planned replacement with working times, the age and periodic replacement policies, where the unit is replaced at a planned time T and at the Nth random working

Applied Reliability Engineering and Risk Analysis: Probabilistic Models and Statistical Inference, First Edition. Edited by Ilia B. Frenkel, Alex Karagrigoriou, Anatoly Lisnianski and Andre Kleyner.

time, were discussed (Chen *et al.* 2010a, 2010b). Furthermore, such a notion of "random maintenance" was applied to a parallel system with random number of units (Nakagawa and Zhao 2012).

It has been assumed in all policies that units are maintained or replaced before failure at some amount of quantities, for example, age, periodic time, usage number, damage level, etc., whichever occurs first. These policies would be reasonable if failures are serious and sometimes may incur heavy production losses. By considering the cases when replacement costs suffered for failures would be estimated to be not so high and the factor of working time, age and periodic replacement policies, where the unit is replaced at a planned operating time T or at a random working cycle Y, whichever occurs last, were proposed (Zhao and Nakagawa 2012a), it also indicated that policies proposed in (Chen *et al.* 2010a, 2010b) would cause frequent and unnecessary replacement which may incur production losses under the assumption of "whichever occurs first". Such a notion of "whichever occurs last" was applied in a cumulative damage model, where the unit is replaced before failure at a planned time T and at a damage level Z, and an optimal Z for a given T was discussed (Zhao *et al.* 2011).

On the other hand, when a job has a variable working cycle or processing time, it would be better to do maintenance or replacement after the job is just completed even though the maintenance time has arrived (Sugiura *et al.* 2004). From such a viewpoint, by considering both planned time and random working time, the age and periodic replacement policies which are done at the first completion of some working times over a planned time T were proposed (Chen *et al.* 2010a, 2010b), and an optimal age replacement policy was derived in (Zhao and Nakagawa 2012b). A representative practical example for such a policy is to maintain a database or to perform a backup of data when a transaction is processing its sequences of operations, because it is necessary to guarantee ACID (atomicity, consistency, isolation, and durability) properties of database transactions, especially for a distributed transaction across a distributed database, that is, if any part of transaction fails, the entire transaction fails and the database state is left unchanged (Haerder and Reuter 1983; Lewis *et al.* 2002).

From the above considerations, we give the following assumptions and summarize four minimal repair policies: It is assumed that an operating unit works for a job with random working cycles, and undergoes minimal repair at each failure between replacements. As preventive replacement, (1) the unit is replaced at periodic times kT (Nakagawa 2005), which is called standard replacement; (2) the unit is replaced at a total operating time T or at a random working cycle Y, whichever occurs first, which is called replacement first; (3) the unit is replaced at a total operating time T or at a random working cycle Y, whichever occurs last, which is called replacement last; (4) the unit is replaced at the first completion of some working cycles over time T, which is called replacement over time. Policies (2) and (3) are the extended ones of standard replacement and Policy (4) is a modified standard replacement. Among policies (2), (3) and (4), we also define that replacement done at Y is a kind of random replacement, that is, replacement first, last, and over time are combined policies with periodic and random replacement.

We first of all summarize optimal replacement policies for the above four replacement policies in the following sections. Second, we compare each policy with one another when the working cycle is exponential and the periodic and random replacement costs are the same. It is shown theoretically that standard replacement is the best one of the four, as estimated previously. In those cases, we can determine replacement over time is better than replacement first and last. We also discuss whether either replacement first or last is better than the other according to the ratio of replacement cost to minimal repair cost. Third, under the assumption of replacement, costs after the completion of working cycles might be lower than those at periodic times, we discuss theoretically and numerically that if by how much the cost of random replacement is lower than that of periodic one, if replacement first, last and over time are the same as the standard replacement.

14.2 Four Policies

14.2.1 Standard Replacement

A new unit begins to operate for a job at time 0, and it undergoes minimal repair when it fails. Suppose that the unit is replaced at periodic times kT $(k = 1, 2, \dots)$, independently of its age, and any unit becomes as good as new after replacement. The repair and replacement times are negligible. It is assumed that the failure time X of the unit has a density function $f(t)$ and a general distribution $F(t) \equiv \Pr\{X \le t\}$ with finite mean. Furthermore, the unit has the failure rate $h(t) \equiv f(t)/\overline{F}(t)$ and the cumulative hazard rate $H(t) \equiv \int_0^t h(u)\mathrm{d}u$, where $\overline{\Phi}(t) \equiv 1 - \Phi(t)$ for any function $\Phi(t)$, and $h(t)$ is assumed to increase strictly.

The expected cost rate is (Barlow and Proschan 1965; Nakagawa 2005)

$$C_S(T) = \frac{c_M H(T) + c_T}{T}, \tag{14.1}$$

where c_M = minimal repair cost at each failure and c_T = replacement cost at time T. If $\int_0^\infty t\mathrm{d}h(t) > c_T/c_M$, then an optimal T_S^* $(0 < T_S^* < \infty)$ which minimizes $C_S(T)$ is given by a finite and unique solution of the equation

$$\int_0^T [h(T) - h(t)]\mathrm{d}t = \frac{c_T}{c_M}, \tag{14.2}$$

and the resulting cost rate is

$$C_S(T_S^*) = c_M h(T_S^*). \tag{14.3}$$

14.2.2 Replacement First

The unit works for a job with a random cycle Y, which has a general distribution $G(t) \equiv \Pr\{Y \le t\}$ with a finite mean $1/\theta$ and is independent of the failure time X. Suppose that the unit is replaced at a total operating time T or at a random working cycle Y, whichever occurs first. Then, the expected cost rate is (Zhao and Nakagawa 2012a)

$$C_F(T) = \frac{c_M \int_0^T \overline{G}(t)h(t)\mathrm{d}t + c_R}{\int_0^T \overline{G}(t)\mathrm{d}t}, \tag{14.4}$$

where c_R = replacement cost at time T or Y and c_M is given in (14.1). If $\int_0^\infty \overline{G}(t)[h(\infty) - h(t)]\mathrm{d}t < \infty$, then an optimal T_F^* $(0 < T_F^* < \infty)$ which minimizes $C_F(T)$ is given by a finite and unique solution of the equation

$$\int_0^T \overline{G}(t)[h(T) - h(t)]\mathrm{d}t = \frac{c_R}{c_M}, \tag{14.5}$$

and the resulting cost rate is

$$C_F(T_F^*) = c_M h(T_F^*). \tag{14.6}$$

14.2.3 Replacement Last

Underlying the same conditions as 14.2.2, suppose that the unit is replaced at a total operating time T or at a random working cycle Y, whichever occurs last. Then, the expected cost rate is (Zhao and Nakagawa 2012a)

$$C_L(T)\frac{c_M[H(T) + \int_T^\infty \overline{G}(t)\mathrm{d}t] + c_R}{T + \int_T^\infty \overline{G}(t)\mathrm{d}t}, \tag{14.7}$$

where c_M and c_R are given in (14.4). If $\int_0^\infty t\mathrm{d}h(t) > c_R/c_M$, then an optimal T_L^* $(0 < T_L^* < \infty)$ which minimizes $C_L(T)$ is given by a finite and unique solution of the equation

$$\int_0^T [h(T) - h(t)]\mathrm{d}t - \int_T^\infty \overline{G}(t)[h(t) - h(T)]\mathrm{d}t = \frac{c_R}{c_M}, \tag{14.8}$$

and the resulting cost rate is

$$C_L(T_L^*) = c_M h(T_L^*). \tag{14.9}$$

14.2.4 Replacement Over Time

The unit works for a job with successive random cycles Y_j, each of which has an identical distribution $G(t) \equiv \Pr\{Y_j \le t\}$ with a finite mean $1/\theta$. It is assumed that $G^{(j)}(t)$ $(j = 1, 2, \cdots)$ denotes the $j-$ fold Stieljes convolution of $G(t)$ with itself and $G^{(0)}(t) \equiv 1$ for $t \ge 0$. Suppose that the unit is replaced at the first completion of Y_j $(j = 1, 2, \cdots)$ over time T. Then, the expected cost rate is (Chen *et al.* 2010b)

$$C_O(T) = \frac{c_M\{H(T) + \int_T^\infty \overline{G}(t)h(t)\mathrm{d}t + \int_0^T \left[\int_T^\infty \overline{G}(u-t)h(u)\mathrm{d}u\right]\mathrm{d}M(t)\} + c_R}{T + \int_T^\infty \overline{G}(t)\mathrm{d}t + \int_0^T \left[\int_T^\infty \overline{G}(u-t)\mathrm{d}u\right]\mathrm{d}M(t)}, \tag{14.10}$$

where $M(t) \equiv \sum_{j=1}^\infty G^{(j)}(t)$, and c_M and c_R are given in (14.4). An optimal T_O^* which minimizes $C_O(T)$ satisfies

$$\int_0^\infty \theta\overline{G}(t)\Bigg(Th(T+t) - H(T) + \int_T^\infty \overline{G}(u)[h(T+t) - h(u)]\mathrm{d}u$$
$$+ \int_0^T \left\{\int_T^\infty \overline{G}(u-x)[h(T+t) - h(u)]\mathrm{d}u\right\}\mathrm{d}M(x)\Bigg)\mathrm{d}t = \frac{c_R}{c_M}, \tag{14.11}$$

whose largest bracket is proved to increase strictly with T for any $t > 0$. Therefore, if

$$\lim_{T\to\infty} \int_0^\infty \theta \overline{G}(t)[Th(T+t) - H(T)]\mathrm{d}t > \frac{c_R}{c_M},$$

then there exists a finite and unique T_O^* $(0 < T_O^* < \infty)$ which satisfies (14.11), and the resulting cost rate is

$$C_O(T_O^*) = c_M \int_0^\infty \theta \overline{G}(t) h(T_O^* + t)\mathrm{d}t. \tag{14.12}$$

14.3 Comparisons of Optimal Policies

Note that if the failure rate $h(t)$ increases strictly to ∞, then all optimal and finite policies of the above four replacements exist uniquely. When $c_T = c_R$ and $G(t) = 1 - \mathrm{e}^{-\theta t}$ $(0 < \theta < \infty)$, we compare four optimal policies as follows:

14.3.1 Comparisons of T_S^* and T_F^*, T_L^*, T_O^*

From (14.5), an optimal T_F^* satisfies,

$$\int_0^T \mathrm{e}^{-\theta t}[h(T) - h(t)]\mathrm{d}t = \frac{c_T}{c_M}, \tag{14.13}$$

and increases with θ from T_S^* to ∞, i.e., $T_F^* > T_S^*$. Thus, from (14.3) and (14.6), $C_S(T_S^*) < C_F(T_F^*)$. In this case, standard replacement is better than replacement first.

Similarly, from (14.8), an optimal T_L^* satisfies,

$$\int_0^T [h(T) - h(t)]\mathrm{d}t - \int_T^\infty \mathrm{e}^{-\theta t}[h(t) - h(T)]\mathrm{d}t = \frac{c_T}{c_M}, \tag{14.14}$$

and decreases with θ to T_S^*, i.e., $T_L^* > T_S^*$. Thus, from (14.3) and (14.9), $C_S(T_S^*) < C_L(T_L^*)$. In this case, standard replacement is better than replacement last.

The expected cost rate of the replacement over time is, from (14.10),

$$C_O(T) = \frac{c_M[H(T) + \int_0^\infty \mathrm{e}^{-\theta t} h(t+T)\mathrm{d}t] + c_T}{T + 1/\theta}. \tag{14.15}$$

From (14.11), an optimal T_O^* satisfies

$$T\int_0^\infty \theta \mathrm{e}^{-\theta t} h(t+T)\mathrm{d}t - H(T) = \frac{c_T}{c_M}, \tag{14.16}$$

and the resulting cost rate is

$$C_O(T_O^*) = c_M \int_0^\infty \theta \mathrm{e}^{-\theta t} h(t + T_O^*)\mathrm{d}t = \frac{c_M H(T_O^*) + c_T}{T_O^*}. \tag{14.17}$$

Clearly, for $0 < T < \infty$,

$$T\int_0^\infty \theta \mathrm{e}^{-\theta t} h(t+T)\mathrm{d}t - H(T) > Th(T) - H(T),$$

which follows that $T_O^* < T_S^*$. In addition, because T_S^* minimizes the expected cost rate $C_S(T)$ in (14.1), from (14.17), $C_O(T_O^*) > C_S(T_S^*)$. Thus, standard replacement is better than replacement over time.

In three cases, when the replacement costs of four policies are the same, i.e., $c_T = c_R$, standard replacement is better than the other ones.

14.3.2 Comparisons of T_O^* and T_F^*, T_L^*

Compare T_O^* and T_F^* which satisfy (14.16) and (14.13), respectively. From (14.16) and (14.13),

$$T\int_0^\infty \theta \mathrm{e}^{-\theta t}h(t+T)\mathrm{d}t - H(T) - \int_0^T \mathrm{e}^{-\theta t}[h(T)-h(t)]\mathrm{d}t$$

$$= T\int_0^\infty \theta \mathrm{e}^{-\theta t}[h(t+T)-h(T)]\mathrm{d}t + \int_0^T (1-\mathrm{e}^{-\theta t})[h(T)-h(t)]\mathrm{d}t > 0,$$

and hence, $T_O^* < T_F^*$. Thus, we compare $C_O(T_O^*)$ in (14.17) and $C_F(T_F^*)$ in (14.6) and determine which policy is better.

For example, when $H(t) = \lambda t^2$, i.e., $h(t) = 2\lambda t$, from (14.17),

$$C_O(T_O^*) = 2c_M\lambda\left(T_O^* + \frac{1}{\theta}\right), \tag{14.18}$$

and from (14.6),

$$C_F(T_F^*) = 2c_M\lambda T_F^*. \tag{14.19}$$

Thus, if $T_O^* + 1/\theta < T_F^*$, then replacement over time is better than replacement first, and vice versa.

Compare T_O^* and T_L^* which satisfy (14.16) and (14.14), respectively. From (14.16) and (14.14),

$$T\int_0^\infty \theta \mathrm{e}^{-\theta t}h(t+T)\mathrm{d}t - H(T) - \int_0^T [h(T)-h(t)]\mathrm{d}t + \int_T^\infty \theta \mathrm{e}^{-\theta t}[h(t)-h(T)]\mathrm{d}t$$

$$= T\int_0^\infty \theta \mathrm{e}^{-\theta t}[h(t+T)-h(T)]\mathrm{d}t + \int_T^\infty \theta \mathrm{e}^{-\theta t}[h(t)-h(T)]\mathrm{d}t > 0,$$

and hence, $T_O^* < T_L^*$. Thus, we compare $C_O(T_O^*)$ in (14.17) and $C_L(T_L^*)$ in (14.9) and determine which policy is better.

For example, when $h(t) = 2\lambda t$, from (14.9),

$$C_L(T_L^*) = 2c_M\lambda T_L^*. \tag{14.20}$$

Thus, if $T_O^* + 1/\theta < T_L^*$, then replacement over time is better than replacement last.

In this case of $h(t) = 2\lambda t$, when $T_O^* + 1/\theta < T_i^*$ $(i = F, L)$, replacement over time is better than both replacement first and last.

14.3.3 *Comparisons of T_F^* and T_L^**

From (14.13) and (14.14),

$$Q(T) \equiv \int_0^T [h(T) - h(t)]\mathrm{d}t - \int_T^\infty \mathrm{e}^{-\theta t}[h(t) - h(T)]\mathrm{d}t - \int_0^T \mathrm{e}^{-\theta t}[h(T) - h(t)]\mathrm{d}t$$

$$= \int_0^T (1 - \mathrm{e}^{-\theta t})[h(T) - h(t)]\mathrm{d}t - \int_T^\infty \mathrm{e}^{-\theta t}[h(t) - h(T)]\mathrm{d}t, \tag{14.21}$$

which increases strictly from $-\int_0^\infty \mathrm{e}^{-\theta t} h(t)\mathrm{d}t$ to ∞. Thus, there exists a finite and unique T_P $(0 < T_P < \infty)$ which satisfies $Q(T) = 0$ and T_P decreases with θ from ∞ to 0.

Denote that

$$L(T_P) \equiv \int_0^{T_P} \mathrm{e}^{-\theta t}[h(T_P) - h(t)]\mathrm{d}t, \tag{14.22}$$

which decreases with θ from ∞ to 0. Then, from (14.6) and (14.9), if $L(T_P) \geq c_T/c_M$ then $T_F^* \leq T_L^*$, and hence, replacement first is better than replacement last. Conversely, if $L(T_P) < c_T/c_M$ then $T_F^* > T_L^*$, and hence, replacement last is better than replacement first. In other words, the units should be replaced earlier as the replacement cost and θ are smaller.

14.4 Numerical Examples 1

Suppose that $c_T = c_R$, the failure time X has a Weibull distribution $F(t) = 1 - \mathrm{e}^{-t^2}$ and a random working cycle Y has an exponential distribution $G(t) = 1 - \mathrm{e}^{-\theta t}$. Tables 14.1 and 14.2 present optimal T_F^*, T_L^*, and T_S^*, which satisfy (14.13), (14.14) and (14.2), respectively, and their cost rates for $1/\theta$ and c_T/c_M. Clearly, T_F^* and T_L^* become T_S^* when $1/\theta = \infty$ and $1/\theta = 0$, respectively. In addition, Table 14.1 gives T_P which satisfies $Q(T) = 0$ in (14.21) and $L(T_P)$ in (14.22).

Tables 14.1 and 14.2 indicate as follows:

(a) Both T_F^* and T_L^* increase with c_T/c_M. When c_T/c_M increases, the replacement time should be longer to lessen a high replacement cost, and replacement last is much better than replacement first, especially for small $1/\theta$. For example, when $1/\theta = 0.5$ and $c_T = 2c_M$, $T_L^* = 1.424$ is much less than $T_F^* = 2.497$.

(b) When $L(T_P) \geq c_T/c_M$, $T_F^* \leq T_L^*$ and replacement first is better than replacement last, and conversely, when $L(T_P) < c_T/c_M$, $T_F^* > T_L^*$ and replacement last is better than replacement first. For example, when $1/\theta = 0.5$, $L(T_P) = 0.286$, and hence, $T_F^* = 0.525 < T_L^* = 0.593$ for $c_T/c_M = 0.20$, and $T_L^* = 0.778 < T_F^* = 0.921$ for $c_T/c_M = 0.50$.

(c) Optimal T_F^* decreases to T_S^* with $1/\theta$ and T_L^* increases from T_S^* with $1/\theta$, because the unit is replaced at $\min\{T_F^*, Y\}$ for replacement first, and at $\max\{T_L^*, Y\}$ for replacement last. Furthermore, replacement first is better than replacement last as $1/\theta$ becomes larger. For example, when $c_T/c_M = 0.10$, if $1/\theta = 0.1$, then replacement last is better than replacement first, and if $1/\theta \geq 0.5$, then replacement first is better than replacement last. When $c_T/c_M = 0.50$ and $1/\theta = 1.0$, $T_F^* = 0.801 < 1/\theta = 1.0 < T_L^* = 1.084$, and optimal replacement times are equal nearly to $1/\theta$. Table 14.4 also

Table 14.1 Optimal T_F^* and T_S^* and their cost rates

c_T/c_M	$1/\theta = 0.1$		$1/\theta = 0.5$		$1/\theta = 1.0$		$1/\theta = \infty$	
	T_F^*	$C_F(T_F^*)/c_M$	T_F^*	$C_F(T_F^*)/c_M$	T_F^*	$C_F(T_F^*)/c_M$	T_S^*	$C_S(T_S^*)/c_M$
0.01	0.120	0.240	0.103	0.206	0.102	0.204	0.100	0.200
0.02	0.184	0.368	0.148	0.296	0.145	0.290	0.141	0.282
0.05	0.347	0.694	0.242	0.484	0.233	0.466	0.224	0.448
0.10	0.600	1.200	0.353	0.706	0.334	0.668	0.316	0.632
0.20	1.100	2.200	0.525	1.050	0.483	0.966	0.447	0.894
0.50	2.600	5.200	0.921	1.842	0.801	1.602	0.707	1.414
1.00	5.100	10.200	1.474	2.948	1.198	2.396	1.000	2.000
2.00	10.100	20.200	2.497	4.994	1.841	3.682	1.414	2.828
5.00	25.100	50.200	5.500	11.000	3.469	6.938	2.236	4.472
T_P	0.130		0.650		1.230			
$L(T_P)$	0.011		0.286		1.045			

Table 14.2 Optimal T_L^* and their cost rates

c_T/c_M	$1/\theta = 0.1$		$1/\theta = 0.5$		$1/\theta = 1.0$		$1/\hat{\theta}$
	T_L^*	$C_L(T_L^*)/c_M$	T_L^*	$C_L(T_L^*)/c_M$	T_L^*	$C_L(T_L^*)/c_M$	
0.01	0.125	0.250	0.458	0.916	0.905	1.810	0.093
0.02	0.156	0.312	0.466	0.932	0.909	1.818	0.132
0.05	0.228	0.456	0.488	0.976	0.920	1.840	0.209
0.10	0.318	0.636	0.525	1.050	0.939	1.878	0.296
0.20	0.447	0.894	0.593	1.186	0.976	1.952	0.418
0.50	0.707	1.414	0.778	1.556	1.084	2.168	0.661
1.00	1.000	2.000	1.031	2.062	1.253	2.506	0.935
2.00	1.414	2.828	1.424	2.848	1.556	3.112	1.322
5.00	2.236	4.472	2.237	4.474	2.281	4.562	2.090

presents $1/\hat{\theta}$ such that two expected costs of replacement first and last are the same: From $Q(T_P) = 0$ in (14.21), θT_P is given by

$$\frac{(\theta T_P)^2}{2} - \theta T_P + 1 = 2e^{-\theta T_P}.$$

Using θT_P and $L(T_P)$ in (14.22), $\hat{\theta}$ is computed for c_T/c_M by

$$2\left(\frac{1}{\hat{\theta}}\right)^2 \left[\frac{(\theta T_P)^2}{2} - e^{-\theta T_P}\right] = \frac{c_T}{c_M}.$$

These values of $1/\hat{\theta}$ increases with c_T/c_M. This indicates that when $c_T/c_M = 0.10$, if $1/\theta <= 0.296$, then replacement last is better than replacement first, and vice versa.

14.5 Comparisons of Policies with Different Replacement Costs

It has been shown that when $c_T = c_R$, the standard replacement is better than the other ones. In general, the random replacement cost would be lower than the periodic one because the job is not suspended by replacement actions. We compute the modified random replacement cost when the expected cost rates of four replacement policies are the same.

14.5.1 Comparisons of T_S^* and T_F^*, T_L^*

First, compute T_S^* which satisfy (14.2). Next, compute c_{RF} and c_{RL}, which satisfy, from (14.5) and (14.8), respectively.

$$\int_0^{T_S^*} \overline{G}(t)[h(T_S^*) - h(t)]\mathrm{d}t = \frac{c_{RF}}{c_M},$$

$$\int_{T_S^*}^{\infty} \overline{G}(t)[h(t) - h(T_S^*)]\mathrm{d}t = \frac{c_T - c_{RL}}{c_M}.$$

For example, when $h(t) = 2t$ and $G(t) = 1 - \mathrm{e}^{-\theta t}$,

$$T_S^* = \sqrt{\frac{c_T}{c_M}}, \tag{14.23}$$

and c_{RF} and c_{RL} are respective solution of the equations:

$$\frac{2}{\theta^2}[\theta T_S^* - (1 - \mathrm{e}^{-\theta T_S^*})] = \frac{c_{RF}}{c_M}, \tag{14.24}$$

$$\frac{2}{\theta^2}\left[\frac{(\theta T_S^*)^2}{2} - \mathrm{e}^{-\theta T_S^*}\right] = \frac{c_{RL}}{c_M}. \tag{14.25}$$

Clearly, c_{RF} is positive, however, c_{RL} might be negative in which cases, replacement last is never better than standard replacement.

14.5.2 Comparisons of T_S^* and T_O^*

Compute T_O^* which satisfies, from (14.11),

$$T\int_0^{\infty} \theta \mathrm{e}^{-\theta t} h(t + T)\mathrm{d}t - H(T) = \frac{c_R}{c_M}.$$

Thus, from (14.3) and (14.17), if

$$h(T_S^*) > \int_0^{\infty} \theta \mathrm{e}^{-\theta t} h(t + T_O^*)\mathrm{d}t \tag{14.26}$$

or

$$\frac{c_M H(T_S^*) + c_T}{T_S^*} > \frac{c_M H(T_O^*) + c_R}{T_O^*},$$

then replacement over time is better than standard replacement. For example, when $h(t) = 2t$,

$$T_S^* = \sqrt{\frac{c_T}{c_M}}, \quad T_O^* = -\frac{1}{\theta} + \sqrt{\frac{1}{\theta^2} + \frac{c_R}{c_M}}.$$

Thus, if

$$T_S^* > T_O^* + \frac{1}{\theta} \quad \text{or} \quad \frac{c_T}{c_M} > \frac{1}{\theta^2} + \frac{c_R}{c_M},$$

then replacement overtime is better than standard replacement. It can be clearly seen that when $c_R/c_M = c_T/c_M - 1/\theta^2$, the two replacements are the same.

14.6 Numerical Examples 2

We show the same numerical examples as those in Section 14.4 when $c_T = c_R$, $F(t) = 1 - \mathrm{e}^{-t^2}$ and $G(t) = 1 - \mathrm{e}^{-\theta t}$. Table 14.3 presents optimal T_O^* and their cost rates $C_O(T_O^*)/c_M$ for $1/\theta$ and c_T/c_M. Optimal T_O^* increases with both c_T/c_M and θ. Compared to T_S^* in Table 14.4, $T_O^* + 1/\theta > T_S^*$, however, $T_O^* + 1/\theta$ approaches to T_S^* as c_T/c_M and θ increase. For example, when $c_T/c_M = 5.00$ and $1/\theta = 0.1$, $T_O^* + 1/\theta = 2.238 > T_S^* = 2.236$.

Compared to T_F^* in Table 14.1, replacement over time becomes better than replacement first as both c_T/c_M and θ increase. On the other hand, compared to T_L^*, $C_L(T_L^*) < C_O(T_O^*)$, however, $T_O^* + 1/\theta$ are a little greater than T_L^*.

When $c_T > c_R$, Table 14.4 presents c_{RF} and c_{RL} given in (14.24) and (14.25), respectively. Values of c_{RF} exist for all $1/\theta$ and c_T/c_M, however, the differences between c_{RF} and c_T become much smaller as c_T/c_M and θ are smaller. That is, if θ and c_T/c_M are small, standard replacement and replacement first become almost the same policy because the unit is replaced mainly at time T. Values of c_{RL} exist for large c_T/c_M and θ. In other words, replacement last cannot be rather than standard replacement when c_T/c_M and θ are small. Furthermore, if c_T/c_M and θ are large, c_{RL} becomes equal to c_T, i.e., the two replacements are almost the same, because the unit is replaced mainly

Table 14.3 Optimal T_O^* and their cost rates

c_T/c_M	$1/\theta = 0.1$		$1/\theta = 0.5$		$1/\theta = 1.0$	
	T_O^*	$C_O(T_O^*)/c_M$	T_O^*	$C_O(T_O^*)/c_M$	T_O^*	$C_O(T_O^*)/c_M$
0.01	0.041	0.283	0.010	1.020	0.005	2.010
0.02	0.073	0.346	0.020	1.039	0.010	2.020
0.05	0.145	0.490	0.048	1.095	0.025	2.049
0.10	0.232	0.663	0.092	1.183	0.049	2.098
0.20	0.358	0.917	0.171	1.342	0.095	2.191
0.50	0.614	1.428	0.366	1.732	0.225	2.450
1.00	0.905	2.010	0.618	2.236	0.414	2.828
2.00	1.318	2.835	1.000	3.000	0.732	3.464
5.00	2.138	4.477	1.791	4.583	1.450	4.899

Table 14.4 Values of c_{RF}/c_M and c_{RL}/c_M

c_T/c_M	$1/\theta = 0.1$		$1/\theta = 0.5$		$1/\theta = 1.0$	
	c_{RF}/c_M	c_{RL}/c_M	c_{RF}/c_M	c_{RL}/c_M	c_{RF}/c_M	c_{RL}/c_M
0.01	0.007	0.003	0.009	–	0.010	–
0.02	0.013	0.015	0.018	–	0.019	–
0.05	0.027	0.048	0.043	–	0.046	–
0.10	0.044	0.099	0.082	–	0.090	–
0.20	0.070	0.200	0.152	–	0.173	–
0.50	0.121	0.500	0.329	0.378	0.400	–
1.00	0.180	1.000	0.568	0.932	0.736	0.264
2.00	0.263	2.000	0.944	1.970	1.315	1.541
5.00	0.427	5.000	1.742	4.944	2.686	4.786

at time T. For example, when $1/\theta = 1.0$ and $c_T/c_M = 0.5$, if $c_{RL}/c_M = 0.378$, i.e., $c_{RL}/c_T \times 100\% = 18.9\%$, the two replacements are almost the same.

14.7 Conclusion

We have compared the standard periodic replacement policy and three modified random replacement policies with minimal repair at failures. It has been clearly shown as estimated previously that when the replacement cost for each policy is the same, standard replacement is the best among them. It has been proved in what cases replacement over time is better than replacement first and last. It is of great interest that replacement last is better than replacement first when the ratio of replacement cost to minimal repair is less than some value. If the replacement cost were lower, replacement last should be adopted in practical fields. We have discussed numerically if by how much the random replacement cost is lower than the periodic one, then the replacement first, last, and over time are better than the periodic one. From such results, the random replacement should be used more in practical fields from an economical and environmental viewpoint.

Finally, as extended replacement policies, we suppose that the unit is replaced before failure at time T or at the Nth working time ($N = 1, 2, \cdots$), whichever occurs first. Then, the expected cost rate is

$$C_F(T, N) = \frac{c_M \int_0^T [1 - G^{(N)}(t)]h(t)\mathrm{d}t + c_R}{\int_0^T [1 - G^{(N)}(t)]\mathrm{d}t}. \tag{14.27}$$

Clearly, when $N = \infty$, $C_F(T, \infty) = C_S(T)$ in (14.1), and when $N = 1$, $C_F(T, 1) = C_F(T)$ in (14.4).

Next, suppose that the unit is replaced before failure at time T or at the Nth working time ($N = 0, 1, 2, \cdots$), whichever occurs last. Then, the expected cost rate is

$$C_L(T, N) = \frac{c_M\{H(T) + \int_T^\infty [1 - G^{(N)}(t)]h(t)\mathrm{d}t\} + c_R}{T + \int_T^\infty [1 - G^{(N)}(t)]\mathrm{d}t}. \tag{14.28}$$

Clearly, when $N = 0$, $C_F(T, 0) = C_S(T)$ in (14.1), and when $N = 1$, $C_L(T, 1) = C_L(T)$ in (14.7).

It would be interesting, using methods similar to those proposed in this chapter, to further study and compare both analytically and numerically optimal policies which minimize the expected cost rate $C_F(T, N)$ in (14.27) and $C_L(T, N)$ in (14.28).

Acknowledgements

This work is partially supported by National Natural Science Foundation of China 71171110; Natural Science Foundation of Jiangsu Province of China BK2010555; Grant-in-Aid for Scientific Research (C) of Japan Society for the Promotion of Science under Grant No. 22500897 and No. 24530371.

References

Barlow, R.E. and F. Proschan. 1965. *Mathematical Theory of Reliability*. New York: Wiley.

Chen, M., S. Mizutani and T. Nakagawa. 2010a. Random and age replacement policies. *International Journal of Reliability, Quality and Safety Engineering* 17 (1): 27–39.

Chen, M., S. Nakamura and T. Nakagawa. 2010b. Replacement and preventive maintenance models with random working times. *IEICE Transactions on Fundamentals of Electronics, Communications and Computer Sciences* E93-A: 500–507.

Haerder, T. and A. Reuter. 1983. Principles of transaction-oriented database recovery. *ACM Computing Surveys* 15 (4): 287–317.

Lewis, P.M., A.J. Bernstein and M. Kifer. 2002. *Databases and Transaction Processing: An Application-Oriented Approach*. Boston: Addison-Wesley.

Nakagawa, T. 2005. *Maintenance Theory of Reliability*. London: Springer-Verlag.

Nakagawa, T. and X. Zhao. 2012. Optimization problems of a parallel system with a random number of units. *IEEE Transactions on Reliability* 61 (2): 543–548.

Nakagawa, T., X. Zhao and W.Y. Yun. 2011. Optimal age replacement and inspection policies with random failure and replacement times. *International Journal of Reliability, Quality and Safety Engineering* 18(5): 405–416.

Pinedo, M. 2012. *Scheduling: Theory, Algorithms and Systems*. New York: Springer.

Stadje, W. 2003. Renewal analysis of a replacement process. *Operations Research Letters* 31 (1): 1–6.

Sugiura, T., S. Mizutani and T. Nakagawa. 2004. Optimal random replacement policies. In: *Proceedings of the 10th ISSAT International Conference on Reliability and Quality in Design*, ed. by H. Pham and S. Yamada. Las Vegas, Nevada, pp. 99–103.

Tadj, L., M.S. Ouali, S. Yacout, and D. Ait-Kadi. 2011. *Replacement Models with Minimal Repair*. London: Springer-Verlag.

Zhao, X. and T. Nakagawa. 2012a. Optimization problems of replacement first or last in reliability theory. *European Journal of Operational Research* 223 (1): 141–149.

Zhao, X. and T. Nakagawa. 2012b. A summary of newly proposed age replacement policies. In *Proceedings of the 18th ISSAT International Conference on Reliability and Quality in Design*, ed. by H. Pham. Boston, Massachusetts, pp. 106–110.

Zhao, X., K. Nakayama, and S. Nakamura. 2011. Cumulative damage models with replacement last. In *Software Engineering, Business Continuity, and Education*, ed. T.-H. Kim, H. Adeli, H.-k. Kim, H.-j. Kang, K. J. Kim, A. Kiumi, B.-H. Kang. Volume 257, Berlin: Springer-Verlag, pp. 338–345.

15

Random Evolution of Degradation and Occurrences of Words in Random Sequences of Letters

Emilio De Santis and Fabio Spizzichino
Department of Mathematics, University La Sapienza, Italy

15.1 Introduction

In our discussion, attention will be concentrated on some remarkable features of a very special class of irreducible, time-homogeneous, finite-space Markov chains in discrete time and on related modeling interest.

For a given positive integer number k, we denote by $\mathcal{M}_k$ the class of irreducible Markov chains $\{X_0, X_1, \ldots\}$ on the state space $E_k \equiv \{0, 1, \ldots, k\}$ with time-homogeneous transition probabilities satisfying the following conditions:

$$p_{i,i+1} > 0 \text{ for } i = 0, 1, \ldots, k-1, \tag{15.1}$$

$$p_{i,j} = 0 \text{ for } i = 0, \ldots, k-2, j > i+1. \tag{15.2}$$

The term *skip-free on the right* has sometimes been used to designate such a special class (see in particular Irle and Gani (2001), Ferreira and Pacheco (2005), De Santis and Spizzichino (2012b)). The states 0 and k are respectively interpreted as the *minimal* and the *maximal* states in E_k (or vice versa, depending on the type of application at hand). One typically assumes that 0 is the initial state of the chain and looks, as an object of main interest, at the first-passage time

$$T_k = \inf\{n \geq k | X_n = k\},$$

namely *the waiting time until the chain reaches the maximal level k.*

Applied Reliability Engineering and Risk Analysis: Probabilistic Models and Statistical Inference, First Edition.
Edited by Ilia B. Frenkel, Alex Karagrigoriou, Anatoly Lisnianski and Andre Kleyner.

Despite the restrictiveness determined by the above conditions, such Markov chains emerge in the modeling of different situations, in numerous applied fields. In particular they manifest, in a completely spontaneous form, in the analysis of waiting times to occurrences of *words* in random sampling of *letters* from an *alphabet*.

Separately from the Markov chain approach, however, some useful and attracting ideas and methods have been developed in the treatment of words' occurrences. In particular, the concept of *leading number* has an important role in such an analysis. This concept was initially introduced by John Conway (see Li (1980) and references therein) and then developed and extended in successive papers (see e.g. De Santis and Spizzichino (2012a), De Santis and Spizzichino (2012c) and Robin and Daudin (1999)).

We think that it is fruitful to combine the two approaches. In order to illustrate such a point of view, we will then present and discuss some specific results. Typical problems related with the analysis of T_k are the computation of the expected value $\mathbb{E}(T_k|X_0 = 0)$, the computation of the probability distribution of T_k, and stochastic comparisons between T_k and T'_k for two different chains $\{X_0, X_1, \ldots\}$ and $\{X'_0, X'_1, \ldots\}$ belonging to the class $\mathcal{M}_k$. Among these problems, in our discussion we will only consider the issue of stochastic comparisons.

Furthermore, it can generally be interesting to compare problems and combine methods arising from different fields where chains of the type $\mathcal{M}_k$ are encountered. In particular we want to point out that some of the ideas, suggested from words' occurrences, can be usefully exported into other fields. Among such fields, we will specifically consider the one of reliability and maintenance.

For our purposes we will first review some relevant aspects related to words' occurrences. Then we show some reliability models that can be recasted in the afore-described Markov chain setting.

Finally we will present and briefly discuss some results concerning stochastic comparisons between the first-passage times T_k and T'_k for two different, but *comparable*, Markov chains in $\mathcal{M}_k$.

We claim that the approach based on words' occurrences can turn out to be useful even when the considered Markov models in $\mathcal{M}_k$, do not necessarily correspond to the occurrence of specific words.

More precisely, the chapter is structured as follows. In Section 15.2 we recall the setting of words' occurrences and point out different aspects and special features of this theme. In particular we recall the concept of leading number. In Section 15.3, we present three different reliability models in discrete time that can be described by Markov chains in the class $\mathcal{M}_k$ and that, under suitable conditions, can be isomorphic to problems related with words' occurrences. Section 15.4 will be devoted to discussing the interest of the words' setting in the analysis of stochastic comparisons between first-passage times for Markov chains in the class $\mathcal{M}_k$.

15.2 Waiting Times to Words' Occurrences

Let $\mathcal{A}_N \equiv \{a_1, \ldots, a_N\}$ be an arbitrary set of cardinality $|\mathcal{A}_N| = N$.

In the present frame we will consider $\mathcal{A}_N$ as an *alphabet* and its elements $a_1, \ldots, a_N$ will be called *letters*. A finite, ordered sequence $\mathbf{w}$ containing k letters $w_1, \ldots, w_k$ of

$\mathcal{A}_N$ (with possible repetitions), namely $\mathbf{w} \equiv (w_1, \ldots, w_k) \in (\mathcal{A}_N)^k$, will be looked at as a *word* of length k over $\mathcal{A}_N$.

Assume now that, at any unit time, a letter is drawn randomly from $\mathcal{A}_N$. Denoting by α_n the letter drawn at time n, we thus obtain a sequence $\alpha_1, \alpha_2, \ldots$ of i.i.d., uniformly distributed, $\mathcal{A}_N$-valued, random variables. Namely we have, for any n and any choice of $i, j, a_{i_1}, \ldots, a_{i_{n-2}}$,

$$P(\alpha_n = a_j | \alpha_1 = a_{i_1}, \ldots, \alpha_{n-2} = a_{i_{n-2}}, \alpha_{n-1} = a_i) = \frac{1}{N}. \tag{15.3}$$

For a fixed word of length k, $\mathbf{w} \equiv (w_1, \ldots, w_k) \in (\mathcal{A}_N)^k$, a main object of interest in this setting is the study of the probability distribution $\pi_{\mathbf{w}}$ of the discrete random variable $T_{\mathbf{w}}$ defined as follows: $T_{\mathbf{w}}$ is the waiting time until, for the first time, we see the occurrence of $\mathbf{w}$ along the observation of the random sequence $\alpha_1, \alpha_2, \ldots$. More formally:

$$T_{\mathbf{w}} = \inf\{n \geq k | \alpha_{n-k+1} = w_1, \ldots, \alpha_n = w_k\}. \tag{15.4}$$

In the following subsections we will review two different methods, that have been considered in the literature for the analysis of the probability distribution $\pi_{\mathbf{w}}$.

One method is based on the study of the absorption behavior of a suitably defined Markov chain. The other method takes into account the repetition structure of the word $\mathbf{w}$ and it is based on the concept of *leading number*. As we will see next, in fact, the latter is a characteristic that collects the information, about such a repetition structure, that is relevant in deriving $\pi_{\mathbf{w}}$.

The present scheme of times to occurrences of words may emerge in several different fields such as communication theory and molecular biology. As to the interest of probabilistic analysis of repetition structures in DNA sequences, see for example, Shukla and Srivastava (1985) and references cited therein.

For conceptual and notational simplicity, in this section we will mainly consider the i.i.d. *uniform* case, defined by the condition (15.3), for the process of letters' sampling. For the purpose of dealing with some of the afore-mentioned applications, however, the need may arise to extend the analysis to the general, that is, non-necessarily uniform, i.i.d. case. In such a case, the letters are drawn randomly from $\mathcal{A}_N$ according to a prescribed probability distribution $\mathbf{p} \equiv (p_1, \ldots, p_N)$ over $\mathcal{A}_N$, namely

$$P(\alpha_n = a_j | \alpha_1 = a_{i_1}, \ldots, \alpha_{n-2} = a_{i_{n-2}}, \alpha_{n-1} = a_i) = p_j \tag{15.5}$$

for all n, all $i, j, i_1, \ldots, i_{n-2}$.

Below we will illustrate the two methods for the special case when $\mathbf{p}$ is the uniform distribution. There is not much difficulty, however, in extending both methods, respectively based on the Markov chains and on the leading numbers, to the cases of non-necessarily uniform distribution. The condition of i.i.d. letters is, on the contrary, more difficult to remove.

15.2.1 *The Markov Chain Approach*

A method for analyzing the probability distribution of $T_{\mathbf{w}}$ can be based on the study of the Markov chain $\{X_0, X_1, X_2, \ldots\}$ defined by the following positions. Define the random variables $X_0, X_1, X_2, \ldots$ with

$$P(X_0 = 0) = 1.$$

For $n \geq 1$

$$X_n = 0 \text{ if } (\alpha_{n-u+1}, \ldots, \alpha_n) \neq (w_1, \ldots, w_u) \quad \textit{for} \text{ each } u = 1, 2, \ldots, k$$

For $h = 1, 2, \ldots, k-1$,

$$X_n = h$$

$$\text{if } (\alpha_{n-h+1}, \ldots, \alpha_n) = (w_1, \ldots, w_h), (\alpha_{n-h}, \ldots, \alpha_n) \neq (w_1, \ldots, w_{h+1}).$$

Finally,

$$X_n = k$$

$$\text{if } (\alpha_{n-k+1}, \ldots, \alpha_n) = (w_1, \ldots, w_k).$$

Thus, for any $n = 1, 2, \ldots, X_n$ is a $\{0, 1, \ldots, k\}$-valued random variable and it is easy to check that $X_0, X_1, X_2, \ldots$ is a time-homogeneous Markov Chain belonging to $\mathcal{M}_k$, that is, besides its state space being $E_k = \{0, 1, \ldots, k\}$, the properties (15.1), (15.2) are also verified.

The value taken by X_n is known at time n and counts how many letters of the word **w** already appeared, in the most recent part of the sequence, in a way adapt for the formation of **w**. The event $\{X_n = k\}$ means that **w** has occurred at time n.

It is then immediately seen that, for this chain, the waiting time until the first occurrence of **w** coincides with the first-passage time of the chain $X_0, X_1, X_2, \ldots$ (starting from the initial state 0) in the maximal state k. More formally, one can write

$$T_{\mathbf{w}} = T_k$$

with $T_{\mathbf{w}}$ defined as in (15.4) and

$$T_k = \inf\{n \geq k | X_n = k\}.$$

For this reason, the probability distribution of $T_{\mathbf{w}}$ can be studied in terms of the equation for absorbing times of Markov chains with absorbing states, by taking into account the special structure determined by the conditions (15.1), (15.2).

In the case when the sampling distribution $\mathbf{p}$ is uniform over $\mathcal{A}_N$, all the non-zero elements of the transition matrix must be equal to $\frac{1}{N}$, except the elements of the type $p_{h,0}$, for $h = 0, 1, \ldots, k-1$, which are of the form $\frac{d}{N}$ with d a suitable integer, $0 \leq d \leq N-1$. The set of non-zero elements, furthermore, must satisfy some compatibility conditions, depending on the structure of the word **w**.

15.2.2 Leading Numbers and Occurrences Times

We again consider a fixed word $\mathbf{w} = (w_1, \ldots, w_k) \in (\mathcal{A}_N)^k$ and assume that the probability distribution $\mathbf{p}$ is uniform over $\mathcal{A}_N$, that is, the letters in the alphabet $\mathcal{A}_N$ are equiprobable. Even in such a special case, the probability distribution, $\pi_{\mathbf{w}}$, of $T_{\mathbf{w}}$ is still influenced by the repetition structure of **w**. A remarkable fact in this respect is that the information, that is contained in such a repetition structure and that is relevant for determining $\pi_{\mathbf{w}}$, is summarized by a simple, special characteristic of $\mathbf{w}$, called *leading number*. The latter (see in particular Li (1980) and Chen *et al.* (2009) and the citation to J. Conway contained in De Santis and Spizzichino (2012c)) is defined as the binary vector

$$\varepsilon_{\mathbf{w}} \equiv (\varepsilon_{\mathbf{w}}(1), \varepsilon_{\mathbf{w}}(2), \ldots, \varepsilon_{\mathbf{w}}(k)),$$

or simply $\varepsilon \equiv (\varepsilon(1), \varepsilon(2), \ldots, \varepsilon(k))$, where, for $u = 1, 2, \ldots, k$,

$$\varepsilon(u) = \mathbf{1}\{(w_{k-u+1}, \ldots, w_k) = (w_1, \ldots, w_u)\}.$$

In other words $\varepsilon(u) = 1$ if and only if the two "sub-words," formed with the first u letters and the last u letters of $\mathbf{w}$, do coincide. For example, if $N = 26$, $\mathcal{A}_N$ is the English alphabet, $k = 11$, and $\mathbf{w} =$ (*abracadabra*), then

$$\varepsilon_{\mathbf{w}} \equiv (1, 0, 0, 1, 0, 0, 0, 0, 0, 0, 1).$$

Notice that, for a binary vector of length k, $\varepsilon = (\varepsilon(1), \ldots, \varepsilon(k))$, the condition $\varepsilon(k) = 1$ is necessary for ε being the leading number associated to a word.

The importance of the notion of leading number in the problem at hand is motivated by the following two facts. It can be proved that

$$\mathbb{E}(T_{\mathbf{w}}) = \sum_{u=1}^{k} \varepsilon_{\mathbf{w}}(u) \cdot N^u \tag{15.6}$$

(see e.g. the "fair-game" argument in Williams (1991)). Furthermore one has the following result (see Li (1980), Blom and Thorburn (1982), Robin and Daudin (1999)).

Proposition 15.1. *If* $\mathbf{w}', \mathbf{w}'' \in (\mathcal{A}_N)^k$ *are such that* $\varepsilon_{\mathbf{w}'} = \varepsilon_{\mathbf{w}''}$, *then* $T_{\mathbf{w}'} \stackrel{d}{=} T_{\mathbf{w}''}$, *namely*

$$\pi_{\mathbf{w}'} = \pi_{\mathbf{w}''}.$$

In the i.i.d. non-uniform case, defined by the condition (15.5), the arguments above can be extended (see, in particular, Chen and Zame 1979, and Li 1980) by replacing the quantity $\sum_{u=1}^{k} \varepsilon_{\mathbf{w}}(u) \cdot N^u$ with

$$\frac{\varepsilon(k) + \sum_{i=2}^{k} \varepsilon(k+1-i) \prod_{j=k-i+2}^{k} p_j}{\prod_{j=1}^{k} p_j}. \tag{15.7}$$

15.3 Some Reliability-Maintenance Models

We start this section by presenting three different models of ageing and degradation that can be naturally described in terms of Markov chains belonging to the class $\mathcal{M}_k$.

15.3.1 Model 1 (Simple Machine Replacement)

We start by considering the classical situation of machines' replacement at pre-established age.

In such a situation $C_1, C_2, \ldots$ are identical devices designed to work in one and the same position in a system S.

C_1 is installed at step 0, C_2 is installed in the place of C_1 as soon as C_1 exits its perfection-state or it reaches its *age* $k, \ldots$, C_r is replaced by C_{r+1} as soon as it exits its perfection-state or it reaches its *age* k.

Let $D_1, D_2, \ldots$, be $\{1, 2, \ldots\}$-valued i.i.d. random variables where D_r represents the number of times (cycles) in which C_r is able to work in the perfection-state and, for $h = 0, 1, \ldots$, set

$$\rho_h \equiv P\{D_1 = h + 1 | D_1 > h\}.$$

Consider, moreover, the random variable

$$Y_r = \min(D_r, k),$$

that is, the number of times in which C_r remains in operations. Then the random variable defined by

$$\mathcal{T}_m = \sum_{r=1}^{m} Y_r$$

can be seen as the instant of m-th replacement, and the discrete-time counting process

$$R_n = \sup\{m | \mathcal{T}_m \leq n\}.$$

counts the total number of replacements within time n, for $n = 1, 2, \ldots$.

We consider now the sequence of the random variables $X_0, X_1, \ldots$ defined by the following positions

$$X_0 = 0,$$
$$\ldots$$
$$X_n = n - \mathcal{T}_{R_n}$$

Under these positions, we see that X_n is the *age* of the device working at time (cycle) n.

$X_0, X_1, \ldots$ is an homogeneous Markov chain belonging to the class $\mathcal{M}_k$. More precisely, the only positive transition probabilities are given as follows. For $h = 0, 1, \ldots, k - 1$,

$$p_{h,h+1} = \rho_h; \; p_{h,0} = 1 - p_{h,h+1} = 1 - \rho_h.$$

Furthermore,

$$p_{k,0} = 1.$$

The first-passage time T_k in the state k has the meaning of the number of times (cycles) until maturation of age k for the working device, in other words until the first replacement, due to planned maintenance rather than to a failure.

15.3.2 Model 2 (Random Reduction of Age)

The model that we consider here comes out as a natural generalization of the previous one.

This time we have a device C which is installed at step 0, and a repair action is made at the end at pre-established number of missions (cycle).

Repair can have beneficial effects on the device, but it is observed that the amounts of such effects are random and possibly null.

We assume the existence of some quality characteristic of the device which is strictly related to its age, namely to the number of cycles for which the device itself would have been working in absence of maintenance. In other words, we assume that the age of the device is physically measurable.

We assume, furthermore, that the (random) benefit of a repair on the device can be measured in terms of a reduction of the age of it. As an example, one may consider the case when the device is the battery of a cellular phone. In the latter case, the perfection state is a pre-assigned duration-time of its charge, the reduced quality is measured by the reduction of the charge-duration, and such an observable characteristic is proportional to the *age,* that is, to the amount of use. We conceive that the failure of the device is defined by a very short and pre-assigned charge-duration.

In this respect, we furthermore assume that the probability distribution of *age-reduction* at repair only depends on the age h reached by the device, at the moment of the repair action. Setting again

$$X_0 = 0$$

$$\cdots$$

$$X_n = \text{age of device at cycle } n,$$

we still have that $X_0, X_1, \ldots$ is a Markov chain and that it belongs to the class $\mathcal{M}_k$.

15.3.3 Model 3 (Random Number of Effective Repairs in a Parallel System)

This time we deal with a parallel system S made with repairable components $C_1, C_2, \ldots, C_k$. Such components are all similar and their life-lengths are measured in continuous time. We assume that, when C_r $(r = 1, 2, \ldots, k)$ fails, a repair is immediately attempted but it is not sure that reparation is effective. The repair action is, furthermore, attempted for all the failed components that we may find in the system at that time.

A difference with the above model is that the components do not undergo degradation. In fact, each component can be only in one of the two states: *on* or *off.* When an off component is repaired, such a maintenance action can only be *effective* or *non-effective*. Then, there is no room for talking about *age* of a component. Degradation is, instead, present at the system level and it is measured in terms of the number of off components.

When C_r fails, and other $(h - 1)$ off (non-repaired) components are present in the system, all the h failed components, as already mentioned, are maintained $(0 \leq h \leq k - 1)$.

In this respect, we assume that the number of components, which remain non-repaired under any of such operations, is a $\{0, 1, \ldots, h\}$-valued random variable with probability distribution depending on h (and only on h).

Also in this context, we define a discrete-time Markov chain as follows.

$$X_0 = 0$$

$$X_n = \text{number of failed components in the system at the time of the } n\text{-th failure.}$$

$X_0, X_1, \ldots$ is then, once again, a Markov chain belonging to the class $\mathcal{M}_k$, with

$$p_{k,k} = 1.$$

We point out that we have here a Markov chain in discrete time since we count the number of failed components, present in the system, after any repair action.

The absorption time

$$T_k = \inf\{n \geq k | X_n = k\}$$

has the meaning of number of components' failures that will cause the failure of the system S and then we can be interested in deriving its probability distribution.

15.3.4 Degradation and Words

We saw that, for all the three cases above, the stochastic model for the time-evolution of degradation can be described in terms of an appropriate Markov chain belonging to the class $\mathcal{M}_k$. In view of the arguments of Section 15.2, it can be useful to recognize when such a Markov chain can coincide with one corresponding to a word $\mathbf{w}$ of length k (for some suitable choice of N and $\mathbf{p}$). When this is the case, in fact, we can exploit the Proposition 15.1, according to which the leading number $\varepsilon_{\mathbf{w}}$ (or the appropriate generalization of it, in the non-uniform case) gives us crucial information about the probability distribution of T_k and the quantity in (15.7) immediately gives us the expected value $\mathbb{E}(T_k)$.

Let us first consider Model 1. If the distribution of the random variables $D_1, D_2, \ldots$ is geometric, then the associated Markov chain coincides with the one associated to a word $\mathbf{w}$ of length k, of the special form

$$w_1 = a, \ldots, w_k = a, \tag{15.8}$$

for some fixed letter $a \in \mathcal{A}_N$. The leading number of $\mathbf{w}$ is

$$\varepsilon_{\mathbf{w}} = (1, 1, \ldots, 1).$$

If $\rho_h = \frac{1}{N}$, for some integer N and all $h = 1, 2, \ldots$, then the distribution $\mathbf{p}$, over $\mathcal{A}_N$, can be taken as uniform. A non-uniform distribution is on the contrary needed when $\rho_h = \rho$, for all $h = 1, 2, \ldots$ and for some quantity ρ not of the form $\frac{1}{N}$.

When the distribution of $D_1, D_2, \ldots$ is not geometric it is impossible to find a corresponding word $\mathbf{w}$.

We note that, for these special models, the condition $p_{h,0} = 1 - p_{h,h+1}$ holds and the description in terms of Markov chains in the class $\mathcal{M}_k$ is quite a classical topic in applied probability. Furthermore, the analogy between the problem arising for the occurrence of the word $\mathbf{w}$ in (15.8) and the one arising in the context of Reliability has been already noted in the literature. Concerning the present remark, see in particular Feller (1968) and the examples devoted to " Success Runs" and " Machine Replacement" in Brémaud (1999).

Also for Model 2 and Model 3, one can select some special cases where the corresponding Markov chains coincide with those associated to suitable words. In particular, for any arbitrary word $\mathbf{w}$ we can find an infinite number of corresponding degradation models of the type 2, by suitably choosing an alphabet $\mathcal{A}_N$ and a distribution $\mathbf{p}$. The triple $(\mathbf{w}, \mathcal{A}_N, \mathbf{p})$ uniquely determines a transition matrix of the form as in Model 2.

It is clear that, given an arbitrary model of the type 2, we cannot generally find a corresponding $(\mathbf{w}, \mathcal{A}_N, \mathbf{p})$. In particular the following conditions are necessary.

For any j, $0 \leq j \leq k-1$, if $p_{h,j} \neq 0$ for some h, $j \leq h \leq k-1$, then

$$p_{h,j} = p_{j-1,j}.$$

In other words, all the non-null transition probabilities $p_{h,j}$ with $j \leq h$ necessarily coincide. Furthermore, in case

$$p_{h',h'+1} \neq p_{h'',h''+1}, \tag{15.9}$$

for all pairs $0 \leq h' < h'' \leq k-1$, then

$$\begin{aligned} p_{h,1} &= p_{1,1} = \gamma, \text{ for } h = 0, 1, \ldots, k-1 \\ p_{h,0} &= 1 - \gamma - p_{h,h+1}. \end{aligned} \tag{15.10}$$

In fact, any word $\mathbf{w}$ for which (15.9) holds, must necessarily be formed by all different letters, and this implies (15.10).

15.4 Waiting Times to Occurrences of Words and Stochastic Comparisons for Degradation

As discussed in the previous section, Markov chains in the class $\mathcal{M}_k$ can generally be interpreted in terms of evolution in discrete time of levels of degradation for appropriate reliability structures.

On the other hand, recognizing that a given Markov chain $\{X_n\}_{n=0,1,\ldots}$ admits corresponding triples $(\mathbf{w}, \mathcal{A}_N, \mathbf{p})$ can be of applied interest in some situations. In fact the identities (15.6), (15.7), and Proposition 15.2.1 can be exploited in different forms when we face the computation of the expected value or of the probability distribution for the first-passage time T_k of $\{X_n\}_{n=0,1,\ldots}$. We notice in this respect that the smaller the value of N (number of letters in the alphabet) the smaller the computational complexity arising in such computations. In particular, problems are remarkably simplified for words consisting of only two letters. In the case $N = 2$ we write, say, $\mathcal{A}_2 = \{a, b\}$ and a probability distribution $\mathbf{p}$ over $\mathcal{A}_2$ is obviously determined by the only scalar quantity, $p_1 = P(a)$.

As has been remarked above, not all Markov chains in the class $\mathcal{M}_k$ admit corresponding triples $(\mathbf{w}, \mathcal{A}_N, \mathbf{p})$.

We want, however, to point out that the arguments presented in Section 15.2 can still be very useful in the more general treatment of Markov chains in the class $\mathcal{M}_k$. Here we are in particular interested in methods to establish stochastic comparisons of the type

$$T'_k \leq_{st} T''_k, \tag{15.11}$$

for the first-passage times T'_k, T''_k of two different Markov chains in the class $\mathcal{M}_k$. The symbol $\leq_{st}$ denotes the *usual stochastic order*: for two scalar random variables X, Y, the condition $X \leq_{st} Y$ then means

$$P(X > t) \leq P(Y > t), \forall t \in \mathbb{R}.$$

For our purposes, one can, in particular, use the following two results, concerning stochastic comparisons between first-passage times for two *comparable* Markov chains belonging to $\mathcal{M}_k$.

Let $\mathbf{X}'$ and $\mathbf{X}''$ be two such Markov chains, with transition matrices $P' = (p'_{i,j})_{0 \le i,j \le k}$ and $P'' = (p''_{i,j})_{0 \le i,j \le k}$, respectively. Let, furthermore, $\pi'_0 = (\pi'_0(i))_{i=1,\ldots,k}$, $\pi''_0 = (\pi''_0(i))_{i=1,\ldots,k}$ denote their initial distributions and let T'_h and T''_h be the corresponding first-passage times to the state h, for $h = 1, \ldots, k$.

Proposition 15.2. *Under the conditions*

$$\sum_{j=r}^{k} p'_{i,j} \ge \sum_{j=r}^{k} p''_{i,j} \text{ for } i = 1, \ldots, k, \text{ for } r = 1, \ldots, k,$$

$$\boldsymbol{\pi}'_0 \ge_{st} \boldsymbol{\pi}''_0,$$

one has the stochastic comparisons

$$T'_k \le_{st} T''_k.$$

Even if this result may sound rather intuitive, it still needs to be proven formally, see Irle and Gani (2001), Ferreira and Pacheco (2005). See also De Santis and Spizzichino (2012b), where the possibility is mentioned of a method of proof, different from the previous one and developed therein for proving a more complex result of stochastic comparison. Such a method is based on a special coupling construction. Two Markov chains are paired in the following sense: when the *faster* chain reaches an upper level, it stops and waits until the other one reaches the same level (*stop and go* behavior). In this respect, see also Ferreira and Pacheco (2005).

The same method of proof can be used for the following result, also presented in De Santis and Spizzichino (2012b).

Proposition 15.3. *Let both the initial distributions* $\boldsymbol{\pi}'_0$, $\boldsymbol{\pi}''_0$ *be degenerate on the state* 0. *Suppose furthermore that there exists an intermediate state m* ($1 \le m \le k-1$) *such that*

$$T'_m \le_{st} T''_m,$$

and, for $i = m, \ldots, k-1$

$$p''_{i,0} + p''_{i,i+1} = 1, \quad p''_{i,i+1} \le p'_{i,i+1}.$$

Then $T'_i \le_{st} T''_i$ *for* $i = m, \ldots, k$.

Let us now come back to the comparisons of the type (15.11). For this purpose, the Propositions 15.2, 15.3 can be exploited in various ways.

As a basic fact we point out that, for a given Markov chain belonging to $\mathcal{M}_k$, one can suitably contruct a word $\mathbf{w}$ such that

$$T_{\mathbf{w}} \le_{st} T_k, \tag{15.12}$$

or, alternatively,

$$T_k \le_{st} T_{\mathbf{w}} \tag{15.13}$$

hold. Concerning such an issue we just mention the following, very simple, constructions.

Let a transition matrix $\hat{P} \equiv (\hat{p}_{i,j})$ belonging to $\mathcal{M}_k$ be given and consider the word $\mathbf{w} = \mathbf{a} \equiv (a, \ldots, a)$ of length k on the alphabet $\mathcal{A}_2 = \{a, b\}$.

In order to let (15.13) hold, we select the distribution $\mathbf{p}^{(>)}$ by setting

$$p_1^{(>)} \equiv P(\alpha_n = a) = \min\{\hat{p}_{h,h+1} : h \in \{0, \ldots, k-1\}\}.$$

The transition matrix for the Markov chain associated with the word $\mathbf{a}$ is then

$$P^{(>)} = \begin{pmatrix} 1 - p_1^{(>)} & p_1^{(>)} & 0 & \cdots \\ \vdots & 0 & p_1^{(>)} & 0 \\ 1 - p_1^{(>)} & \vdots & \ddots & p_1^{(>)} \\ 0 & 0 & \cdots & 1 \end{pmatrix}. \tag{15.14}$$

It is rather intuitive, and in any case easy to check by Proposition 15.2, that the inequality (15.13) holds. The construction of a distribution $\mathbf{p}^{(<)}$ in order to get (15.12) can be made in an analogous way. However, this time we should assume the possibility of a " perfect repair at any age." In other words we consider the quantity

$$\sigma_0 = \min_{0 \le h \le k-1} (\hat{p}_{h,0}),$$

and assume $\sigma_0 > 0$. We then fix the quantities $p_1^{(<)}, p_2^{(<)}$ as follows

$$p_1^{(<)} = P(\alpha_n = a) = 1 - \sigma_0, \; p_2^{(<)} = P(\alpha_n = b) = \sigma_0. \tag{15.15}$$

As it is easy to see, the transition matrix associated with the triple $(\mathbf{a}, A_2, \mathbf{p}^{(<)})$ can be obtained from the original transition matrix $\hat{P}$ just by transporting, on each row, some mass from left to right. Therefore, also in this case, the inequality (15.12) can be readily obtained by applying Proposition 15.2.

For brevity's sake, here we limit ourselves only to this easy construction based on the word $\mathbf{a}$ containing only one letter. However, other more efficient constructions can be made on the basis of the specific elements of $\hat{P}$.

As mentioned above, arguments based on words' occurrences can be implemented in various ways to obtain comparisons of the type (15.11) for first-passages times for two Markov chains belonging to $\mathcal{M}_k$, not necessarily associated in a direct way to words' occurrences.

An interesting possibility is based on the following argument. Let $\mathbf{X}'$ and $\mathbf{X}''$ be two such Markov chains, with transition matrices $P' = (p'_{i,j})_{0 \le i,j \le k}$ and $P'' = (p''_{i,j})_{0 \le i,j \le k}$, respectively. Suppose that we are able to find $\mathcal{A}_n$ with $\mathbf{p} = (\frac{1}{N}, \frac{1}{N}, \ldots, \frac{1}{N})$ and two different words $\mathbf{w}', \mathbf{w}'' \in (\mathcal{A}_N)^k$ such that

$$T'_k \le_{st} T_{\mathbf{w}'}, T_{\mathbf{w}''} \le_{st} T''_k, \tag{15.16}$$

$$\varepsilon_{\mathbf{w}'} = \varepsilon_{\mathbf{w}''}. \tag{15.17}$$

Once such $\mathbf{w}', \mathbf{w}''$ have been found, one easily gets (15.11) by taking into account the transitivity property of the $\le_{st}$ ordering and by using Proposition 15.1. The earlier Propositions 15.2 and 15.3, on the other hand, can be exploited in the search for $\mathbf{w}', \mathbf{w}''$ satisfying (15.16) and (15.17) at a time.

Furthermore, in the case when we are only interested in comparisons of the type

$$\mathbb{E}(T_k^{'}) \leq \mathbb{E}(T_k^{''}), \tag{15.18}$$

the condition (15.17) can be replaced by the weaker condition $\mathbb{E}(T_{\mathbf{w}'}) \leq \mathbb{E}(T_{\mathbf{w}''})$ in view of the Formulas (15.6) or (15.7). Remember, in this respect, that $X \leq_{st} Y$ implies $\mathbb{E}(X) \leq \mathbb{E}(Y)$.

As one other hint, one can exploit the connection with words at an intermediate level l, $1 \leq l \leq k-1$. Consider, for example, the two matrices

$$P^{'} = \begin{pmatrix} \frac{1}{2} & \frac{1}{2} & 0 & 0 & 0 \\ \frac{1}{2} & 0 & \frac{1}{2} & 0 & 0 \\ 0 & 0 & \frac{1}{2} & \frac{1}{2} & 0 \\ 1-\alpha & 0 & 0 & 0 & \alpha \\ 0 & 0 & 0 & 0 & 1 \end{pmatrix}, \quad P^{''} = \begin{pmatrix} \frac{1}{2} & \frac{1}{2} & 0 & 0 & 0 \\ 0 & \frac{1}{2} & \frac{1}{2} & 0 & 0 \\ 0 & \frac{1}{2} & 0 & \frac{1}{2} & 0 \\ 1-\alpha & 0 & 0 & 0 & \alpha \\ 0 & 0 & 0 & 0 & 1 \end{pmatrix}. \tag{15.19}$$

We note that, here, one has $k = 4$. Furthermore the first-passage times T'_3 and T''_3 can respectively be seen as the occurrences times $T_{\mathbf{w}'}$ and $T_{\mathbf{w}''}$ for the two words $\mathbf{w}' = (b, b, a)$ and $\mathbf{w}'' = (a, b, b)$, belonging to the alphabet $\mathcal{A}_2 = \{a, b\}$, under the uniform distributions $\mathbf{p}' = \mathbf{p}'' = (\frac{1}{2}, \frac{1}{2})$. In fact the elements of the transition matrices Q' and Q'', corresponding to such two words respectively, satisfy the relations

$$q^{'}_{i,j} = p^{'}_{i,j}, q^{''}_{i,j} = p^{''}_{i,j}, \text{ for } i = 0, 1, 2, \;\; j = 0, 1, 2, 3.$$

Since $\mathbf{w}'$ and $\mathbf{w}''$ share the same leading number (0, 0, 1), we can say that T'_3 and T''_3 are identically distributed. Thus we can conclude that T'_4 and T''_4 also have the same distribution by taking into account Proposition 15.3.

Similarly to the previous case also, this argument can be extended, by combining Proposition 15.1, Proposition 15.2, and Proposition 15.3 in various ways, to prove comparisons of the type (15.11) or comparisons of the weaker type (15.18).

Taking into account the discussion in the Section 15.3, such results have an evident interest in the choices between two different devices or two different maintenance policies.

15.5 Conclusions

In the chapter we pointed out some analogies between words' occurrences and random degradation of devices or systems. Our interest in such an analysis is not only based on an intellectual curiosity. We argue in fact that, looking at some peculiar aspects of words' occurrences, one can reflect on *useful* probabilistic aspects of waiting times. In particular our attention has been focused on times to failures of systems, where maintenance and random reduction of age can oppose the degradation process. Furthermore, the time-honored literature concerning reliability and degradation may stimulate cross-fertilization between the analysis of first-passage times for an important class of Markov chains and special ideas that had arisen in the field of word occurrences.

Acknowledgements

We like to thank the Editors for inviting us to contribute to this volume dedicated to the centennial of the birth of Boris Gnedenko. Our work has been partially supported in the frame of Research Projects "Modelli e Algoritmi Stocastici: Convergenza ed Ottimizzazione" 2009 and "Modelli Stocastici e Applicazioni alla Fisica e alla Finanza" 2010 of Università La Sapienza, Roma.

References

Blom G. and D. Thorburn. 1982. How many random digits are required until given sequences are obtained? *Journal of Applied Probability*, 19(3), 518–531.

Brémaud P. 1999. *Markov Chains. Gibbs Fields, Monte Carlo Simulation, and Queues*. New York:, Springer.

Chen R. and A. Zame. 1979. On fair coin-tossing games. *Journal of Multivariate Analysis*. 9(1): 150–156.

Chen R., A. Zame, and B. Rosenberg. 2009. On the first occurrence of strings. *Electronic Journal of Combinatorics*, 16(1):1–16.

De Santis E. and F. Spizzichino. 2012a. First occurrence of a word among the elements of a finite dictionary in random sequences of letters. *Electronic Journal of Probability*, 17: 1–9.

De Santis E. and F. Spizzichino. 2012b. Stochastic comparisons between first-passage times for Markov chains. *arXiv:1210.1116 [math.PR]*:1–18.

De Santis E. and F. Spizzichino. 2012c. Waiting for ABRACADABRA. Occurrences of words and leading numbers. In *Imagine Math: Between Culture and Mathematics*, ed. M Emmer. Milan: Springer, pp. 175–185.

Feller W. 1968. *An Introduction to Probability Theory and its Applications. Vol. I*. New York: John Wiley & Sons.

Ferreira F. and A. Pacheco. 2005. Level-crossing ordering of semi-Markov processes and Markov chains. *Journal of Applied Probability*, 42(4): 989–1002.

Irle A. and J. Gani J. 2001. The detection of words and an ordering for Markov chains. *Journal of Applied Probability*, 38A: 66–77.

Li S.-Y. R. 1980. A martingale approach to the study of occurrence of sequence patterns in repeated experiments. *Annals of Probability*, 8(6): 1171-1176.

Robin S. and J.J. Daudin. 1999. Exact distribution of word occurrences in a random sequence of letters. *Journal of Applied Probability*, 36(1): 179–193.

Shukla, R. and R.C. Srivastava. 1985. The statistical analysis of direct repeats in nucleic acid sequences. *Journal of Applied Probability*, 22(1): 15–24.

Williams D. 1991. *Probability with Martingales*. Cambridge: Cambridge University Press.

16

Occupancy Times for Markov and Semi-Markov Models in Systems Reliability

Alan G. Hawkes,[1] Lirong Cui[2] and Shijia Du[2]
[1]*School of Business, Swansea University, UK*
[2]*School of Management & Economics, Beijing Institute of Technology, China*

16.1 Introduction

Because of their simplicity, there is a long history of Markov processes, and to some extent semi-Markov processes, being much used in modelling the reliability of systems. It is impossible to list all aspects of their use or to construct a comprehensive bibliography in this chapter. We shall therefore concentrate primarily on general aspects of occupancy times when studying the reliability of systems that are described by finite-state continuous-time Markov or semi-Markov processes. The general models are essentially the same as those used to describe the activity of biological ion channels, and it turns out that many of the results published in the reliability literature were already known in the ion-channel literature. Occupancy times have various forms in the reliability field such as the waiting time of arriving in some class, age of systems, hitting times, residual lifetimes, and so forth. Particularly with multi-class partitions for the state spaces of Markov or semi-Markov processes, many kinds of occupancy times are of interest to reliability engineers and theoretical workers.

In general, there are two kinds of occupancy times. One is so-called 'integer-segment occupancy times' which consist of some integer segments of durations in some subsets, and they have been widely considered. Another is 'non-integer segment occupancy times' which consist of durations by a given time t: for example, in reliability systems, interval availability is defined by the fraction of time during which a system is in operation over

Applied Reliability Engineering and Risk Analysis: Probabilistic Models and Statistical Inference, First Edition.
Edited by Ilia B. Frenkel, Alex Karagrigoriou, Anatoly Lisnianski and Andre Kleyner.

a finite observation period, which requires the occupancy time of operational states by time t. Another example is the occupancy time in a given interval $[a, b]$. In the chapter, the first kind of problems will be studied, but the second one also needs some attention.

We shall not discuss the many articles dealing with details of specific practical applications. Nor shall we discuss the important problem of large systems. These often occur in computer/communication systems, where the number of states of the system can run into millions and it is impractical to define the model parameters, let alone carry out an analysis. There is, however, a substantial and important literature about calculating bounds on statistics, such as various availabilities, based on partial knowledge of the system.

The underlying stochastic processes used in the chapter are semi-Markov processes including its special situation: Markov processes. We therefore begin with a general definition and establish some notation to be used.

Suppose a process $X(t)$ can take only a finite number n of possible values $\{1, 2, \cdots, n\}$, called the states of the process, and that the process jumps from one state to another at times $0 = S_0 < S_1 < \cdots < S_r < \cdots$. The increment $T_k = S_k - S_{k-1}$ is the sojourn time in state $J_{k-1}, k = 1, 2, \ldots$. Thus, $X(t) = J_{k-1}$ $\forall t$ in the interval $[S_{k-1}, S_k)$.

Define a matrix $\mathbf{G}(t)$, called the semi-Markov kernel, whose ijth element is

$$G_{ij}(t) = P((J_k = j) \cap (T_k \le t) | J_{k-1} = i), \quad t > 0.$$

and denote its Laplace-Stieltjes transform by

$$\mathbf{G}^*(s) = \int_0^\infty \exp(-st) d\mathbf{G}(t).$$

$\{J_k\}$ is a homogeneous Markov chain with transition probability matrix $\mathbf{P} = \mathbf{G}(\infty) = \mathbf{G}^*(0)$. The function $\mathbf{G}(t)$, together with an initial probability vector $\boldsymbol{\pi}_0$ with elements $P(J_0 = i), i = 1$ to n, completely determines the behaviour of the process. The process $\{(J_r, T_r)\}$ is clearly Markov and gives an equivalent description for the process $X(t)$, sometimes called a semi-Markov process, and the process $\{(J_k, S_k)\}$, sometimes called a Markov-renewal process. The three processes are therefore just different ways of looking at the same thing, which we will refer to simply as a semi-Markov process.

The sojourn times in state i that end with a transition to state j have cumulative distribution function $F_{ij}(t) = P(T_k \le t | J_k = j, J_{k-1} = i\} = G_{ij}(t)/p_{ij}$, provided $p_{ij} \ne 0$. The overall distribution of sojourns in state i has cumulative distribution $F_i(t) = \sum_j G_{ij}(t)$. In some cases it may be useful to allow transitions from state i to state i, so $G_{ii}(t)$, and therefore also p_{ii}, are not necessarily zero: this will not usually be the case, however. If $G_{ii}(t) = 0$ and $G_{ij}(t) = q_{ij}(1 - e^{-\lambda_{i\cdot}t})/\lambda_{i\cdot}, t > 0, j \ne i, \forall i$, then the semi-Markov process is a finite-state continuous-time Markov process with transition rate matrix $\mathbf{Q}$, and the transition matrix $\mathbf{P}(t) = \exp(\mathbf{Q}t) = e^{\mathbf{Q}t}$ has the ijth element given by $p_{ij}(t) = P(X(s+t) = j | X(s) = i)$. Note that we only consider continuous-time homogeneous Markov processes.

The rest of the chapter is organized as follows. Occupancy times for Markov models for system reliability are discussed in Section 16.2.2 together with a brief comment on reward processes. In Section 16.3, similar problems, focused on joint distributions and cumulative distributions, are considered for semi-Markov models. Similar distribution problems for time interval omission situations are given briefly in Section 16.4. In Section 16.5, some numerical examples are presented to illustrate the results obtained in the previous sections. Finally, the conclusions are summarized.

16.2 Markov Models for Systems Reliability

We suppose that at any point in time a system must be in one of a set of states, $\mathcal{S}$, and each state is classified as being in subset $\mathcal{O}$ if the system is operational and subset $\mathcal{F}$ if the system is failed, so $\mathcal{S} = \mathcal{O} \cup \mathcal{F}$. Denote the state occupied at time t by $X(t)$.

For convenience, we label the states so that the operational set $\mathcal{O}$ consists of states 1 to $n_{\mathcal{O}}$ and $\mathcal{F}$ consists of the remaining $n_{\mathcal{F}}$ states $(n_{\mathcal{O}+1}, n_{\mathcal{O}+2}, \cdots, n)$, $n = n_{\mathcal{O}} + n_{\mathcal{F}}$. Then it is useful to partition the transition rate matrix as

$$\mathbf{Q} = \begin{pmatrix} \mathbf{Q}_{\mathcal{OO}} & \mathbf{Q}_{\mathcal{OF}} \\ \mathbf{Q}_{\mathcal{FO}} & \mathbf{Q}_{\mathcal{FF}} \end{pmatrix}. \tag{16.1}$$

$\mathbf{Q}_{\mathcal{OF}}$ contains transition rates from operational to failed states; $\mathbf{Q}_{\mathcal{FO}}$ contains transition rates from failed to operational; $\mathbf{Q}_{\mathcal{OO}}$ and $\mathbf{Q}_{\mathcal{FF}}$ contain transition rates between states within the operational set and the failed set, respectively.

In this chapter we assume the process is irreducible, so that the system will always return from $\mathcal{F}$ to $\mathcal{O}$ by virtue of some repair/maintenance function. Then the process can be seen as an alternating sequence of sojourns in the subset $\mathcal{O}$ (operational periods or uptimes) and sojourns in the subset $\mathcal{F}$ (failed periods or downtimes).

Our purpose is to consider some results concerning this sequence of intervals. The foundation publication in the reliability literature that addressed this problem, to which all subsequent work refers, seems to be that of Rubino and Sericola (1989). They showed that, if the underlying process, $X(t)$, is Markov, there is an embedded semi-Markov process, $Y(t)$, with the same state space for which the transitions occur only at those points in time, $\{S_k\}$, where $X(t)$ switches between $\mathcal{O}$ and $\mathcal{F}$, or vice versa. The intervals $T_k = S_k - S_{k-1}$ are the durations of, alternately, uptimes and downtimes. $J_{k-1} = X(S_{k-1})$ is the state in which the sojourn begins and $J_k = X(S_k)$ is the state into which the system jumps at the end of the sojourn. The underlying Markov process $X(t)$ may move between operational states during an operational sojourn and between failed states during a failed sojourn but the embedded semi-Markov process $Y(t)$ is defined as remaining constant throughout a sojourn, remaining in the same state in which the sojourn began. Thus $Y(t) = J_{k-1},\ S_{k-1} \le t < S_k$.

Because the operational and failed intervals are alternating, the semi-Markov kernel takes the form

$$\mathbf{G}(t) = \begin{pmatrix} \mathbf{0} & \mathbf{G}_{\mathcal{OF}}(t) \\ \mathbf{G}_{\mathcal{FO}}(t) & \mathbf{0} \end{pmatrix}, \tag{16.2}$$

where we quote probability densities rather than cumulative distributions

$$d\mathbf{G}_{\mathcal{OF}}(t)/dt = \exp(\mathbf{Q}_{\mathcal{OO}}t)\mathbf{Q}_{\mathcal{OF}};\ d\mathbf{G}_{\mathcal{FO}}(t)/dt = \exp(\mathbf{Q}_{\mathcal{FF}}t)\mathbf{Q}_{\mathcal{FO}},\quad t > 0. \tag{16.3}$$

See Colquhoun and Hawkes (1982: Equations 1.16 and 1.20).

The Markov chain $\{J_k\}$ embedded at the switching points has transition probability matrix

$$\mathbf{P} = \begin{pmatrix} \mathbf{0} & \mathbf{P}_{\mathcal{OF}} \\ \mathbf{P}_{\mathcal{FO}} & \mathbf{0} \end{pmatrix}, \tag{16.4}$$

where

$$\mathbf{P}_{\mathcal{OF}} = \mathbf{G}_{\mathcal{OF}}(\infty) = -\mathbf{Q}_{\mathcal{OO}}^{-1}\mathbf{Q}_{\mathcal{OF}};\ \mathbf{P}_{\mathcal{FO}} = \mathbf{G}_{\mathcal{FO}}(\infty) = -\mathbf{Q}_{\mathcal{FF}}^{-1}\mathbf{Q}_{\mathcal{FO}}. \quad (16.5)$$

If we consider the alternate points in the above process, when the system enters the $\mathcal{O}$ subset, we have a Markov chain with transition probability matrix $\mathbf{P}_{\mathcal{OF}}\mathbf{P}_{\mathcal{FO}}$. Let $\boldsymbol{\pi}_{r-1}^{(\mathcal{O})}$ be the state occupancy probability row vector at the start of the rth operational period and let $\boldsymbol{\alpha} = (\boldsymbol{\alpha}_{\mathcal{O}}\ \boldsymbol{\alpha}_{\mathcal{F}})$ be the initial occupancy probability vector of the underlying Markov process, $X(t)$, partitioned appropriately, then

$$\boldsymbol{\pi}_0^{(\mathcal{O})} = \boldsymbol{\alpha}_{\mathcal{O}} + \boldsymbol{\alpha}_{\mathcal{F}}\mathbf{P}_{\mathcal{FO}};\ \boldsymbol{\pi}_r^{(\mathcal{O})} = \boldsymbol{\pi}_{r-1}^{(\mathcal{O})}\mathbf{P}_{\mathcal{OF}}\mathbf{P}_{\mathcal{FO}},\ r \geq 1.$$

The cumulative distribution of the duration of the rth operational period; its Laplace-Stieltjes transform; and its probability density are then given, respectively, by

$$\boldsymbol{\pi}_{r-1}^{(\mathcal{O})}\mathbf{G}_{\mathcal{OF}}(t)\mathbf{1}_{\mathcal{F}};\ \boldsymbol{\pi}_{r-1}^{(\mathcal{O})}(s\mathbf{I} - \mathbf{Q}_{\mathcal{OO}})^{-1}\mathbf{Q}_{\mathcal{OF}}(t)\mathbf{1}_{\mathcal{F}};\ \boldsymbol{\pi}_{r-1}^{(\mathcal{O})}\exp(\mathbf{Q}_{\mathcal{OO}}t)\mathbf{Q}_{\mathcal{OF}}(t)\mathbf{1}_{\mathcal{F}}, \quad (16.6)$$

where $\mathbf{1}_{\mathcal{F}}$ is a column vector of $n_{\mathcal{F}}$ ones.

If $\boldsymbol{\pi}^{(\mathcal{O})}$ is the equilibrium vector of this chain, then the steady-state distribution of operational times has cumulative distribution; its Laplace-Stieltjes transform; and its probability density given, respectively, by

$$\boldsymbol{\pi}^{(\mathcal{O})}\mathbf{G}_{\mathcal{OF}}(t)\mathbf{1}_{\mathcal{F}};\ \boldsymbol{\pi}^{(\mathcal{O})}(s\mathbf{I} - \mathbf{Q}_{\mathcal{OO}})^{-1}\mathbf{Q}_{\mathcal{OF}}(t)\mathbf{1}_{\mathcal{F}};\ \boldsymbol{\pi}^{(\mathcal{O})}\exp(\mathbf{Q}_{\mathcal{OO}}t)\mathbf{Q}_{\mathcal{OF}}(t)\mathbf{1}_{\mathcal{F}}. \quad (16.7)$$

Note that, by differentiating the Laplace-Stieltjes transform in (Equation 16.6), we can obtain the moments of the rth operational period, $T_{\mathcal{O}r}$. In particular, using $\mathbf{Q}_{\mathcal{OF}}\mathbf{1}_{\mathcal{F}} = -\mathbf{Q}_{\mathcal{OO}}\mathbf{1}_{\mathcal{O}}$,

$$E(T_{\mathcal{O}r}) = -\boldsymbol{\pi}_{r-1}^{(\mathcal{O})}\mathbf{Q}_{\mathcal{OO}}^{-1}\mathbf{1}_{\mathcal{O}};\ \ E(T_{\mathcal{O}r}^2) = \boldsymbol{\pi}_{r-1}^{(\mathcal{O})}\mathbf{Q}_{\mathcal{OO}}^{-2}\mathbf{1}_{\mathcal{O}}.$$

Similar results hold for the equilibrium distribution, replacing $\boldsymbol{\pi}_{r-1}^{(\mathcal{O})}$ by $\boldsymbol{\pi}^{(\mathcal{O})}$.

If the $n_{\mathcal{O}}$ eigenvalues, $\{\lambda_i\}$ of $-\mathbf{Q}_{\mathcal{OO}}$ are distinct then, from the spectral expansion of $-\mathbf{Q}_{\mathcal{OO}}$, it follows that the above distributions, expressed as probability densities, are linear combinations of exponentials

$$\sum_{j=1}^{n_{\mathcal{O}}} a_j e^{-\lambda_j t},\quad t > 0.$$

Distributions of failed periods can be obtained in a similar way.

It is often the case that not all operational states can be reached directly from the set of failed states, corresponding to columns in $\mathbf{P}_{\mathcal{FO}}$ that contain only zeros: the corresponding rows of $\mathbf{G}_{\mathcal{OF}}(t)$ are in effect redundant because the system cannot begin an operational sojourn in such states. Those that can be thus reached, say $m_{\mathcal{O}}(\leq n_{\mathcal{O}})$ in number, are called *operational gateway states*. Similarly, there will be a set of *failed gateway states*, $m_{\mathcal{F}}(\leq n_{\mathcal{F}})$ in number, in which it is possible to begin a sojourn in the set of $\mathcal{F}$ states.

Then the semi-Markov process embedded at the switching points may be defined in terms of a reduced state space. See Ball *et al.* (1991) for further details. This may have an advantage in computing results for some specific examples but does little to simplify the general formulae. In formulating the general theory, we will, therefore, use the full state space.

In reliability studies it is sometimes useful to attach a value, or reward rate $r(i)$, to each state i and usually $r(i) = 0$ if $i \in \mathcal{F}$. Then the total reward over some interval (a, b) is given by $\int_a^b r(X(t))dt$. In particular, if r is an indicator function, $r(i) = 1 \leftrightarrow i \in \mathcal{O}$, the reward is just total operational time over the interval, the instantaneous availability at time t is $P(r(X(t)) = 1)$, the probability $P(\frac{1}{t}\int_0^t r(X(u))du \le x)$ is an interval availability distribution function, etc. Sometimes you might want to specify a lesser value for some states that are only partly operational: for example, if a road is reduced to single-lane operation by road works, you can still use it but it is not totally satisfactory.

Distributions of rewards can be found by appropriate modification of the time scales within each state (imagine a clock running faster or slower depending on which state the system is in) so that the total reward is equivalent to total operational time on the modified clock: see, for example, Rubino and Sericola (1993). Therefore, from here on, we will not discuss rewards explicitly.

Hawkes *et al.* (2011) built a model for the dynamics of a system represented by a finite-state Markov process operating under two alternating regimes. The distributions of uptimes and the numbers of completed regimes within an uptime, including the special case of constant regime durations, were given as the solutions to the occupancy time distributions in some aggregated stochastic processes.

16.3 Semi-Markov Models

More generally, suppose now that the underlying process $X(t)$ is a semi-Markov process with kernel partitioned as

$$\mathbf{\Gamma}(t) = \begin{pmatrix} \mathbf{\Gamma}_{\mathcal{OO}}(t) & \mathbf{\Gamma}_{\mathcal{OF}}(t) \\ \mathbf{\Gamma}_{\mathcal{FO}}(t) & \mathbf{\Gamma}_{\mathcal{FF}}(t) \end{pmatrix} \tag{16.8}$$

and Laplace-Stieltjes transform $\mathbf{\Gamma}^*(s) = \int_0^\infty e^{-st} d\mathbf{\Gamma}(t)$. Then the embedded process $Y(t)$ is still a semi-Markov process whose kernel has the form given by (Equation 16.2), but now $\mathbf{G}_{\mathcal{OF}}(t)$ and $\mathbf{G}_{\mathcal{FO}}(t)$ are matrices whose Laplace-Stieltjes transforms are,

$$\mathbf{G}^*_{\mathcal{OF}}(s) = \int_0^\infty e^{-st} d\mathbf{G}_{\mathcal{OF}}(t) = \sum_{n=0}^\infty \{\mathbf{\Gamma}^*_{\mathcal{OO}}(s)\}^n \mathbf{\Gamma}^*_{\mathcal{OF}}(s) = \{\mathbf{I} - \mathbf{\Gamma}^*_{\mathcal{OO}}(s)\}^{-1}\mathbf{\Gamma}^*_{\mathcal{OF}}(s);$$

$$\mathbf{G}^*_{\mathcal{FO}}(s) = \int_0^\infty e^{-st} d\mathbf{G}_{\mathcal{FO}}(t) = \sum_{n=0}^\infty \{\mathbf{\Gamma}^*_{\mathcal{FF}}(s)\}^n \mathbf{\Gamma}^*_{\mathcal{FO}}(s) = \{\mathbf{I} - \mathbf{\Gamma}^*_{\mathcal{FF}}(s)\}^{-1}\mathbf{\Gamma}^*_{\mathcal{FO}}(s). \tag{16.9}$$

The first of these arises because the sojourn in $\mathcal{O}$ may consist of the convolution of any number n of transitions within $\mathcal{O}$ followed by a final transition into $\mathcal{F}$. The second case is similar. Equations (16.2), (16.4), (16.6) and (16.7) still hold, except that the partitions of the transition matrix of the embedded Markov chain may now be given by

$$\mathbf{P}_{\mathcal{OF}} = \mathbf{G}_{\mathcal{OF}}(\infty) = \int_0^\infty d\mathbf{G}_{\mathcal{OF}}(t) = \mathbf{G}^*_{\mathcal{OF}}(0); \quad \mathbf{P}_{\mathcal{FO}} = \int_0^\infty d\mathbf{G}_{\mathcal{FO}}(t) = \mathbf{G}^*_{\mathcal{FO}}(0). \tag{16.10}$$

When there is an equilibrium probability vector $\boldsymbol{\pi}^{(\mathcal{O})}$ for entry into the operational state, it is the solution of

$$\boldsymbol{\pi}^{(\mathcal{O})}\mathbf{P}_{\mathcal{OF}}\mathbf{P}_{\mathcal{FO}} = \boldsymbol{\pi}^{(\mathcal{O})} \quad \text{together with} \quad \boldsymbol{\pi}^{(\mathcal{O})}\mathbf{1}_{\mathcal{O}} = 1.$$

16.3.1 Joint Distributions of Operational and Failed Times

The joint distribution of the first m operational times and the first m failed times can be given in terms of a multivariate Laplace-Stieltjes transform as

$$E\left(\exp\left(\sum_{r=1}^{m}\left(T_{\mathcal{O}r}s_{\mathcal{O}r} + T_{\mathcal{F}r}s_{\mathcal{F}r}\right)\right)\right) = \boldsymbol{\pi}_0^{(\mathcal{O})}\prod_{r=1}^{m}\mathbf{G}^*_{\mathcal{OF}}(s_{\mathcal{O}r})\mathbf{G}^*_{\mathcal{FO}}(s_{\mathcal{F}r})\mathbf{1}_{\mathcal{O}}, \tag{16.11}$$

where $s_{\mathcal{O}r}$, $s_{\mathcal{F}r}$ are the Laplace-Stieltjes transform parameters corresponding, respectively, to the duration of the rth operational period, $T_{\mathcal{O}r}$, and the duration of the r th failed period, $T_{\mathcal{F}r}$. Here, and in all further equations, we assume that the system starts at the beginning of an operational period in a state with initial probability row vector $\boldsymbol{\pi}_0^{(\mathcal{O})}$.

For the Markov case we can replace the Laplace-Stieltjes transform by the joint density function

$$\boldsymbol{\pi}_0^{(\mathcal{O})}\prod_{r=1}^{m}\exp(\mathbf{Q}_{\mathcal{OO}}t_{\mathcal{O}r})\mathbf{Q}_{\mathcal{OF}}\exp(\mathbf{Q}_{\mathcal{FF}}t_{\mathcal{F}r})\mathbf{Q}_{\mathcal{FO}}\mathbf{1}_{\mathcal{O}}. \tag{16.12}$$

Equivalent results are given in the reliability literature by Csenki (1991, 1992). Csenki also has a somewhat complicated result for dealing with a situation with three or more classes, with simpler results in some special cases.

It can be useful to look at the relationship between neighbouring intervals, such as the mth operational time and the failure time that follows. They have joint Laplace-Stieltjes transform

$$\boldsymbol{\pi}_0^{(\mathcal{O})}(\mathbf{P}_{\mathcal{OF}}\mathbf{P}_{\mathcal{FO}})^{m-1}\mathbf{G}^*_{\mathcal{OF}}(s_{\mathcal{O}m})\mathbf{G}^*_{\mathcal{FO}}(s_{\mathcal{F}m})\mathbf{1}_{\mathcal{O}}, \tag{16.13}$$

obtained simply by setting all the other Laplace-Stieltjes parameters in Equation (16.11) to zero. More usefully, the joint equilibrium distribution for any operational time and its following failure time, as given by the joint Laplace-Stieltjes transform, is

$$\boldsymbol{\pi}^{(\mathcal{O})}\mathbf{G}^*_{\mathcal{OF}}(s_{\mathcal{O}})\mathbf{G}^*_{\mathcal{FO}}(s_{\mathcal{F}})\mathbf{1}_{\mathcal{O}}. \tag{16.14}$$

Again, the Laplace-Stieltjes transforms may be replaced by probability densities where they can be explicitly calculated.

Once we have joint distributions we can consider summary statistics to express the relationship between variables, notably correlations. For example, the joint moment of adjacent intervals can be found from differentiating (16.14)

$$E\left(T_{\mathcal{O}}T_{\mathcal{F}}\right) = \boldsymbol{\pi}^{(\mathcal{O})}\left\{\left(\partial\mathbf{G}^*_{\mathcal{OF}}/\partial s_{\mathcal{O}}\right)\left(\partial\mathbf{G}^*_{\mathcal{FO}}/\partial s_{\mathcal{F}}\right)\right\}_{s_{\mathcal{O}}=s_{\mathcal{F}}=0}\mathbf{1}_{\mathcal{O}}.$$

Correlations can occur between various operational times and failures times if there are at least two $\mathcal{O}$ states and at least two $\mathcal{F}$ states. See, for example, Colquhoun and Hawkes (1987), Ball and Rice (1989).

The marginal distribution of the mth operational sojourn (alternatively, failed sojourn) has Laplace-Stieltjes transform given by setting $s_{\mathcal{F}m} = 0$ (alternatively, $s_{\mathcal{O}m} = 0$) in Equation (16.13) giving

$$\pi_0^{(\mathcal{O})}(\mathbf{P}_{\mathcal{O}\mathcal{F}}\mathbf{P}_{\mathcal{F}\mathcal{O}})^{m-1}\mathbf{G}^*_{\mathcal{O}\mathcal{F}}(s_{\mathcal{O}m})\mathbf{1}_{\mathcal{F}}, \text{ for operational sojourns;}$$

$$\pi_0^{(\mathcal{O})}(\mathbf{P}_{\mathcal{O}\mathcal{F}}\mathbf{P}_{\mathcal{F}\mathcal{O}})^{m-1}\mathbf{P}_{\mathcal{O}\mathcal{F}}\mathbf{G}^*_{\mathcal{F}\mathcal{O}}(s_{\mathcal{F}m})\mathbf{1}_{\mathcal{O}}, \text{ for failed sojourns.} \tag{16.15}$$

Note that the first of these results has been simplified by using the fact that $\mathbf{P}_{\mathcal{F}\mathcal{O}}\mathbf{1}_{\mathcal{O}} = \mathbf{1}_{\mathcal{F}}$. The distribution of the mth operational period for semi-Markov processes is discussed in Rubino and Sericola (1993), including results equivalent to Equation (16.15).

The equilibrium distributions of operational sojourns and failed sojourns are given by Laplace-Stieltjes transforms

$$\pi^{(\mathcal{O})}\mathbf{G}^*_{\mathcal{O}\mathcal{F}}(s)\mathbf{1}_{\mathcal{F}}, \text{ for operational sojourns;}$$

$$\pi^{(\mathcal{F})}\mathbf{G}^*_{\mathcal{F}\mathcal{O}}(s)\mathbf{1}_{\mathcal{O}}, \text{ for failed sojourns,} \tag{16.16}$$

where

$$\pi^{(\mathcal{F})} = \pi^{(\mathcal{O})}\mathbf{P}_{\mathcal{O}\mathcal{F}}.$$

Mean operational time and mean failed time can be found by differentiating the transforms in (16.15) or (16.16), as appropriate. For example, the mean equilibrium operational time is

$$\pi^{(\mathcal{O})}\{-\partial\mathbf{G}^*_{\mathcal{O}\mathcal{F}}(s)/\partial s\}_{s=0}\mathbf{1}_{\mathcal{F}}.$$

16.3.2 Distribution of Cumulative Times

If in Equation (16.11) we let $s_{\mathcal{O}r} = s_{\mathcal{O}}$ and $s_{\mathcal{F}r} = s_{\mathcal{F}}, \forall r$, we get the Laplace-Stieltjes transform of the joint distribution of total operational time and total failed time over the first m cycles of operation and failure,

$$\pi_0^{(\mathcal{O})}(\mathbf{G}^*_{\mathcal{O}\mathcal{F}}(s_{\mathcal{O}})\mathbf{G}^*_{\mathcal{F}\mathcal{O}}(s_{\mathcal{F}}))^m\mathbf{1}_{\mathcal{O}}. \tag{16.17}$$

If we now set $s_{\mathcal{F}} = 0$ (or alternatively $s_{\mathcal{O}} = 0$) in Equation (16.17), we get the Laplace-Stieltjes transform of the distribution of total operational time (respectively total failure time) over the first m cycles as

$$\pi_0^{(\mathcal{O})}(\mathbf{G}^*_{\mathcal{O}\mathcal{F}}(s_{\mathcal{O}})\mathbf{P}_{\mathcal{F}\mathcal{O}})^m\mathbf{1}_{\mathcal{O}} \text{ for operational times;}$$

$$\pi_0^{(\mathcal{O})}(\mathbf{P}_{\mathcal{O}\mathcal{F}}\mathbf{G}^*_{\mathcal{F}\mathcal{O}}(s_{\mathcal{F}}))^m\mathbf{1}_{\mathcal{O}} \text{ for failure times.} \tag{16.18}$$

The distribution of the total operational time up to and including the m th operational period is discussed in Rubino and Sericola (1992).

Cui *et al.* (2007) introduced a Markov system in which the states were classified as up (U), down (D) or changeable (C): if a visit to C followed a sojourn in U, it was classified as up, if it followed a sojourn in D, it was classified as down. Zheng *et al.* (2008) considered the instantaneous availability of semi-Markov repairable systems with history-dependent up and down states. For the same semi-Markov model, Wang and Cui (2011) provided the distribution of the time to the first system failure and the Laplace-Stieltjes transforms of several time distributions in a cycle, such as the up and down time, the total time the system is in the up, down and changeable states.

16.4 Time Interval Omission

Recently interest has been shown in problems in which it may be reasonable to ignore short breaks in operation, so that an effective operational period might be considered to consist of a number of sojourns in $\mathcal{O}$ separated by sojourns in $\mathcal{F}$ that last for less than some specified critical constant ξ, say. For example, in the operation of a gas platform, short interruptions of several minutes can probably be covered by inventory but longer delays, say, a day or two, are likely to cause operational difficulties and supply contract penalties. Similarly, penalties for downtimes longer than a few minutes may be written into a contract for computer services. At the same time, very short operational periods may not be worth considering.

Similar problems have already been solved in the ion-channel literature. In that case the processes operate so fast that the recording equipment is unable to see very short intervals, of a few microseconds in duration, so that open sojourns and shut sojourns appear to be fewer in number and of greater duration than they should be: these are called apparent open and shut times. This is called the time interval omission (TIO) problem.

To use general terminology that covers various applications, we define an apparent sojourn in $\mathcal{O}$ (or *e*-sojourn: *e* for extended) as starting with a sojourn in $\mathcal{O}$ of duration at least $\xi_{\mathcal{O}}$ that may be extended by a sequence of $\mathcal{O}$ sojourns separated by short $\mathcal{F}$ sojourns of duration less than $\xi_{\mathcal{F}}$. Similarly, define an *e*-sojourn in $\mathcal{F}$ that starts with a sojourn in $\mathcal{F}$ of duration at least $\xi_{\mathcal{F}}$ that may be extended by a sequence of $\mathcal{F}$ sojourns separated by short $\mathcal{O}$ sojourns of duration less than $\xi_{\mathcal{O}}$. Then there is a semi-Markov process embedded at the start of each *e*-sojourn, having a semi-Markov kernel whose Laplace transform will be denoted

$$ {}^{e}\mathbf{G}^*(s) = \begin{pmatrix} \mathbf{0} & {}^{e}\mathbf{G}^*_{\mathcal{OF}}(s) \\ {}^{e}\mathbf{G}^*_{\mathcal{FO}}(s) & 0 \end{pmatrix}. $$

Ball and Sansom (1988) obtained these Laplace Transforms when the underlying process is a Markov process. Hawkes *et al.* (1990) obtained exact expressions for the corresponding probability densities. These are fine for small to moderate values of time t, but can be numerically unstable for large t. Asymptotic approximations were found (Jalali and Hawkes 1992; Hawkes *et al.* 1992), that are extremely accurate for values of t from very large right down to very small. For small t the exact results are readily obtainable, so that the distributions are obtained over the whole range. If the true distribution (without TIO) is a mixture of k exponentials, then the approximation to the distribution of apparent times allowing for TIO is also a mixture of k exponentials: the time constants are, however, different.

Colquhoun *et al.* (1996) studied the effect of TIO on joint distributions of neighbouring apparent open and shut times. They also used the joint distribution of a complete series of intervals to calculate the likelihood, using an equation similar to Equation (16.12) with the *e*-distributions instead of the standard Markov ones, and thus estimated the model parameters.

The TIO problem was also studied for an underlying semi-Markov framework by Ball *et al.* (1991, 1993) who gave formal expressions for the Laplace transform of the semi-Markov kernel of the sequence of apparent sojourns.

These ideas have been used for simple reliability models, but concentrating on availability measures: see Zheng *et al.* (2006), Bao and Cui (2010). There is great scope for further applications to reliability models of the missing interval properties described earlier.

16.5 Numerical Examples

In order to illustrate the results given in the previous sections, numerical results are given in this section for two simple models: one Markov and one semi-Markov.

For a Markov model, we assume the underlying process $X(t)$ has a state space $\mathcal{S} = \{1, 2, 3\} = \{1, 2\} \cup \{3\}$, that is, the operational and failed subsets are $\mathcal{O} = \{1, 2\}$, $\mathcal{F} = \{3\}$. The transition rate matrix $\mathbf{Q}$ is assumed to be

$$\mathbf{Q} = \begin{pmatrix} \mathbf{Q}_{\mathcal{O}\mathcal{O}} & \mathbf{Q}_{\mathcal{O}\mathcal{F}} \\ \mathbf{Q}_{\mathcal{F}\mathcal{O}} & \mathbf{Q}_{\mathcal{F}\mathcal{F}} \end{pmatrix} = \left(\begin{array}{cc|c} -5 & 1 & 4 \\ 2 & -4 & 2 \\ \hline 4 & 2 & -6 \end{array}\right).$$

Then the results given in Section 16.2 can be computed.

The Markov chain $\{J_k\}$ embedded at the switching points has the transition matrix

$$\mathbf{P} = \begin{pmatrix} \mathbf{0} & \mathbf{P}_{\mathcal{O}\mathcal{F}} \\ \mathbf{P}_{\mathcal{F}\mathcal{O}} & \mathbf{0} \end{pmatrix} = \begin{pmatrix} 0 & 0 & 1 \\ 0 & 0 & 1 \\ 2/3 & 1/3 & 0 \end{pmatrix}.$$

Thus, the duration of failed periods have probability density $6e^{-6t}, t > 0$, with mean $1/6$ and the following operating period begins in state 1 or state 2 with respective probabilities $\left(2/3 \;\; 1/3\right)$, which is also the equilibrium entry probability $\pi^{(\mathcal{O})}$. The distribution of operating times can be found from

$$\exp(\mathbf{Q}_{\mathcal{O}\mathcal{O}} t)\mathbf{Q}_{\mathcal{O}\mathcal{F}} = \begin{pmatrix} 2e^{-3t} + 2e^{-6t} \\ 4e^{-3t} - 2e^{-6t} \end{pmatrix},$$

the top and bottom elements of which are the probability densities of operational times starting, respectively, from state 1 or state 2. The corresponding means are, respectively $5/18$, $7/18$ so the equilibrium distribution of operational periods has mean

$$\pi^{(\mathcal{O})} \begin{pmatrix} 5/18 \\ 7/18 \end{pmatrix} = 17/54.$$

For the semi-Markov model, we assume the same state space as above and that cumulative transition functions are given by

$$\mathbf{\Gamma}(t) = \left(\begin{array}{cc|c} 0 & 0.4\left\{1 - (1 + 2t)\, e^{-2t}\right\} & 0.6\{1 - e^{-5t}\} \\ 0.5\{1 - e^{-t}\} & 0 & 0.5\{1 - (1 + 4t)e^{-4t}\} \\ \hline 0.2\left\{1 - \left(1 + 3t + \dfrac{9}{2}t^2\right) e^{-3t}\right\} & 0.8\{1 - e^{-2t}\} & 0 \end{array}\right),$$

$$\mathbf{P} = \begin{pmatrix} 0 & 0.4 & 0.6 \\ 0.5 & 0 & 0.5 \\ 0.2 & 0.8 & 0 \end{pmatrix} \quad \text{and } \mathbf{\Gamma}^*(s) = \left(\begin{array}{cc|c} 0 & \dfrac{1.6}{(s+2)^2} & \dfrac{3}{s+5} \\ \dfrac{0.5}{s+1} & 0 & \dfrac{8}{(s+4)^2} \\ \hline \dfrac{5.4}{(s+3)^3} & \dfrac{1.6}{s+2} & 0 \end{array}\right)$$

Then the kernel $\mathbf{G}(t)$ for the embedded process $Y(t)$ can be obtained: here we give Laplace-Stieltjes transforms from (16.9) as

$$\mathbf{G}^*_{\mathcal{OF}}(s) = \frac{(s+1)(s+2)^2}{(s+1)(s+2)^2 - 0.8} \begin{pmatrix} \dfrac{3}{s+5} + \dfrac{12.8}{(s+2)^2(s+4)^2} \\ \dfrac{1.5}{(s+1)(s+5)} + \dfrac{8}{(s+4)^2} \end{pmatrix};$$

$$\mathbf{G}^*_{\mathcal{FO}}(s) = \begin{pmatrix} \dfrac{5.4}{(s+3)^3} & \dfrac{1.6}{2+s} \end{pmatrix},$$

so that $d\mathbf{G}_{\mathcal{FO}}(t)/dt = \begin{pmatrix} 2.7t^2e^{-3t} & 1.6e^{-2t} \end{pmatrix}$ and $\mathbf{P}_{\mathcal{OF}} = \begin{pmatrix} 1 \\ 1 \end{pmatrix}$, $\mathbf{P}_{\mathcal{FO}} = \begin{pmatrix} 0.2 & 0.8 \end{pmatrix}$.

The derivative of $\mathbf{G}_{\mathcal{FO}}$, obtained from the inverse transform of $\mathbf{G}^*_{\mathcal{OF}}(s)$ is more complicated.

From equations in Section 16.3, some numerical results for distributions are obtained. Here we just take several distributions as examples, which are as follows.

In Equation (16.15), when $m = 2$, the marginal probability density function (p.d.f.) of the second operational sojourn is shown in Figure 16.1.

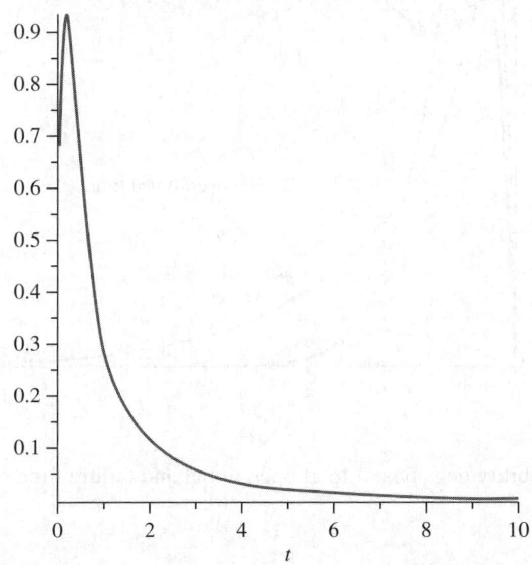

Figure 16.1 The marginal probability density of the *second* operational sojourn

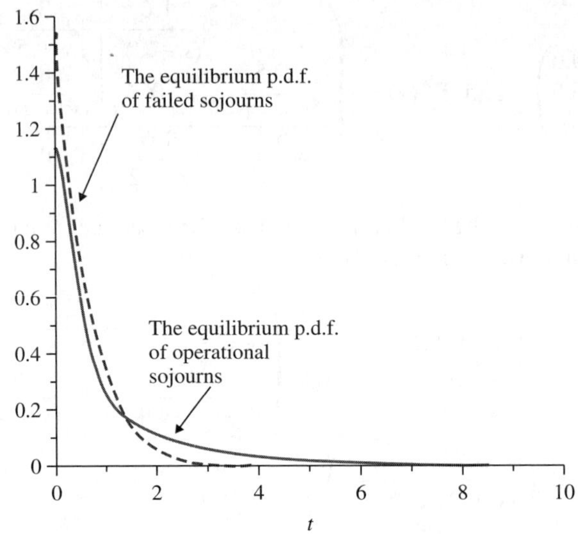

Figure 16.2 The equilibrium probability densities of operational and failed sojourns

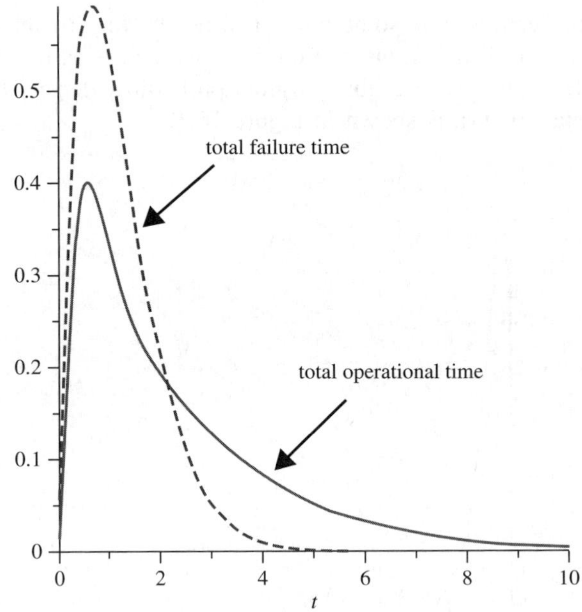

Figure 16.3 The probability densities of total operational and failure time over the first *two* cycles

For equilibrium situations, the equilibrium probability densities of operational sojourns and failed sojourns in Equation (16.16) are shown in Figure 16.2.

In Equation (16.18), when $m = 2$, the probability densities of total operational time and total failure time over the first m cycles are shown in Figure 16.3.

16.6 Conclusion

In this chapter we have presented some general results for certain probability distributions of occupancy times in terms of Laplace-Stieltjes transforms. The advantages are that the results are elegant and moments can usually be easily obtained by differentiation. The probability densities themselves are often difficult, or impossible, to express in closed form except when the underlying process is Markov. Densities can, however, be computed using standard techniques for numerical inversion of Laplace-Stieltjes transforms, see, for example, Den Iseger (2006), or using software such as Maple or Mathematica.

We have pointed out the similarity of some Markov and semi-Markov models for reliability and ion-channel theory. In particular, we think there is scope for further applications of the theory of omitted intervals, already well developed in ion channel literature, to reliability systems. Unfortunately, space restrictions have limited the attention we could give to this topic in this chapter. We have also had to omit discussion of systems that suffer terminal failure.

Acknowledgements

This work is partially supported by NSFC (71071020) and Beijing Excellent Doctoral Dissertation Instructor Project of Humanities and Social Sciences (YB20091000701).

References

Ball, F. G., R. K. Milne and G. F. Yeo. 1991. Aggregated semi-Markov processes incorporating time interval omission. *Advances in Applied Probability* 23 (4): 772–797.

Ball, F. G., R. K. Milne and G. F. Yeo. 1993. On the exact distribution of observed open times in single ion channel models. *Journal of Applied Probability* 30 (3): 529–537.

Ball, F. G. and J. A. Rice. 1989. A note on single-channel autocorrelation functions. *Mathematical Biosciences* 97 (1): 17–26.

Ball, F. G. and M. S. P. Sansom. 1988. Aggregated Markov processes incorporating time interval omission. *Advances in Applied Probability* 20: 546-572.

Bao, X. Z. and L. R. Cui. 2010. An analysis of availability for series Markov repairable system with neglected or delayed failures. *IEEE Transactions on Reliability* 59 (4): 734–743.

Colquhoun, D. and A. G. Hawkes. 1982. On the stochastic properties of bursts of single ion channel openings and of clusters of bursts. *Philosophical Transactions of the Royal Society of London B* 200: 1–59.

Colquhoun, D., A. G. Hawkes and K. Srodsinski. 1996. Joint distributions of apparent open and shut times of single ion channels and the maximum likelihood fitting of mechanisms. *Philosophical Transactions of the Royal Society of London: Mathematical, Physical and Engineering Science A* 354: 2555–2590.

Csenki, A. 1991. The joint distribution of sojourn times in finite semi-Markov processes. *Stochastic Processes and Their Applications* 39 (2): 287–299.

Csenki, A. 1992. The joint distribution of sojourn times in finite Markov processes. *Advances in Applied Probability* 24 (1): 141–160.

Cui, L. R., H. J. Li and J. L. Li. 2007. Markov repairable systems with history-dependent up and down states. *Stochastic Models* 23 (4): 665–681.

Hawkes, A. G., L. R. Cui and Z. H. Zheng. 2011. Modeling the evolution of system reliability performance under alternative environments. *IIE Transactions* 43 (11): 761–772.

Hawkes, A. G., A. Jalali and D. Colquhoun. 1990. The distributions of the apparent open times and shut times in a single channel record when brief events cannot be detected. *Philosophical Transactions of the Royal Society of London: Physical Sciences and Engineering A* 332: 511–538.

Hawkes, A. G., A. Jalali and D. Colquhoun. 1992. Asymptotic distributions of apparent open times and shut times in a single channel record allowing for the omission of brief events. *Philosophical Transactions of the Royal Society of London B* 337: 383–404.

Iseger, P. D. 2006. Numerical transform inversion using Gaussian quadrature. *Probability in the Engineering and Informational Sciences* 20 (1): 1–44.

Jalali, A. and A. G. Hawkes 1992. Generalised eigenproblems arising in aggregated Markov processes allowing for time interval omission. *Advances in Applied Probability* 24 (2): 302–321.

Rubino, G. and B. Sericola. 1989. Sojourn times in finite Markov processes. *Journal of Applied Probability* 26 (4): 744–756.

Rubino, G. and B. Sericola. 1992. Interval availability analysis using operational periods. *Performance Evaluation* 14 (304): 257–272.

Rubino, G. and B. Sericola. 1993. Sojourn times in semi-Markov reward processes: application to fault-tolerant systems modelling. *Reliability Engineering and Systems Safety* 41 (1): 1–4.

Wang, L. Y. and L. R. Cui. 2011. Aggregated semi-Markov repairable systems with history-dependent up and down states. *Mathematical and Computer Modelling* 53 (5–6): 883–895.

Zheng, Z. H., L. R. Cui and A. G. Hawkes. 2006. A study on a single-unit Markov repairable system with repair time omission. *IEEE Transactions on Reliability* 55 (2): 182–188.

Zheng, Z. H., L. R. Cui, and H. J. Li. 2008. Availability of semi-Markov repairable systems with history-dependent up and down states. In: *Proceedings of the 3rd Asian International Workshop*. Taipei, Taiwan, pp. 186–193.

17

A Practice of Imperfect Maintenance Model Selection for Diesel Engines

Yu Liu, Hong-Zhong Huang, Shun-Peng Zhu and Yan-Feng Li
School of Mechanical, Electronic, and Industrial Engineering, University of Electronic Science and Technology of China, China

17.1 Introduction

Maintenance activities have been extensively conducted in engineering practices to enhance reliability and performance of degraded systems, restore failed systems to working condition, and extend as much as possible the remaining life of systems. An effective maintenance planning is highly desirable from an industry perspective as it can reduce not only the frequency of failure of failure-prone systems, but also maintenance expenditures (Wang 2002). Before developing a decision model for maintenance scheduling of engineered systems, the impact of maintenance activities on the physical condition of maintained systems must first be characterized. According to its efficiency, maintenance can be classified into one of the following five situations (Pham and Wang 1996; Wu 2011): better than new, perfect, imperfect, minimal, and worse. The "better than new" maintenance changes the condition of a repairable item into a state with less failure intensity than a brand-new identical one. Perfect maintenance (e.g. complete replacement) is able to restore the repaired system back to a brand new ("as good as new") condition, whereas a minimal maintenance (e.g. correction of a minor flaw) assumes a maintenance action can only leave the system in the same condition as it was before failure ("as bad as old"). Worse maintenance results in a negative impact on the system's condition due to ineffective maintenance. Nevertheless, *imperfect maintenance* which can only restore the maintained systems to a condition somewhere

Applied Reliability Engineering and Risk Analysis: Probabilistic Models and Statistical Inference, First Edition.
Edited by Ilia B. Frenkel, Alex Karagrigoriou, Anatoly Lisnianski and Andre Kleyner.

between "as good as new" and "as bad as old" has long been recognized as the set of the most common maintenance activities in practices.

Numerous mathematical models have been put forward in the literature to characterize the efficiency of imperfect maintenance from various perspectives. A comprehensive survey of imperfect maintenance models for binary state systems components can be found in (Pham and Wang 1996; Wu and Zuo 2010; Wu 2011). Additionally, as many engineered systems possess multiple intermediate states between perfect functioning and complete failure, several attempts have been made to investigate imperfect maintenance models for multi-state systems as of late (Liu and Huang 2010; Soro *et al.* 2010).

As every imperfect maintenance model characterizes impacts of maintenance activities via its own assumption of underlying physics, appropriate imperfect maintenance models must be chosen before constructing maintenance decision models. Due to the lack of a comprehensive understanding of the true physics and mechanism by which maintained components/systems are governed, especially in the case where a sophisticated system is maintained by replacing and fixing multiple components subject to a variety of failure modes, the data collected from system operation and maintenance become the only available information that can be utilized to select an appropriate imperfect maintenance model. Therefore, it requires a tool to examine whether the pre-assumed imperfect maintenance model is adequate enough to characterize the effect of maintenance. The goodness-of-fit (GOF) test is a mathematically rigorous way to validate the adequacy of model in static sense. However, to the best of our knowledge, the GOF test for imperfect maintenance models has rarely been investigated in the literature. In addition, if more than one imperfect maintenance model passes the GOF test, it is oftentimes necessary to choose the best model among multiple competitive candidate models. This issue is known as the model selection problem. Without any knowledge of underlying physics, the model which has the best fitness to the data from real systems will be chosen.

To address the two aforementioned issues, a novel GOF test method along with a tailored Bayesian model selection method was proposed in our recent work (Liu *et al.* 2012). The unique features of the GOF test method are twofold: First, by pooling all the failure data into a u-space, the adequacy of the imperfect maintenance model can be validated by checking the uniformity of u-pooling data. Second, the tailored Bayesian model selection method identifies the most adequate imperfect maintenance model among several competitive candidates while considering the uncertainty arising from the parameters' estimation. As demonstrated in the designed test problems, the new method has an exceptional capability to identify the best imperfect maintenance model even with limited failure data. In this chapter, the proposed imperfect maintenance model selection method is applied to a preventive maintenance scheduling problem of diesel engines to further demonstrate the use of our method to engineering practices. The framework presented in this chapter provides a general guideline for practitioners to use the imperfect maintenance model selection method in a broader range of applications.

The remainder of this chapter is organized as follows: The proposed imperfect maintenance model selection method is first reviewed in Section 17.2. Detailed procedures of applying the proposed method to an engineering example of diesel engines are presented in Section 17.3, and this is followed by a brief closure in Section 17.4.

17.2 Review of Imperfect Maintenance Model Selection Method

The proposed imperfect maintenance model selection method consists of six steps as illustrated in Figure 17.1. In Step 1, the operation and failure data of repairable systems in question are collected. The unknown parameters of candidate imperfect maintenance models are estimated by using the maximum likelihood estimation (MLE) in Step 2. The likelihood function for parameter estimation will be detailed in Section 17.2.1. The proposed GOF test method is conducted in Step 3. By pooling all the failure data into a u-space, the adequacy of the candidate imperfect maintenance models can be validated by checking whether or not the pooled u values are uniformly distribution. The tailored Bayesian model selection method is performed in Step 4 if more than one competitive model passes the GOF test under a pre-specified confidence level (say, 95%), and then one can obtain posterior probabilities of the hypothesis that failure data are from one of competitive imperfect maintenance models. The model possessing the highest posterior probability is regarded as the best model and can be identified in Step 5. The best model chosen from the candidates or all the competitive models along with their posterior probabilities is implemented in the subsequent maintenance decision in Step 6. As uncertainty from both parameters estimation and model selection is incorporated in the maintenance decision, several proposed criteria for decision-making can be used to identify the optimal maintenance strategy. The ensuing sections review more details of formulations and algorithms of this method with the assumption that systems are minimally repaired upon failure and preventively maintained at pre-specified time intervals. The preventive maintenance (PM) herein is treated as imperfect maintenance, and failure data are collected from N identical repairable systems. Some notations are listed as follows:

t_k^p: Time interval between the $(k-1)$th and the kth PMs
$\lambda_k(t)$: Failure intensity function in the kth PM cycle
$n_{l,k}$: Number of failures of the lth repairable system in the kth PM cycle

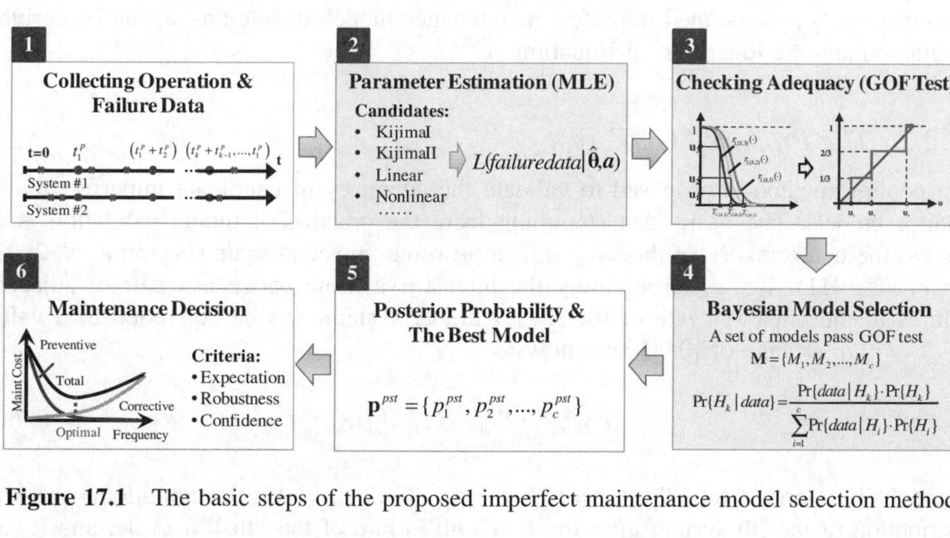

Figure 17.1 The basic steps of the proposed imperfect maintenance model selection method

$t^f_{l,(k,i)}$: Time instant of the ith failure of the lth system in the kth PM cycle
$S_{l,(k,i)}(t)$: Conditional survival distribution of the lth repairable system after the $(i-1)$, the failure of the kth PM cycle
$T_{l,(k,i)}$: Inter-arrival time between the $(i-1)$th and the ith failures of the kth PM cycle for the lth repairable system
$\Pr\{H_k|data\}$: Posterior probability of hypothesis H_k
$\Pr\{\widehat{a}|H_i\}$: Distribution of estimated parameters $\widehat{a}$ of the ith imperfect maintenance model.

17.2.1 Estimation of the Parameters

Before conducting the proposed GOF test, unknown parameters in candidate imperfect maintenance models need to be first of all estimated. The MLE can be used here to infer the unknown parameters of each model. Suppose that $n_{l,k}$ failures are observed in the kth PM cycle of the lth system, and the corresponding failure time instants are denoted as $t^f_{l,(k,1)}, t^f_{l,(k,2)}, \ldots, t^f_{l,(k,n_{l,k})}$, where $0 < t^f_{l,(k,1)} < t^f_{l,(k,2)} < \ldots < t^f_{l,(k,n_{l,k})} < t^p_k$. If the system is discarded at the end of the Hth PM, the likelihood function can be formulated as:

$$L(failure\ data|\boldsymbol{\theta}, \boldsymbol{a}) = \prod_{l=1}^{N}\prod_{k=1}^{H}\left(\prod_{i=1}^{n_{l,k}}\left(\lambda_k\left(t^f_{l,(k,i)}\right)\exp\left(-\int_{t^f_{l,(k,i-1)}}^{t^f_{l,(k,i)}}\lambda_k(t)dt\right)\right),\right.$$
$$\left.\exp\left(-\int_{t^f_{l,(k,n_{l,k})}}^{t^p_k}\lambda_k(t)dt\right)\right) \quad (17.1)$$

where $\boldsymbol{\theta}$ denotes the parameters of initial failure intensity function, and it is assumed to be known in advance or can be estimated based on failure data collected in the first PM repair cycle where imperfect PM are not involved. The estimators of the unknown parameters of a pre-assumed imperfect maintenance model, denoted as $\hat{\mathbf{a}}$, can be derived by maximizing the logarithm of Equation (17.1).

17.2.2 The Proposed GOF Test

A u-pooling method is proposed to validate the adequacy of candidate imperfect maintenance models. The basic idea stemming from the practice of model validation is to express the conformance of theory and data on some universal scale (Ferson *et al.* 2008; Liu *et al.* 2011). $T_{l,(k,i)}$, representing the inter-arrival time between a pair of adjacent failures in the kth PM cycle of the lth repairable system, can be converted to a value $u_{l,(k,i)}$ within a range of [0, 1] as follows:

$$u_{l,(k,i)} = S_{l,(k,i)}(T_{l,(k,i)}), \quad (17.2)$$

where the inter-arrival time $T_{l,(k,i)} = t^f_{l,(k,i)} - t^f_{l,(k,i-1)}$, $S_{l,(k,i)}(\cdot)$ is the conditional survival distribution of the lth system after the $(i-1)$th failure of the kth PM cycle, and it can be written as:

$$S_{l,(k,i)}(t') = \exp\left(-\int_{t^f_{l,(k,i-1)}}^{t'+t^f_{l,(k,i-1)}} \lambda_k(t|\widehat{a})dt\right), \tag{17.3}$$

where t' represents the time elapsed since the last minimal repair, and $\lambda_k(t|\widehat{a})$, with estimators $\widehat{a}$ from MLE, is the failure intensity function of the kth PM cycle of a pre-assumed imperfect maintenance model. An illustration of this conversation is shown in Figure 17.2(a) where three inter-arrival times $T_{l,(k,1)}$, $T_{l,(k,2)}$, and $T_{l,(k,3)}$ are transformed into u values $u_{l,(k,1)}$, $u_{l,(k,2)}$, and $u_{l,(k,3)}$ with respect to the corresponding conditional survival distribution. It should be noted that if the collected failure data is censored (or suspended due to the next PM), the corresponding truncated conditional survival distribution will replace Equation (17.3) to attain u values. Based upon the probability integral transform theorem, if all the inter-arrival times $T_{l,(k,i)}$ are distributed according to their respective conditional survival distribution $S_{l,(k,i)}$, all the u values converted from $T_{l,(k,i)}$ constitute a standard uniform distribution. By pooling all the u values into a u-space with a range of [0, 1] × [0, 1] as shown in Figure 17.2(b), the adequacy of the pre-assumed imperfect maintenance model can be validated by checking the uniformity of the u values. Several well-established GOF test methods, such as Kolmogorov-Smirnov (K-S), Anderson-Darling, and Cramér-von Mises, can be used to detect evidence of significant disagreement between the distribution of the u values and the standard uniform distribution with a pre-set confidence level.

17.2.3 Bayesian Model Selection

Model selection is of great significance if more than one candidate model passes the aforementioned GOF test. Suppose a finite set **M** consists of all the competitive imperfect maintenance models, for example, $\mathbf{M} = \{M_1, M_2, \ldots, M_c\}$, which are not rejected in the GOF test. According to the basis of Bayesian model selection method, one can define a set of hypothesis H_k as follows:

H_k : The failure data are from the $M_k(k = 1, 2, \ldots, c)$ imperfect maintenance model.

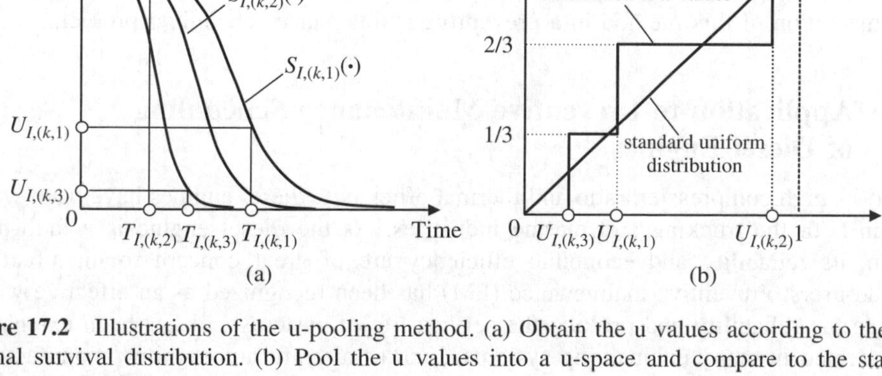

Figure 17.2 Illustrations of the u-pooling method. (a) Obtain the u values according to the conditional survival distribution. (b) Pool the u values into a u-space and compare to the standard uniform distribution

Given the failure data set, the posterior probability of the hypothesis H_k is defined by:

$$\Pr\{H_k|data\} = \frac{\Pr\{data|H_k\} \cdot \Pr\{H_k\}}{\sum_{i=1}^{c} \Pr\{data|H_i\} \cdot \Pr\{H_i\}}, \tag{17.4}$$

where $\Pr\{H_i\}$ is the prior preference of the hypothesis H_k; $\Pr\{data|H_i\}$ is the likelihood function under the hypothesis H_k which can be further written as:

$$\Pr\{data|H_i\} = \int \Pr\{data|\widehat{\boldsymbol{a}}, H_i\} \Pr\{\widehat{\boldsymbol{a}}|H_i\} d\widehat{\boldsymbol{a}}, \tag{17.5}$$

where $\Pr\{data|\widehat{\boldsymbol{a}}, H_i\}$ is the likelihood function of the ith imperfect maintenance model with estimator $\widehat{\boldsymbol{a}}$ and expressed by Equation (17.1). $\Pr\{\widehat{\boldsymbol{a}}|H_i\}$, quantifying the uncertainty associated with the parameters' estimation, is the distribution of the estimated parameters $\widehat{\boldsymbol{a}}$ of the ith imperfect maintenance model. To derive $\Pr\{\widehat{\boldsymbol{a}}|H_i\}$, the bootstrap method can be used where multiple copies of identical repairable systems exist. The basic procedures of the bootstrap method in our specific problem are as follows:

1. Obtain the initial N sets of failure data from N identical repairable systems.
2. Randomly choose N sets of failure data by re-sampling with a replacement.
3. Estimate the unknown parameter $\boldsymbol{a}$ of the ith imperfect maintenance model by the MLE method introduced in Section 17.2.1.
4. Repeat Step 2 and Step 3 p times (typically 500 to 5000).
5. Compute the empirical distribution of the estimators $\widehat{\boldsymbol{a}}$ based on the p sets of MLE results.

After deriving $\Pr\{\widehat{\boldsymbol{a}}|H_i\}$, the posterior probability of each hypothesis can be computed by Equation (17.4). The imperfect maintenance model with the highest posterior probability and strongly dominating the others is identified as the most adequate model to fit the failure data and will be used in maintenance decision. On the other hand, if the posterior probabilities of several competitive models are close to one another and none of models possesses a very strong domination, all these competitive models along with the associated posterior probability will have to be considered in the maintenance decision. In Section 17.3, a practical example of diesel engines will be presented to demonstrate the implementation of this method in a preventive maintenance scheduling problem.

17.3 Application to Preventive Maintenance Scheduling of Diesel Engines

Due to its high compression ratio and thermal efficiency, diesel engines have been widely used in both the trucking and marine industries. As the diesel engine is a fault-prone system, its reliability and economic efficiency are of great concern to manufacturers and end-users. Preventive maintenance (PM) has been recognized as an effective way to enhance the reliability and prolong the service life of systems and has been extensively implemented in many engineering systems, such as aircraft, automobiles, manufacturing systems, and it has also been adopted by diesel engine manufacturers. However, an inappropriate plan of PM may cause under-maintenance or over-maintenance, which will lead to further unexpected failures and a waste of resources respectively. Scheduling an effective PM plan for diesel engines is a priority for the manufacturers.

The diesel engine studied (the Model 6G) has six cylinders and is extensively used in coaches. Based on functional and physical logics, the diesel engine can be decomposed into six major subsystems, namely, the cylinder and crankcase subsystem, the fuel subsystem, the valve subsystem, the starting and charging subsystem, the lubrication subsystem, and the cooling subsystem. Any malfunction of a subsystem will lead to the failure of the diesel engine, and therefore these subsystems can be regarded as serially connected from reliability engineering point of view. With the aim of reducing maintenance costs over the promised warranty period (300,000 km), the optimal PM sequence for each individual subsystem needs to be determined. As the mileage is a more suitable measure than the work time in the reliability analysis of vehicle engines, it should be noted that we herein use 10,000 km as the unit for both parameters' estimation and maintenance decisions.

17.3.1 Initial Failure Intensity Estimation

Before choosing appropriate imperfect maintenance models and constructing maintenance decision models, the initial failure intensity function which characterizes the failure behavior of each individual subsystem without interventions from PM needs to be identified. Therefore, a group of ten diesel engines were put into use without executing any PM, and every diesel engine was minimally repaired upon failures. Based on the failure data collected up to 400,000 km, the initial failure intensity function (e.g. power law, log-linear, etc.) of these six subsystems along with unknown parameters can be identified, based on the maximum likelihood estimation. Due to the limitations of this chapter, the details of estimating the initial failure intensity function are not elaborated here. But the general procedures can be found in (Rigdon and Basu 2000). The initial failure intensity functions and the corresponding estimators of unknown parameters are listed in Table 17.1. The average cumulative numbers of failures observed from the ten diesel engines studied set against the estimators from the identified initial failure intensity function are shown in Figure 17.3 for the valve subsystem and the lubrication subsystem. It appears that the estimated initial failure intensity functions are in good agreement with the collected failure data.

17.3.2 Imperfect Maintenance Model Selection

During the lifetime of a diesel engine, the six subsystems will be preventively maintained at pre-determined time intervals. The PM can be treated as imperfect maintenance as it is unable to restore the maintained subsystems to a completely new condition, and the repairs upon failures are viewed as minimal repairs. To quantify the PM efficiency and choose appropriate imperfect maintenance models for each subsystem, another group of ten studied diesel engines were preventively maintained at un-optimized periodic PM intervals: 100,000 km (for the cylinder and crankcase subsystem, the valve subsystem, and the fuel subsystem) and 50,000 km (for the starting and charging subsystem, the lubrication subsystem, and the cooling subsystem). The failures occurred within PM intervals were

Table 17.1 The initial failure intensity functions and estimators of unknown parameters

Subsystem	Type of failure intensity function and estimated parameters	Estimated initial failure intensity function
Cylinder and crankcase subsystem	Power law ($\alpha = 0.04$, $\beta = 1.32$)	$\lambda_1(t) = 0.0528t^{0.32}$
Valve subsystem	Bounded intensity process [11] ($\alpha = 0.63$, $\beta = 34.0$)	$\lambda_1(t) = 0.63\left[1 - e^{(-\frac{t}{34.0})}\right]$
Fuel subsystem	Log-linear ($\alpha = -3.5$, $\beta = 0.07$)	$\lambda_1(t) = e^{-3.5+0.07t}$
Starting and charging subsystem	Log-linear ($\alpha = -2.8$, $\beta = 0.073$)	$\lambda_1(t) = e^{-2.8+0.073t}$
Lubrication subsystem	Power law ($\alpha = 0.051$, $\beta = 1.51$)	$\lambda_1(t) = 0.077t^{0.51}$
Cooling subsystem	Log-linear ($\alpha = -3.2$, $\beta = 0.12$)	$\lambda_1(t) = e^{-3.2+0.12t}$

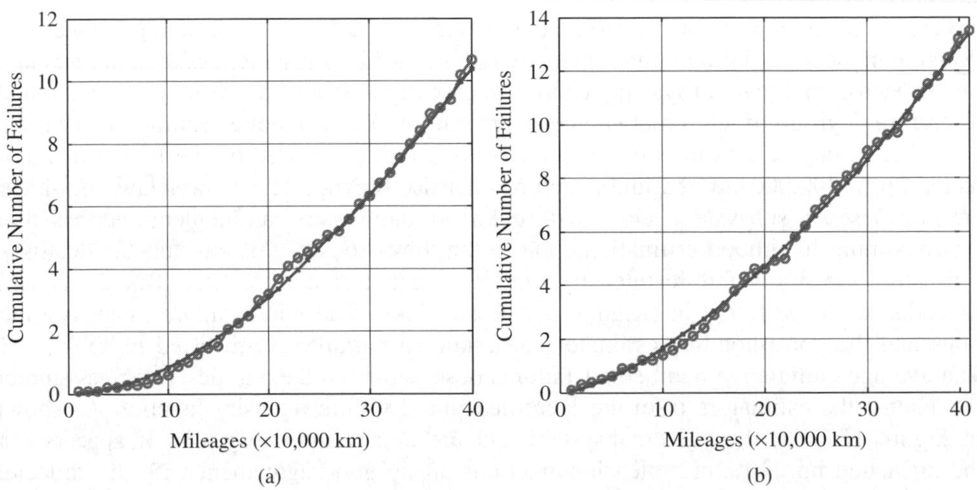

Figure 17.3 The observed average cumulative number of failures vs. estimations from the initial failure intensity function: (a) the valve subsystem; (b) the lubrication subsystem

recorded for each subsystem. One sample of the failure data set collected from the value subsystem of the studied diesel engines is tabulated in Table 17.2.

Four imperfect preventive maintenance models with different assumptions of underlying physics are used as candidates in our work, and the corresponding mathematical relationships of the failure intensity functions before and after imperfect PMs are shown in Table 17.3. For any subsystem, PMs periodically performed on this subsystem are the same over the lifetime, therefore, in our study, we assume $a_i = a$ for $i = 1, 2, \ldots, k$. It is noted that for the subsystems with the power law initial failure intensity function,

Table 17.2 An example of the failure data set collected from the fuel subsystem of a diesel engine

Preventive maintenance cycle	Failure data	u values (for Model I)
2nd PM cycle: 10.0 $\sim$ 20.0 ($\times$10,000 km)	11.6, 19.0	0.17, 0.90
3rd PM cycle: 20.0 $\sim$ 30.0 ($\times$10,000 km)	23.0, 29.0	0.42, 0.93

Table 17.3 The corresponding failure intensity functions for candidate imperfect maintenance models

Imperfect PM models	Relation of failure intensity functions
Model I (Kijima Type I Model (Kijima, 1989))	$\lambda_k(t) = \lambda_1\left(t + \sum_{i=1}^{k-1} a_i t_i^p\right)$
Model II (Kijima Type II Model (Kijima, 1989))	$\lambda_k(t) = \lambda_1\left(t + \sum_{i=1}^{k-1}\left(t_i^p \prod_{j=o}^{k-1} a_j\right)\right)$
Model III (Nakagawa, 1988)	$\lambda_k(t) = \lambda_1(t)\prod_{i=1}^{k-1} a_i$
Model IV (Nonlinear PM Model (Wu and Zuo, 2010))	$\lambda_k(t) = \lambda_1\left(t\prod_{i=1}^{k-1} a_i\right)$

Note: * t denotes the time elapsed in each PM cycle.

Models III and IV are exactly same from mathematical point of view and will be treated as one candidate model for these subsystems.

Following the proposed framework as shown in Figure 17.1, the adequacy of the candidate for the imperfect PM models is validated by using the proposed GOF test. After inferring the unknown parameters of imperfect PM models via the MLE, all the failure data are pooled into the u-space according to the corresponding conditional survival density derived from the initial failure intensity function and the estimated imperfect maintenance model. Figure 17.4 plots the empirical distribution of u values under different pre-assumed imperfect maintenance modes (the stepwise line with dots) versus the standard uniform distribution (the diagonal line) for the fuel subsystem. The u values with respect to failure data tabulated in Table 17.2 are given in the last column of Table 17.2. As indicated by Figure 17.4, Models I, II, and III are closer to the standard uniform distribution than Model IV. By applying the K-S test, we conclude that Model IV cannot pass the uniformity test under a 95% confidence level.

The proposed Bayesian model selection method is used to further identify the most adequate model among these competitive models (Models I, II, and III). Without any knowledge, the prior preference $\Pr\{H_i\}$ for each competitive model is the same and set to $1/c$, that is, $\Pr\{H_1\} = \Pr\{H_2\} = \Pr\{H_3\} = 1/3$ for the fuel subsystem. The distribution of estimated parameters $\Pr\{\widehat{a}|H_i\}$ in Equation (17.5) is gained via the bootstrap method with $p = 500$. The posterior probabilities of competitive models can be computed by

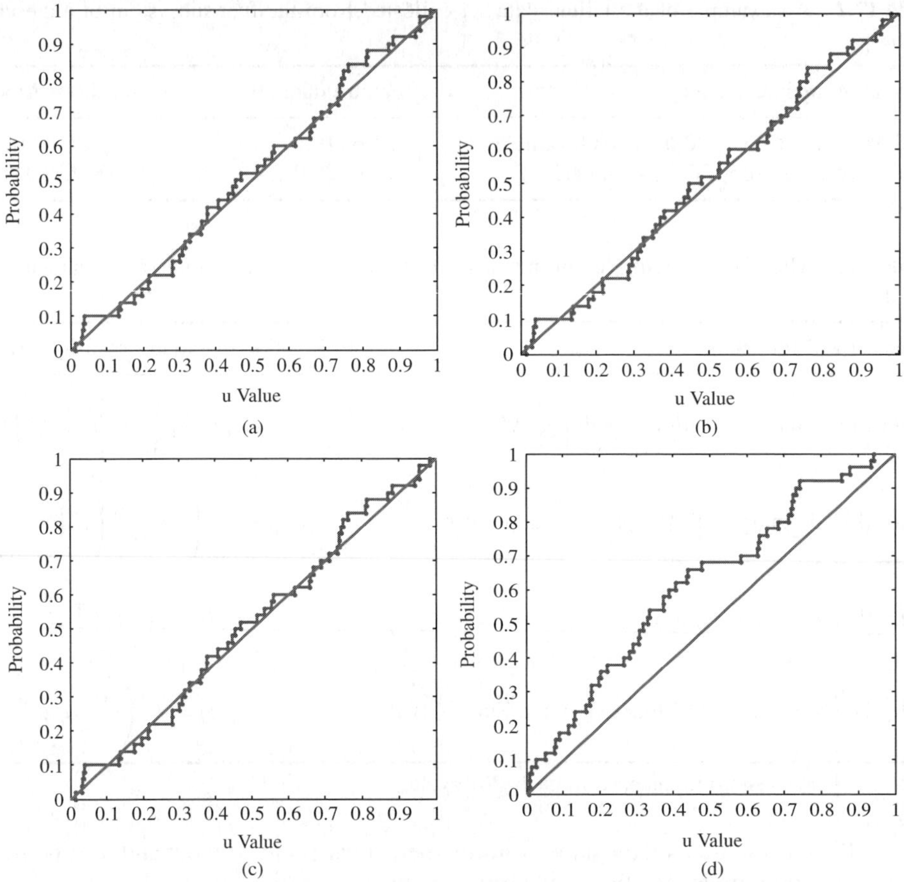

Figure 17.4 Empirical distribution of u values vs. the standard uniform distribution. (a) Model I ($\widehat{a} = 0.59$); (b) Model II ($\widehat{a} = 0.69$); (c) Model III ($\widehat{a} = 1.38$); (d) Model IV ($\widehat{a} = 1.58$)

Equation (17.4), and they are 36.6%, 27.4%, and 36.0% for Model I, Model II, and Model III respectively. Following the same procedure as the fuel subsystem, the posterior probabilities of competitive imperfect maintenance models for all the subsystems can be obtained, and the results are presented in Table 17.4 where the most adequate models with the highest posterior probability are highlighted. As observed from Table 17.4, Models I, II, and III are equally competitive in most cases and no one greatly dominates the other two. Since the initial failure intensity functions of cylinder and crankcase subsystem and the lubrication subsystem are power law type, Model III and Model IV are completely identical and regarded as one candidate model. Model IV fails to pass the GOF test for three subsystems.

17.3.3 Implementation in Preventive Maintenance Decision-Making

The posterior probabilities of competitive imperfect maintenance models quantify the uncertainty associated with model selection. As exemplified in our earlier work

Table 17.4 The posterior probability of competitive imperfect maintenance models (%)

Subsystems	Model I	Model II	Model III	Model IV
Cylinder and crankcase subsystem	30.6	**40.0**	29.4	
Valve subsystem	**37.4**	28.5	34.1	–
Fuel subsystem	**36.6**	27.4	36.0	–
Starting and charging subsystem	29.4	**38.9**	31.7	–
Lubrication subsystem	20.4	15.3	**64.3**	
Cooling subsystem	29.2	17.8	**30.0**	23.0

Note: * – means that the imperfect maintenance model cannot pass the GOF test under a 95% confidence level.

(Liu *et al.*, 2012) incorporating such uncertainty in maintenance decision will achieve more accurate results even if none of the candidate models is the true model. Before constructing the maintenance decision model, some basic assumptions are given as follows:

1. PMs are performed on subsystem l with a sequence of time interval $\mathbf{T}_{PM}^{l} = \{T_{PM}^{l,1}, T_{PM}^{l,2}, \ldots, T_{PM}^{l,NP_l}\}$ where NP_l is the total number of PMs for subsystem l, and the cost for preventively maintaining a subsystem l is c_p^l. Any PM will involve an extra fixed cost c_{pf} which includes some basic cost, like the assembly cost. But if several subsystems are maintained simultaneously, such cost will be charged only once. This economic dependency is very common in engineering practices (Dekker *et al.* 1997).
2. Any failure of subsystem l between two adjacent PMs is minimally repaired with a repair cost c_f^l and a production delay cost c_d^l.
3. The promised warranty duration of the diesel engine is $T_L = 300{,}000$ kilometers.
4. Compared with the operation time, the time for PMs and minimal repairs is negligible;
5. The objective is to minimize the expected average maintenance cost per unit time $C(\mathbf{T}_{PM}^1, \mathbf{T}_{PM}^2, \ldots, \mathbf{T}_{PM}^6)$ of the warranty period. Herein, the unit time is 10,000 km.

Let $\mathbf{p}_l^{pst} = \{p_{l,1}^{pst}, p_{l,2}^{pst}, \ldots, p_{l,c_l}^{pst}\}$ denote the posterior probabilities of competitive models for subsystem l, and c_l the number of competitive models which pass the GOF test. If one of these models greatly dominates the others, say, $p_{l,i}^{pst} \geq 99\%$, this model can be used alone as the imperfect maintenance model, otherwise all the competitive models along with their posterior probabilities will be involved in maintenance decision. Hence, the expected average maintenance cost per unit time $C(\mathbf{T}_{PM}^1, \mathbf{T}_{PM}^2, \ldots, \mathbf{T}_{PM}^6)$ can be formulated as:

$$C(\mathbf{T}_{PM}^1, \mathbf{T}_{PM}^2, \ldots, \mathbf{T}_{PM}^6) = \frac{\sum_{l=1}^{6}\left(\left(c_f^l + c_d^l\right)\cdot \sum_{i=1}^{c_l} p_{l,i}^{pst} \cdot \left(\sum_{j=1}^{NP_l}\int_0^{T_{PM}^{l,j}} \lambda_{l,j}^i(t|\widehat{\boldsymbol{a}})\,dt + \int_0^{T_L - \sum_{j=1}^{NP_l} T_{PM}^{l,j}} \lambda_{l,NP_l+1}^i(t|\widehat{\boldsymbol{a}})dt\right)\cdot \Pr\{\widehat{\boldsymbol{a}}|H_i\} + c_p^l \cdot (NP_l - 1)\right) + c_{pf}\cdot N_{PM}}{T_L} \tag{17.6}$$

where $\lambda_{l,j}^i(\cdot)$ is the failure intensity function of subsystem l in the jth PM cycle by using the imperfect maintenance model i. $\Pr\{\widehat{\boldsymbol{a}}|H_i\}$ from the proposed bootstrap method

Table 17.5 The cost associated with maintenance activities of subsystems

Subsystems	c_f^l	c_p^l	c_d^l	c_{pf}
Cylinder and crankcase subsystem	\$50.0	\$12.0		
Valve subsystem	\$40.0	\$8.0		
Fuel subsystem	\$38.0	\$12.0	\$100.0	\$12.0
Starting and charging subsystem	\$28.0	\$6.0		
Lubrication subsystem	\$36.0	\$10.0		
Cooling subsystem	\$35.0	\$8.0		

corresponds to the uncertainty associated with the parameters estimation. If more than one subsystem is preventively maintained at the same time, the extra fixed cost c_{pf} will be counted once. The last term of the numerator characterizes the economic dependency among PMs of subsystems. N_{PM} refers to the number for the extra fixed cost to be paid for the entire system and one has $N_{PM} \le \sum_l NP_l$. Therefore, it is better to combine the PMs of all the subsystems as much as possible, and the PM cost will be greatly reduced if more than one subsystem is maintained at the same time. The cost of maintenance activities is listed in Table 17.5.

Uncertainties from both model selection and parameters estimation will be eventually spread to the quantity of interest, and thereby $C(\mathbf{T}_{PM}^1, \mathbf{T}_{PM}^2, \ldots, \mathbf{T}_{PM}^6)$ is a random quantity. To identify the optimal maintenance strategy for the random quantity of interest, decision-makers have to define criteria depending on their preferences. Three criteria, namely, expectation criterion, robustness criterion, and confidence criterion, were proposed in our previous work (Liu *et al.* 2012). In this study, we chose the expectation criterion as our preference and identify the optimal PM plan $\{\mathbf{T}_{PM}^{1*}, \mathbf{T}_{PM}^{2*}, \ldots, \mathbf{T}_{PM}^{6*}\}$ that possesses the minimal expectation value of the average maintenance cost per unit time $C_M(\mathbf{T}_{PM}^{1*}, \mathbf{T}_{PM}^{2*}, \ldots, \mathbf{T}_{PM}^{6*})$. It can be written as

$$C_M^*(\mathbf{T}_{PM}^{1*}, \mathbf{T}_{PM}^{2*}, \ldots, \mathbf{T}_{PM}^{6*}) = \min_{\substack{\mathbf{T}_{PM}^1=\{T_{PM}^{1,1}, T_{PM}^{1,2}, \ldots, T_{PM}^{1,NP_1}\} \\ \vdots \\ \mathbf{T}_{PM}^6=\{T_{PM}^{6,1}, T_{PM}^{6,2}, \ldots, T_{PM}^{6,NP_6}\}}} \{E[C(\mathbf{T}_{PM}^1, \mathbf{T}_{PM}^2, \ldots, \mathbf{T}_{PM}^6)]\} \tag{17.7}$$

It is noteworthy that the number of decision variables in Equation (17.7) is equal to the total number of PMs of all the subsystems (i.e. $\sum_l NP_l$), and it cannot be pre-determined before conducting optimization. In addition, due to the economic dependency, these decision variables have complex relationships with the quantity of interest, and traditional gradient-based optimization methods may not be able to find the global optimal solution of the entire system. In our work, to reduce the computation burden and the number of decision variables, a two-stage optimization strategy is developed to solve the resulting optimization problem. The idea behind this strategy is that the number of decision variables (i.e. the number of PMs in the promised warranty period) for each individual subsystem is first determined by conducting an optimization for the PM sequence of each subsystem respectively without considering economic dependency. The results, along with the numbers of decision variables determined in the first stage optimization, will be

for $N_l = 1$ to NP_{max}

  Obtain optimal $T_{PM}^{l,1}*,\ldots,T_{PM}^{l,N_l}*$ by solving $(\partial C_M(T_{PM}^{l,1},\ldots,T_{PM}^{l,N_l})/\partial T_{PM}^{l,i})|_{T_{PM}^{l,i}\;T_{PM}^{l,i}*} = 0$

  if $(C_M^*(T_{PM}^{l,1}*,\ldots,T_{PM}^{l,N_l-1}*) > C_M^*(T_{PM}^{l,1}*,\ldots,T_{PM}^{l,N_l}*))$

    $N_l^* = N_l,\ \mathbf{T}_{PM}^{l}* = \{T_{PM}^{l,1}*,\ldots,T_{PM}^{l,N_l^*}*\}$

  end

end

Figure 17.5 The search for the first stage optimization

used as the initial solution for the second stage optimization which seeks the optimal PM sequence of each subsystem in a joint manner when considering economic dependency. Figure 17.5 illustrates the search for the first stage optimization for each individual subsystem, where $NP_{\max}$ is the maximum value for NP_i, and it is set to 30 in our case.

The search process only takes few seconds for each individual subsystem. The optimized PM sequence for each subsystem is presented in Table 17.6, where the integer values indicate PM time. For example, 15 means that the PM is conducted at the mileage of 150,000 km. With such a set of PM sequences, the expectation value of the average maintenance cost of the entire system is \$97.01 per 10,000 km, and from the entire system perspective, 12 PMs are required to be performed during the warranty period. The integer-coded genetic algorithm (GA) is utilized in the second stage to further jointly optimize the PM sequences for all the subsystems with considering economic dependency in PM activities. As the total number of PMs for all the subsystems is 22, after executing

Table 17.6 The optimized PM plan for the diesel engines over the warranty period*

Subsystems	PM time (per unit: 10,000 kilometers)											
Respectively optimized PM plan for each subsystems (Stage 1)												
Subsystem 1			11					20				
Subsystem 2	6		11		15		19			23		26
Subsystem 3				14						23		
Subsystem 4			11			18					24	
Subsystem 5		8		14			19			23		26
Subsystem 6		8			15				21			26
Optimized PM plan with considering economic dependency (Stage 2)												
Subsystem 1			11				19					
Subsystem 2		8	11	14			19			23		26
Subsystem 3				14						23		
Subsystem 4			11				19					26
Subsystem 5		8		14			19			23		26
Subsystem 6		8		14			19					26

* Subsystems 1 to 6 refer to the cylinder and crankcase subsystem to the cooling subsystem respectively.

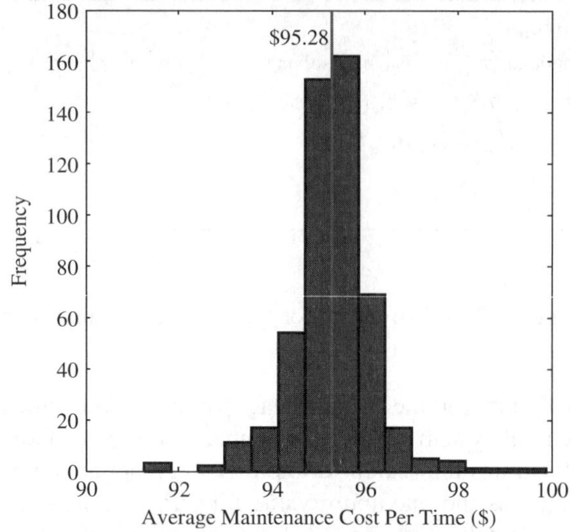

Figure 17.6 Average maintenance cost per unit time

the first stage optimization, the number of decision variables used in the ensuing GA optimization is also 22 and the optimal results in the first stage optimization are put into the initial population. The roulette-wheel selection scheme together with crossover and mutation operations is used to generate a new population generation in each optimization iteration. The GA is executed 10 times via MATLAB® on a PC with Intel Core(TM) Duo 2GHz CPU and 4GB RAM, and the average computation time for a single run is 580.32 sec. The optimal results gained from the second stage optimization are given in Table 17.6. As seen in Table 17.6, even though the number of PMs for each individual subsystem does not change, the number of PMs for the entire system reduces from 12 to 6, leading to a reduction of maintenance expenditure. With this PM plan, the corresponding expectation value of the average maintenance cost is $95.28 per 10,000 km. The new PM schedule achieves cost savings as the expectation value of the average maintenance cost is $101.09 per 10,000 km for the original un-optimized periodic PM intervals (see Section 17.3.2).

Figure 17.6 plots the histogram of the average maintenance cost per unit time with the optimized PM schedule, and it demonstrates the variation of cost due to the uncertainties from the model selection and the parameters estimation.

17.4 Conclusion

In this chapter, an imperfect maintenance model selection method is introduced with the aim of choosing appropriate models that can better characterize the efficiency of maintenance activities performed on repairable systems. The adequacy of candidate models is validated by using the proposed u-pooling method, whereas the tailored Bayesian model selection identifies the most adequate model and the posterior probabilities of competitive models. An example of scheduling the preventive maintenance plan for diesel engines is presented to demonstrate a step-by-step guideline for implementing the proposed method in engineering practices by incorporating uncertainties from both model selection and parameters estimation. Since PM activities have widespread applications in the automobile industry, manufacturing, aviation, and military, etc., the proposed method can be further applied to these fields as well. Additionally, our research focus is to appropriately choose imperfect maintenance models to better quantify the maintenance efficiency. The method therefore can be used in all situations, from preventive maintenance to condition-based and predictive maintenance, where the maintained components/systems are in an intermediate condition between "as good as new" and "as bad as old."

As many engineered systems possess multiple intermediate states between perfectly working and complete failure, the imperfect maintenance model selection for multi-state repairable systems merits further research.

Acknowledgement

The authors greatly acknowledge grant supports from the National Natural Science Foundation of China under contract numbers 71101017 and 51075061, the Specialized Research Fund for the Doctoral Program of Higher Education of China under contract numbers 20110185120014 and 20120185110032, and the Fundamental Research Funds for the Central Universities under contract number ZYGX2011J084.

References

Dekker, R., R. E. Wildeman, and F. A. van der Duyn Schouten. 1997. A review of multi-component maintenance models with economic dependence. *Mathematical Methods of Operation Research* 45 (3): 411–435.

Ferson, S., W. L. Oberkampf and L. Ginzburg. 2008. Model validation and predictive capability for the thermal challenge problem. *Computer Methods in Applied Mechanics and Engineering* 197 (29–32): 2408–2430.

Kijima, M. 1989. Some results for repairable systems with general repair. *Journal of Applied Probability* 26 (1): 89–102.

Liu, Y., W. Chen, P. Arendt and H. Z. Huang. 2011. Toward a better understanding of model validation metrics. *Journal of Mechanical Design* 133 (7): 071005.

Liu, Y. and H. Z. Huang. 2010. Optimal replacement policy for multi-state system under imperfect maintenance. *IEEE Transactions on Reliability* 59 (3): 483–495.

Liu, Y., H. Z. Huang and X. Zhang. 2012. A data-driven approach to selecting imperfect maintenance models. *IEEE Transactions on Reliability* 61 (1): 101–112.

Nakagawa, T. 1988. Sequential imperfect preventive maintenance policies. *IEEE Transactions on Reliability* 37 (3): 295–298.

Pham, H. and H. Z. Wang. 1996. Imperfect maintenance. *European Journal of Operational Research* 94 (3): 425–438.

Pulcini, G. 2001. A bounded intensity process for the reliability of repairable equipment. *Journal of Quality Technology* 33 (4): 480–492.

Rigdon, S. E. and A. P. Basu. 2000. *Statistical Methods for the Reliability of Repairable Systems*. New York: John Wiley & Sons.

Soro, I. W., M. Nourelfath and D. Aït-Kadi. 2010. Performance evaluation of multi-state degraded systems with minimal repairs and imperfect preventive maintenance. *Reliability Engineering and System Safety* 95 (2): 65–69.

Wang, H. Z. 2002. A survey of maintenance policies of deteriorating systems. *European Journal of Operational Research* 139 (3): 469–489.

Wu, S. 2011. Preventive maintenance models: a review. In *Replacement Models with Minimal Repair*, ed. L. Tadj, M.-S. Ouali, S. Yacout and D. Aït-Kadi. London: Springer, pp. 129–140.

Wu, S. and M. J. Zuo. 2010. Linear and nonlinear preventive maintenance models. *IEEE Transactions Reliability* 59 (1): 242–249.

18

Reliability of Warm Standby Systems with Imperfect Fault Coverage

Rui Peng,[1] Ola Tannous,[2] Liudong Xing[3] and Min Xie[4]
[1]*University of Science & Technology Beijing, Beijing, China*
[2]*Reliability and Quality Engineering Department, Electro Motive Diesels, Chicago, IL, USA*
[3]*University of Massachusetts Dartmouth, Dartmouth, MA, USA*
[4]*Department of Systems Engineering & Engineering Management, City University of Hong Kong, Hong Kong, China*

18.1 Introduction

It is crucial for a mission-critical or a safety-critical system to be designed with fault tolerance in order to achieve high reliability (Coll-Mayor *et al.* 2004; Iwayemi *et al.* 2010; Wu and Zhou 2011). Standby sparing is a common fault-tolerance technique, and there are three types: hot, cold and warm standby sparing, to adapt to different situations (Tannous *et al.* 2011b). Hot standby SParing (HSP) is used as a failover mechanism to provide reliability in system configurations. The hot spare is active and connected as part of a working system. This type of sparing is generally used for applications where the recovery time is critical. For Cold standby SParing (CSP), the spare component is powered up only when the online component fails and needs to be replaced. CSP is typically used for applications where energy consumption is critical. Warm standby SParing (WSP) compromises the energy consumption and the recovery time; the spare components are partially powered up when the primary component is operational and it is fully powered up only after the primary component fails. For WSP systems, the standby components have time-dependent

Applied Reliability Engineering and Risk Analysis: Probabilistic Models and Statistical Inference, First Edition.
Edited by Ilia B. Frenkel, Alex Karagrigoriou, Anatoly Lisnianski and Andre Kleyner.

failure behavior; they have different failure rates, in general, different time-to-failure distributions before and after they are used to replace the on-line faulty components.

Even in the case where a system is designed with adequate redundancy, a single uncovered component fault may propagate through the system and lead to the overall system failure (Dobson *et al.* 2007; Pepyne 2007; Aranya and Marija 2011). This behavior is known as imperfect fault coverage (IFC), see Bouricius *et al.* (1969), Arnold (1973) and Xing (2007). Due to the IFC, the system reliability cannot increase unconditionally with the increase in the system redundancy (Amari *et al.* 2004; Levitin 2008; Myers 2008). Peng (2011) studied the optimal structure of a series-parallel system with a consideration of different kinds of fault coverage models. Incorporating the IFC is a challenging task, in particular regarding the reliability analysis of a WSP system due to the aforementioned time-dependent failure behavior of the standby components. Some researchers have studied the availability of a WSP system with repair distribution and IFC (Ke *et al.* 2008a, 2008b, 2010). But their work is restricted to cases where the WSP system has only one spare. Some other studies of WSP with the IFC are restricted to cases where the failure time of each system component follows an exponential distribution (Ke *et al.* 2008c; Hsu *et al.* 2009; Lee *et al.* 2009). In order to overcome these limitations, this chapter presents a binary decision diagram (BDD) based approach to evaluating the reliability of a general WSP with IFC.

Section 18.2 reviews recent works on the reliability analysis of warm standby systems. Section 18.3 presents the BDD-based approach with illustrative examples. Section 18.4 summarizes this chapter.

18.2 Literature Review

Wang *et al.* (2007) deal with the reliability and sensitivity analysis of a repairable system with M operating components, S warm standby components, and R unreliable service stations. Failure times and service times of operating or standby components are assumed to be exponentially distributed. While the service station is working, it is subject to breakdowns according to a Poisson process. When the station breaks down, it requires repair at a repair facility, where repair times follow the negative exponential distribution. Explicit expressions for the reliability function of the system and the MTTF are developed using Laplace transform techniques.

Bagdonavicius *et al.* (2008) study the mathematical formulation of fluent switching from warm to hot conditions of standby components using the well-known Sedyakin's and accelerated failure time models. Non-parametric estimators of cumulative distribution function and mean failure time of a redundant system with several standby components are proposed. Goodness-of-fit tests for two given models are given.

Cha *et al.* (2008) propose a general method for modeling standby systems and derive the system performance measures. The cold, hot and warm standby systems with components of exponential distribution, which were studied in the literature, are shown to be special cases of the proposed model. An optimal allocation problem for a standby system is also discussed.

Ke *et al.* (2008a) propose a procedure to construct the membership functions of the system characteristics of a redundant repairable system with two primary components and one standby, in which the coverage factor is the same for an operating component failure as that for a standby component failure. Times to failure and times to repair of the operating and standby components are assumed to follow fuzzified exponential distributions. The alpha-cut approach is used to extract from the fuzzy repairable system a family of conventional crisp intervals for the desired system characteristics, determined with a set of parametric nonlinear programs using their membership functions.

Ke *et al.* (2008b) study the system characteristics of a repairable system from a Bayesian viewpoint with different types of priors assumed for unknown parameters, in which the system consists of one active component and one standby component. The detection of standby, the coverage factor and reboot delay of failed components are possibly considered. Time to failure of the components is assumed to follow an exponential distribution. Time to repair and time to reboot of the failed components also follow exponential distributions. When time to failure, time to repair and time to reboot have uncertain parameters, a Bayesian approach is adopted to evaluate the system characteristics. Monte Carlo simulation is used to derive the posterior distribution for the mean time to system failure and the steady-state availability. Similarly, Ke *et al.* (2008c) study the system characteristics of a redundant repairable system with two primary components and one standby from a Bayesian viewpoint with different types of priors assumed for unknown parameters, in which the coverage factor is the same for an operating component failure as that for a standby component failure. Lee *et al.* (2009) also studied the system performance measures of a repairable system with two active components and one warm standby. There is a failure probability associated with switching from a standby state to an active state.

Li *et al.* (2009) investigated the heterogeneity of components with proportional hazard rates in a redundant system. The total number of those standbys surviving the failure time of some active component is derived, and the algorithm to determine the optimal number of standbys is also discussed.

Jenab and Rashidi (2009) present a flow-graph-based method to calculate time-to-failure data and failure probability of an aircraft environmental control system (ECS). The obtained data from the model may be used for maintenance optimization that employs the failure limit strategy for the ECS. The model incorporates detectable failures such as hardware failures, critical human errors, common-cause failures, maintenance categories, and switch activation methods. A numerical example is also presented to demonstrate the application of the model.

Papageorgiou and Kokolakis (2010) investigate a parallel (2, n-2)-system where two components start their operation simultaneously and any one of them is replaced instantaneously upon its failure by one of the (n-2) warm standbys. The system reliability is evaluated by recursive relations.

Levitin and Amari (2010) present an approximation algorithm for evaluating the cumulative distribution function (cdf) of time-to-failure for k-out-of-n systems with shared spare components. The working and spare components can have different arbitrary distributions of time-to-failure. Different spare components can be in different standby modes: cold, hot and warm. The algorithm is based on a universal generating function approach.

Yun and Cha (2010) consider a standby system with two components in which the first component (component 1) starts its operation in an active state and the other component (component 2) is in a cold standby state at the starting point. After a specified time s (switching time), the state of component 2 is changed to a warm standby state and, as soon as the operating component 1 fails, the state of component 2 is changed to the active state. If component 1 fails before time s, the system fails. Components can fail in both active and warm standby states. A general method for modeling the standby system is adopted and system performance measures (system reliability and mean life) based on the proposed model are derived.

Ke *et al.* (2010) studied the statistical inferences of an availability system with imperfect coverage. The system consists of two active components and one warm standby. The time-to-failure and time-to-repair of the components are assumed to follow an exponential and a general distribution respectively. The coverage factors for an active-component failure and for a standby-component failure are assumed to be the same. A consistent and asymptotically normal estimator of availability is constructed for such a repairable system. Based on this estimator, interval estimation and testing hypothesis are performed. To implement the simulation inference for the system availability, two repair-time distributions, namely, lognormal and Weibull, are adopted and three types of Weibull distributions characterized by their shape parameters are considered. Finally, all simulation results are displayed in appropriate tables and curves to highlight the performance of the statistical inference procedures.

Li and Meng (2011) study a warm standby repairable system consisting of two dissimilar components and one repairman. It is assumed that the working time distributions and the repair time distributions of the two components are both exponential, and component 1 is given priority in use. The system fails immediately once the switch fails. Some important reliability indices and some steady state system indices are derived using Markov process theory and the Laplace transform.

Eryilmaz (2011) studies a coherent system consisting of active components and equipped with warm standby components. In particular, the study investigates the random quantity which denotes the number of surviving warm standby components at the time of system failure. The distribution of the corresponding random variable is presented in terms of system signature and its potential utilization with a certain optimization problem is discussed.

Hsu *et al.* (2011) statistically examine an availability system with reboot delay, standby switching failures and an unreliable repair facility, which consists of two active components and one warm standby. The time-to-failure and the reboot time are assumed to be exponentially distributed. The repair time of the service station and the time-to-repair of components are assumed to be generally distributed. A consistent and asymptotically normal estimator of availability of such a repairable system is developed. Based on this estimator, interval estimation and testing hypothesis are performed by using logit transformation. To implement the simulation inference for the system availability, two repair-time distributions, lognormal and Weibull distributions, are used. Three Weibull distributions characterized by distinct shape parameters are considered. Finally, all simulation results are displayed as appropriate tables and curves to reveal the performance of the statistical inference procedures.

Tannous *et al.* (2011a) propose a combinatorial approach based on sequential binary decision diagrams to evaluate the reliability of WSP systems with an arbitrary number of spares.

Figure 18.1 illustrates the general structure of a WSP gate with n spares, where A represents the primary component and S_i represents a warm-standby component. All the spares are used in the specified order (from left to right) as long as it is still operational.

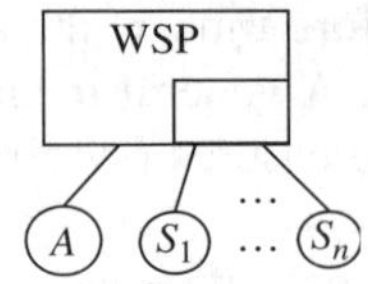

Figure 18.1 WSP Gate with n-spares

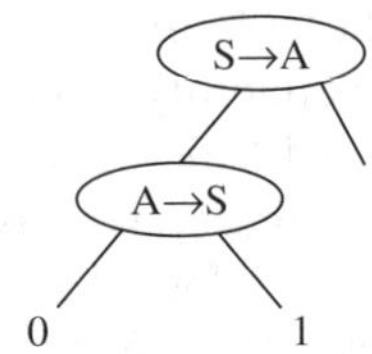

Figure 18.2 Sequential BDD for WSP with one spare

Figure 18.2 illustrates the sequential BDD for a WSP with one spare, where "→" indicates the failure order of the primary component and the spare. The approach is applicable to any type of time-to-failure distributions for the system components. However, imperfect fault coverage is not considered in this work.

Tannous *et al.* (2011b) present two optimization solution methodologies, respectively based on genetic algorithm and integer programming to achieve optimal design of a series-parallel system with warm standby redundancy. Both methodologies are illustrated using examples. Similarly, this work does not incorporate the effects of imperfect fault coverage.

18.3 The BDD-Based Approach

Binary decision diagram (BDD) was initially developed as a tool for validating VLSI circuitry design by Bryant (Bryant 1986). BDD is a direct acyclic graph based on Shannon decomposition:

$$f = x \cdot f_{x=1} + \overline{x} \cdot f_{x=0} = ite(x, f_{x=1}, f_{x=0}) \tag{18.1}$$

where f represents a Boolean expression for a set of Boolean random variables X and x being a member of X, the $f_{x=1}$ after $x\cdot$ is the Shannon co-factor of function f with all instances of x replaced by 1, and the $f_{x=0}$ after $\overline{x}\cdot$ is the Shannon co-factor of function f with all instances of x replaced by 0. The two terminal nodes labelled "1" and "0" in the BDD represent the system being in the failure and operational states respectively. The total failure probability of the system can be calculated as the sum of probabilities of all the disjoint paths that lead to the sink node "1." The BDD is generated via a bottom-up traversal of the fault tree, applying the following manipulation rules (Bryant 1986):

$$\begin{aligned} G \Diamond H &= ite\ (x, G_1, G_2) \Diamond ite\ (y, H_1, H_2) \\ &= \begin{cases} ite\ \left(x, G_1 \Diamond H_1, G_2 \Diamond H_2\right) & index(x) = index(y) \\ ite\ (x, G_1 \Diamond H, G_2 \Diamond H) & index(x) < index(y) \\ ite\ (y, G \Diamond H_1, G \Diamond H_2) & index(x) > index(y) \end{cases} \end{aligned} \tag{18.2}$$

where G and H represent two Boolean expressions corresponding to the traversed sub-fault trees, the index() represents the order of the Boolean variable in the input list, and $\Diamond$ represents the logical operation (AND, OR).

The BDD method has been used to analyse static systems. Some additional rules need to be applied to address the time dependency of warm spare failures and the effect of imperfect fault coverage during the model construction and evaluation, as detailed in the following subsections.

18.3.1 The BDD Construction

An individual component is represented by a BDD as shown in Figure 18.3. The BDD is constructed iteratively by combining the BDD representing $A, S_1, \ldots, S_n$ in sequence. Besides the traditional manipulation rules represented by (18.2), the following additional rules need to be applied:

1. Since the system fails in case of any global failure (i.e. failure revealed due to the imperfect fault coverage) regardless the status of any other component, the right child of X_G (which can be either A_G, $S_{iG}(\alpha)$, or $S_{iG}(\lambda)$) is always 1.
2. If the primary component does not fail, only global failure of subsequent spares can cause the system to fail. Thus, following the left branch of $S_{iL}(\lambda)$ or A_L, there will only be $S_{jG}(\alpha)$, where $j > i$.

18.3.2 System Unreliability Evaluation

The unreliability of a warm standby system can be evaluated as the sum of probabilities of all the disjoint paths from the root node to sink node "1" in the BDD model. For a warm standby system, the failure probability of a standby component while in the online operational mode is affected by the switch-over time. In this study, the effect is modeled

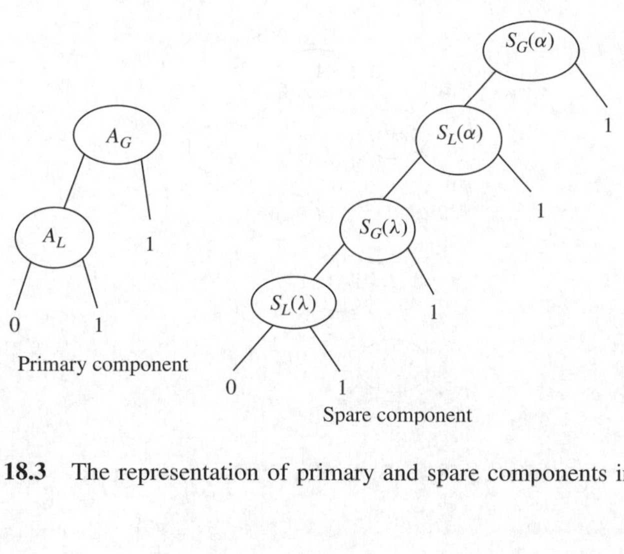

Figure 18.3 The representation of primary and spare components in BDD

by accelerated failure time model (AFTM) with the concept of cumulative exposure model (CEM) (Nelson 1990). Accordingly, the probability that both primary component A and the spare S fail locally is given by (18.3) or (18.4) where the spare fails before the primary component with a reduced failure rate α or the spare fails after the primary component with a full failure rate λ, respectively:

$$\Pr\{A_L, S_L(\alpha)\} = \Pr\{S_L \to A_L\} = \int_0^t \int_0^{\tau_1} f_{\lambda A_L}(\tau_1) f_{\alpha SL}(\tau_2) d\tau_2 d\tau_1 \tag{18.3}$$

$$\begin{aligned}\Pr\{A_L, S_L(\lambda)\} &= \Pr\{A_L \to S_L\} \\ &= \int_0^t \int_{\tau_1}^t f_{\lambda A_L}(\tau_1) f_{\lambda SL}(\tau_3 - \tau_1 + \phi_S \tau_1) \left(1 - \int_0^{\tau_1} f_{\alpha S}(\tau_2)\, d\tau_2\right) d\tau_3 d\tau_1\end{aligned} \tag{18.4}$$

where α_{SL} and α_{SG} are respectively the reduced local and global failure rates of the spare component; λ_{SL} and λ_{SG} are respectively the full local and global failure rates of the spare component; λ_S is the standby component's total failure rate, which is the sum of λ_{SL} and λ_{SG}; λ_{AL} and λ_{AG} are respectively the local and global failure rates of the primary component. To consider the switch-over time, the acceleration factor φ_s is used. The probability density function (pdf) f can be any distribution.

18.3.3 Illustrative Examples

This section considers a warm standby system with one spare for illustration. The BDD for a WSP with one spare can be constructed by combining the BDD of the primary and the BDD of the spare, as shown in Figure 18.4.

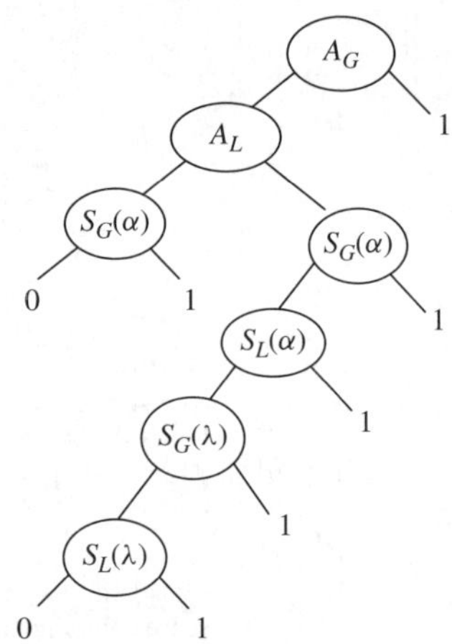

Figure 18.4 BDD of a WSP with one spare

The system unreliability, denoted by UR, can be obtained by adding up the probabilities of all the paths leading to 1-terminal as

$$\begin{aligned} UR &= \Pr\{A_G\} + \Pr\{A_L \cdot [S(\alpha) + \neg S(\alpha) \cdot S(\lambda)]\} + \Pr\{\neg A_G \cdot \neg A_L \cdot S_G(\alpha)\} \\ &= \int_0^t f_{A_G\lambda}(\tau_1)\, d\tau_1 \\ &\quad + \int_0^t f_{A_L\lambda}(\tau_1) \left[\int_0^{\tau_1} f_{S\alpha}(\tau_2)\, d\tau_2 + \left(1 - \int_0^{\tau_1} f_{S\alpha}(\tau_2)\, d\tau_2\right) \right. \\ &\quad \left. \times \int_{\tau_1}^t f_{S\lambda}(\tau_2 - \tau_1 + \phi_S \tau_1)\, d\tau_2 \right] d\tau_1 \\ &\quad + \int_0^t f_{SG\alpha}(\tau_1) d\tau_1 \left[1 - \int_0^t f_{A\lambda}(\tau_2)\, d\tau_2 \right] \end{aligned} \tag{18.5}$$

Consider a special case where the failure times of all the components follow exponential distribution with the following failure rates: $\lambda_{AL} = 0.18$/year, $\lambda_{AG} = 0.02$/year, $\alpha_{SG} = 0.01$/year, $\lambda_{SG} = 0.03$/year, $\alpha_{SL} = 0.09$/year, $\lambda_{SL} = 0.27$/year. If the system is supposed to work for 5 years, the system unreliability can be obtained as 0.4457.

We also consider the case where the system components have Weibull time-to-failure distribution. The *pdf* function f is given by $(k/\lambda)(x/\lambda)^{k-1}e^{-(x/\lambda)^k}$, where k is the shape parameter, and λ is the scale parameter. The parameter values used for analysis are: $k_{AL} = k_{AG} = 2$, $k_{S\alpha G} = k_{S\alpha l} = 1.5$, $k_{S\alpha G} = k_{S\lambda G} = 0.9$, $\lambda_{AL} = 0.27$/year, $\lambda_{AG} = 0.03$/year, $\alpha_{SG} = 0.01$/year, $\lambda_{SG} = 0.02$/year, $\alpha_{SL} = 0.09$/year, $\lambda_{SL} = 0.18$/year. The acceleration factor in the standby mode is assumed to be $\varphi_s = 1.5$. If the system is supposed to work for 5 years, the system unreliability can be obtained as 0.25777.

Examples with more than one warm standby component are not presented here due to space limitation.

18.4 Conclusion

This chapter studied the reliability of a warm standby system with the consideration of imperfect fault coverage. For WSP systems, the standby components have different failure rates before and after they are used to replace the on-line faulty components. Furthermore, an uncovered component fault may cause the entire system to fail despite enough redundancy being provided. It is a challenging task to incorporate imperfect fault coverage into the reliability analysis of WSP systems. The existing approaches have various limitations, such as assuming exponential time-to-failure distributions for all the system components or being applicable to cases with only one spare. This chapter presented a combinatorial BDD-based approach for the reliability analysis of WSP systems with IFC after a review of recent works on reliability analysis of WSP systems. The BDD-based approach can work well for WSP systems with any number of spares having arbitrary type of time-to-failure distributions.

Acknowledgments

The research is partially supported by the China NSFC under grant number 71231001 and US National Science Foundation under grants 0832594 and 1107868.

References

Amari, S., H. Pham and G. Dill. 2004. Optimal design of k-out-of-n: G subsystems subjected to imperfect fault-coverage. *IEEE Transactions on Reliability* 53 (4): 567–575.

Aranya, C. and D. Mariya. 2011. *Control and Optimization Methods for Electric Smart Grids*. New York: Springer.

Arnold, T. F. 1973. The concept of coverage and its effect on the reliability model of a repairable system. *IEEE Transactions on Computers* C-22 (3): 251–254.

Bagdonavicius, V., I. Masiulaityte and M. Nikulin. 2008. Statistical analysis of redundant systems with 'warm' stand-by units. *Stochastics: An International Journal of Probability and Stochastic Processes* 80 (2–3): 115–128.

Bouricius, W. G., V. Carter and P. R. Schneider. 1969. Reliability modeling techniques for self-repairing computer systems. In *Proceedings of the 1969 24th National Conference*, ACM'69: 295–309.

Bryant, R. 1986. Graph based algorithms for Boolean function manipulation. *IEEE Transactions on Computers* C-35 (8), 677–691.

Cha, J.H., J. Mi and W. Y. Yun. 2008. Modelling a general standby system and evaluation of its performance. *Applied Stochastic Models in Business and Industry* 24 (2), 159–169.

Coll-Mayor, D., R. Picos and E. Garcia-Moreno. 2004. State of the art of the virtual utility: the smart distributed generation network. *International Journal of Energy Research* 28 (1): 65–80.

Dobson, I., B. A. Carreras, V. E. Lynch and D. E. Newman. 2007. Complex systems analysis of series of blackouts: cascading failure, critical points, and self-organization. *Chaos* 17 (2): 026103.

Eryilmaz, D. 2011. The behavior of warm standby components with respect to a coherent system. *Statistics and Probability Letters* 81 (8): 1319–1325.

Hsu, Y., J. C. Ke and T. H. Liu. 2011. Standby system with general repair, reboot delay, switching failure and unreliable repair facility: A statistical standpoint. *Mathematics and Computers in Simulation* 81 (11): 2400–2413.

Hsu, Y., S. L. Lee and J. C. Ke. 2009. A repairable system with imperfect coverage and reboot: Bayesian and asymptotic estimation. *Mathematics and Computers in Simulation* 79 (7): 2227–2239.

Iwayemi, A., P. Yi, P. Liu and C. Zhou. 2010. A perfect power demonstration system. *Innovative Smart Grids Technologies*: 1–7.

Jenab, K. and K. Rashidi. 2009. Operational reliability assessment of an aircraft environmental control system. *Reliability Engineering & System Safety* 94 (2): 456–462.

Ke, J. C., H.I. Huang and C. H. Lin. 2008a. A redundant repairable system with imperfect coverage and fuzzy parameters. *Applied Mathematical Modelling* 32 (12): 2839–2860.

Ke, J. C., S. L. Lee and Y. L. Hsu. 2008b. On a repairable system with detection, imperfect coverage and reboot: Bayesian approach. *Simulation Modelling Practice and Theory* 16 (3): 353–367.

Ke, J. C., S. L. Lee and Y. L. Hsu. 2008c. Bayesian analysis for a redundant repairable system with imperfect coverage. *Communications in Statistics: Simulation and Computation* 37 (5): 993–1004.

Ke, J. C., Z. L. Su, K. H. Wang and Y. L. Hsu. 2010. Simulation inferences for an availability system with general repair distribution and imperfect fault coverage. *Simulation Modelling Practice and Theory* 18 (3): 338–347.

Lee, S. L., J. C. Ke and Y. L. Hsu. 2009. Bayesian assessing for a repairable system with standby imperfect switching and reboot delay. *International Journal of Systems Science* 40 (11): 1149–1159.

Levitin, G. 2008. Optimal structure of multi-state systems with uncovered failures. *IEEE Transactions on Reliability* 57 (1): 140–148.

Levitin, G. and S.V. Amari. 2010. Approximation algorithm for evaluating time-to-failure distribution of k-out-of-n system with shared standby elements. *Reliability Engineering and Systems Safety* 95 (4): 396–401.

Li, X. H., R. F. Yan and M. J. Zuo. 2009. Evaluating a warm standby system with components having proportional hazard rates. *Operations Research Letters* 37 (1): 56–60.

Li, Y. and X. Y. Meng. 2011. Reliability analysis of a warm standby repairable system with priority in use. *Applied Mathematical Modelling* 35 (9): 4295–4303.

Myers, A. 2008. Achievable limits on the reliability of k-out-of-n: G systems subject to imperfect fault coverage. *IEEE Transactions on Reliability* 57 (2): 349–354.

Nelson, A. 1990. *Accelerated Testing: Statistical Models, Test Plans and Data Analyses*. New York: John Wiley & Sons.

Papageorgiou, E. and G. Kokolakis. 2010. Reliability analysis of a two-unit general parallel system with (n-2) warm standbys. *European Journal of Operational Research* 201 (3): 821–827.

Peng, R. 2011. Some extensions to reliability modeling and optimization of networked systems. PhD thesis. National University of Singapore.

Pepyne, D. L. 2007. Topology and cascading line outages in power grids. *Journal of Systems Science and Systems Engineering* 16 (2): 202–221.

Tannous, O., L. Xing and J. B. Dugan. 2011a. Reliability analysis of warm standby systems using sequential BDD. In *Proceedings of the 57th Annual Reliability & Maintainability Symposium*, 1–7. January 24–27, 2011. Florida, USA.

Tannous, O., L. Xing, R. Peng, M. Xie and S. H. Ng. 2011b. Redundancy allocation for series-parallel warm-standby systems. In *Proceedings of the IEEE International Conference on Industrial Engineering and Engineering Management* (IEEM), 1261–1265. December 6–9, 2011. Singapore.

Wang, K. H., J. B. Ke and W. C. Lee. 2007. Reliability and sensitivity analysis of a repairable system with warm standbys and R unreliable service stations. *The International Journal of Advanced Manufacturing Technology* 31 (11–12): 1223–1232.

Wu, D. and C. Zhou. 2011. Fault-tolerant and scalable key management for smart grids. *IEEE Transactions on Smart Grid* 2 (2): 375–381.

Xing, L. 2007. Reliability evaluation of phased-mission systems with imperfect fault coverage and common-cause failures. *IEEE Transactions on Reliability* 56 (1): 58–68.

Yun, W. Y. and J. H. Cha. 2010. Optimal design of a general warm standby system. *Reliability Engineering & System Safety* 95 (8): 880–886.

[illegible]

[illegible]

[illegible] 2011 [illegible]

[illegible]

[illegible]

[illegible]

[illegible]

[illegible]

Part Four

Statistical Inference in Reliability

19

On the Validity of the Weibull-Gnedenko Model

Vilijandas Bagdonavičius,[1] Mikhail Nikulin[2] and Ruta Levuliene[1]
[1]*Department of Mathematical Statistics, Vilnius University, Lithuania*
[2]*IMB, Bordeaux Victor Segalen University, France*

19.1 Introduction

The two-parameter Weibull distribution is a continuous probability distribution with the cumulative distribution function

$$F(t) = 1 - e^{-(t/\theta)^{\nu}}, \quad t \geq 0, \tag{19.1}$$

where $\theta > 0$ is the scale parameter and $\nu > 0$ is the shape parameter.

This distribution is named after a Swedish engineer Waloddi Weibull (1887–1979) who popularized its use for reliability analysis, especially for metallurgical failure modes. Weibull's first paper on the subject was published in 1939 (Weibull 1939), but the method did not attract much attention until the 1950s after his 1951 paper (Weibull 1951), where he described in detail the wide applicability of the distribution: modeling the yield strength and fatigue life of steel, the fiber strength of cotton, the size distribution of fly ash, the length of cyrtoidae, etc. After this extremely successful paper this distribution became commonly known as "*Weibull distribution*."

Nevertheless (see Stoyan 2011, for discussion), this distribution was first used by (Rosin *et al.* 1933) to describe the size distribution of particles. However, in its modern form (19.1) had first appeared in (Bennett 1936). Therefore, before publication of Weibull's paper the name "*RRS distribution*" or "*RRSB distribution*" was used on several occasions.

Apart from real data modeling the distribution (19.1) appeared in papers written by mathematicians considering limiting distribution for extreme order statistics. Fisher and Tippett (1928) showed that the asymptotic distribution of the maximum of a sample

Applied Reliability Engineering and Risk Analysis: Probabilistic Models and Statistical Inference, First Edition.
Edited by Ilia B. Frenkel, Alex Karagrigoriou, Anatoly Lisnianski and Andre Kleyner.

of independent and identically distributed random variables after proper normalization follows one of three possible distributions. Following Fisher and Tippett, Boris Vladimirovich Gnedenko (Gnedenko 1943) was the first to present a mathematically rigorous theory of extremes. He gave the necessary and sufficient conditions to the normalized maximum of a sample to converge in distribution to any of three possible limiting distributions given by (Fisher and Tippett 1928). This implies the necessary and sufficient conditions for the normalized minimum of a sample to converge in distribution to the distribution (19.1).

Let us consider more closely the result of Fisher-Tippett-Gnedenko: let $X_1, \dots, X_n$ be a sequence of independent and identically distributed random variables, $X_{n,n} = max_{1\le i\le n} X_i$. If a sequence of pairs of real numbers (b_n, σ_n) exists such that $\sigma_n > 0$ and $\lim_{n\to\infty} P\{(X_{n,n} - b_n)/\sigma_n \le x\} = G(x)$, where G is a non-degenerate cumulative distribution function, then the limiting distribution belongs to either the Gumbel, the Fréchet or the Weibull family, that is, the function G has one of the three following forms:

$$\Phi_\alpha(x) = \begin{cases} e^{-x^{-\alpha}}, & x > 0, \\ 0, & x \le 0, \end{cases} \quad \Psi_\alpha(x) = \begin{cases} e^{-(-x)^\alpha}, & x \le 0, \\ 1, & x > 0, \end{cases} \quad \Lambda(x) = e^{-e^{-x}}, x \in \mathbf{R},$$

here $\alpha > 0$.

In the third case this result implies that the distribution of $X_{n,n}$ for large n is approximated by the cumulative distribution function

$$e^{-e^{-(x-b_n)/\sigma_n}}.$$

Taking into account that $minX_i = -max(-X_i)$, the distribution of $X_{1,n}$ for large n is approximated by the cumulative distribution function

$$1 - e^{-e^{(x-\mu_n)/\sigma_n}}, \quad \mu_n = -b_n.$$

Taking $T_i = e^{X_i}$ we find that the distribution of $T_{1,n}$ for large n is approximated by the cumulative distribution function

$$1 - e^{-e^{(\ln t - \mu_n)/\sigma_n}} = 1 - e^{-(t/\theta_n)^{\nu_n}}, \quad t \ge 0,$$

where $\theta_n = e^{\mu_n}$, $\nu_n = 1/\sigma_n$. So the distribution of the minimum of positive random variables under the conditions given by (Gnedenko 1943), is approximated by the Weibull-Gnedenko distribution if n is large.

Therefore, we conclude that the importance of the distribution (19.1) for a large quantity of examples of real data models was first shown by Waloddi Weibull and the necessary and sufficient conditions under which this distribution is the limiting distribution of the "weakest chain" were given by Boris Vladimirovich Gnedenko. So we refer to the distribution (19.1) as "Weibull-Gnedenko" distribution.

The purpose of this chapter is to survey the statistical methods for the validation of the two-parameter Weibull-Gnedenko distribution from type II censored samples. A modification of the powerful Zhang test (Zhang 2002) from complete to type II censored data is proposed. We also present simple explicit formulas for modified chi-squared tests from censored data with data-driven choice of partitioning which are discussed in the recent papers (Bagdonavičius and Nikulin 2011) and (Bagdonavičius *et al.* 2012). We analyze powers of tests against the most commonly used alternatives to the Weibull-Gnedenko distribution.

Let T be an absolutely continuous random variable with the c.d.f. F_T and the p.d.f. f. We consider the hypothesis

$$H_{WG} : F_T(t) = 1 - e^{-(t/\theta)^\nu}, t \geq 0, \tag{19.2}$$

stating that the random variable T has the Weibull-Gnedenko distribution with unknown parameters θ and ν.

Suppose that data is *complete*, that is, i.i.d. random variables

$$T_1, T_2, \ldots, T_n \tag{19.3}$$

are observed, or *type II censored*, that is, only first r order statistics

$$T_{1,n} \leq T_{2,n} \leq \ldots \leq T_{r,n} \tag{19.4}$$

are observed.

The following tests are most commonly discussed in the literature (see, for example, Greenwood and Nikulin 1996): chi-squared tests based on differences of observed and expected numbers of failures in intervals of some partition of $[0, \infty)$; modified Kolmogorov-Smirnov, Cramér-von-Mises, Andersen-Darling tests for composite hypotheses based on the differences of non-parametric and parametric (obtained under the hypothesis H_{WG}) estimators of the cumulative distribution function; tests based on the differences of logarithms of order statistics, such as Mann-Scheuer-Fertig and Tiku-Singh tests; correlation tests based on empirical correlation coefficients between logarithms of order statistics and their "expected values."

We would like to draw attention to the not so widely used but powerful general tests of (Zhang, 2002) which can be applied to test the Weibull hypothesis in the case of complete data. In the next section we propose "integrated likelihood ratio test" which is a modification of Zhang's test from complete to type II censored data.

19.2 Integrated Likelihood Ratio Test

Suppose that data are complete. Let us test the simple hypothesis

$$H_0 : F(t) = G(t), \quad \text{for all } t \in \mathbf{R}, \tag{19.5}$$

where G is a completely specified function of t.

Zhang (2002) proposed a class of powerful tests for such hypotheses. His idea was to begin with tests for narrower hypotheses in the form of

$$H_{0t} : F(t) = G(t) \tag{19.6}$$

where t is fixed. He used the indicators $\mathbf{1}_{\{T_i \leq t\}}$ of the events $\{T_i \leq t\}$ to construct test statistics for H_{0t}. For such restricted data the hypothesis H_{0t} is equivalent to the hypothesis testing the value of the Bernoulli random variable parameter. Likelihood ratio test statistics are used to test these hypotheses and test statistics for the hypothesis H_0 are written as functionals of these simple likelihood ratio statistics.

Zhang's statistics can be modified to test the hypothesis H_{WG}.

We shall use Zhang's idea to construct tests when data are type II censored. Set

$$U_{i,n}(t) = \mathbf{1}_{A_{i,n}(t)}, \quad i = 0, \ldots, r, \tag{19.7}$$

where

$$A_{i,n}(t) = \{T_{i,n} \le t, T_{i+1,n} > t\}, \quad i = 0, \ldots, r-1, \quad T_{0,n} = 0, \quad A_{r,n}(t) = \{T_{r,n} \le t\}.$$

Note that

$$A_{i,n}(t) \cap A_{l,n}(t) = \emptyset, \quad i \neq l, \quad i, l = 0, \ldots, r, \quad \cup_{i=0}^{r} A_{i,n}(t) = \Omega,$$

where Ω is a certain event.

Let us use the data

$$U_{0,n}(t), U_{1,n}(t), \ldots, U_{r,n}(t) \tag{19.8}$$

for hypothesis H_{0t} testing.

The probabilities of the events $A_{i,n}(t)$ are

$$P\{A_{i,n}(t)\} = Q_i(F(t); n), \quad i = 0, \ldots, r. \tag{19.9}$$

where

$$Q_i(x; n) = C_n^i x^i (1-x)^{n-i}, \quad i = 0, \ldots, r-1 \quad Q_r(x; n) = I_x(r, n-r+1), \tag{19.10}$$

and

$$I_x(a, b) = \frac{\Gamma(a+b)}{\Gamma(a)\Gamma(b)} \int_0^x t^{a-1}(1-t)^{b-1} dt, \quad 0 < x < 1; \quad a > 0, \quad b > 0,$$

is the cumulative distribution function of the beta distribution with a and b degrees of freedom. If $r = n$, then

$$Q_n(i; n) = I_x(n, 1) = x^n.$$

The likelihood function based on the data (19.8) is

$$L(F(t); n) = Q_{N_n(t)}(F(t); n), \tag{19.11}$$

where

$$N_n(t) = \sum_{i=0}^{r} i U_{i,n}(t)$$

is the number of failures in the interval $[0, t]$.

The likelihood ratio statistic is

$$LR(t) = 2 \ln \frac{Q_{N_n(t)}(\hat{F}_n(t); n)}{Q_{N_n(t)}(G(t); n)}. \tag{19.12}$$

The estimator $\hat{F}_n(t) = \frac{N_n(t)}{n}$ for $t \le T_{r,n}$ coincides with the empirical distribution function.

Please note that

$$LR(t) = 2\left(N_n(t) \ln \frac{\hat{F}_n(t)}{G(t)} + (n - N_n(t)) \ln \frac{1 - \hat{F}_n(t)}{1 - G(t)} \right), \tag{19.13}$$

if $N_n(t) = 1, \ldots, r-1$, and

$$LR(t) = 2 \ln \frac{I_{\hat{F}_n(t)}(r, n-r+1)}{I_{G(t)}(r, n-r+1)}, \tag{19.14}$$

if $N_n(t) = r$. If $r = n$, then (19.13) holds for any values of $N_n(t)$.

Using the continuity correction we define $\hat{F}_n^-(T_{i,n}) = (i - 0.5)/n)$,

$$LR^-(t) = 2\left((i - 0.5) \ln \frac{i - 0.5}{nG(t)} + (n - i + 0.5) \ln \frac{n - i + 0.5}{n(1 - G(t))} \right), \tag{19.15}$$

if $N_n(t) = 1, \ldots, r-1$, and

$$LR^-(t) = 2 \ln \frac{I_{(i-0.5)/n}(r, n-r+1)}{I_{G(t)}(r, n-r+1)}, \tag{19.16}$$

if $N_n(t) = r$. Let now introduce test statistics for the hypothesis H_0. Integrating weighted statistics $LR(t)$ we can obtain various tests (depending on the choice of the weights) for the hypothesis H_0.

Integrating $LR^-(t)$ on $[0, T_{r,n}]$ with respect to

$$dw(t) = \frac{d\hat{F}_n(t)}{n\hat{F}_n^-(t)(1 - \hat{F}_n^-(t))},$$

results in the following statistic

$$\frac{1}{n}\int_0^{T_{r,n}} LR^-(t)dw(t) = 2n \sum_{i=1}^{r-1} \left(\frac{1}{n - i + 0.5} \ln \frac{i - 0.5}{nG(T_{i,n})} + \frac{1}{i - 0.5} \ln \frac{n - i + 0.5}{n(1 - G(T_{i,n}))} \right)$$
$$+ \frac{2n}{(r - 0.5)(n - r + 0.5)} \ln \frac{I_{(r-0.5)/n}(r, n-r+1)}{I_{G(T_{r,n})}(r, n-r+1)}.$$

Now we modify this statistic to test the Weibull-Gnedenko hypothesis H_{WG}. Let $\hat{\theta}$ and $\hat{\nu}$ be the ML estimators of the parameters θ and ν maximizing the log-likelihood function

$$\ell(\theta, \nu) = (\nu - 1)\sum_{i=1}^{r} \ln T_{i,n} - \nu r \ln \theta + r \ln \nu - \frac{1}{\theta^\nu}\left(\sum_{i=1}^{r} T_{i,n}^\nu + (n - r)T_{r,n}^\nu \right).$$

Set

$$U_{i,n} = 1 - e^{-(T_{i,n}/\hat{\theta})^{\hat{\nu}}}, \quad i = 1, \ldots, r, \quad U_{0,n} = 0.$$

Let us define the "integrated likelihood ratio statistic" as

$$ILR_{r,n} = 2n\sum_{i=1}^{r-1}\left(\frac{1}{n-i+0.5}\ln\frac{i-0.5}{nU_{i,n}} + \frac{1}{i-0.5}\ln\frac{n-i+0.5}{n(1-U_{i,n})}\right) + \frac{2n}{(r-0.5)(n-r+0.5)}\ln\frac{I_{(r-0.5)/n}(r,n-r+1)}{I_{U_{r,n}}(r,n-r+1)}. \tag{19.17}$$

The distribution of the random vector $(U_{1,n}, \ldots, U_{r,n})^T$ does not depend on unknown parameters under the hypothesis H_{WG} (see Bagdonavičius *et al.* 2011), so the distribution of the test statistic $ILR_{r,n}$ and the critical values of this statistic do not depend on unknown parameters too.

Let us denote by $ILR_\alpha(r,n)$ the α-critical value of the $ILR_{r,n}$ statistic. The hypothesis H_{WG} is rejected with the significance level α if $ILR_{r,n} > ILR_\alpha(r,n)$.

It is not difficult to obtain asymptotic distribution of the test statistic, however, an asymptotic result is not useful because the rate of convergence to the limiting distribution (as $r, n \to \infty$) is very slow even for complete data, as mentioned in Zhang (2002). This is not a problem because the critical values are easily found by simulation for reasonable values of r and n.

19.3 Tests based on the Difference of Non-Parametric and Parametric Estimators of the Cumulative Distribution Function

Michael (1983) used the arcsin transformation to stabilize the variances of the plotted points on the probability graphs associated with the Kolmogorov-Smirnov test. The modified Michael statistic (Castro-Kuriss 2011) has the form

$$D_{r,n} = \max_{1\le i\le r}\left\{\frac{2}{\pi}\left|arcsin\sqrt{\frac{i-0.5}{n}} - arcsin\sqrt{U_{i,n}}\right|\right\} \tag{19.18}$$

and has greater power than the Kolmogorov-Smirnov statistic.

Cramér-von-Mises and Anderson-Darling statistics $nC_{r,n}$ and $nA_{r,n}$ are modified in the case of the composite hypothesis and type II censored case as follows:

$$nC_{r,n} = \frac{r}{12n^2} + \sum_{i=1}^{r}\left(U_{i,n} - \frac{2i-1}{2n}\right)^2 + \frac{n}{3}\left(U_{r,n} - \frac{r}{n}\right)^3, \tag{19.19}$$

$$nA_{r,n} = -\frac{1}{n}\sum_{i=1}^{r}(2i-1)\ln\frac{U_{i,n}}{1-U_{i,n}} - 2\sum_{i=1}^{r}\ln\,(1-U_{i,n}) - \frac{1}{n}[(n-r)^2\ln\,(1-U_{r,n}) - r^2\ln\,U_{r,n} + n^2U_{r,n}]. \tag{19.20}$$

These expressions are obtained in the following way. Let us define the modified ω^2-

$$\omega^2_{r,n} = n\int_0^{Y_{r,n}}\left(\frac{i}{n} - y\right)^2\psi(y)dy,$$

where $\psi(x) = 1$ or $\psi(x) = 1/(x(1-x))$.

Let $g(x)$ and $h(x)$ be the primitives of $x\psi(x)$ and $\psi(x)$, respectively, and let

$$k(y) = \int_0^y x^2\psi(x)dx.$$

The value $h(0)$ is not defined for the function $\psi(x) = 1/(x(1-x))$. We have

$$n\int_0^{Y_{r,n}}\left(\frac{i}{n}-y\right)^2\psi(y)dy = n\int_0^{Y_{1,n}} y^2\psi(y)dy + n\sum_{i=1}^{r-1}\int_{U_{i,n}}^{U_{i+1,n}}\left(\frac{i}{n}-y\right)^2\psi(y)dy$$

$$= n\sum_{i=1}^{r-1}\frac{i^2}{n^2}[h(U_{i+1,n}) - h(U_{i,n})] - 2n\sum_{i=1}^{r-1}\frac{i}{n}[g(U_{i+1,n}) - g(U_{i,n})] + nk(U_{r,n})$$

$$= \frac{1}{n}\sum_{i=1}^{r}((i-1)^2 - i^2)h(U_{i,n}) + 2\sum_{i=1}^{r} g(U_{i,n}) + \frac{r^2}{n}h(U_{r,n}) - 2rg(U_{r,n}) + nk(U_{r,n})$$

$$= -\frac{1}{n}\sum_{i=1}^{r}(2i-1)h(U_{i,n}) + 2\sum_{i=1}^{r} g(U_{i,n}) + \frac{r^2}{n}h(U_{r,n}) - 2rg(U_{r,n}) + nk(U_{r,n}).$$

If $\psi(y) \equiv 1$, then

$$g(y) = \frac{y^2}{2}, \quad h(y) = y, \quad k(y) = \frac{y^3}{3},$$

$$nC_{r,n} = -\frac{1}{n}\sum_{i=1}^{r}(2i-1)U_{i,n} + \sum_{i=1}^{r} U_{i,n}^2 + \frac{r^2}{n}U_{r,n} - rU_{r,n}^2 + n\frac{U_{r,n}^3}{3}$$

$$= \sum_{i=1}^{r}\left[U_{i,n} - \frac{i-0.5}{n}\right]^2 - \frac{1}{n^2}\sum_{i=1}^{r}(i-0.5)^2 + \frac{r^2}{n}U_{r,n} - rU_{r,n}^2 + n\frac{U_{r,n}^3}{3}$$

$$= \sum_{i=1}^{r}\left[U_{i,n} - \frac{i-0.5}{n}\right]^2 + \frac{r}{12n^2} + \frac{n}{3}\left(U_{r,n} - \frac{r}{n}\right)^3.$$

If $\psi(y) = 1/(y(1-y))$, then

$$g(y) = -\ln(1-y), \quad h(y) = \ln\frac{y}{1-y}, \quad k(y) = -y - \ln(1-y),$$

$$nA_{r,n} = -\frac{1}{n}\sum_{i=1}^{r}(2i-1)\ln\frac{U_{i,n}}{1-U_{i,n}} - 2\sum_{i=1}^{r}\ln(1-U_{i,n})$$

$$+ \frac{r^2}{n}\ln\frac{U_{r,n}}{1-U_{r,n}} + 2r\ln(1-U_{r,n}) - nU_{r,n} - n\ln(1-U_{r,n})$$

$$= -\frac{1}{n}\sum_{i=1}^{r}(2i-1)\ln\frac{U_{i,n}}{1-U_{i,n}} - 2\sum_{i=1}^{r}\ln(1-U_{i,n})$$

$$-\frac{1}{n}[-r^2\ln U_{r,n} + r^2\ln(1-U_{r,n})$$

$$-2rn\ln(1-U_{r,n}) + n^2U_{r,n} + n^2\ln(1-U_{r,n})]$$

$$= -\frac{1}{n}\sum_{i=1}^{r}(2i-1)\ln\frac{U_{i,n}}{1-U_{i,n}} - 2\sum_{i=1}^{r}\ln(1-U_{i,n})$$

$$-\frac{1}{n}[(n-r)^2\ln(1-U_{r,n}) - r^2\ln U_{r,n} + n^2U_{r,n}].$$

The critical values of the statistics (19.18)–(19.20) are easily calculated by simulation and the hypothesis H_{WG} is rejected if these statistics take large values.

19.4 Tests based on Spacings

The hypothesis H_{WG} is equivalent to the hypothesis that the random variable $X = \ln T$ has distribution with the c.d.f.

$$F_X(x) = 1 - e^{-e^{\frac{x-\mu}{\sigma}}}, \quad x \in \mathbf{R},$$

where $\mu = \ln\theta, \sigma = 1/\nu$.

Set $X_{i,n} = \ln T_{i,n}$, $r_1 = [r/2]$. Let us define the following two statistics as

$$MSF_{r,n} = \frac{\sum_{i=r_1+1}^{r-1} M_{i,n}(X_{i+1,n} - X_{i,n})}{\sum_{i=1}^{r-1} M_{i,n}(X_{i+1,n} - X_{i,n})}, \quad TS_{r,n} = \frac{2\sum_{i=1}^{r-2}(r-i+1)M_{i,n}(X_{i+1,n} - X_{i,n})}{(r-2)\sum_{i=1}^{r-1} M_{i,n}(X_{i+1n} - X_{i,n})},$$

where $M_{i,n}$ are constants (depending on i and n).

Let us define $Z_{i,n} = (X_{i,n} - \mu)/\sigma$ which is the ith order statistic from a simple sample $Z_1, \ldots, Z_n$ with the cumulative distribution function of Z_i being $F(x) = 1 - e^{-e^x}$. This function does not depend on unknown parameters.

Based on the equality $X_{i+1,n} - X_{i,n} = \sigma(Z_{i+1,n} - Z_{i,n})$ and the fact that under the hypothesis H_{WG} the distribution of $(Z_{1,n}, \ldots, Z_{r,n})$ does not depend on unknown parameters, we can demonstrate that the distribution of either of the statistics $MSF_{r,n}$ and $TS_{r,n}$ also does not depend on unknown parameters.

The asymptotic distribution (as $i/n \to p \in (0,1)$) of the difference $n(Z_{i+1,n} - Z_{i,n})$ is exponential with the mean $-1/((1-p)\ln(1-p))$. So the distribution of each term $M_{i,n}(X_{i+1,n} - X_{i,n})$ is approximated by the same law if

$$M_{i,n} = -\left(1 - \frac{i}{n}\right)\ln\left(1 - \frac{i}{n}\right).$$

If n is small (in this case, tests have poor power and are not very useful), the weights $M_{i,n}$ could be replaced by $1/E(Z_{i+1,n} - Z_{i,n})$ to have equal means of all terms.

Mann *et al.* (1973) and Tiku and Singh (1981) give tests based on the statistics $MSF_{r,n}$ and $TS_{r,n}$, respectively. These tests generalize the well-known Gnedenko's F-test for exponentiality (Gnedenko *et al.* 1965) in the following way. The F-statistic of Gnedenko is

$$F = \frac{(r-r_1)\sum_{i=1}^{r_1}(n-i+1)(X_{i+1,n} - X_{i,n})}{r_1\sum_{i=r_1+1}^{r}(n-i+1)(X_{i+1,n} - X_{i,n})}.$$

If the lifetime distribution is exponential with hazard rate λ, then the random variables $(n-i+1)(X_{i+1,n}-X_{i,n})$, $i=1,\ldots,r$, are i.i.d. and each has exponential distribution with hazard rate λ. So under the hypothesis of exponentiality the Gnedenko's F-statistic has the Fisher F-distribution with $2r_1$ and $2(r-r_1)$ degrees of freedom. The Gnedenko test is powerful when the distribution under the alternative hypothesis has an increasing (or decreasing) hazard rate. If the hazard rate increases (decreases), then the statistic F tends to take greater (smaller) values than in the case of an exponential distribution.

If the Weibull-Gnedenko hypothesis H_{WG} is tested, then a modification of the Gnedenko statistic can be used by replacing the $(n-i+1)$ weight with the $M_{i,n}$ weight for $i=1,\ldots,r$. Then the random variables $M_{i,n}(X_{i+1,n}-X_{i,n})$ have asymptotic exponential distribution. For large n the distribution of such modified statistic is approximated by the same Fisher distribution. The only difference between the Gnedenko statistic modified this way and the Mann-Scheuer-Fertig statistic is as follows: in the modified Gnedenko statistic the first spacings appear in the numerator and the last spacings appear in the denominator; in the Mann-Scheuer-Fertig statistic the last spacings appear in the numerator and all spacings appear in the denominator. Therefore for large n the distribution of the Mann-Scheuer-Fertig statistic is approximated by the beta distribution.

The hypothesis H_{WG} may be tested using any of the statistics $MSF_{r,n}$ or $TS_{r,n}$. Let us denote by $MSF_\alpha(r,n)$ and $TS_\alpha(r,n)$ the α critical values of the distribution of the respective statistics $MSF_{r,n}$ and $TS_{r,n}$. The critical values $S_\alpha(r,n)$ and $TS_\alpha(r,n)$ are easily computed by simulation.

If the statistic $S_{r,n}$ is used, then the hypothesis H_{WG} is rejected with the significance level α if

$$MSF_{r,n} < S_{1-\alpha/2}(r,n) \quad \text{or} \quad MSF_{r,n} > S_{\alpha/2}(r,n).$$

If the statistic $TS_{r,n}$ is used, then the hypothesis H_{WG} is rejected with the significance level α if

$$TS_{r,n} < TS_{1-\alpha/2}(r,n) \quad \text{or} \quad TS_{r,n} > TS_{\alpha/2}(r,n).$$

19.5 Chi-Squared Tests

Bagdonavičius and Nikulin (2011) presented the modified chi-squared tests from arbitrary censored data with data-driven choice of partitioning.

Here we give simple explicit formulas for test statistic for type II censored samples.

Let us denote the ML estimators of the parameters θ and ν by $\hat{\theta}$ and $\hat{\nu}$. These estimators maximize the loglikelihood function

$$\ell(\theta,\nu) = (\nu-1)\sum_{i=1}^{r} \ln\, T_{i,n} - \nu r \ln\, \theta + r \ln\, \nu - \frac{1}{\theta^\nu}\left(\sum_{i=1}^{r} T_{i,n}^\nu + (n-r)T_{r,n}^\nu\right).$$

Let us divide the interval $[0, T_{r,n}]$ into k smaller intervals $I_j=(a_{j-1},a_j]$, $a_0=0$, $a_k=T_{r,n}$.

Set

$$b_0=0, \quad b_i=\sum_{l=1}^{i} U_{l,n} + (n-i)U_{i,n},\ i=1,\ldots,r.$$

For fixed j $(j = 1, \ldots, k)$ find the smallest number $i = i(j, U_{1,n}, \ldots, U_{r,n})$ verifying the inequalities

$$b_{i(j)-1} \leq \frac{j}{k} b_r \leq b_{i(j)}$$

(to save space, let us skip $U_{1,n}, \ldots, U_{r,n}$ in the expression of i). Then the limits of the intervals $I_j = (a_{j-1}, a_j]$ will be defined as

$$a_j = \hat{\theta} \left(\frac{\frac{j}{k} b_r - \sum_{l=1}^{i(j)-1} U_{l,n}}{n - i(j) + 1} \right)^{1/\hat{\nu}}, \quad j = 1, \ldots, k-1, \quad a_k = T_{r,n}.$$

As was noted in Section 19.2, the distribution of the random vector $(U_{1,n}, \ldots, U_{r,n})^T$ does not depend on unknown parameters under the hypothesis H_{WG}. Let us define $D_j = \sum_{s=1}^{r} \mathbf{1}_{\{T_{sn} \in I_j\}}$ the number of failures in the interval I_j. Note that the distribution of $(\mathbf{1}_{\{T_{1,n} \in I_j\}}, \ldots, \mathbf{1}_{\{T_{r,n} \in I_j\}})^T$ and consequently of $(D_1, \ldots, D_k)^T$ is parameter-free under the hypothesis H because the inequality $T_{i,n} \in I_j$ is equivalent to the inequality

$$\frac{\frac{j-1}{k} b_r - \sum_{l=1}^{i(j-1)-1} U_{l,n}}{n - i(j-1) + 1} < U_{sn} \leq \frac{\frac{j}{k} b_r - \sum_{l=1}^{i(j)-1} U_{l,n}}{n - i(j) + 1},$$

where i, b_r are functions of $(U_{1,n}, \ldots, U_{r,n})^T$ and the distribution of the random vector $(U_{1,n}, \ldots, U_{r,n})^T$ is parameter-free.

Let us define the statistic

$$Y^2(k, r, n) = \sum_{j=1}^{k} \frac{(D_j - r/k)^2}{D_j} + Q(k, r, n),$$

where (we skip (k, r, n) in the following expressions)

$$Q = \frac{W^2}{g}, \quad W = \frac{1}{\sqrt{n}} \sum_{j=1}^{k} \frac{C_j (D_j - r/k)}{D_j}, \quad C_j = \frac{1}{\hat{\nu}} \sum_{i=1}^{r} \mathbf{1}_{\{T_{i,n} \in I_j\}} (1 + \ln\ U_{i,n}).$$

$$g = \frac{1}{n \hat{\nu}^2} \sum_{i=1}^{r} (1 + \ln\ U_{i,n})^2 - \frac{1}{n} \sum_{j=1}^{k} D_j^{-1} C_j^2.$$

The distribution of the statistic $Y^2(k, r, n)$ is parameter-free because this statistic is a function of $(U_{1,n}, \ldots, U_{r,n})^T$.

High values of Y^2 indicate a poor fit and thus lead to the rejection of the hypothesis. The critical values are easily computed by simulation.

If n and r are large, the distribution of the statistic Y^2 is approximated by the chi square distribution with $k - 1$ degrees of freedom.

19.6 Correlation Test

Taking into account that the values of $X_{i,n} = \ln\ T_{i,n}$ are concentrated around

$$a_{i,n} = \ln\left[-\ln\left(1 - \frac{i}{n+1}\right)\right],$$

the square of the empirical correlation coefficient

$$R^2_{r,n} = \frac{\left[\sum_{i=1}^{r}(X_{i,n} - \overline{X}_{r,n})(a_{i,n} - \overline{a}_{r,n})\right]^2}{\sum_{i=1}^{r}(X_{i,n} - \overline{X}_{r,n})^2 \sum_{i=1}^{r}(a_{i,n} - \overline{a}_{r,n})^2}, \quad \overline{X}_{r,n} = \frac{1}{r}\sum_{i=1}^{r} X_{i,n}, \quad \overline{a}_{r,n} = \frac{1}{r}\sum_{i=1}^{r} a_{i,n},$$

should be large under the Weibull hypothesis. Let us define (Gerlach 1979)

$$CO_{r,n} = n(1 - R^2_{r,n}).$$

The distribution of this statistic is parameter-free under the hypothesis H_{WG} because the order statistics $X_{i,n}$ have the form $X_{i,n} = \sigma Z_{i,n} + \mu$ and the distribution of $(Z_{1,n}, \ldots, Z_{r,n})^T$ is parameter-free.

High values of $CO_{r,n}$ indicate a poor fit and thus lead to the rejection of the hypothesis.

19.7 Power Comparison

Here we investigate the power of the proposed tests against the most commonly used alternatives to the Weibull-Gnedenko distribution:

1. lognormal distribution $LN\ (\mu, \sigma)$;
2. loglogistic distribution $LL(\mu, \sigma)$;
3. gamma distribution $G(a, \lambda)$.

The distribution of test statistics does not depend on unknown parameters under the null hypotheses but depends on unknown parameters under the alternatives. We used parameter values for alternative distributions under which the powers are smaller than 1 in the case of samples of size n.

We obtained critical values of the following statistics by simulation (100,000 simulations were used for each test statistic with fixed r and n, levels of censoring were 10% and 25%):

1. integrated likelihood ratio ($ILR_{r,n}$);
2. Cramér-von-Mises ($nC_{r,n}$);
3. Anderson-Darling ($nA_{r,n}$);
4. Michael ($D_{r,n}$);
5. Mann-Scheuer-Fertig ($MSF_{r,n}$);
6. Tiku-Singh($TS_{r,n}$);
7. chi-squared ($Y^2(k, r, n)$);
8. correlation ($CO_{r,n}$).

Table 19.1 shows the critical values of the new integrated likelihood ratio statistic. These values do not depend on the unknown parameters as was shown in Section 19.2.

Powers of tests based on 10,000 simulations were computed for censored samples with 10% and 25%levels of censoring. The level of significance was 0.05. The optimal number (giving the highest power) of grouping intervals k was used for chi squared test power analysis.

Powers of the tests against lognormal, loglogistic and gamma alternatives are given in Tables 19.2, 19.3 and 19.4, respectively. The powers of three most powerful tests are written in bold for each alternative and fixed values of r and n.

In the case of lognormal and gamma alternatives, the proposed ILR test and the Tiku-Singh test are the most powerful. More specifically, under lognormal alternative, the ILR

Table 19.1 Critical values $ILR_\alpha(r, n)$

p	α	n						
		40	50	60	80	100	150	200
0.9	0.1	5.8	6.2	6.6	7.1	7.5	8.2	8.8
	0.05	7.3	7.8	8.3	8.9	9.5	10.4	11.1
0.75	0.1	4.9	5.4	5.7	6.2	6.7	7.4	7.9
	0.05	6.3	6.9	7.2	7.9	8.5	9.5	10.1

Table 19.2 Powers of tests against lognormal distribution LN (0, 5)

p	Test	n						
		40	50	60	80	100	150	200
	$ILR_{r,n}$	**0.373**	**0.484**	**0.576**	**0.717**	**0.820**	**0.954**	**0.996**
	$nC_{r,n}$	0.271	0.336	0.396	0.516	0.619	0.818	0.958
	$nA_{r,n}$	0.287	0.361	0.437	0.569	0.678	0.870	0.977
	$D_{r,n}$	**0.344**	**0.434**	**0.514**	**0.648**	**0.757**	**0.912**	**0.988**
0.9	$MSF_{r,n}$	0.295	0.358	0.428	0.545	0.648	0.836	0.963
	$TS_{r,n}$	**0.397**	**0.487**	**0.574**	**0.716**	**0.811**	**0.950**	**0.995**
	$Y^2(k, r, n)$	0.247	0.326	0.393	0.535	0.647	0.848	0.970
	$CO_{r,n}$	0.048	0.056	0.067	0.099	0.134	0.266	0.739
	Optimal k	3	4	5	6	6	7	7
	$ILR_{r,n}$	**0.215**	**0.298**	**0.364**	**0.489**	**0.595**	**0.787**	**0.895**
	$nC_{r,n}$	0.169	0.214	0.260	0.340	0.414	0.596	0.743
	$nA_{r,n}$	0.166	0.218	0.269	0.362	0.447	0.644	0.793
	$D_{r,n}$	**0.204**	**0.276**	**0.327**	**0.440**	**0.538**	**0.727**	**0.848**
0.75	$MSF_{r,n}$	0.181	0.236	0.282	0.371	0.447	0.623	0.751
	$TS_{r,n}$	**0.231**	**0.299**	**0.364**	**0.475**	**0.583**	**0.774**	**0.889**
	$Y^2(k, r, n)$	0.165	0.215	0.261	0.341	0.431	0.622	0.764
	$CO_{r,n}$	0.028	0.025	0.028	0.036	0.043	0.084	0.144
	Optimal k	3	4	4	5	5	5	6

Table 19.3 Powers of tests against loglogistic distribution $LL(1, 3)$

p	Test	n						
		40	50	60	80	100	150	200
	$ILR_{r,n}$	**0.285**	**0.342**	**0.397**	**0.489**	0.576	0.735	0.840
	$nC_{r,n}$	0.272	0.327	0.385	0.483	**0.578**	**0.753**	**0.868**
	$nA_{r,n}$	0.278	**0.337**	**0.396**	**0.501**	**0.601**	**0.779**	**0.888**
	$D_{r,n}$	**0.281**	0.336	0.386	0.468	0.544	0.694	0.803
0.9	$MSF_{r,n}$	0.247	0.299	0.368	0.475	0.555	0.732	0.857
	$TS_{r,n}$	**0.349**	**0.427**	**0.505**	**0.618**	**0.714**	**0.866**	**0.946**
	$Y^2(k, r, n)$	0.203	0.266	0.301	0.392	0.496	0.670	0.796
	$CO_{r,n}$	0.049	0.047	0.047	0.048	0.059	0.098	0.131
	Optimal k	3	4	4	4	4	5	5
	$ILR_{r,n}$	**0.120**	**0.155**	**0.174**	**0.214**	0.252	0.347	0.412
	$nC_{r,n}$	0.117	0.151	0.167	**0.215**	**0.264**	**0.386**	**0.475**
	$nA_{r,n}$	0.107	0.139	0.159	0.209	**0.254**	**0.380**	**0.476**
	$D_{r,n}$	**0.119**	**0.158**	**0.174**	0.213	0.247	0.334	0.396
0.75	$MSF_{r,n}$	0.118	0.146	0.164	0.208	0.250	0.375	0.464
	$TS_{r,n}$	**0.156**	**0.210**	**0.220**	**0.284**	**0.345**	**0.496**	**0.591**
	$Y^2(k, r, n)$	0.103	0.123	0.138	0.169	0.219	0.309	0.373
	$CO_{r,n}$	0.029	0.027	0.029	0.027	0.028	0.027	0.031
	Optimal k	3	3	4	4	4	5	5

Table 19.4 Powers of tests against the gamma distribution $G(8, 1)$

p	Test	n						
		40	50	60	80	100	150	200
	$ILR_{r,n}$	**0.182**	**0.234**	**0.285**	**0.346**	**0.447**	**0.627**	**0.749**
	$nC_{r,n}$	0.142	0.165	0.194	0.255	0.305	0.445	0.570
	$nA_{r,n}$	0.143	0.174	0.209	0.277	0.334	0.492	0.631
	$D_{r,n}$	**0.174**	**0.218**	**0.269**	**0.330**	**0.399**	**0.566**	**0.688**
0.9	$MSF_{r,n}$	0.152	0.177	0.223	0.281	0.344	0.497	0.620
	$TS_{r,n}$	**0.194**	**0.242**	**0.298**	**0.379**	**0.471**	**0.662**	**0.786**
	$Y^2(k, r, n)$	0.128	0.159	0.190	0.242	0.311	0.450	0.580
	$CO_{r,n}$	0.019	0.018	0.019	0.024	0.024	0.043	0.069
	Optimal k	3	4	4	5	5	5	6
	$ILR_{r,n}$	**0.128**	**0.154**	**0.180**	**0.238**	**0.294**	**0.422**	**0.521**
	$nC_{r,n}$	0.095	0.116	0.130	0.171	0.208	0.297	0.382
	$nA_{r,n}$	0.089	0.110	0.130	0.174	0.217	0.316	0.413
	$D_{r,n}$	**0.114**	**0.151**	**0.169**	**0.223**	**0.269**	**0.387**	**0.468**
0.75	$MSF_{r,n}$	0.103	0.126	0.143	0.187	0.227	0.335	0.423
	$TS_{r,n}$	**0.121**	**0.161**	**0.183**	**0.239**	**0.291**	**0.446**	**0.542**
	$Y^2(k, r, n)$	0.088	0.118	0.135	0.170	0.203	0.294	0.369
	$CO_{r,n}$	0.016	0.013	0.014	0.014	0.013	0.016	0.023
	Optimal k	3	4	4	4	4	5	6

test is the most powerful for samples of size over 50 and the Tiku-Singh test is most powerful for the sample sizes $n = 40, 50$. It holds true for both censoring percentages. Under gamma alternative Tiku-Singh test is the best, the ILR test is second, being slightly less powerful. Michael's test took the third place under both alternative distributions.

In the case of loglogistic alternative the situation is slightly different. Clearly the Tiku-Singh test has the highest power. For sample sizes 40, 50, 60, the second is the ILR test. For the sample sizes of 80 and above, the Anderson-Darling test takes the second place.

19.8 Conclusion

We conclude that the proposed ILR test has only one serious challenger: the Tiku-Singh test. There are situations where the ILR test is the best, and also situations (which are more frequent) where the Tiku-Singh test is the best.

Please note that the ILR test and the Tiku-Singh test can only be applied for complete and type II censored samples. Under more general censoring schemes the modified chi-squared test can be applied (see Bagdonavičius *et al.* 2011).

References

Bagdonavičius, V., Kruopis, J., and Nikulin, M. 2011. *Nonparametric Tests for Censored Data*, London: ISTE/WILEY.

Bagdonavičius, V., Levuliene, R., and Nikulin, M. 2012. Chi-squared goodness-of-fit test for parametric accelerated failure time models. *Communications in Statistics: Theory and Methods* (accepted for publication).

Bagdonavičius, V., and Nikulin, M. 2011. Chi-squared goodness-of-fit test for right censored data. *International Journal of Applied Mathematics and Statistics* 24: 30–50.

Bennett, J. G. 1936. Broken coal. *Journal of the Institute of Fuel* 10: 22–39.

Castro-Kuriss, C. 2011. On a goodness-of-fit test for censored data from a location-scale distribution with applications. *Chilean Journal of Statistics* 2(1): 115–136.

Fisher, R. A., and Tippett, L. H. C. 1928. Limiting forms of the frequency distribution of the largest and smallest member of a sample, in *Proceedings of the Cambridge Philosophical Society* 24(2): 180–190.

Gerlach, B. 1979. A consistent correlation-type goodness-of-fit test with application to the two-parameter Weibull distribution. *Statistics*, 10 (3): 427–452.

Gnedenko, B.V. 1943. Sur la distribution limite du terme maximum d'une série aléatoire, *Annals of Mathematics* 44: 423–453.

Gnedenko, B. V., Belyaev, Y. K. and Soloviev, A. D. 1965. *Mathematical Methods in Reliability*. Moscow: Nauka (in Russian)

Greenwood, P. E., and Nikulin, M. S. 1996. *A Guide to Chi-Squared Testing*. New York: John Wiley & Sons.

Mann, N. R., Scheuer, M., and Fertig, K. W. 1973. A new goodness-of-fit test for the two parameter Weibull or extreme-value distribution with unknown parameters. *Communication in Statistics* 2: 283–400.

Michael, J. R. 1983. The stabilized probability plot. *Biometrika* 70(1): 11–17.

Rosin, P., Rammler, E., and Sperling, K. 1933. *Korngrossenprobleme des Kohlenstaubes und ihre Bedeutung für die Vermahlung Berichte der Technisch-Wirtschaftlichen Sachverstndigenaus schusse des Reichskohlenrats*. Bericht C52, pp.1–25.

Stoyan, D. 2011. Weibull, RRSB or extreme-value theorists? *Metrika* 76 (2): 153–159.

Tiku, M. L., and Singh, M. 1981. Testing the two parameter Weibull distribution. *Communication in Statistics A*. 10(9): 907–917.

Weibull, W. 1939. The phenomenon of rupture in solids. *Proceedings of the Royal Swedish Institute for Engineering Research*, No 153, Stockholm.

Weibull, W. 1951. A statistical distribution function of wide applicability. *Journal of Applied Mechanics* 18 (3): 293–297.

Zhang, J. 2002. Powerful goodness-of-fit tests based on the likelihood ratio. *Journal of the Royal Statistical Society. B* 64 (2): 281–294.

20

Statistical Inference for Heavy-Tailed Distributions in Reliability Systems

Ilia Vonta[1] and Alex Karagrigoriou[2]
[1]*Department of Mathematics, School of Applied Mathematical and Physical Sciences, National Technical University of Athens, Greece*
[2]*Department of Mathematics and Statistics, University of Cyprus, Cyprus*

20.1 Introduction

The distributions which are widely used in physics and engineering, viz., the normal Gaussian distribution, the exponential (Boltzmann) distribution, and combinations of these, are sufficient to describe many empirical distributions of physical characteristics of natural disasters and of losses due to disasters. The Normal law is of universal applicability possibly due to the reason provided by Poincare who observed that physicists rely on it due to its mathematical justification and mathematicians due to its physical nature. On the other hand, the exponential law has attracted considerable attention as a typical physical distribution. One explanation for this is the fact that physical phenomena like biological mechanisms are quite complex and therefore one needs to rely on distributions with easily determined parameters. Earthquakes, hurricanes and other rare and extreme events were solely relying on the exponential law. Similar examples can be found in economics and reliability. In reliability in particular, the interest lies in the occurrence of rather exceptional events which are associated with the tail part of a probability distribution.

All the above-mentioned rare events and the rate at which they occur, are related to the shape and the heaviness of the tail of the generating mechanism that produces such events. At the same time both mathematical and physical arguments can also be advanced to explain the prevalence of heavy-tailed distributions. There is a broad class of distributions which are called heavy or fat or long-tailed or subexponential distributions.

Applied Reliability Engineering and Risk Analysis: Probabilistic Models and Statistical Inference, First Edition.
Edited by Ilia B. Frenkel, Alex Karagrigoriou, Anatoly Lisnianski and Andre Kleyner.

A sub-class of heavy-tailed distributions is the class of power-tailed distributions known as Pareto, Lévy or Lévy-Pareto distributions. The mathematical importance of the latter class lies on the asymptotic theory of stable laws. More specifically it is related to the second theorem in extreme value theory which provides the asymptotic tail distribution of a random variable X, with unknown cumulative distribution function F. Unlike the first theorem (the Fisher-Tippett-Gnedenko theorem, Fisher and Tippett 1928; Gnedenko 1943), the second, known as the Gnedenko-Pickands-Balkema-de Haan limit theorem (Balkema and de Haan 1974; Pickands 1975), focuses on the conditional distribution above a threshold, say, t. Indeed, if F^t and f^t are the excess conditional distribution and the excess conditional density functions of the random variable $X^t = [X|X > t]$ or $[X - t|X > t]$, then F^t is well approximated by the Generalized Pareto distribution (GPD) as $t \to \infty$, with cumulative distribution function (cdf) given by

$$\lim_{t\to\infty} F^t(y) = F(y) = \begin{cases} 1 - \left(1 + \dfrac{ky}{\sigma}\right)^{-1/k}, & k \neq 0 \\ 1 - \exp^{\,-y/\sigma}, & k = 0 \end{cases} \tag{20.1}$$

where $y \geq 0$ if $k > 0$ and $0 \leq y \leq -\sigma/k$ if $k < 0$. The distribution reduces to the exponential distribution for $k = 0$, to the Pareto for $k < 0$ and to the Uniform distribution in $[0, \sigma]$ for $k = -1$.

The importance of the above result lies in the fact that rare or extreme events of phenomena (that exceed a certain threshold) such as electrical or power massive failures and physical disasters (e.g. earthquakes or hurricanes) obey the power-law even if the phenomenon itself obeys a distribution different from a power-law distribution.

It is quite often in reliability that we talk about a situation where a device is *wearing out* or *aging*. Both terms refer to cases where the reliability function decreases as a function of time, after a certain threshold time t. This is the case of the increasing failure rate (IFR) distribution. If the reverse occurs, we are dealing with a decreasing failure rate (DFR) distribution. In both instances we have, as opposed to the exponential distribution, varying failure (or hazard) rates and consequently heavier (and lighter) tails with the former attracting much more attention. It is therefore quite important to define and analyze heavy-tailed distributions and provide statistical tools to investigate their presence in reliability, economics, telecommunications, etc. For this purpose we intend to rely on the so-called measures of divergence and establish distances appropriate for both tails of the distribution, namely the *residual lifetime* as well as the *past lifetime* of events. These measures will then be used for statistical inference.

The chapter is organized as follows. In the next two sections we provide the definitions, examples and some properties and characterizations of heavy-tailed distributions. In Section 20.4 we discuss measures of divergence and in Section 20.5 we explain how these measures can be implemented as statistical tools for the assessment of heavy-tailed distributions. Finally, in Section 20.6 a series of Monte Carlo simulations are used to explore the capabilities and the general performance of the proposed techniques.

20.2 Heavy-Tailed Distributions

In probability theory, heavy-tailed distributions are probability distributions whose tails are not exponentially bounded, that is, they have heavier tails than the exponential distribution.

In the literature the term *leptokurtic distributions* also appears which indicates that such distributions have kurtosis greater than 0 and their tail tends to zero at a rate slower than the normal distribution. In many applications it is the right tail of the distribution that is of interest, but a distribution may have a heavy left tail, or both tails may be heavy. There is a broad class of distributions called heavy-, long-, fat- power-tailed or sub-exponential distributions. The classification is associated with their behavior for large values. In all instances, the reliability function $R(x)$ tends to 0 at a rate slower than that of the exponential distribution.

Let X be a random variable (r.v.) representing the time for a process to complete and let $F(x) = P(X \leq x)$ the cumulative distribution function of X. It is customary to denote by $f(x)$ the probability density function (pdf) of X and by $R(x)$ (or $\overline{F}(x)$) the reliability (or survival) function of X which is be given by

$$R(x) = \overline{F}(x) = 1 - F(x) = P(X > x).$$

Another function that is frequently used is the failure (or hazard) function which is given by

$$h(x) = \frac{f(x)}{\overline{F}(x)}.$$

Definition 20.1. We define a distribution F to be (right-)heavy-tailed if and only if

$$\int_{\mathbb{R}} e^{\lambda x} f(x)dx = \int_{\mathbb{R}} e^{\lambda x} dF(x) = \infty, \qquad \text{for all} \quad \lambda > 0. \tag{20.2}$$

If on the other hand, the above quantity is finite for *some* $\lambda > 0$ then it is said that the distribution F is *light-tailed.*

The corresponding definition for functions says that a function $g(x)$ is said to be *heavy-tailed function* if and only if

$$\limsup_{x\to\infty} g(x)e^{\lambda x} = \infty, \qquad \text{for all} \quad \lambda > 0.$$

It should be noted that a distribution F is heavy-tailed if and only if the corresponding reliability function $R(x)$ is a heavy-tailed function. Note that if F is heavy-tailed, then the pdf $f(x)$ is a heavy-tailed function but the converse is *not* true. The piecewise continuous density

$$f(x) + \sum_{n=1}^{\infty} \mathbb{I}\{x \in [n, n + 2^{-n}]\}$$

which is not a heavy-tailed function but the corresponding F is a heavy-tailed distribution, is discussed in Foss *et al.* (2011, p. 9).

Finally, for the so-called hazard function $h(x)$ we have:

$$\liminf_{x\to\infty} h(x)/x = \liminf_{x\to\infty} \left\{ -\frac{1}{x} \ln \overline{F}(x) \right\} = 0.$$

The definitions for the long-tailed and the power-tailed distributions are given as follows.

Definition 20.2. A distribution F is called long-tailed if $R(x) > 0$ for all x and for any fixed t we have

$$\lim_{x\to\infty} \frac{R(x+t)}{R(x)} = \lim_{x\to\infty} P[X > x + t | X > x] = 1.$$

Definition 20.3. A distribution F is called power-tailed if for large x

$$R(x) \sim \frac{c}{x^a}$$

for $c > 0$ and $a > 0$.

An alternative definition indicates that F is power-tailed if

$$\lim_{x\to\infty} x^k R(x) = \begin{cases} 0 & k < a \\ \infty & k > a. \end{cases} \tag{20.3}$$

The power-tailed class is also known as the Pareto, Lévy or Lévy-Pareto distribution. Observe that all moments of order $k > a$ are infinite so that if $a < 1$ then both the mean and the variance are infinite. We point out that power-tailed distributions are both long- and heavy-tailed but the reverse is not necessarily true. Furthermore, any long-tailed distribution is heavy-tailed but the reverse is not necessarily true (see Shortle *et al.* 2004; Bashkansky and Gadrich 2012). The function

$$f(x) = \sum_{n=1}^{\infty} 2^{-n} \mathbb{I}\{2^{n-1} < x \le 2^n\}$$

is long-tailed but not a long-tailed distribution (see Foss *et al.* 2011, p. 19).

Finally, we provide the definition for the sub-exponential distributions. Subexponentiality is defined in terms of convolutions of probability distributions. For two i.i.d. random variables X_1 and X_2 with common distribution function F, the convolution of F with itself, is defined, using Lebesgue-Stieltjes integration, by:

$$F^{*2}(x) = P[X_1 + X_2 \le x] = \int_0^{\infty} F(x-y)dF(y).$$

The n-fold convolution, F^{*n} is defined in the same way.

Definition 20.4. A distribution F in the positive half-line is sub-exponential if

$$\liminf_{x\to\infty} \frac{\overline{F^{*2}}(x)}{\overline{F}(x)} = 2.$$

The generalization to $n > 2$ is natural. All subexponential distributions are both long-tailed and heavy-tailed distributions but the converse is not true.

Heavy-tailed distributions can be characterized by their moments. Indeed, if a distribution F on the positive half-line $\mathbb{R}^+$ does not have all its moments finite, namely

$$\int_0^{\infty} x^k dF(x) = \infty, \quad \text{for some} \quad k > 0,$$

then F is a heavy-tailed distribution. Naturally, this property does not hold for light-tailed distributions. Note that the earlier characterization result can be extended to generalized moments:

Theorem 20.1. Let a non-negative r.v. X with heavy-tailed distribution F. Let a function $g(x)$ s.t. $g(x) \to \infty$ as x tends to ∞. Then, there exists appropriate monotone concave function $\varphi(x)$ s.t.

$$E(e^{g(X)}) < \infty \quad \text{and} \quad E(e^{g(X)+\varphi(X)}) = \infty.$$

The book by Bingham *et al.* (1987) is considered classic in the theory of heavy-tailed distributions. Large deviations for heavy-tailed distributions have recently been investigated by Konstantinides and Loukissas (2011) and in a PhD thesis (in Greek) by Loukissas (2012).

20.3 Examples of Heavy-Tailed Distributions

The Pareto distribution of the first kind (also known as Pearson type VI distribution), is the most popular of the power-tailed distributions. The distribution was originally used to describe the allocation of wealth. The tail function, namely the reliability function of the Pareto distribution of the first kind, is given by

$$R(x) = \left(\frac{x_{min}}{x}\right)^a, \quad x > x_{min}$$

where $x_{min} > 0$ is the minimum value of X and $a > 0$ the shape parameter.

If the Pareto of the first kind is shifted by $x_{\min}$ units so that the shifted variable is defined for $x > 0$, then the resulting distribution is known as Pareto of the second kind or Lomax distribution which finds its applications, among other areas, in actuarial science. Its reliability function is given by

$$R(x) = \overline{F}(x) = \left(\frac{k}{x+k}\right)^a, \quad x > 0$$

where $k > 0$ the scale parameter and $a > 0$ the shape parameter.

The Pareto distribution and its generalization, the Generalized Pareto distribution (GPD) given in (20.1), have extensive applications in various fields, like wind engineering, where they are implemented for the statistical analysis of extreme wind speeds produced by storms (Holmes and Moriarty 1999) or earthquake engineering where GPD is used for the tail part of the distribution or earthquake ground-motion values (Huyse *et al.* 2010) or the reliability of a software system (Kantam and Subba Rao 2009). In hydrology and climate science, the Pareto distribution is used to explain environmental extremes such as annually maximum one-day rainfalls and river discharges. Bashkansky and Gadrich (2012) state that the popularity and great applicability of the Pareto distribution could be attributed to its special properties of scale invariance and self-similarity.

For the case of the Pareto distribution, the heavy tail can be revealed if one considers the logarithm of the reliability function. Then, clearly the resulting log-reliability function is increased linearly with log x:

$$\log\,(R(x)) = \log\,(x_{\min})^a - a \log\, x.$$

The above setting can be potentially useful for estimating purposes. Indeed, using the empirical distribution function the left-hand side can be estimated through the available data and the least squares estimators of the slope of the regression line a and the $y-$intercept $\log\ (x_{\min})^a$ can easily be obtained.

Although the above distribution is right-tailed, there are distributions that are two-tailed. One such distribution is a special case of the Pearson type VII distribution, known as *Cauchy distribution* (also known as Lorentz distribution), the probability density function of which is given by

$$f(x) = \frac{k}{\pi\left((x-a)^2 + k^2\right)}$$

where $k > 0$ is the scale parameter, and $a \in \mathbb{R}$ is the location parameter. The simplest Cauchy distribution is the standard Cauchy distribution which is defined as the distribution of a random variable that is the ratio of two independent standard normal random variables, the pdf of which is given by

$$f(x) = \frac{1}{\pi(1+x^2)}.$$

It is well known that for the Cauchy distribution both the mean and the variance are undefined and in addition, it has no moment generating function.

The Cauchy distribution is related to the so-called *signal-to-noise ratio* (S/N) which plays a pivotal role in the performance quality in reliability control. Indeed, the S/N ratio is a representative example of the *ratio distribution*, namely the distribution of the ratio of two random variables. Cauchy is such a *ratio distribution* since, as mentioned above, it can be defined as the ratio of two independent standard normal random variables (it should be pointed out though that the converse is *not* true). For more details, see Bashkansky and Gadrich (2012) and Johnson *et al.* (1994, p. 313). Since the tail of the Cauchy distribution is heavier than that of the Pareto distribution, it is regarded as a *super-heavy-tailed* distribution.

A related distribution which is, however, right-tailed is the log-Cauchy distribution. If X follows a Cauchy distribution, then the random variable $Y = e^X$ follows the log-Cauchy distribution. Equivalently, if Y follows the log-Cauchy distribution, then the r.v. $X = \log\ Y$ has a Cauchy distribution.

Another one-tailed distribution is *the lognormal distribution*. As in the case of the Cauchy distribution, if X is a random variable with a normal distribution, then $Y = e^X$ has a log-normal distribution; likewise, if Y is log-normally distributed, then $X = \log\ Y$ has a normal distribution. The reliability function of the lognormal distribution is given by

$$R(x) = \overline{F}(x) = \Phi\left(\frac{\log\ x - \mu}{\sigma}\right)$$

where μ is the mean, σ the standard deviation, and $\Phi(\cdot)$ the cdf of the standard normal distribution.

The lognormal distribution is the most commonly used life distribution model for many high technology applications. The lognormal distribution is quite popular in reliability and maintainability engineering. For instance, in reliability analysis, it is often used to model times to repair a maintainable system. It has been also proposed that coefficients of friction and wear may be treated as having a lognormal distribution (Steele, 2008). It is also used to describe failure due to chemical reactions or degradation, such as corrosion, migration, or diffusion, which is common with semiconductor failure and time to fracture in metals that are subject to the growth of fatigue cracks. Furthermore, it has been demonstrated that it describes dust concentration in industrial areas and mineral concentration in geology. Many more applications have been reported in telecommunications, quality control, survival analysis, etc. (see Johnson *et al.* 1994, p. 238).

The Burr system of distributions was constructed in 1942 by Irving W. Burr (1942). The reliability function of *the Burr type XII distribution* is given by

$$R(x) = \overline{F}(x) = \left(\frac{k}{x^{\tau} + k}\right)^{a}$$

for parameters $a, k, \tau > 0$. Clearly for $\tau = 1$ it becomes the Lomax distribution (Pareto of the second kind). The log-logistic distribution is also a special case of the Burr distribution. It is also called the *transformed Pareto distribution* since if X follows the Burr distribution, then the transformed random variable $Y = X^{1/\tau}$ follows the Pareto distribution with parameters a and k. The Burr distribution can be developed as a mixture of Weibull distributions with a gamma mixing distribution. Indeed, let Λ be a Gamma random variable with parameters a and k. Let another r.v. X be such that the conditional distribution of X given $\Lambda = \lambda$ is Weibull with parameters τ and λ. Then, the unconditional distribution of X is a Burr distribution with the reliability function given above.

The Burr distribution, together with the lognormal distribution and the Inverse Gaussian are the most popular distributions with non-monotone failure rates. The usual failure function of these distributions increases over time from 0 to a maximum and thereafter decreases as $t \to \infty$ which is of some concern since the risk of failure does not, in typical cases, decrease as time increases. It should be, however, pointed out that although the hazard function of both the lognormal and the Burr distribution decrease to 0 as time increases, the same is not true for the Inverse Gaussian distribution. This feature makes the Inverse Gaussian more attractive than its two competitors as far as the tail of the distribution is concerned.

We close this section with the most popular heavy-tailed distribution, the *Weibull distribution*, the reliability function of which is given by

$$R(x) = \overline{F}(x) = e^{-(x/k)^{a}}$$

where $k > 0$ the scale parameter and $a > 0$ the shape parameter. Although the distribution is popular due to its mathematically straightforward form, the inference is not so easy due to the fact that the sufficient statistic for the shape parameter a is $n-$dimensional. For $a = 1$ it reduces to the Exponential distribution and for $a = 2$ becomes the Rayleigh distribution. Notice that for $a > 1$ it becomes an IFR distribution and for $a < 1$ becomes a DFR distribution. In the former case it is used to model wear-out (old aging) and in the latter to model wear-in (infant mortality) for mechanical and technical units and systems. Since it does not have a *bath-tub-shaped failure rate*, one needs more than one Weibull to model different, in shape, stages of a unit's life. To overcome this defect, a generalization of the Weibull distribution has been developed. More specifically, the exponentiated Weibull distribution with cdf given by

$$F(x) = \left(1 - e^{-(x/k)^{a}}\right)^{\gamma}$$

has been proposed by Zacks (1984) and Mudholkar and Srivastava (1993). This distribution which exhibits a bath-tub-shaped failure rate, has been extensively investigated in recent years (see, e.g., Mudholkar *et al.* 1995).

The Weibull distribution, like the Pareto distribution is related to the theory of extremes. Indeed, the well-known Fisher-Tippett-Gnedenko first theorem (Fisher and Tippett 1928; Gnedenko 1943) of the extreme value theory deals with the distribution of the maximum of a sequence of independent and identically distributed random variables. If one considers the reverse of this result, then, under conditions given by Gnedenko (1943), the distribution of the minimum of positive random variables is approximated by the Weibull distribution (see Chapter 19 in the present volume by Bagdonavičius *et al.* (2014) where the term *Weibull-Gnedenko distribution* is used). The Weibull distribution is used in industrial engineering to represent manufacturing and delivery times, in weather forecasting to describe the distribution of wind speed and in wireless communications for the modeling of fading channels. Many more applications can be found in reliability and industrial engineering, survival analysis, hydrology, etc. The special case of the Rayleigh distribution is often used to describe measurement data in the field of communications engineering and in the life testing of electrovacuum devices.

20.4 Divergence Measures

Measures of information are powerful statistical tools with diverse applicability. In this section we will focus on a specific type of information measures, known as measures of discrepancy (distance or divergence) between the lifetimes of two items X and Y with pdf's f and g. Furthermore, we will explore ways to measure the discrepancy between (i) the distribution of the lifetime of an item X as deduced from an available set of data; and (ii) the distribution of a hypothesized distribution that is believed to be the generating mechanism that produced the set of data at hand. Although we will present general measures available for new items, we will pay special attention to measures appropriate for measuring used items, which will be appropriate for measuring the discrepancy between the tail-part of the involved distributions. More specifically we will provide ways to quantify the divergence between the residual lives as well as the past lifetimes which are associated with the tail heaviness of the distributions.

For historical reasons we present first Shannon's Entropy (Shannon 1948) given by

$$I^S(X) \equiv I^S(f) = -\int f \ln f \, d\mu = E_f[-\ln \ f],$$

where X is a random variable with density function $f(x)$ and μ a probability measure on $\mathbb{R}$. Shannon's entropy was introduced and used during World War II in Communication Engineering. Shannon derived the discrete version of $I^S(f)$ where f is a probability mass function and named it *entropy* because of its similarity to thermodynamics entropy. The continuous version was defined by analogy. For a finite number of points, Shannon's entropy measures the expected information of a signal transferred without noise from a source X with density $f(x)$ and is related to Kullback-Leibler divergence (Kullback and Leibler 1951) through the following expression:

$$I^S(f) = I^S(h) - I_X^{KL}(f, h)$$

where h is the density of the uniform distribution and the Kullback-Leibler divergence between two densities $f(x)$ and $g(x)$ is given by

$$I_X^{KL}(f,g) = \int f \ln(f/g) d\mu = E_f[\ln(f/g)]. \tag{20.4}$$

Many generalizations of Shannon's Entropy were hereupon introduced. Rényi's (1961) entropy as extended by Liese and Vajda (1987) is given by

$$I^{R_{lv},a}(X) \equiv I^{R_{lv},a}(f) = \frac{1}{a(a-1)} \ln E_f[f]^{a-1}, \quad a \neq 0, 1.$$

For more details on entropy measures, the reader is referred to Mathai and Rathie (1975), Nadarajah and Zografos (2003) and Zografos and Nadarajah (2005).

A measure of divergence is used as a way to evaluate the distance (divergence) between any two functions f and g associated with the variables X and Y. Among the most popular measures of divergence are the Kullback-Leibler measure of divergence given in (20.4) and Csiszar's φ-divergence family of measures (Csiszar 1963; Ali and Silvey 1966) given by

$$I_{f,g}^{\varphi} = \int_0^{\infty} g(x)\varphi\left(\frac{f(x)}{g(x)}\right) dx. \tag{20.5}$$

The class of Csiszar's measures includes a number of widely used measures that can be recovered for appropriate choices of the function φ. Both these measures are used to measure the distance between the entire lifespan of the associated random variables X and Y. When the function φ is defined as

$$\varphi(u) = u \log u \quad \text{or} \quad \varphi(u) = u \log u + 1 - u \tag{20.6}$$

then the above measure reduces to the Kullback-Leibler measure given in (20.4). If

$$\varphi(u) = \frac{1}{2}(1-u)^2, \tag{20.7}$$

Csiszar's measure yields the Pearson's chi-square divergence (also known as Kagan's divergence; (Kagan 1963). If

$$\varphi(u) \doteq \varphi_1(u) = (u^{a+1} - u - a(u-1)/(a(a+1)) \tag{20.8}$$

we obtain the Cressie and Read power divergence (Cressie and Read 1984), $a \neq 0, -1$.

Another function that we usually consider is

$$\varphi(u) \doteq \varphi_2(u) = 1 - \left(1 + \frac{1}{a}\right)u + \frac{u^{1+a}}{a}, \quad a \neq 0. \tag{20.9}$$

This function is related to a recently proposed measure of divergence (BHHJ power divergence, Basu *et al.* (1998)) while it is a special case of the BHHJ family of divergence measures proposed by Mattheou *et al.* (2009)

$$I_X^{BHHJ} = \int g^{1+\alpha}(x)\varphi\left(\frac{f(x)}{g(x)}\right) dx, \qquad \alpha \geq 0. \tag{20.10}$$

Appropriately chosen functions $\varphi(\cdot)$ give rise to special measures mentioned above while for $\alpha = 0$ the BHHJ family reduces to the Csiszar's family. For more details on divergence measures, see Cavanaugh (2004), Pardo (2006), Toma (2009), and Toma and Broniatowski (2011). Specifically for robust inference based on divergence measures see Basu *et al.* (2011) and a recent paper by Patra *et al.* (2012) on the power divergence and the density power divergence families. The problem of robust parameter estimation for the 3-parameter Generalized Pareto distribution given in (20.1) with $y - \theta$ in place of y, is investigated in Chapter 21 of this volume by Ghosh *et al.* (2014).

It was quite recently that the need to measure the distance between the tails of two distributions, surfaced. In reliability we might know the current age t of a mechanical or technical system which is a piece of information that needs to be taken into consideration when two systems are compared. Ebrahimi and Kirmani (1996) achieved this by replacing the distribution functions f and g of the random variables X and Y in the Kullback-Leibler divergence measure given in (20.4), by the distributions of their *residual lifetimes*. Di Crescenzo and Longobardi (2004) define a dual measure of divergence which constitutes a distance between *past life distributions*. Vonta and Karagrigoriou (2010) focus on the distance between lifetimes and propose generalized measures of divergence between residual lives of two items that have both survived up to some time t as well as between past lives, based on Csiszar's family of measures. It should finally be noted that in a recent paper Sachlas and Papaioannou (2012) worked on entropies rather than measures and investigated the residual and past entropies with applications to actuarial science.

To formulate the problem, let X and Y be absolutely continuous, non-negative random variables that describe the lifetimes of two items. Let $f(x)$, $F(x)$ and $\overline{F}(x)$ (or $R(x)$) be the density function, the cumulative distribution function and the reliability function of X respectively. Let also $g(x)$, $G(x)$ and $\overline{G}(x)$ (or $S(x)$) be the density function, the cumulative distribution function and the reliability function of Y respectively. Without loss of generality we assume that the support of f and g is $(0, +\infty)$.

Recently Vonta and Karagrigoriou (2010) proposed two new measures of discrepancy which are based on Csiszar's φ-divergence family, namely, the *φ-distance between residual lifetimes*

$$I_{X,Y}^{\varphi}(t) = \int_t^{\infty} \frac{g(x)}{\overline{G}(t)} \varphi\left(\frac{f(x)/\overline{F}(t)}{g(x)/\overline{G}(t)}\right) dx, \qquad t > 0 \tag{20.11}$$

and the *φ-distance between past lifetimes*

$$\overline{I}_{X,Y}^{\varphi}(t) = \int_0^{t} \frac{g(x)}{G(t)} \varphi\left(\frac{f(x)/F(t)}{g(x)/G(t)}\right) dx, \qquad t > 0 \tag{20.12}$$

where the function φ belongs to a class of functions Φ with properties

(i) $\varphi(x)$ is continuous, differentiable and convex for $x \geq 0$
(ii) $\varphi(1) = 0$
(iii) $\varphi'(1) = 0$.

From the above assumptions we deduce that $\varphi(x) \geq 0$ for all $x > 0$ and $\varphi'(x) > 0$ for $x > 1$ and $\varphi'(x) < 0$ for $x < 1$.

Observe that special cases of the above measures of divergence are the ones associated with the Kullback-Leibler measure with $\varphi(x) = x \ln x$. Indeed, Ebrahimi and Kirmani (1996) introduced a measure of discrepancy between the residual lives of X and Y at time t as follows

$$I_{X,Y}(t) = \int_t^{\infty} \frac{f(x)}{\overline{F}(t)} \ln \left(\frac{f(x)/\overline{F}(t)}{g(x)/\overline{G}(t)} \right) dx, \qquad t > 0 \tag{20.13}$$

and Di Crescenzo and Longobardi (2004) introduced the dual measure for the distance between past lifetimes, given by

$$\overline{I}_{X,Y}(t) = \int_0^t \frac{f(x)}{F(t)} \ln \left(\frac{f(x)/F(t)}{g(x)/G(t)} \right) dx, \qquad t > 0. \tag{20.14}$$

For a 2-tail distribution (like the Cauchy distribution) one may wish to investigate the behavior of both tails simultaneously. This is possible if, for φ Csiszar's class, we consider two time points t_1 and t_2 with $t_1 < t_2$ and define the 2-tail $\varphi-$ distance by

$$I\overline{I}_{X,Y}(t_1, t_2) = \int_{\mathbb{R}-[t_1,t_2]} \frac{g(x)}{\overline{G}(t_2) + G(t_1)} \varphi \left(\frac{f(x)/\left(\overline{F}(t_2) + F(t_1)\right)}{g(x)/(\overline{G}(t_2) + G(t_1))} \right) dx, \qquad t_1 > 0. \tag{20.15}$$

Although the φ-family of measures in (20.11), (20.12), and (20.15) is a very broad family of measures, the Rényi measure does not belong to it. As a result, we propose here, for $a > 0$ and $a \neq 1$, the Rényi distance between residual lives X and Y is given by

$$\mathcal{R}^a_{X,Y}(t) = \frac{1}{1-a} \ln \left(\int_t^{\infty} \left(\frac{f(x)}{\overline{F}(t)} \right)^a \left(\frac{g(x)}{\overline{G}(t)} \right)^{1-a} dx, \right) \tag{20.16}$$

the Rényi distance between past lives X and Y is given by

$$\overline{\mathcal{R}}^a_{X,Y}(t) = \frac{1}{1-a} \ln \left(\int_0^t \left(\frac{f(x)}{F(t)} \right)^a \left(\frac{g(x)}{G(t)} \right)^{1-a} dx \right) \tag{20.17}$$

and the Rényi 2-tail distance is given by

$$\overline{\mathcal{R}}^a_{X,Y}(t_1, t_2) = \frac{1}{1-a} \ln \left(\int_{\mathbb{R}-[t_1,t_2]} \left(\frac{f(x)}{\overline{F}(t_2) + F(t_1)} \right)^a \left(\frac{g(x)}{\overline{G}(t_2) + G(t_1)} \right)^{1-a} dx \right). \tag{20.18}$$

Another type of divergence measure, the interval distance between the random lifetimes X and Y at interval t_1, t_2 is the Kullback-Leibler discrimination measure between the truncated lives $[X|t_1 < X < t_2]$ and $[Y|t_1 < Y < t_2]$ and is given by (Misagh and Yari 2012)

$$\overline{ID}_{X,Y}(t_1, t_2) = \int_{t_1}^{t_2} \frac{f(x)}{F(t_2) - F(t_1)} \ln \left(\frac{f(x)/\left(F(t_2) - F(t_1)\right)}{g(x)/(G(t_2) - G(t_1))} \right) dx, \qquad t_1 > 0. \tag{20.19}$$

As expected, natural generalizations be to the Rényi interval distance and Csiszar's interval φ-distance can easily established:

$$\overline{\mathcal{IR}}^{a}_{X,Y}(t_1,t_2) = \frac{1}{1-a}\ln\left(\int_{t_1}^{t_2}\left(\frac{f(x)}{F(t_2)-F(t_1)}\right)^{a}\left(\frac{g(x)}{G(t_2)-G(t_1)}\right)^{1-a}dx\right) \quad (20.20)$$

and

$$\overline{ID}^{\varphi}_{X,Y}(t) = \int_{t_1}^{t_2}\frac{g(x)}{G(t_2)-G(t_1)}\varphi\left(\frac{f(x)/\left(F(t_2)-F(t_1)\right)}{g(x)/(G(t_2)-G(t_1))}\right)dx, \qquad t_1 > 0. \quad (20.21)$$

20.5 Hypothesis Testing

Let a continuous random variable X describe the lifetime of an individual or a technical system with a distribution function F. Let F^* be one of the following four conditional cdf's, of past life F_t, residual lifetime F^t, 2-tail lifetime F_2, and interval lifetime $F_{1,2}$. Let also f^* be the probability density function of the random variable X^* associated with F^* so that, for example, $f_t(x) = \frac{f(x)}{F(t)}$ is the probability distribution function of the random variable $X_t = [X|X < t]$ associated with $F_t(x)$ and $f^t(x) = \frac{f(x)}{\bar{F}(t)}$ is the probability distribution function of the random variable $X^t = [X|X > t]$ associated with $F^t(x)$. In what follows we use the notation F^* which refers to any of the above four conditional distributions. Assume that $X_1^*, \ldots, X_n^*$ is a random sample with conditional distribution F^*.

Consider also a parametric family of distributions $\{F_\theta\}_{\theta\in\Theta}$ where Θ is an open subset of $\mathbb{R}^{M_0}$ and the parameter θ is unknown. Let $F_\theta^*(x)$ denote the conditional distribution function of a random variable Z^*.

We would like now to test the hypothesis that the data set $X_1^*, \ldots, X_n^*$ follows a specific past life distribution, that is, we would like to test the null hypothesis

$$H_0 : F^* = F_\theta^*. \quad (20.22)$$

To test the above null hypothesis we propose partitioning the interval of interest into M equal subintervals $E_i, i = 1, \ldots, M$. Note that in the case of the 2-tail measure the interval of interest is $[0, t_1] \cup [t_2, \infty)$.

Let $\hat{p}^* = (\hat{p}_1^*, \ldots, \hat{p}_M^*)^T$ be the relative frequencies in the subintervals E_i, $i = 1, \ldots, M$, and $p^*(\boldsymbol{\theta}) = (p_1^*(\boldsymbol{\theta}), \ldots, p_M^*(\boldsymbol{\theta}))^T$ be the probabilities corresponding to the intervals E_i, $i = 1, \ldots, M$ under the null hypothesis, that is, $p_i^*(\boldsymbol{\theta})$ are defined as

$$p_i^*(\boldsymbol{\theta}) = \int_{\frac{i-1}{M}t}^{\frac{i}{M}t} f_\theta^*(x)dx, \quad i = 1, \ldots, M.$$

Since the parameter θ under the null, is unknown, a consistent estimator, say, $\hat{\boldsymbol{\theta}}_\varphi$ will be used. For possible choices of this estimator, see the following comments.

Observe that in (20.11), (20.12), (20.15)–(20.18), (20.20) and (20.21) the integral will be replaced by summation, and the quantities $p_i^*(\hat{\boldsymbol{\theta}}_\varphi)$ and $\hat{p}_i^*$ will play the roles of the (appropriate) conditional distributions of g^* and f^*.

We will test the null hypothesis based on the φ-divergence test statistic defined as

$$\begin{aligned} T_n^{\varphi}(\hat{\boldsymbol{\theta}}_{\varphi}) = T_n^{\varphi}(\hat{p}^*, p^*(\hat{\boldsymbol{\theta}}_{\varphi})) &= \frac{2n}{\varphi''(1)} D_{\varphi}(\hat{p}^*, p^*(\hat{\boldsymbol{\theta}}_{\varphi})) \\ &= \frac{2n}{\varphi''(1)} \sum_{i=1}^{M} p_i^*(\hat{\boldsymbol{\theta}}_{\varphi}) \varphi\left(\frac{\hat{p}_i^*}{p_i^*(\hat{\boldsymbol{\theta}}_{\varphi})}\right). \end{aligned} \tag{20.23}$$

We present here the asymptotic distribution of the test statistic introduced earlier.

Theorem 20.2. Under the null hypothesis given in (20.22) and regularity conditions as in Pardo (2006), the test statistic defined in (20.23) follows asymptotically a χ^2 distribution with $M - M_0 - 1$ degrees of freedom as $n \to \infty$. Furthermore, the estimator $\hat{\theta}_{\varphi}$ is unique in a neighborhood of θ_0, consistent for θ_0 and asymptotically normally distributed.

The proof is omitted but related results can be found in Pardo (2006) and Mattheou and Karagrigoriou (2010).

According to the previous theorem, the null hypothesis is rejected if the value of the test statistic is over the upper $\alpha-$percentile of the χ^2 distribution with the appropriate degrees of freedom:

$$T_n^{\varphi}(\hat{\theta}_{\varphi}) > \chi^2_{M-M_0-1;\alpha}$$

where M_0 the dimension of the unknown parameter θ.

Comments.

(i) The partition of the relevant interval into M subintervals could be accomplished in different ways, for example, by using the quantiles of the null distribution.

(ii) The estimation of the parameter θ is found from

$$\begin{aligned} \hat{\theta}_{\varphi} &= \arg \inf_{\theta \in \Theta \subseteq \mathbb{R}^{M_0}} D_{\varphi}(\hat{p}^*, p^*(\theta)) \\ &= \arg \inf_{\theta \in \Theta \subseteq \mathbb{R}^{M_0}} \left(\sum_{i=1}^{M} p_i^*(\theta) \varphi\left(\frac{\hat{p}_i^*}{p_i^*(\theta)}\right)\right). \end{aligned} \tag{20.24}$$

Note that the φ function used in (20.24) could be different from the one used in (20.23). If φ is chosen to be the one given in (20.6), then the maximum likelihood estimator of θ will be obtained. For testing purposes, a consistent estimator of θ is sufficient for the asymptotic theory (Theorem 20.2) to be valid.

(iii) It should be noted that the asymptotic distribution of the test statistic in (20.23) could be established also under the alternative hypothesis for specific type of alternatives (local contiguous alternatives or Pitman (1949) transition (local) alternatives). For details, see Mattheou and Karagrigoriou (2010) and Vonta and Karagrigoriou (2013).

(iv) Such tests as the above, for the case of the left-tail, have also been examined in Chouchoumis 2010.

20.6 Simulations

In this section we perform a small simulation study to show the behavior of the proposed tests. More specifically, we present results for the comparison of the behavior of the following test statistics: the Kullback-Leibler test statistic, Pearson's X^2 test statistic, the Cressie-Read test statistic and the φ_2- test statistic which are associated with the φ functions given in (20.6)–(20.9). For the Cressie and Read function which depends on a parameter a, we will consider the best choice of a which is equal to 2/3 (Read and Cressie 1988). For the φ_2 function we will examine two different values of the parameter a, namely, $a = 10/9$ and $a = 0.8$. The choice 10/9 constitutes the best choice of a for the φ_2 function (Tsanousa 2010).

We generated each time 20000 random samples for different sample sizes, namely $n = 20$, 50, 100, and 200 from an appropriate residual distribution function $F^t_{\theta_0}$ and for t such that

$$F_{\theta_0}(t) = P(X \leq t) = 0.25$$

where θ_0 is the true value of the parameter of the distribution. All the tests are conducted at a significance level $a = 0.05$ with $M = 3$ subintervals. In Tables 20.1–20.3 we present the size and the power of the test when the null distribution is exponential with true parameter $\theta_0 = 1$ and the alternative is Gamma with scale parameter equal to 1 and shape parameter equal to 0.5 and 4 respectively.

Our results show that in general all tests have good sizes for large sample sizes as well as for small sample sizes. All sizes are fairly close to the nominal 5% level. In terms of power we observe that all five tests have a very similar behavior for large sample sizes which shows that they are equivalent. Although the differences between them are minimal, we must indicate that the φ_2 test with $a = 0.8$ appears to have a slight edge

Table 20.1 Distribution $Exp^t(1)$

n	KL	X^2	CR $a = 2/3$	$\varphi_2 - Test$ $a = 10/9$	$\varphi_2 - Test$ $a = 0.8$
20	0.05425	0.03750	0.04975	0.03325	0.05875
50	0.05495	0.04275	0.04915	0.03715	0.05340
100	0.05440	0.04705	0.04980	0.04475	0.05160
200	0.05560	0.05230	0.05325	0.04980	0.05460

Table 20.2 Power of the test $Exp^t(1)$ versus G^t (0.5, 1)

n	KL	X^2	CR $a = 2/3$	$\varphi_2 - Test$ $a = 10/9$	$\varphi_2 - Test$ $a = 0.8$
20	0.07435	0.03730	0.06040	0.03095	0.07465
50	0.08020	0.06510	0.07010	0.04500	0.07860
100	0.07345	0.07820	0.07915	0.07715	0.07930
200	0.09810	0.10660	0.10385	0.10630	0.10415

Table 20.3 Power of the test $Exp^t(1)$ versus $G^t(4, 1)$

n	KL	X^2	CR $a = 2/3$	$\varphi_2 - Test$ $a = 10/9$	$\varphi_2 - Test$ $a = 0.8$
20	0.51265	0.57195	0.56470	0.57190	0.56470
50	0.82865	0.82940	0.83360	0.83625	0.83395
100	0.94805	0.94510	0.94630	0.94310	0.94720
200	0.98375	0.98130	0.98265	0.98030	0.98295

over its competitors. The subject of powers of the different tests, however, needs further investigation in order to safely conclude which test is actually better.

20.7 Conclusion

In this work we focus on the class of heavy-tailed distributions and discuss their significance in reliability engineering. Relying on divergence measures we evaluate the residual and past lifetimes of events which are associated with the tail part of the distribution of events and propose a class of goodness of fit tests designed for heavy-tailed distributions. We establish the asymptotic distribution of the proposed test statistic and through some simulations we show that the proposed tests behave in a similar fashion and perform quite satisfactory with minimal differences between them.

References

Ali, S. M. and S. D. Silvey. 1966. A general class of coefficients of divergence of one distribution from another. *Journal of the Royal Statistical Society* B28 (1): 131–142.

Bagdonavićius, V., M. S. Nikulin and R. Levuliene. 2014. On validity of the Weibull-Gnedenko model. In *Applied Reliability Engineering and Risk Analysis: Probabilistic Models and Statistical Inference*, ed. I. Frenkel, A. Karagrigoriou, A. Lisnianski and A. V. Kleyner. Chichester: John Wiley & Sons.

Balkema, A. A. and L. de Haan. 1974. Residual life time at great age. *Annals of Probability* 2 (5): 792–804.

Bashkansky, E. and T. Gadrich. 2012. Reliability of continuous-state systems in view of heavy-tailed distributed performance features. In *Recent Advances in System Reliability*, ed. A. Lisnianski and I. Frenkel. London, Springer, pp. 295–306.

Basu, A., Harris, I. R., Hjort, N. L. and M. C. Jones. 1998. Robust and efficient estimation by minimising a density power divergence. *Biometrika* 85 (3): 549–559.

Basu, A., Shioya, H. and C. Park. 2011. *Statistical Inference: The Minimum Distance Approach*. Boca Raton FL: Chapman & Hall/CRC.

Bingham, N. H., Goldie, C. M. and J. L. Teugels. 1987. *Regular Variation*. Cambridge: Cambridge University Press.

Burr, I. W. 1942. Cumulative frequency functions. *Annals of Mathematical Statistics* 13 (2): 215–232.

Cavanaugh, J. E. 2004. Criteria for linear model selection based on Kullback's symmetric divergence. *Australian and New Zealand Journal of Statistics* 46 (2): 257–274.

Chouchoumis, I. 2010. φ-divergence measures and their applications to survival analysis and reliability. Master's thesis. Athens, National Technical University of Athens (in Greek).

Cressie, N. and T. R. C. Read. 1984. Multinomial goodness-of-fit tests. *Journal of the Royal Statistical Society* B46 (3): 440–464.

Csiszar, I. 1963. Eine Informationstheoretische Ungleichung und ihre Anwendung auf den Beweis der Ergodizität on Markhoffschen Ketten. *Publ. of the Math. Inst. of the Hungarian Academy* 8: 95–108.

Di Crescenzo, A. and M. Longobardi. 2004. A measure of discrimination between past lifetime distributions. *Statistics and Probability Letters* 67 (2): 173–182.

Ebrahimi, N. and S. N. U. A. Kirmani. 1996. A measure of discrimination between two residual life-time distributions and its applications. *Annals of the Institute of Statistical Mathematics* 48 (2): 257–265.

Fisher, R. A. and L. H. C. Tippett. 1928. Limiting forms of the frequency distribution of the largest and smallest member of a sample. *Mathematical Proceedings of the Cambridge Philosophical Society* 24 (2): 180–190.

Foss, S., Korshunov, D. and A. Zachary. 2011. *An Introduction to Heavy-Tailed and Subexponential Distributions*. New York: Springer.

Ghosh, A., Maji, A. and A. Basu. 2014. Robust inference based on divergence. In *Applied Reliability Engineering and Risk Analysis: Probabilistic Models and Statistical Inference*, ed. I. Frenkel, A. Karagrigoriou, A. Lisnianski and A. Kleyner. Chichester: John Wiley & Sons.

Gnedenko, B.V. 1943. Sur la distribution limite du terme maximum d'une serie aleatoire. *Annals of Mathematics* 44 (3): 423–453.

Holmes, J.D. and W. W. Moriarty. 1999. Application of the generalized Pareto distribution to extreme value analysis in wind engineering. *Journal of Wind Engineering and Industrial Aerodynamics* 83 (1–3): 1–10.

Huyse, L., R. Chen and J. A. Stamatakos. 2010. Application of generalized Pareto distribution to constrain uncertainty in peak ground accelerations. *Bulletin of the Seismological Society of America* 100 (1): 87–101.

Johnson, N. L., S. Kotz and N. Balakrishnan. 1994. *Continuous Univariate Distributions*, Vol. 1. New York: John Wiley & Sons.

Kagan, A. M. 1963. On the theory of Fisher's amount of information. *Soviet Mathematics. Doklady* 4: 991–993.

Kantam, R. R. L and R. Subba Rao. 2009. Pareto distribution: a software reliability growth model. *International Journal of Performability Engineering* 5 (3): 275–281.

Konstantinides, D. G. and F. Loukissas. 2011. Precise large deviations for sums of negatively dependent random variables with common long tailed distributions. *Communications in Statistics: Theory and Methods* 40 (19–20): 3663–3671.

Kullback, S. and R. Leibler. 1951. On information and sufficiency. *The Annals of Mathematical Statistics* 22 (1): 79–86.

Liese, F. and I. Vajda. 1987. *Convex Statistical Distances*. Leipzig: Teubner.

Loukissas, F. 2012. The exact large deviations in heavy-tailed distributions. PhD Thesis. Dept. of Statistics and Actuarial-Financial Mathematics. Aegean, Greece: University of the Aegean, Greece (in Greek).

Mathai, A. and P. N. Rathie. 1975. *Basic Concepts in Information Theory and Statistics: Axiomatic Foundations and Applications*. New Delhi: John Wiley & Sons.

Mattheou, K. and A. Karagrigoriou. 2010. A new family of divergence measures for tests of fit. *Australian and New Zealand Journal of Statistics* 52 (2): 187–200.

Mattheou, K., S. Lee and A. Karagrigoriou. 2009. A model selection criterion based on the BHHJ measure of divergence. *Journal of Statistical Planning and Inference* 139 (1): 228–235.

Matusita, K. 1967. On the notion of affinity of several distributions and some of its applications, *Annals of the Institute of Statistical Mathematics* 19 (1): 181–192.

Misagh, F. and G. Yari. 2012. Interval entropy and informative distance, *Entropy*, 14: 480–490.

Mudholkar, G.S. and D.K. Srivastava. 1993. Exponentiated Weibull family for analyzing bathtub failure-rate data. *IEEE Transactions on Reliability* 42 (2): 299–302.

Mudholkar, G.S., D.K. Srivastava and M. Freimer. 1995. The exponentiated Weibull family: A reanalysis of the bus-motor-failure data. *Technometrics* 37 (4): 436–445.

Nadarajah, S. and K. Zografos. 2003. Formulas for Rényi information and related measures for univariate distributions. *Information Sciences* 155 (1–2): 119–138.

Pardo, L. 2006. *Statistical Inference Based on Divergence Measures*. Boca Raton, FL: Chapman & Hall/CRC.

Patra, S., A. Maji, A. Basu and L. Pardo. 2012. The power divergence and the density power divergence families: the mathematical connection. *Sankhya B* (To appear).

Pickands, J. 1975. Statistical inference using extreme order statistics. *The Annals of Statistics* 3 (1): 119–131.

Pitman, E. 1949. *Notes on Nonparametric Statistical Inference*. New York: Columbia Univ., mimeo.

Read, T. R. C. and N. Cressie. 1988. *Goodness-of-Fit Statistics for Discrete Multivariate Data*. New York: Springer-Verlag.

Rényi, A. 1961. On measures of entropy and information. In *Proceedings of the Fourth Berkeley Symposium on Mathematical Statistics and Probability* June 20–July 30, 1960. 547-561. Berkeley, California: University of California Press.

Sachlas, A. and T. Papaioannou. 2012. Residual and past entropy in actuarial science and survival models. *Methodology and Computing in Applied Probability*. (To appear).

Shannon, C. E. 1948. A mathematical theory of communication. *Bell System Technical Journal* 27: 379–423 & 623–656.

Shortle J. F., P. H. Brill, M. J. Fischer, D. Gross and D. M. B. Masi. 2004. An algorithm to compute the waiting time distribution for the M/G/1 queue. *INFORMS Journal on Computing* 16 (2): 152–161.

Steele, C. 2008. Use of the lognormal distribution for the coefficients of friction and wear. *Reliability Engineering and System Safety* 93 (10): 1574–1578.

Toma, A. 2009. Optimal robust M-estimators using divergences. *Statistics and Probability Letters* 79 (1): 1–5.

Toma, A. and M. Broniatowski. 2011. Dual divergence estimators and tests: robustness results. *Journal of Multivariate Analysis* 102 (1): 20–36.

Tsanousa, A. 2010. Statistical inference via φ-measures of divergence. Master's thesis. Athens, National Technical University of Athens (in Greek).

Vonta, F. and A. Karagrigoriou. 2010. Generalized measures of divergence in survival analysis and reliability. *Journal of Applied Probability* 47 (1): 216–234.

Vonta, I. and A. Karagrigoriou. 2013. Goodness of fit tests via measures of divergence for censored data, *Journal of Statistical Computation and Simulation*, DOI.org/10.1080/00949655.2012.733396 (to appear).

Zacks, S. 1984. Estimating the shift to wear-out of systems having exponential-weibull life distributions. *Operations Research* 32 (3): 741–749.

Zhang, J. 2002. Powerful goodness-of-fit tests based on the likelihood ratio. *Journal of the Royal Statistical Society* B64 (2): 281–294.

Zografos, K. and S. Nadarajah. 2005. Survival exponential entropies. *IEEE Transactions on Information Theory* 51 (3): 1239–1246.

21

Robust Inference based on Divergences in Reliability Systems

Abhik Ghosh, Avijit Maji and Ayanendranath Basu
Indian Statistical Institute, India

21.1 Introduction

The minimum distance approach is an useful technique in parametric inference. The idea is a natural one; the discrepancy between the data and the model is quantified through a suitable divergence which is then minimized over the parameter space. Many such methods have been particularly useful because of the strong robustness properties that they inherently possess. Minimum distance methods in the literature can be divided into two classes: (i) those based on distances between the empirical and the model distribution functions, such as the Kolmogorov-Smirnov and Cramér von-Mises statistics, and (ii) those based on distances between a nonparametric density estimate and the model density, such as the chi-square and the Bregman divergences. Although the history of Pearson's chi-square (Pearson 1900) goes far back, the use of density-based divergences in robust statistical inference is much more recent. Beran's (1977) paper appears to be the starting point for this area of research. Since then the literature has grown substantially. The monographs by Vajda (1989), Pardo (2006) and Basu *et al.* (2011) provide a unified description of the development of this field.

Our main interest in the use of the minimum distance technique is on account of their inherent robustness properties. Many density-based minimum distance estimators, in addition, have full (or very high) asymptotic efficiency, which makes them very useful estimators in practice. Such estimators have also demonstrated the limitations of the influence function in quantifying their robustness, which will be further illustrated here.

Applied Reliability Engineering and Risk Analysis: Probabilistic Models and Statistical Inference, First Edition. Edited by Ilia B. Frenkel, Alex Karagrigoriou, Anatoly Lisnianski and Andre Kleyner.

The discussion that we present in this chapter will describe the power divergence family (Cressie and Read 1984) and density power divergence family (Basu *et al.* 1998) under a common framework along the lines of Ghosh *et al.* (2012) and show that both families are part of a larger superfamily.

The rest of this chapter is organized as follows. Sections 21.2 and 21.3 describe the power divergence (PD) and the density power divergence (DPD) families and, apart from discussing their robustness properties, talk about the interconnection between the families. Section 21.4 ties in these families through a larger superfamily, and Section 21.5 talks about applications. Section 21.6 has some concluding remarks.

In this chapter we will use the term "density function" for both discrete and continuous models. We also use the term "distance" loosely, to refer to any divergence which is nonnegative and is equal to zero if and only if its arguments are identically equal.

21.2 The Power Divergence (PD) Family

Density-based minimum distance inference is largely dominated by the class of chi-square distances, referred to as ϕ-divergences, f-divergences, or disparities (Csiszár 1963; Ali and Silvey 1966; Lindsay 1994; Pardo 2006). The power divergence family (Cressie and Read 1984) is perhaps its most prominent subclass. We begin with a discrete probability model $\mathcal{F}_\theta = \{F_\theta : \theta \in \Theta \subseteq \mathbb{R}^p\}$. The original Cressie-Read paper developed the power divergence family in the context of multinomial goodness-of-fit testing. However, the family has been used successfully by a host of subsequent authors to produce robust and efficient estimators under parametric settings; see Basu *et al.* (2011) for an extended discussion.

To exploit the structural geometry, we follow Lindsay's (1994) disparity approach to describe the PD family. Let $X_1, \cdots, X_n$ denote n independent and identically distributed observations from a discrete distribution G. Without loss of generality, let the support of G and the parametric model $\mathcal{F}_\theta$ be $\mathcal{X} = \{0, 1, 2, \cdots\}$. Denote the relative frequency of the value x in above sample by $r_n(x)$. We assume that both G and $\mathcal{F}_\theta$ belong to $\mathcal{G}$, the class of all distributions having densities with respect to the appropriate measure. Let f_θ be the model density function. We estimate the parameter by choosing the model element that provides the closest match to the data. The separation between the probability vectors $\underline{r} = (r_n(0), r_n(1), \cdots)^T$ and $\underline{f_\theta} = (f_\theta(0), f_\theta(1), \cdots)^T$ will be quantified by the class of disparities.

Definition 21.1. Let C be a thrice differentiable, strictly convex function on $[-1, \infty)$, satisfying $C(0) = 0$. Let the Pearson residual at the value x be defined by

$$\delta(x) = \delta_n(x) = \frac{r_n(x)}{f_\theta(x)} - 1.$$

Then the disparity between $\underline{r}$ and $\underline{f_\theta}$ generated by C is defined by

$$\rho_C(r_n, f_\theta) = \sum_{x=0}^{\infty} C(\delta(x)) f_\theta(x). \tag{21.1}$$

The strict convexity of C and Jensen's inequality immediately imply that the disparity defined in (1) is nonnegative; it equals zero only when $r_n = f_\theta$, identically. For notational simplicity, we will write the expression on the right-hand side of Equation (21.1) as $\sum C(\delta) f_\theta$ whenever the context is clear, and use similar notation throughout the rest of this chapter.

Specific forms of the function C generate many well-known disparities. For example, $C(\delta) = (\delta + 1)\log(\delta + 1) - \delta$ generates the well-known likelihood disparity (LD) given by

$$\text{LD}(r_n, f_\theta) = \sum [r_n \log(r_n/f_\theta) + (f_\theta - r_n)] = \sum r_n \log(r_n/f_\theta). \tag{21.2}$$

The symmetric opposite of the above is the Kullback-Leibler divergence which has the form

$$\text{KLD}(r_n, f_\theta) = \sum [f_\theta \log(f_\theta/r_n) + (r_n - f_\theta)] = \sum f_\theta \log(f_\theta/r_n). \tag{21.3}$$

It corresponds to $C(\delta) = \delta - \log(\delta + 1)$. The (twice, squared) Hellinger distance (HD) has the form

$$\text{HD}(r_n, f_\theta) = 2\sum \left[r_n^{1/2} - f_\theta^{1/2}\right]^2 \tag{21.4}$$

and has $C(\delta) = 2((\delta + 1)^{1/2} - 1)^2$. The Pearson's chi-square (divided by 2) is defined as

$$\text{PCS}(r_n, f_\theta) = \sum \frac{(r_n - f_\theta)^2}{2f_\theta}, \tag{21.5}$$

where $C(\delta) = \delta^2/2$, and the Neyman's chi-square (divided by 2) is defined as

$$\text{NCS}(r_n, f_\theta) = \sum \frac{(r_n - f_\theta)^2}{2r_n}, \tag{21.6}$$

where $C(\delta) = \delta^2/[2(\delta + 1)]$.

Arguably, the best-known subfamily of the disparities is the power divergence family (Cressie and Read 1984) which is indexed by a real parameter λ, and has the form

$$\text{PD}_\lambda(r_n, f_\theta) = \frac{1}{\lambda(\lambda + 1)} \sum r_n \left[\left(\frac{r_n}{f_\theta}\right)^\lambda - 1\right]. \tag{21.7}$$

Note that for values of $\lambda = 1, 0, -1/2, -1$ and -2 the Cressie-Read form in Equation (21.7) generates the PCS, the LD, the HD, the KLD and the NCS respectively. The LD and KLD are actually the continuous limits of the expression on the right-hand side of (21.7) as $\lambda \to 0$ and -1 respectively. The measure HD ($\lambda = -1/2$) is the only symmetric measure within this family, and the only one that is linked to a metric. Divergences within this family that are equally spaced on either side of $\lambda = -1/2$ in the λ scale are symmetric opposites of each other obtained by interchanging r_n and f_θ. Thus, one gets the KLD ($\lambda = -1$) by interchanging r_n and f_θ in the expression for LD ($\lambda = 0$), and the NCS ($\lambda = -2$) by interchanging r_n and f_θ in the expression for the PCS ($\lambda = 1$). In fact one can write $\text{PD}_\lambda(r_n, f_\theta)$ alternatively as

$$\text{PD}^*_\beta(r_n, f_\theta) = \frac{4}{1 - \beta^2} \sum r_n \left[1 - \left(\frac{f_\theta}{r_n}\right)^{(1+\beta)/2}\right]. \tag{21.8}$$

Under this formulation $\mathrm{PD}_\lambda = \mathrm{PD}^*_\beta$, with $\beta = -(1+2\lambda)$. This class generates a symmetric measure (the HD) for $\beta = 0$, and the choices β and $-\beta$ generate distances which are symmetric opposites of each other when $\beta \neq 0$. Jimenez and Shao (2001) have referred to PD^*_β and $\mathrm{PD}^*_{(-\beta)}$ as adjoints of each other. The measure HD is self adjoint.

The power divergence family can be alternatively expressed as

$$\mathrm{PD}_\lambda(r_n, f_\theta) = \sum \left\{ \frac{1}{\lambda(\lambda+1)} r_n \left[\left(\frac{r_n}{f_\theta} \right)^\lambda - 1 \right] + \frac{f_\theta - r_n}{\lambda + 1} \right\}, \tag{21.9}$$

which makes all the terms in the summand on the right-hand side nonnegative. The $C(\cdot)$ function for the Cressie-Read family of power divergence under this formulation is given by

$$C_\lambda(\delta) = \frac{(\delta+1)^{\lambda+1} - (\delta+1)}{\lambda(\lambda+1)} - \frac{\delta}{\lambda+1}.$$

See Basu *et al.* (2011) for a discussion of several other disparity subfamilies.

21.2.1 Minimum Disparity Estimation

The minimum disparity estimator (MDE) $\hat{\theta}$ of θ based on ρ_C is defined by the relation

$$\rho_C(r_n, f_{\hat{\theta}}) = \min_{\theta \in \Theta} \rho_C(r_n, f_\theta) \tag{21.10}$$

provided such a minimum exists. The maximum likelihood estimator (MLE) $\hat{\theta}_{\mathrm{ML}}$ of θ is also an MDE. Under the parametric set-up, the MLE of θ maximizes (with respect to θ)

$$\log \prod_{i=1}^n f_\theta(X_i) = \sum_{i=1}^n \log f_\theta(X_i),$$

or, equivalently,

$$n \sum_{x=0}^\infty r_n(x) \log f_\theta(x) = n \sum r_n \log f_\theta. \tag{21.11}$$

The measure LD in (21.2), when multiplied by $(-n)$, is separated from the expression in (21.11) only by a term which is independent of θ. Thus, the likelihood disparity is minimized by the MLE of θ; thus the class of MDEs includes the MLE under discrete models.

Under differentiability of the model, the MDE solves the estimating equation

$$-\nabla \rho_C(r_n, f_\theta) = \sum (C'(\delta)(\delta+1) - C(\delta)) \nabla f_\theta = 0, \tag{21.12}$$

where ∇ represents the gradient with respect to θ. Letting $A(\delta) = C'(\delta)(\delta+1) - C(\delta)$, the estimating equation for θ has the form

$$-\nabla \rho_C(r_n, f_\theta) = \sum A(\delta) \nabla f_\theta = 0. \tag{21.13}$$

We can standardize the function $A(\delta)$, without changing the estimating properties of the disparity, so that $A(0) = 0$ and $A'(0) = 1$. This standardized function $A(\delta)$ is called

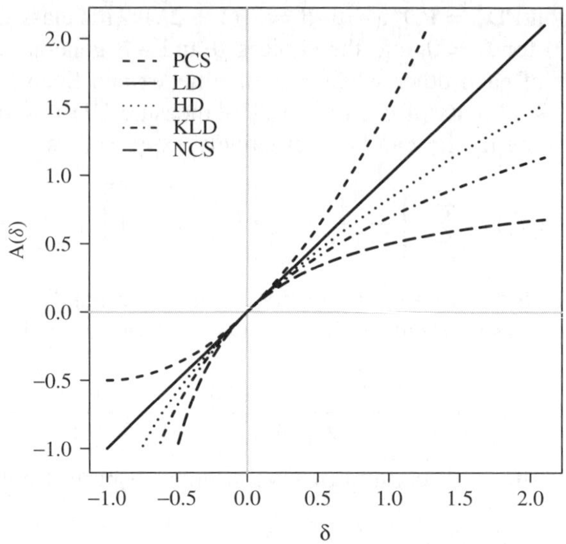

Figure 21.1 Residual Adjustment Functions for five common disparities

the residual adjustment function (RAF) of the disparity. The different properties of the minimum disparity estimators are governed by the form of the function $A(\delta)$. The residual adjustment function for the Cressie-Read family of divergences is given by

$$A_\lambda(\delta) = \frac{(\delta + 1)^{\lambda+1} - 1}{\lambda + 1}. \tag{21.14}$$

It is easy to see that the RAF for likelihood disparity is linear, given by $A_{\mathrm{LD}}(\delta) = \delta$.

21.2.2 The Robustness of the Minimum Disparity Estimators (MDEs)

An element x of the sample space having a large positive value of $\delta(x)$ is considered to be an outlier relative to the model; in this case the observed proportion $r_n(x)$ at x is large compared to what is predicted by the model. To get a stable estimate of θ, we aim to downweight observations having such values of δ. This requires that $A(\delta)$ should curve sharply down for increasing (positive) δ. With the LD as the basis for comparison, we need to explore how the other RAFs depart from linearity. Note that the conditions $A(0) = 0$ and $A'(0) = 1$ guarantee that all RAFs are tangential to the line $A_{\mathrm{LD}}(\delta) = \delta$ at the origin.

Figure 21.1 shows the graphs of the RAFs of the five common disparities within the PD family. Notice that the RAFs for the HD, the KLD and the NCS all provide strong downweighting for large δ (relative to the LD); on the other hand, the PCS actually magnifies the effect of large δ outliers and hence generates estimators that are worse than the MLE in terms of robustness. We refer to the quantity $A_2 = A''(0)$ as the estimation curvature of the disparity. It is a measure of local robustness, with negative values of A_2 being preferred; $A_2 = 0$ for the LD. A negative value for the A_2 parameter is achieved for the HD, the KLD, the NCS (Figure 21.1), and all other members of the Cressie-Read family with $\lambda < 0$.

The influence function of an estimator is a useful indicator of its asymptotic efficiency, as well as of its classical first-order robustness. To obtain the influence function of the MDE based on the disparity ρ_C, we consider the ϵ contaminated version of the true density g given by $g_\epsilon(x) = (1-\epsilon)g(x) + \epsilon\chi_y(x)$. Similarly $G_\epsilon(x) = (1-\epsilon)G(x) + \epsilon\wedge_y(x)$. [Here $\chi_y(x)$ is the indicator function which takes the value 1 at $x = y$, and $\wedge_y(x)$ is the degenerate distribution which puts all its mass at the point y.] The minimum distance functional $T(G)$ is the minimizer of $\rho_C(g, f_\theta)$; $\theta_\epsilon = T(G_\epsilon)$, the minimizer of $\rho_C(g_\epsilon, f_\theta)$, satisfies

$$\sum_x A(\delta_\epsilon(x))\nabla f_{\theta_\epsilon}(x) = 0, \tag{21.15}$$

where $\delta_\epsilon(x) = g_\epsilon(x)/f_{\theta_\epsilon}(x) - 1$. The influence function $\phi_G(y) = T'(y)$ of the functional T at the distribution G is the derivative of θ_ϵ evaluated at $\epsilon = 0$. Let $u_\theta(x) = \nabla \ln\ f_\theta(x)$ be the score function of the model. A routine differentiation of (21.15) gives the form below.

Theorem 21.1. [Lindsay (1994, Proposition 1)]. Let $T(G) = \theta^g$ be the minimum disparity functional corresponding to the disparity $\rho_C(\cdot,\cdot)$; let the associated estimating function be $\sum_x A(\delta(x))\nabla f_\theta(x)$. Then the influence function of T at the distribution G has the form

$$T'(y) = J_g^{-1}\{A'(\delta(y))u_{\theta^g}(y) - E_g[A'(\delta(X))u_{\theta^g}(X)]\},$$

where

$$J_g = E_g[u_{\theta^g}(X)u_{\theta^g}^T(X)A'(\delta(X))] - \sum_x A(\delta(x))\nabla_2 f_{\theta^g}(x), \tag{21.16}$$

for $\delta(x) = g(x)/f_{\theta^g}(x) - 1$. Here $A'(\delta)$ is the first derivative of $A(\delta)$, and $\nabla_2 f_{\theta^g}$ is the second derivative of f_θ with respect to θ evaluated at θ^g.

Under the conditions of the above theorem, when the distribution G belongs to the model, so that the density $g(x) = f_\theta(x)$ for some $\theta \in \Theta$, we get $\theta^g = \theta$, $\delta(x) = 0$ for all x, and the MDE corresponding to the estimating equation $\sum A(\delta(x))\nabla f_\theta(x) = 0$ has influence function $T'(y) = I^{-1}(\theta)u_\theta(y)$, where $I(\theta)$ is the Fisher information matrix of the model at θ. Note that this is also the influence function of the MLE at the model. Thus all MDEs (including those within the Cressie-Read family of power divergences) have the same influence function at the model as the MLE, as is necessary if these estimators are to be asymptotically fully efficient. Thus, the influence functions of these MDEs are not useful indicators when describing their robustness.

21.2.3 Asymptotic Properties

Expanding the estimating Equation (21.13) in a Taylor series around $\delta = 0$, we get

$$-\nabla\rho_C(r_n, f_\theta) = \sum A(\delta)\nabla f_\theta = \sum\left\{\delta + \frac{A_2}{2}\delta^2 + \ldots\right\}\nabla f_\theta = 0. \tag{21.17}$$

Note that the leading term in the estimating function of any disparity is the same as that of the LD; under appropriate conditions one can expect similar behavior for the minimum disparity estimating equation and the maximum likelihood score equation. This gives some intuitive justification of the asymptotic equivalence of any MDE and the MLE.

Lindsay (1994) proved the consistency and asymptotic distribution of the MDEs. It was shown that, under the conditions $A1 - A7$ of Basu *et al.* (2011, pp. 60–61), the minimum disparity estimators have the following asymptotic properties [Lindsay (1994, Theorem 33)]:

1. The minimum disparity estimating equation (21.13) has a consistent sequence of roots $\hat{\theta}_n$.
2. $n^{1/2}(\hat{\theta}_n - \theta^g)$ has an asymptotic multivariate normal distribution with (vector) mean zero and covariance matrix $J_g^{-1} V_g J_g^{-1}$, where J_g is as defined in Theorem 2.1 and $V_g = Var_g[A'(\delta(X))u_{\theta^g}(X)]$.

When the true distribution G belongs to the model, so that $G = F_\theta$ for some $\theta \in \Theta$ and $\theta^g = \theta$, the asymptotic variance of $n^{1/2}(\hat{\theta}_n - \theta)$ reduces to $I^{-1}(\theta)$. Thus all MDEs have the same asymptotic distribution as that of the MLE at the model and hence have full asymptotic efficiency. The estimation curvature A_2 is also related to the concept of the second-order efficiency (Rao 1961, 1962); for the multinomial models $A_2 = 0$ implies second order efficiency of the MDE. In this case the corresponding RAF has a second order contact with that of the LD at $\delta = 0$. We will take $A_2 = 0$ to be our working definition of second order efficiency of the MDE. For the PD_λ family the estimation curvature for the disparity corresponding to tuning parameter λ is λ.

21.3 Density Power Divergence (DPD) and Parametric Inference

Minimum disparity estimation based on the PD family described in the context of discrete models in the previous section can easily be generalized in case of continuous models also. However, for continuous models one is forced to use some nonparametric smoothing technique such as kernel density estimation to produce a continuous estimate of the true density [see, e.g. Basu *et al.* 2011]. So, the minimum distance estimation method based on disparities inherits all the associated complications in continuous models such as those related to bandwidth selection. In this section we will present another related family of divergences, namely the "Density Power Divergence" family, as a function of a tuning parameter $\alpha \in [0, 1]$ that allows us to avoid the complication of the kernel density estimation even in continuous models.

The motivation behind this development can be appreciated by comparing the estimating equations of the minimum L_2 distance estimator and the MLE. For the case of location models these estimating equations are given by

$$\sum_i u_\theta(X_i) f_\theta(X_i) = 0 \qquad \text{and} \qquad \sum_i u_\theta(X_i) = 0 \tag{21.18}$$

respectively. Note that the estimating equation of the minimum L_2 distance estimator is a weighted likelihood equation with the model density as the weight. This provides an automatic downweighting for the scores $u_\theta(\cdot)$ of the outlying observations relative to the model giving the desired shrinkage.

Sometimes the robustness of the minimum L_2 distance estimator requires a high price in terms of efficiency. In order to provide a better compromise, Basu *et al.* (1998) developed the density power divergence (DPD) family which establishes a smooth bridge between the MLE ($\alpha = 0$) and the minimum L_2 distance estimator ($\alpha = 1$). Note that the two estimating equations in (21.18) can be considered as special cases of a generalized estimating equation

$$\sum_{i=1}^{n} u_\theta(X_i) f_\theta^\alpha(X_i) = 0, \quad \alpha \in [0, 1], \tag{21.19}$$

which gives a density power downweighting, and hence robustness of the likelihood score equation. The degree of downweighting increases with α. The family can be extended for values of α beyond 1, but by then the efficiency of the resulting estimators become very low.

For general models beyond the location model, the estimating equation in (21.19) can be further generalized to obtain an unbiased estimating equation (at the model) as

$$\frac{1}{n}\sum_{i=1}^{n} u_\theta(X_i) f_\theta^\alpha(X_i) - \int u_\theta(x) f_\theta^{1+\alpha}(x)dx = 0, \quad \alpha \geq 0. \tag{21.20}$$

Basu *et al.* (1998) used this form to reconstruct the DPD family. Let $\mathcal{G}$ denote the set of distributions having densities with respect to the dominating measure. Given densities g, f for distributions G and F in $\mathcal{G}$, the density power divergence in terms of a parameter α is

$$\mathrm{DPD}_\alpha(g, f) = \begin{cases} \displaystyle\int \left[f^{1+\alpha} - \left(1 + \frac{1}{\alpha}\right) f^\alpha g + \frac{1}{\alpha} g^{1+\alpha} \right] & \text{for } \alpha > 0, \\ \displaystyle\int g \log(g/f) & \text{for } \alpha = 0. \end{cases} \tag{21.21}$$

Here $\mathrm{DPD}_0(g, f) = \lim_{\alpha \to 0} \mathrm{DPD}_\alpha(g, f)$. The measures are genuine divergences; under the parametric set-up of Section 21.2, one can define the minimum DPD functional $T_\alpha(G)$ at G as

$$\mathrm{DPD}_\alpha(g, f_{T_\alpha(G)}) = \min_{\theta \in \Theta} \mathrm{DPD}_\alpha(g, f_\theta)$$

where g represents the density of G. The corresponding estimator is Fisher consistent. Since $\int g^{1+\alpha}$ is independent of θ, $T_\alpha(G)$ essentially minimizes

$$\int f_\theta^{1+\alpha} - \left(1 + \frac{1}{\alpha}\right) \int f_\theta^\alpha g. \tag{21.22}$$

Notice that in expression (21.22) the density g appears only as a linear term. Thus, given a random sample $X_1, \ldots, X_n$ from G we can approximate the expression (21.22) by replacing G with its empirical estimate G_n. For given α, therefore, the minimum density power divergence estimator (MDPDE) $\hat{\theta}_\alpha$ of θ can be obtained by minimizing

$$\int f_\theta^{1+\alpha} - \left(1 + \frac{1}{\alpha}\right) \int f_\theta^\alpha dG_n = \int f_\theta^{1+\alpha} - \left(1 + \frac{1}{\alpha}\right) \frac{1}{n}\sum_{i=1}^{n} f_\theta^\alpha(X_i) = \frac{1}{n}\sum_{i=1}^{n} V_\theta(X_i) \tag{21.23}$$

over $\theta \in \Theta$, where $V_\theta(x) = \int f_\theta^{1+\alpha} - \left(1 + \frac{1}{\alpha}\right) f_\theta^\alpha(x)$. The minimization of the above expression over θ does not require the use of a nonparametric density estimate. Under differentiability of the model the minimization of the objective function in (21.23) leads to the estimating equation (21.20). Expression (21.23) also shows that the MDPDE is an M-estimator, so that the M-estimation theory applies in determining its asymptotic properties.

The DPD family is a subfamily of the class of Bregman divergences (Bregman 1967). For a convex function B, the Bregman divergence between the densities g and f is given by

$$\int [B(g(x)) - B(f(x)) - \{g(x) - f(x)\}B'(f(x))]\, dx.$$

The choice $B(f) = f^{1+\alpha}$ generates $\alpha\text{DPD}_\alpha(g, f)$. Although Basu *et al.* (1998) considered the density power divergence for $\alpha \geq 0$, recently Patra *et al.* (2013) argued that we can also define the density power divergences with negative tuning parameter α by considering the scaled divergence $(1+\alpha)^{-1}\text{DPD}_\alpha(g, f)$, $\alpha \in \mathbb{R}$. However, there is no immediate visible benefit in robustness or efficiency for such values of α.

21.3.1 Connections between the PD and the DPD

There is an interesting connection between the density power divergence and the power divergence families (Patra *et al.* 2013). Note that we can express the power divergence measure between a generic density g and a model density f_θ as

$$\text{PD}_\lambda(g, f_\theta) = \int \left\{ \frac{1}{\lambda(1+\lambda)} \left[\left(\frac{g}{f_\theta}\right)^{1+\lambda} - \left(\frac{g}{f_\theta}\right) \right] + \frac{1 - g/f_\theta}{1+\lambda} \right\} f_\theta. \tag{21.24}$$

Suppose we wish to modify this measure, preserving the divergence properties, such that the computation of the minimum divergence estimator avoids nonparametric density estimation. To do this we need to eliminate the terms that contain a product of a nonlinear function of g with some function of f_θ. A quick look at Equation (21.24) reveals that we only need to adjust the term $(g/f_\theta)^{1+\lambda}$. As the expression within the parentheses is nonnegative and equals zero only if $g = f_\theta$, we can replace the outer f_θ term in (21.24) by $f_\theta^{1+\lambda}$ and still get a valid divergence that simplifies to

$$\begin{aligned} \int \left\{ \frac{[g^{1+\lambda} - g f_\theta^\lambda]}{\lambda(1+\lambda)} + \frac{f_\theta^{1+\lambda} - g f_\theta^\lambda}{1+\lambda} \right\} &= \frac{1}{1+\lambda} \int \left\{ \frac{1}{\lambda} [g^{1+\lambda} - g f_\theta^\lambda] + f_\theta^{1+\lambda} - g f_\theta^\lambda \right\} \\ &= \frac{1}{1+\lambda} \int \left\{ f_\theta^{1+\lambda} - \left(1 + \frac{1}{\lambda}\right) g f_\theta^\lambda + \frac{1}{\lambda} g^{1+\lambda} \right\}, \end{aligned} \tag{21.25}$$

a scaled version of the measure given in Equation (21.21) for $\lambda = \alpha$. Interestingly, the above operation generates strongly robust divergences starting from divergences that are highly non-robust and vice versa. For example, starting from Pearson's chi-square divergence, we derive the L_2 distance, a measure which has sound robustness credentials.

We can also reverse the above transformation to get the power divergence from the density power divergence by replacing the outer $f_\theta^{1+\alpha}$ term in

$$\mathrm{DPD}_\alpha(g, f_\theta) = \int \left\{ 1 - \left(1 + \frac{1}{\alpha}\right)\frac{g}{f_\theta} + \frac{1}{\alpha}\left(\frac{g}{f_\theta}\right)^{1+\alpha} \right\} f_\theta^{1+\alpha} dx \tag{21.26}$$

with f_θ. After simplification, the measure is easily seen to be equal to the PD_α measure.

Patra *et al.* (2013) considered the general class of divergence given by

$$\rho(g, f_\theta) = \int h(\delta + 1) f_\theta^\beta dx, \tag{21.27}$$

where $\beta > 1$, δ is the Pearson residual defined earlier and $h(y) = \sum_{t \in T} a_t y^t$ for some finite set T with elements in $\mathbb{R}$ and real coefficients $\{a_t\}$ such that $h(.)$ is nonnegative on $[0, \infty)$ and $h(y) = 0$ only when $y = 1$. They found that the restriction that the measure must be a genuine statistical divergence and nonparametric smoothing will be avoided for the purpose of estimation leads to the DPD measure with parameter $\beta - 1$ as the unique solution.

21.3.2 Influence Function of the Minimum DPD estimator

Let $T_\alpha(\cdot)$ represent the minimum density power divergence functional based on the tuning parameter α. A routine differentiation of the estimating equation as in the case of power divergence shows that the influence function for the minimum density power divergence (DPD) estimator is given by

$$T_\alpha'(y) = \mathrm{IF}(y, T_\alpha, G) = J_\alpha^{-1}(\theta)[u_\theta(y) f_\theta^\alpha(y) - \xi_\alpha(\theta)], \tag{21.28}$$

where $\theta = T_\alpha(G)$, and

$$J_\alpha(\theta) = \int u_\theta u_\theta^T f_\theta^{1+\alpha} + \int \{i_\theta - \alpha u_\theta u_\theta^T\}\{g - f_\theta\} f_\theta^\alpha \tag{21.29}$$

and

$$\xi_\alpha(\theta) = \int u_\theta(x) f_\theta^\alpha(x) g(x) dx \tag{21.30}$$

with $i_\theta(x) = -\nabla u_\theta(x)$ being the so-called information function of the model. However, for $g = f_\theta$, the influence function at the model simplifies to

$$\mathrm{IF}(y, T_\alpha, G) = \left[\int u_\theta u_\theta^T f_\theta^{1+\alpha}\right]^{-1} \left\{ u_\theta(y) f_\theta^\alpha(y) - \int u_\theta f_\theta^{1+\alpha} \right\}. \tag{21.31}$$

When J_α and ξ_α are finite, this influence function is bounded whenever $u_\theta(y) f_\theta^\alpha(y)$ is bounded. Unless $\alpha = 0$, this is true for most standard parametric models. This is in direct contrast with density-based minimum distance estimation using the PD family.

In Figure 21.2, we present the influence function of the minimum DPD functional for the Poisson model and the normal model (with known variance) for several different values of $\alpha \geq 0$. The redescending nature of the function for $\alpha > 0$ is clearly evident.

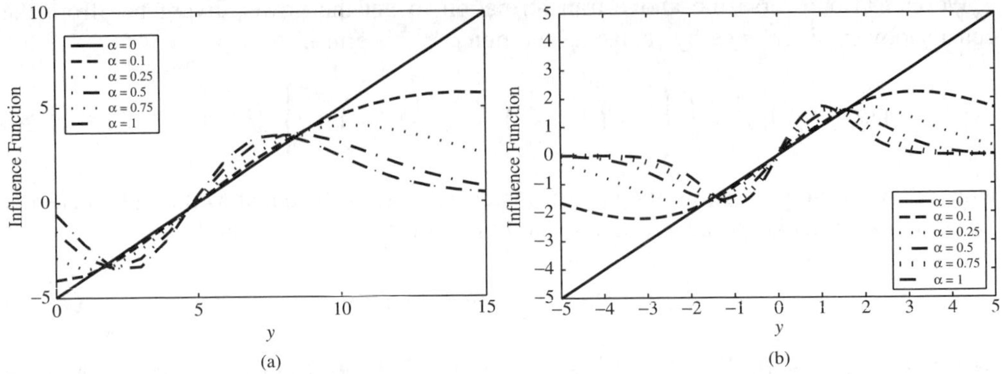

Figure 21.2 Influence function for the MDPDEs of θ under (a) the *Poisson*(θ) model at the *Poisson*(5) distribution and (b) the normal mean under the $N(\theta, 1)$ model at the $N(0, 1)$ distribution

21.3.3 Asymptotic properties of the Minimum DPD estimator

Consider the problem of estimating θ under the parametric model $\mathcal{F}_\theta$ where G is the true data-generating distribution. Let g represent the density function of G, and let $\theta^g = T_\alpha(G)$ be the best fitting parameter. Define

$$K_\alpha(\theta) = \int u_\theta u_\theta^T f_\theta^{2\alpha} g - \xi_\alpha(\theta)\xi_\alpha(\theta)^T. \tag{21.32}$$

Under the conditions $D1 - D5$ of Basu *et al.* (2011, p. 304), the minimum density power divergence estimators (MDPDEs) have the following asymptotic properties: [Basu *et al.* (1998, Theorem 2)].

1. The minimum DPD estimating equation (21.20) has a consistent sequence of roots $\hat{\theta}_\alpha = \hat{\theta}_n$.
2. $n^{1/2}(\hat{\theta}_\alpha - \theta^g)$ has an asymptotic multivariate normal distribution with (vector) mean zero and covariance matrix $J^{-1}KJ^{-1}$, where $J = J_\alpha(\theta^g)$, $K = K_\alpha(\theta^g)$ and $J_\alpha(\theta)$, $K_\alpha(\theta)$ are as in (21.29) and (21.32) respectively, and $\theta^g = T_\alpha(G)$, the best fitting minimum density power divergence functional at G corresponding to tuning parameter α.

When the true distribution G belongs to the model so that $G = F_\theta$ for some $\theta \in \Theta$, the formula for $J = J_\alpha(\theta^g)$, $K = K_\alpha(\theta^g)$ and $\xi = \xi(\theta^g)$ simplifies to

$$J = \int u_\theta u_\theta^T f_\theta^{1+\alpha}, \qquad K = \int u_\theta u_\theta^T f_\theta^{1+2\alpha} - \xi\xi^T, \qquad \xi = \int u_\theta f_\theta^{1+\alpha}. \tag{21.33}$$

A consistent estimate of the asymptotic variance of $n^{1/2}$ times the MDPDE $\hat{\theta}_\alpha$ can be obtained in a sandwich fashion using the form of the influence function in Equation (21.28).

21.4 A Generalized Form: The *S*-Divergence

21.4.1 The Divergence and the Estimating Equation

The DPD family provides a smooth bridge between the likelihood disparity and the L_2 distance and nicely illustrates the trade-off between robustness and efficiency. A natural question is whether one can also construct families of divergences which can similarly connect other members of the PD family with the L_2 distance. One such family has the form

$$\rho(g, f) = \frac{2}{1+\alpha}\left[\int f^{1+\alpha} - 2\int f^{(1+\alpha)/2} g^{(1+\alpha)/2} + \int g^{1+\alpha}\right], \tag{21.34}$$

which generates the HD measure for $\alpha = 0$ and the L_2 distance for $\alpha = 1$. This family, therefore, provides a smooth bridge between the Hellinger distance and the L_2 distance. All these suggest that these families could be the member of a larger superfamily of divergences.

Recently Ghosh *et al.* (2012) have developed a generalized family of density based divergence measures as a function of two parameters α and λ, that serves as this super-family. This family, to be referred to here as the family of S-divergences, contains both the PD and DPD families as subfamilies. This family is defined as

$$\begin{aligned} S_{(\alpha,\lambda)}(g, f) &= \frac{1}{1+\lambda(1-\alpha)}\int\Bigg[\left(f^{1+\alpha} - g^{1+\alpha}\right) \\ &\quad - \frac{(1+\alpha)}{(\alpha-\lambda(1-\alpha))} g^{1+\lambda(1-\alpha)}\left(f^{\alpha-\lambda(1-\alpha)} - g^{\alpha-\lambda(1-\alpha)}\right)\Bigg] \\ &= \frac{1}{A}\int f^{1+\alpha} - \frac{1+\alpha}{AB}\int f^B g^A + \frac{1}{B}\int g^{1+\alpha}, \quad \alpha > 0, \quad -\infty < \lambda < \infty, \end{aligned} \tag{21.35}$$

where $A = 1 + \lambda(1-\alpha)$ and $B = \alpha - \lambda(1-\alpha)$. Note that, $A + B = 1 + \alpha$. When A equals zero, the divergence measure is defined as the limit of the form in (21.35) as $A \to 0$ and is given by

$$S_{(\alpha,\lambda:A=0)}(g, f) = \lim_{A\to 0} S_{(\alpha,\lambda)}(g, f) = \int f^{1+\alpha}\log\left(\frac{f}{g}\right) - \int\frac{(f^{1+\alpha} - g^{1+\alpha})}{1+\alpha}. \tag{21.36}$$

Similarly, if $B = 0$, then the divergence measure was defined to be

$$S_{(\alpha,\lambda:B=0)}(g, f) = \lim_{B\to 0} S_{(\alpha,\lambda)}(g, f) = \int g^{1+\alpha}\log\left(\frac{g}{f}\right) - \int\frac{(g^{1+\alpha} - f^{1+\alpha})}{1+\alpha}. \tag{21.37}$$

Note that for $\alpha = 0$, the class of S-divergences reduces to the PD family with parameter λ; for $\alpha = 1$, $S_{1,\lambda}$ equals the L_2 distance irrespective of the value of λ. On the other hand, $\lambda = 0$ generates the DPD family as a function of α, while $\lambda = -1/2$ generates the family in (21.34). As demonstrated by Ghosh *et al.* (2012), members of the S-divergence family generate genuine divergences for all $\lambda \in \mathbb{R}$ and all nonnegative α.

The case $\lambda = -1/2$ gives us an interesting class of divergence measures. This class of divergences is the same class as described in Equation (21.34), and has been referred to as the S-Hellinger (SHD) family by Ghosh *et al.* (2012) who described some of its special properties. It is also interesting to note that just as the Hellinger distance represents the self adjoint member of the PD family, any other cross-section of the class of S-divergences for a fixed value α has a self adjoint member in $S_{(\alpha,-1/2)}$, which is the point of intersection of this cross-section with the SHD family.

Other special cases are possible for particular values of λ. For example the $S_{(\alpha,\lambda=1)}$ subfamily generates the Pearson chi-square for $\alpha = 0$ and the L_2 distance for $\alpha = 1$. This family is a generalization of the Pearson's chi-square, and is given by

$$S_{(\alpha,\lambda=1)}(g, f) = \frac{1}{(2-\alpha)} \int \left[f^{1+\alpha} - \frac{(1+\alpha)}{(2\alpha-1)} g^{2-\alpha} f^{2\alpha-1} + \frac{(2-\alpha)}{(2\alpha-1)} g^{1+\alpha} \right], \quad (\alpha \neq 1/2),$$

$$S_{(\alpha=1/2,\lambda=1)}(g, f) = \int \left[\frac{2}{3} \left(f^{3/2} - g^{3/2}\right) + g^{3/2} \log(g/f) \right]. \tag{21.38}$$

Let us again consider the parametric class of model distributions $\{F_\theta : \theta \in \Theta \subset \mathbb{R}^p\}$ where we are interested in the estimation of the parameter θ. Let G denote the true distribution (having density g). The minimum S-divergence functional $T_{\alpha,\lambda}(G)$ is defined by the relation

$$S_{(\alpha,\lambda)}(g, f_{T_{\alpha,\lambda}(G)}) = \min_{\theta \in \Theta} S_{(\alpha,\lambda)}(g, f_\theta).$$

Since the $S_{(\alpha,\lambda)}$ class is a family of genuine divergences, the functional $T_{\alpha,\lambda}(G)$ is Fisher consistent. When G is outside the model family, let $\theta^g_{\alpha,\lambda} = T_{\alpha,\lambda}(G)$ represent the best fitting parameter, and f_{θ^g} is the model element closest to g in the S-divergence sense. For simplicity in the notation, we suppress the subscript α, λ for $\theta^g_{\alpha,\lambda}$.

When the aim of the statistician is to perform minimum distance estimation using the family of S-divergences without having recourse to density estimation, we need to choose α and λ so that the parameter A in equation (21.35) is equal to 1. This requires that $\lambda = 0$ or $\alpha = 1$. Thus the DPD family is the only family of divergences within this class that allows parameter estimation without the construction of a nonparametric density estimate. In fact, as shown by Patra *et al.* (2013), the DPD family is the only family of divergences having this property over a class divergences, given by Equation (21.27), which is much larger than the class of S-divergences.

Given the observed data, we estimate the parameter θ by minimizing the divergence $S_{(\alpha,\lambda)}(g, f_\theta)$ over θ, where g is the relative frequency or any density estimate based on the sample data depending on whether the model is discrete or continuous. The estimating equation has the form

$$\int f_\theta^{1+\alpha} u_\theta - \int f_\theta^B g^A u_\theta = 0 \quad \text{or} \quad \int K(\delta(x)) f_\theta^{1+\alpha}(x) u_\theta(x) dx = 0, \tag{21.39}$$

where $\delta(x) = g(x)/f_\theta(x) - 1$ and $K(\delta) = [(\delta+1)^A - 1]/A$. For $\alpha = 0$ one gets $A = 1 + \lambda$, and the function $K(\cdot)$ coincides with the RAF of the PD family in Equation (21.14). Also, note that for any fixed α, the estimating equation of the minimum divergence estimators differ only in the form of the function $K(\cdot)$, so that the robustness properties of the corresponding estimators may at least be partially explained in terms of this function.

21.4.2 *Influence Function of the Minimum S-Divergence estimator*

Let $T_{\alpha,\lambda}$ denote the minimum S-divergence functional at G. As in the previous sections, a straightforward differentiation of the estimating equation shows the influence function to be

$$\text{IF}(y, T_{\alpha,\lambda}, G) = J^{-1}[Au_\theta(y) f_\theta^B(y) g^{A-1}(y) - \xi], \tag{21.40}$$

$$\xi = \xi_\alpha(\theta) = A \int u_\theta f_\theta^B g^A,$$

$$J = J_\alpha(\theta) = A \int u_\theta u_\theta^T f_\theta^{1+\alpha} + \int (i_\theta - Bu_\theta u_\theta^T)(g^A - f_\theta^A) f_\theta^B, \tag{21.41}$$

where $i_\theta(x) = -\nabla[u_\theta(x)]$. However, for $g = f_\theta$, the influence function is exactly the same as that given by Equation (21.31). The remarkable observation here is that the above influence function depends only on α and not on λ. Thus the influence function analysis will predict similar behavior (in terms of robustness) for all minimum S-divergence estimators with the same value of α irrespective of the value of λ. However, actual calculations shows that the robustness properties of these estimators are critically dependent on the choice of the value of λ. This provides a very strong demonstration of the limitation of the influence function approach in describing the robustness properties of the minimum S-divergence estimators. This point has been discussed in further detail in Ghosh *et al.* (2012). Lindsay also observed such limitations of the influence function in the context of minimum disparity estimators. However, in that case all the estimators had unbounded influence. In the present case, the estimators corresponding to $\alpha = 0$ and $\lambda < 0$ are observed to produce highly robust estimates but have unbounded influence functions, while for $\alpha > 0.5$ and $\lambda > 0$ the estimators are seen to be quite nonrobust although the influence function would predict a robust behavior. The above has been observed by the authors in repeated simulations; for brevity, actual numerical results are not presented here.

21.4.3 *Minimum S-Divergence estimators: Asymptotic Properties*

Suppose $X_1, X_2, \ldots, X_n$ are n independent and identically distributed observations from a discrete distribution G modeled by $F = \{F_\theta : \theta \in \Theta \subseteq \mathbb{R}^p\}$ and let the distribution be supported on $\chi = \{0, 1, 2, \cdots\}$. Consider the minimum S-divergence estimator obtained by minimizing the measure $S_{(\alpha,\lambda)}(r_n, f_\theta)$ for $\theta \in \Theta$. Define

$$J_g = E_g[u_{\theta^g}(X) u_{\theta^g}^T(X) K'(\delta_g^g(X)) f_{\theta^g}^\alpha(X)] - \sum_x K(\delta_g^g(x)) \nabla_2 f_{\theta^g}(x) \tag{21.42}$$

$$V_g = V_g[K'(\delta_g^g(X)) f_{\theta^g}^\alpha(X) u_{\theta^g}(X)] \tag{21.43}$$

Ghosh *et al.* (2012) have shown that under suitable conditions [$A1$–$A7$ of Ghosh *et al.* 2012], the minimum S-divergence estimators have the following important asymptotic properties:

1. The S-divergence estimating Equation (21.39) has a consistent sequence of roots $\hat{\theta}_{\alpha,\lambda} = \hat{\theta}_n$.

2. $n^{1/2}(\hat{\theta}_{\alpha,\lambda} - \theta^g)$ has an asymptotic multivariate normal distribution with (vector) mean zero and covariance matrix $J_g^{-1} V_g J_g^{-1}$, where J_g and V_g are as defined earlier.

When the true distribution G belongs to the model so that $G = F_\theta$ for some $\theta \in \Theta$, the formula for the asymptotic variance simplifies to $J^{-1} K J^{-1}$, where J and K are as defined in Equation (21.33). Note that, this asymptotic distribution is independent of the λ parameter and is the same as that of the MDPDE with tuining parameter α.

21.5 Applications

In this article we have described the PD, DPD and the S-Divergence families and discussed their properties and connections. The families have wide-ranging potential for application in different areas of statistics. The PD family has already been quite extensively studied; the S-divergences are too recent to see significant use. Therefore, we present some applications of the density power divergence and the MDPDE. Due to space restrictions, we discuss a particular problem in reliability, and also briefly discuss applications in model selection and survival analysis.

21.5.1 Reliability: The Generalized Pareto Distribution

In this section we will consider minimum DPD estimation of the parameters of the generalized Pareto distribution (GPD) comprising three parameters k (shape parameter), σ (scale parameter) and θ (threshold parameter). The importance of the GPD in the extreme value theory is paramount and it is extensively used in various practical situations (including fields outside reliability). Like the exponential distribution, the GPD is often used to model the tails of another distribution. It allows a continuous range of possible shapes that includes both the exponential and Pareto distributions as special cases. The GPD has three basic forms, each corresponding to a limiting distribution of exceedance data from different classes of underlying distributions.

1. Distributions whose tails decrease exponentially (e.g. normal) lead to shape parameter $k = 0$.
2. Distributions whose tails decrease as a polynomial (e.g. Student's t) lead to a positive shape parameter.
3. Distributions with finite tails, such as the beta, lead to a negative shape parameter.

Here we consider the MLEs and the MDPDEs of the GPD parameters through a robustness example. (See de Zea Bermudez and Kotz 2010, for a discussion and comparison of several other estimation methods.) The forms of the GPD are

$$f(x|k,\sigma,\theta) = \frac{1}{\sigma}\left(1 + k\frac{(x-\theta)}{\sigma}\right)^{-1-\frac{1}{k}} \tag{21.44}$$

for $\theta < x$, when $k > 0$, or for $0 < x < -\sigma/k$ when $k < 0$. For $k = 0$, the density is

$$f(x|0,\sigma,\theta) = \frac{1}{\sigma}e^{-\frac{(x-\theta)}{\sigma}} \tag{21.45}$$

for $\theta < x$. If $k = 0$ and $\theta = 0$, the GPD is equivalent to the exponential distribution. If $k > 0$ and $\theta = \sigma/k$, the GPD is equivalent to the Pareto distribution.

21.5.1.1 Comparison of the Estimation Methods

We first consider the maximum likelihood method. For sample observations $x_1, \ldots, x_n$ from a GPD(k, σ, θ) distribution, the logarithm of the likelihood can be expressed as

$$\log L(k, \sigma, \theta | x_1, \ldots, x_n) = \begin{cases} -n \log \sigma - \left(1 + \frac{1}{k}\right) \sum_{i=1}^{n} \log \left(1 + k \frac{(x_i - \theta)}{\sigma}\right) & \text{if } k \neq 0, \\ -n \log \sigma - \sum_{i=1}^{n} \frac{(x_i - \theta)}{\sigma} & \text{if } k = 0 \end{cases} \tag{21.46}$$

which will then be maximized with respect to k, σ and θ. To find the MDPDEs of the estimators of the model we minimize

$$\int f(x|k, \sigma, \theta)^{1+\alpha}\, dx - \left(1 + \frac{1}{\alpha}\right) \frac{1}{n} \sum_{i=1}^{n} f(x_i|k, \sigma, \theta)^{\alpha}.$$

The integral in this expression can be explicitly evaluated, which simplifies the computation.

We now discuss a small example to demonstrate the robustness benefits due to the use of the MDPDE over the MLE. Consider the mixture 90% GPD ($k = 0.05, \sigma = 1, \theta = 0$) + 10% GPD ($k = 0.95, \sigma = 1, \theta = 0$). The first component is our target and the second component is the contamination. We generated samples of size 50 from this mixture and estimated the parameters assuming a GPD model. The process is repeated 1000 times, and the means of the estimates of k under the different methods are presented in Table 21.1. Clearly large values of α keep the estimates stable. However, the estimates of the other two parameters (not reported here) do not appear to be significantly influenced by the contamination.

21.5.2 Survival Analysis

One of the primary advantages of the minimum DPD method is that it does not require any nonparametric smoothing to produce a data-based estimate of the true density function; the empirical distribution function alone may be used to approximate the appropriate divergence in case of independently and identically distributed data. The method has a natural extension in the context of right censored data. In this situation one can exploit

Table 21.1 Estimate of k for ML method and DPD method ($\alpha = 0.25, 0.5, 0.75, 1$). Samples of size $n = 50$ are generated from 90% GPD ($k = 0.05, \sigma = 1, \theta = 0$) + 10% GPD ($k = 0.95, \sigma = 1, \theta = 0$)

$\hat{k}_{ML}$	$\hat{k}_{DPD(\alpha=0.25)}$	$\hat{k}_{DPD(\alpha=0.50)}$	$\hat{k}_{DPD(\alpha=0.75)}$	$\hat{k}_{DPD(\alpha=1.00)}$
0.1977	0.1505	0.0871	0.0582	0.0498

the well-known Kaplan-Meier estimator of the survival curve and substitute it for the empirical distribution function for the construction of the divergence, which can then be minimized for the estimation of the parameters. For right censored data one observes the (possibly censored) observations and the indicator for censoring. The parameter θ of the target density f_θ is then obtained by minimizing

$$\int f_\theta(x)^{1+\alpha}(x)dx - \left(1 + \frac{1}{\alpha}\right) \int f_\theta(x)^\alpha d\hat{G}_n(x)$$

over θ, where $\hat{G}_n = 1 - \hat{S}_n$, and $\hat{S}_n$ is the Kaplan-Meier estimator of the survival curve. See Basu *et al.* (2006) for some examples and more details of the method.

21.5.3 Model Selection: Divergence Information Criterion

Model selection is the problem of selecting a statistical model, given the data, from a set of candidate models. The two most popular model selection criteria are Akaike information criterion (AIC), developed by Akaike (1973) by using the Kullback-Leibler measure, and the Bayesian information criterion (BIC), attributed to Schwarz (1978), which also exploits the likelihood. In this connection Mattheou *et al.* (2009) proposed the divergence information criterion (DIC) for model selection. As divergence measures are indices of dissimilarity between two populations, they may also naturally be used for the construction of model selection criteria. Such a criterion, based on the density power divergence, has been proposed by Mattheou *et al.* (2009). This model selection criterion constructs an approximately unbiased estimator of the expected overall discrepancy which measures the distance between the true unknown model and the fitted approximating model. Mattheou *et al.* have illustrated the properties of the DIC through a modest simulation study where the method performs satisfactorily in comparison to AIC and BIC.

21.6 Conclusion

We have considered two prominent families of divergences in density-based minimum distance estimation, discussed the connection between them, and provided a common framework for their analysis. Apart from tying in several different isolated results in this field, this chapter also illustrates the failure of the influence function approach to appropriately quantify the robustness of these procedures. Further follow-up research will help us to understand the full potential of the new superfamily of divergences considered in this chapter.

References

Akaike, H. 1973. Information theory and an extension of the maximum likelihood principle. In *Proceedings of the Second International Symposium on Information Theory*, ed. B. N. Petrov and F. Csaki. Akademiai Kaido: Budapest, pp. 267–281.

Ali, S. M. and S. D. Silvey. 1966. A general class of coefficients of divergence of one distribution from another. *Journal of the Royal Statistical Society* B 28 (1): 131–142.

Basu, S., A. Basu and M. C. Jones. 2006. Robust and efficient parametric estimation for censored survival data. *Annals of the Institute of Statistical Mathematics* 58: 341–355.

Basu, A., I. R. Harris, N. L. Hjort and M. C. Jones. 1998. Robust and efficient estimation by minimising a density power divergence. *Biometrika* 85 (3): 549–559.

Basu, A., H. Shioya and C. Park. 2011. *Statistical Inference: The Minimum Distance Approach*. Boca Raton, FL: Chapman & Hall/CRC.

Beran, R. J. 1977. Minimum Hellinger distance estimates for parametric models. *The Annals of Statistics* 5 (3): 445–463.

Bregman, L. M. 1967. The relaxation method of finding the common point of convex sets and its application to the solution of problems in convex programming. *USSR Computational Mathematics and Mathematical Physics* 7 (3): 200–217.

Cressie, N. and T. R. C. Read. 1984. Multinomial goodness-of-fit tests. *Journal of the Royal Statistical Society* B46 (3): 440–464.

Csiszár, I. 1963. Eine informations theoretische Ungleichung und ihre Anwendung auf den Beweis der Ergodizität von Markoffschen Ketten. *Publ. of the Math. Inst. of the Hungarian Academy* 8: 95–108.

de Zea Bermudez, P. and S. Kotz. 2010. Parameter estimation of the generalized Pareto distribution-Part I. *Journal of Statistical Planning and Inference* 140 (6): 1353–1373.

Ghosh, A., A. Maji, A. Basu, and L. Pardo. 2012. *The Robust Parametric Inference based on a New Family of Generalized Density Power Divergence Measures*. Technical Report BIRU/2012/2. Bayesian and Interdisciplinary Research Unit, Indian Statistical Institute, Kolkata, India.

Jimenez, R. and Y. Shao. 2001. On robustness and efficiency of minimum divergence estimators. *Test* 10 (2): 241–248.

Lindsay, B. G. 1994. Efficiency versus robustness: The case for minimum Hellinger distance and related methods. *The Annals of Statistics* 22 (2): 1081–1114.

Mattheou, K., S. Lee and A. Karagrigoriou. 2009. A model selection criterion based on the BHHJ measure of divergence. *Journal of Statistical Planning and Inference* 139 (1): 228–235.

Pardo, L. 2006. *Statistical Inference based on Divergences*. Boca Raton, FL: Chapman & Hall/CRC.

Patra, S., A. Maji, A. Basu and L. Pardo. 2013. The power divergence and the density power divergence families: the mathematical connection. *Sankhya B* (To appear).

Pearson, K. 1900. On the criterion that a given system of deviations from the probable in the case of a correlated system of variables is such that it can be reasonably supposed to have arisen from random sampling. *Philosophical Magazine* Series 5, 50 (302): 157–175.

Rao, C. R. 1961. Asymptotic efficiency and limiting information. In *Proceedings of Fourth Berkeley Symposium on Mathematical Statistics and Probability*, June 20–July 30, 1960. 531–546. Berkeley, California: University of California Press.

Rao, C. R. 1962. Efficient estimates and optimum inference procedures in large samples. *Journal of the Royal Statistical Society* B 24 (1): 46–72.

Schwarz, G. 1978. Estimating the dimension of a model. *The Annals of Statistics* 6 (2): 461–464.

Vajda, I. 1989. *Theory of Statistical Inference and Information*. Dordrecht: Kluwer.

22

COM-Poisson Cure Rate Models and Associated Likelihood-based Inference with Exponential and Weibull Lifetimes

N. Balakrishnan[1,2] and Suvra Pal[1]

[1]*Department of Mathematics and Statistics, McMaster University, Canada*

[2]*Department of Statistics, King Abdulaziz University, Saudi Arabia*

22.1 Introduction

Cure rate models, used for modeling time-to-event data, consist of a surviving fraction and have become increasingly popular in the analysis of data from clinical trials. Recently, the development of new drugs and treatment regimens have resulted in patients living longer with certain types of cancer and heart disease. In cohorts of patients with certain types of cancer, some of the patients show no recurrence of the disease. These patients are said to be cured or long-term survivors, while the remaining patients who experience a recurrence of the disease are said to be susceptibles. The population of interest may thus be regarded as a mixture of these two types of patients. Standard methods of parametric and non-parametric analysis are concerned only with survival and do not address the possibility of cure, and are therefore inappropriate for analyzing such data. The main interest of the investigator conducting a clinical trial is in determining the proportion of patients cured and studying the causes for the failure of the treatment in the uncured group of patients. Estimation of treatment-specific cure rate also provides valuable information that will be of primary interest to both the investigator conducting the trial and the patient being diagnosed.

Applied Reliability Engineering and Risk Analysis: Probabilistic Models and Statistical Inference, First Edition.
Edited by Ilia B. Frenkel, Alex Karagrigoriou, Anatoly Lisnianski and Andre Kleyner.

Rodrigues *et al.* (2009) proposed the Conway-Maxwell Poisson (COM-Poisson) cure rate model by viewing the COM-Poisson distribution as a weighted Poisson distribution (Kokonendji *et al.* 2008). This model includes some of the well-known cure rate models discussed in the literature due to Berkson and Gage (1952), Yakovlev and Tsodikov (1996), and Yin and Ibrahim (2005). The exponential distribution is the simplest model for lifetime data and has been widely used in the analysis of reliability and survival data. Recently, Balakrishnan and Pal (2012) discussed the estimation of the parameters of flexible COM-Poisson cure rate models by considering exponential as the lifetime distribution, see also Balakrishnan and Pal (2013). Balakrishnan and Basu (1995) have presented a book-length account of various developments on theory, methods and applications of the exponential distribution. The Weibull distribution, on the other hand, has considerable flexibility while describing lifetime data and is widely used in many situations; see, for example, Johnson *et al.* (1994). In this work, by considering both the exponential and Weibull distributions for the time-to-event and assuming the number of competing causes related to the event of interest to follow the COM-Poisson distribution, we discuss the steps of the Expectation Maximization (EM) algorithm for the determination of the maximum likelihood estimates (MLEs) of the parameters of this cure rate survival model based on right censored data.

There are many applications of cure rate models in a wide range of fields such as biomedical studies, criminology, finance, demography, reliability, and economics. For example, in biomedical studies, one might be interested in a patient's death, which can occur due to several competing causes or recurrence of a tumor that can happen due to some metastasis-component tumor cells that are left active in an individual after an initial treatment. For further details on cure rate models and their applications, interested readers may refer to Yakovlev *et al.* (1993), Ibrahim *et al.* (2001), Meeker and Escobar (1998), and Hoggart and Griffin (2001).

Let M be an unobservable random variable denoting the number of competing causes related to the occurrence of an event of interest, with probability mass function (p.m.f.) $p_m = P[M = m]$ for $m = 0, 1, 2, \ldots$. Given $M = m$, let W_j $(j = 1, 2, \ldots, m)$ denote the time-to-event due to the j-th competing cause, hereafter referred to as "lifetime". W_js are assumed to be independent random variables, distributed independently of M, with a common distribution function $F(y) = 1 - S(y)$, where $S(y)$ denotes the survival function. In the competing cause scenario, the number of competing causes M and the lifetime W_j associated with a particular cause are latent variables, that is, not observable, but what we actually observe is the minimum lifetime Y among all causes. So, in order to include those individuals who are not susceptible to the event occurrence, the lifetime is defined as

$$Y = \min\{W_0, W_1, W_2, \ldots, W_M\}, \tag{22.1}$$

where W_0 is such that $P[W_0 = \infty] = 1$. This actually leads to a proportion p_0 of the population who are not susceptible to the event occurrence, also termed as the "cure rate."

The rest of this article is organized as follows. In Section 22.2, we discuss applications of cure rate models in reliability. In Section 22.3, we introduce the COM-Poisson cure rate model. The form of the data and the likelihood function are presented in Section 22.4. In Section 22.5, we develop the estimation procedure for estimating the parameters of the COM-Poisson cure rate model and some of its special cases. The asymptotic variance-covariance matrix of the MLEs is derived in Section 22.6. In Sections 22.7 and 22.8, we present the results of the proposed methodology when the lifetime distribution is taken to be exponential and Weibull, respectively. An application of the proposed methodology to a real data on cutaneous melanoma is illustrated in Section 22.9, and finally in Section 22.10, some concluding remarks are made.

22.2 Role of Cure Rate Models in Reliability

It has been mentioned earlier that there are numerous applications of cure rate models in various fields. In this section, we discuss applications of cure rate models in reliability. For instance, in software reliability, consider the computers that are sent back to the dealership due to the failure of their software. Now, after the software problem is fixed and the computers are sent back to the customers, it is quite likely that a substantial proportion of the computers would not undergo software problems again and come back to the dealership. These computers are termed as long-term survivors or cured. Furthermore, the failure of a computer can be due to a number of competing causes, for example, floating point error, bugs, hardware problems, etc. In this case, the retailer would always want the cure rate of the computers to be as high as possible so that customers can keep their faith in the product and hence its estimation is important. In industrial reliability, for instance, an event of interest may be the failure of circuit boards which are subjected to various levels of humidity stress. At low levels of humidity, a proportion of the boards lasts indefinitely and hence can be considered as long-term survivors. Also, the failure of circuit boards can occur due to infant failure or wear-out, thus giving rise to a competing cause scenario.

22.3 The COM-Poisson Cure Rate Model

The COM-Poisson distribution, first introduced by Conway and Maxwell (1961), generalizes the Poisson distribution by adding a parameter to model over-dispersion and under-dispersion. This distribution was revived recently by Shmueli *et al.* (2005) and was discussed further by Boatwright *et al.* (2006) and Kokonendji *et al.* (2009). It is a member of the exponential family and can be viewed as a weighted Poisson distribution. Herein, we take the competing cause random variable M to follow the COM-Poisson distribution with probability mass function

$$P[M = m; \eta, \phi] = \frac{1}{Z(\eta, \phi)} \frac{\eta^m}{(m!)^\phi}, \quad m = 0, 1, 2, \ldots, \tag{22.2}$$

where

$$Z(\eta, \phi) = \sum_{j=0}^{\infty} \frac{\eta^j}{(j!)^\phi}. \tag{22.3}$$

The cured fraction is then given by

$$p_0 = P[M = 0; \eta, \phi] = \frac{1}{Z(\eta, \phi)}.$$

When $\phi = 1$ in (22.2), we obtain the Poisson distribution with mean η. Values of $\phi > 1$ correspond to under-dispersion while $\phi < 1$ correspond to over-dispersion relative to the

Poisson distribution. The COM-Poisson distribution approaches the Bernoulli distribution with parameter $\frac{1}{1+\eta}$ as a limiting case when $\phi \to \infty$. When $\phi = 0$, the COM-Poisson distribution reduces to the geometric distribution with parameter $1 - \eta$ provided $\eta < 1$, and is undefined in this special case when $\eta \geq 1$.

From Rodrigues *et al.* (2009), the long-term survival function of the random variable Y in (22.1) is given by

$$S_p(y) = \frac{Z(\eta S(y), \phi)}{Z(\eta, \phi)}. \tag{22.4}$$

It is to be noted that the function $S_p(y)$ in (22.4) is not a proper survival function as $\lim_{y\to\infty} S_p(y) > 0$. From (22.4), one can easily obtain the improper density function of Y as

$$f_p(y) = -S'_p(y) = \frac{1}{Z(\eta, \phi)} \frac{f(y)}{S(y)} \sum_{j=1}^{\infty} \frac{j\{\eta S(y)\}^j}{(j!)^\phi},$$

where $f(y)$ denotes the common (proper) density function of the lifetime in (22.1).

One can present an alternative expression for $S_p(y)$ by considering the population as a mixture of two types of patients. In this case, we define an indicator variable I, which takes the value 0 if the subject is immune, and 1 if the subject is susceptible. Let $P[I = 0] = p_0$ and $P[I = 1] = 1 - p_0$. Also, let F_p and F_1 denote the cumulative distribution function (c.d.f.) of the overall population and susceptibles, respectively. Then, the c.d.f. of the overall population is given by

$$F_p(y) = (1 - p_0)F_1(y),$$

and the survival function of the overall population is given by

$$S_p(y) = 1 - F_p(y) = p_0 + (1 - p_0)S_1(y), \tag{22.5}$$

where S_1 denotes the survival function of the susceptibles. From (22.5), we can get an expression for $S_1(y)$ as

$$S_1(y) = \frac{S_p(y) - p_0}{1 - p_0},$$

which involves $S_p(y)$ that can be calculated from (22.4). We note that $S_1(y)$ is indeed a proper survival function.

In Table 22.1, we present the survival function, the improper density function and the cured fraction corresponding to the three special cases of the COM-Poisson cure rate model.

22.4 Data and the Likelihood

Let us consider the situation when the lifetime in (22.1) is not completely observed and is subject to right censoring. Let C_i denote the censoring time for the i-th subject. Then, from a group of n subjects in the study, what we observe is $T_i = \min\{Y_i, C_i\}$ and $\delta_i = I(Y_i \leq C_i)$, where $\delta_i = 1$ if Y_i is a lifetime and $\delta_i = 0$ if it is right censored, for $i = 1, 2, \ldots, n$. Let $\boldsymbol{\gamma}$ denote the parameter vector of the distribution of the lifetime variable in (22.1).

Table 22.1 Survival function (S_p), density function (f_p), and cured fraction (p_0) for the three special cases of COM-Poisson cure rate model

Model	$S_p(y)$	$f_p(y)$	p_0
Bernoulli	$\frac{1+\eta S(y)}{1+\eta}$	$\frac{\eta}{1+\eta} f(y)$	$\frac{1}{1+\eta}$
Poisson	$\exp(-\eta(1-S(y)))$	$\eta S_p(y) f(y)$	$\exp(-\eta)$
Geometric	$\frac{1-\eta}{1-\eta S(y)}$	$\frac{\eta}{1-\eta S(y)} S_p(y) f(y)$	$1-\eta$

Now, to introduce the covariate effect, we can either link the parameter η to covariates $\boldsymbol{x}_i$ by the log-linear link function η_i=$\exp(\boldsymbol{x}_i'\boldsymbol{\beta})$, $i = 1, \ldots, n$, or link the cure rate parameter p_0 to covariates $\boldsymbol{x}_i$ by the logistic link function $p_{0i} = \frac{1}{1+\exp(\boldsymbol{x}_i'\boldsymbol{\beta})}$, $i = 1, \ldots, n$, where $\boldsymbol{\beta} = (\beta_0, \beta_1, \ldots, \beta_k)'$ denotes the vector of regression coefficients.

From n pairs of times and censoring indicators $(t_1, \delta_1), \ldots, (t_n, \delta_n)$ as the data, the likelihood function under the non-informative censoring assumption can be written as

$$L(\boldsymbol{\theta}; \boldsymbol{t}, \boldsymbol{\delta}) \propto \prod_{i=1}^{n} \{f_p(t_i; \boldsymbol{\theta})\}^{\delta_i} \{S_p(t_i; \boldsymbol{\theta})\}^{1-\delta_i}, \tag{22.6}$$

where $\boldsymbol{\theta} = (\phi, \boldsymbol{\beta}', \boldsymbol{\gamma}')'$, $\boldsymbol{t} = (t_1, \ldots, t_n)'$, and $\boldsymbol{\delta} = (\delta_1, \ldots, \delta_n)'$. Using the alternative expression for the long-term survival function in (22.5), the likelihood function in (22.6) can also be expressed as

$$L(\boldsymbol{\theta}; \boldsymbol{t}, \boldsymbol{x}, \boldsymbol{\delta}) \propto \prod_{I_1} f_p(t_i, \boldsymbol{x}_i; \boldsymbol{\theta}) \prod_{I_0} \{p_0(\boldsymbol{\theta}_1, \boldsymbol{x}_i) + (1 - p_0(\boldsymbol{\theta}_1, \boldsymbol{x}_i)) S_1(t_i, \boldsymbol{x}_i; \boldsymbol{\theta})\}, \tag{22.7}$$

where $\boldsymbol{\theta}_1 = (\phi, \boldsymbol{\beta}')'$, $I_1 = \{i : \delta_i = 1\}$, $I_0 = \{i : \delta_i = 0\}$, and $\boldsymbol{x}$ denotes the vector of $\boldsymbol{x}_i$ values. Let n_1 denote the cardinality of I_1 and $I^* = I_0 \cup I_1$.

22.5 EM Algorithm

In order to determine the MLE of the unknown parameter $\boldsymbol{\theta}$, we note that the random variable I, for each individual, takes the value 1 if the individual belongs to the set I_1 and is unknown if it belongs to the set I_0. This actually introduces the missing data and the MLE of $\boldsymbol{\theta}$ can then be obtained by employing the EM algorithm; see McLachlan and Krishnan (2008) for a detailed discussion on this topic. Let us denote the complete data by $(t_i, \delta_i, \boldsymbol{x}_i, I_i)$, $i = 1, 2, \ldots, n$, which includes the observed data and the unobserved I_is. The complete data likelihood function is then given by

$$\begin{aligned} L_c(\boldsymbol{\theta}; \boldsymbol{t}, \boldsymbol{x}, \boldsymbol{\delta}, \boldsymbol{I}) &\propto \prod_{I_1} \{f_p(t_i, \boldsymbol{x}_i; \boldsymbol{\theta})\}^{I_i} \\ &\times \prod_{I_0} \{p_0(\boldsymbol{\theta}_1, \boldsymbol{x}_i)\}^{1-I_i} \{(1 - p_0(\boldsymbol{\theta}_1, \boldsymbol{x}_i)) S_1(t_i, \boldsymbol{x}_i; \boldsymbol{\theta})\}^{I_i}, \end{aligned}$$

where $\boldsymbol{I}$ denote the vector of I_i values. Ignoring the constant term, the corresponding log-likelihood function is given by

$$l_c(\boldsymbol{\theta}; \boldsymbol{t}, \boldsymbol{x}, \boldsymbol{\delta}, \boldsymbol{I}) = \sum_{I_1} I_i \log f_p(t_i, \boldsymbol{x}_i; \boldsymbol{\theta}) + \sum_{I_0} (1 - I_i) \log p_0(\boldsymbol{\theta}_1, \boldsymbol{x}_i)$$
$$+ \sum_{I_0} I_i \log((1 - p_0(\boldsymbol{\theta}_1, \boldsymbol{x}_i)) S_1(t_i, \boldsymbol{x}_i; \boldsymbol{\theta})). \tag{22.8}$$

E-Step: In the expectation step (E-step), for a fixed value of the parameter ϕ, the expectation of the complete data log-likelihood function is computed with respect to the distribution of the unobserved $I_i s$, given the current values of the parameter and the observed data $\boldsymbol{O}$, where $\boldsymbol{O} = \{\text{observed } I_i s, (t_i, \delta_i, \boldsymbol{x}_i); i = 1, 2, \ldots, n\}$. It is to be noted that $I_i s$ are Bernoulli random variables and that we have to compute $\pi_i^{(k)} = E(I_i | \boldsymbol{\theta}^{*(k)}, \boldsymbol{O}), i = 1, 2, \ldots, n$, where $\boldsymbol{\theta}^* = (\boldsymbol{\beta}', \boldsymbol{\gamma}')'$ and $\boldsymbol{\theta}^{*(k)}$ denotes the current parameter value at the k-th iteration step. Now, if the i-th observation is censored, we have

$$\begin{aligned}
\pi_i^{(k)} &= P[I_i = 1 | T_i > t_i; \boldsymbol{\theta}^{*(k)}] \\
&= \left. \frac{P[T_i > t_i | I_i = 1] P[I_i = 1]}{P[T_i > t_i]} \right|_{\boldsymbol{\theta}^* = \boldsymbol{\theta}^{*(k)}} \\
&= \left. \frac{(1 - p_0(\boldsymbol{\theta}_1, \boldsymbol{x}_i)) S_1(t_i, \boldsymbol{x}_i; \boldsymbol{\theta})}{S_p(t_i, \boldsymbol{x}_i; \boldsymbol{\theta})} \right|_{\boldsymbol{\theta}^* = \boldsymbol{\theta}^{*(k)}} \\
&= w_i^{(k)},
\end{aligned}$$

whereas if the i-th observation is uncensored, we simply have $\pi_i^{(k)} = I_i = 1$. Thus, the E-step replaces the $I_i s$ in (22.8) by $\pi_i^{(k)}$, where $\pi_i^{(k)} = w_i^{(k)}$ for censored observation and $= 1$ for uncensored observation. We denote the conditional expectation of the complete data log-likelihood function, for a fixed value of ϕ, by $Q(\boldsymbol{\theta}^*, \boldsymbol{\pi}^{(k)})$, where $\boldsymbol{\pi}^{(k)}$ is the vector of $\pi_i^{(k)}$ values.

M-Step: In the maximization step (M-step), for a fixed value of the parameter ϕ, the maximization of the $Q(\boldsymbol{\theta}^*, \boldsymbol{\pi}^{(k)})$ function is carried out with respect to $\boldsymbol{\theta}^*$ over the corresponding parameter space $\boldsymbol{\Theta}^*$, given $\boldsymbol{\pi}^{(k)}$, in order to obtain the improved estimate of $\boldsymbol{\theta}^*$ as

$$\boldsymbol{\theta}^{*(k+1)} = \arg\max_{\boldsymbol{\theta}^* \in \boldsymbol{\Theta}^*} Q(\boldsymbol{\theta}^*, \boldsymbol{\pi}^{(k)}).$$

The E-step and the M-step, for a fixed value of ϕ, are then continued iteratively until convergence to obtain the MLE of the parameter $\boldsymbol{\theta}^*$. In this work, the maximization step is carried out by using the EM gradient algorithm (Lange 1995) as the MLEs of $\boldsymbol{\beta}$ and $\boldsymbol{\gamma}$ do not have explicit expressions. This is a one-step Newton-Raphson method and qualifies as a special case of the generalized EM algorithm (Dempster *et al.* 1977). The parameter ϕ is then finally estimated by employing the profile likelihood technique. In this technique, we fix a set of distinct admissible values of ϕ and for each fixed value of ϕ, we estimate the parameter $\boldsymbol{\theta}^*$. Finally, we take the value of ϕ for which the maximized log-likelihood function is the maximum as the MLE of ϕ.

22.6 Standard Errors and Asymptotic Confidence Intervals

An approximation of the asymptotic variance-covariance matrix of $(\hat{\boldsymbol{\beta}}, \hat{\boldsymbol{\gamma}})$ is obtained by inverting the observed information matrix of $\boldsymbol{\beta}$ and $\boldsymbol{\gamma}$ under the assumption that ϕ is fixed. Although one can obtain the first-order derivatives of the observed data log-likelihood function directly from (22.7), an easier alternative way is to use the complete data log-likelihood function from the EM algorithm and then derive the score function using the method of Louis (1982); see Sy and Taylor (2000). Let $l(\boldsymbol{\theta}; \boldsymbol{t}, \boldsymbol{x}, \boldsymbol{\delta}) = \log L(\boldsymbol{\theta}; \boldsymbol{t}, \boldsymbol{x}, \boldsymbol{\delta})$. For a fixed value of ϕ, the components of the score function are as follows:

$$\frac{\partial}{\partial \beta_l} l(\boldsymbol{\theta}; \boldsymbol{t}, \boldsymbol{x}, \boldsymbol{\delta}) = \left. \frac{\partial}{\partial \beta_l} l_c(\boldsymbol{\theta}; \boldsymbol{t}, \boldsymbol{x}, \boldsymbol{\delta}, \boldsymbol{I}) \right|_{I_i = \pi_i \forall i}$$

for $l = 0, 1, \ldots, k$, and

$$\frac{\partial}{\partial \boldsymbol{\gamma}} l(\boldsymbol{\theta}; \boldsymbol{t}, \boldsymbol{x}, \boldsymbol{\delta}) = \left. \frac{\partial}{\partial \boldsymbol{\gamma}} l_c(\boldsymbol{\theta}; \boldsymbol{t}, \boldsymbol{x}, \boldsymbol{\delta}, \boldsymbol{I}) \right|_{I_i = \pi_i \forall i},$$

where

$$\pi_i = \frac{(1 - p_0(\boldsymbol{\theta}_1, \boldsymbol{x}_i)) S_1(t_i, \boldsymbol{x}_i; \boldsymbol{\theta})}{S_p(t_i, \boldsymbol{x}_i; \boldsymbol{\theta})} = w_i$$

if the i-th observation is censored and $\pi_i = 1$ if the i-th observation is uncensored. We can then take the negative of the second-order derivatives of the function $l(\boldsymbol{\theta}; \boldsymbol{t}, \boldsymbol{x}, \boldsymbol{\delta})$ with respect to $\boldsymbol{\beta}$ and $\boldsymbol{\gamma}$ to obtain the components of the observed information matrix.

Once we obtain the MLEs of $\boldsymbol{\beta}$ and $\boldsymbol{\gamma}$ and the asymptotic variances and covariances of the MLEs, we can then construct asymptotic confidence intervals for the parameters by using the asymptotic normality of the MLEs and the estimate of the asymptotic variance-covariance matrix of the MLEs. To evaluate the accuracy of this asymptotic method, we may examine the coverage probabilities of these confidence intervals based on a Monte Carlo simulation study. Next, we present the details of the associated likelihood inference based on the EM algorithm for the exponential and Weibull lifetime distributions separately.

22.7 Exponential Lifetime Distribution

In this case, we assume the time-to-event W to follow the exponential distribution with

$$F(w; \gamma) = 1 - \exp(-\gamma w) \quad \text{and} \quad f(w; \gamma) = \gamma \exp(-\gamma w)$$

for $w > 0$, $\gamma > 0$ and relate the parameter η to covariates $\boldsymbol{x}_i$ by the log-linear link function $\eta_i = \exp(\boldsymbol{x}_i' \boldsymbol{\beta})$, $i = 1, \ldots, n$. The simplified explicit expressions of the $Q(\boldsymbol{\theta}^*, \boldsymbol{\pi}^{(k)})$ function and the expressions of the first- and second-order derivatives of the $Q(\boldsymbol{\theta}^*, \boldsymbol{\pi}^{(k)})$ function with respect to $\boldsymbol{\theta}^*$ that enable us to carry out the required maximization step for different cure rate models are presented in Appendix A1. Furthermore, the expressions of the components of the observed information matrix for different cure rate models are presented in Appendix B1.

22.7.1 Simulation Study: Model Fitting

An extensive Monte Carlo simulation study is performed in order to evaluate the performance of the proposed estimation method. In this simulation study, we considered two sample sizes: $n = 400$ and $n = 600$ so that we can observe the behavior of the model under moderate and large sample sizes. The total sample is then divided into four groups. The sample size in each group can be viewed as the number of patients in that group. Stratification of patients into different groups allows us to introduce the covariate effect, wherein the patients in group j are assigned a covariate value of j, $j = 1, 2, 3, 4$. As the group category is considered to be the only covariate involved in the model, we have two regression parameters, that is, β_0 and β_1. Again, as the cured proportion is purely a function of the regression parameters, on fixing the cured proportion for any of the two groups, the regression parameters β_0 and β_1 can be calculated. Once the regression parameters are obtained, the cured proportion for the remaining two groups can be automatically determined. In our study, we took the cured proportion to be decreasing with the groups and we fixed the cured proportion for the first and fourth groups. We considered two different choices of cured proportions for Groups $(1, 4)$: $(0.65, 0.25)$ and $(0.40, 0.15)$. We note that for a group having cured proportion p_0, p_0 proportion of patients is always censored. In order to introduce random censoring, we considered the distribution of the censoring times to be exponential with censoring rate α. The parameter α is to be calculated such that the proportion of censoring for a group is p (say). It is clear that α will be different for each group. We considered two different choices of censoring proportions corresponding to Groups 1–4: $(0.80, 0.65, 0.50, 0.35)$ and $(0.50, 0.40, 0.30, 0.20)$. This allows us to observe the behavior of the model under heavy as well as small censoring amounts. Next, we discuss the procedure to generate data from different cure rate models.

Bernoulli cure rate model: For a group with cured proportion p_0 and censoring proportion p, we first generate a uniform random variable U and a censoring time C. If $U \le p_0$, the observed time T is taken to be the censoring time, that is, $T = C$. Otherwise, we generate an exponential random variable W with rate γ, for a specific choice of γ, and then take $T = \min\{W, C\}$.

Poisson cure rate model: For a group with cured proportion p_0 and censoring proportion p, we first generate a censoring time C and a Poisson random variable M with parameter $-\log(p_0)$. If $M = 0$, the observed time T is taken to be the censoring time, that is, $T = C$. Otherwise, we generate M exponential random variables $\{W_1, W_2, \ldots, W_M\}$ with rate γ, for a specific choice of γ, and then take $T = \min\{Y, C\}$, where $Y = \min\{W_1, W_2, \ldots, W_M\}$.

Geometric cure rate model: In this case, the procedure is the same as in the Poisson cure rate model just described, except that the competing cause random variable M is generated from a geometric distribution with parameter p_0.

COM-Poisson cure rate model: In this case, the procedure is once again the same as in the Poisson cure rate model described above, except that the competing cause random variable M is generated from a COM-Poisson distribution with p.m.f. given by (22.2) for a specific choice of the parameter ϕ.

We considered two different choices for the parameter γ: 1 and 3, which allow us to evaluate the behavior of the model under low and high hazard rates. In the simulation study, we simulated data from different cure rate models under different simulation settings described above and the estimation method proposed was employed to estimate the unknown parameters of the model. To examine the performance of the estimates of the parameters, we calculated the bias and mean square error of the estimates. We also calculated the coverage probability of the confidence interval based on the asymptotic normality of MLEs for different nominal confidence levels. All simulations were performed using the R software and the results were based on 1000 Monte Carlo runs.

To find an initial guess of the parameters β_0, β_1 and γ for starting the iterative procedure, we considered the first and fourth groups and calculated the Kaplan-Meier estimates of the cured proportions for these two groups separately. Once the estimates of the cured proportions for these two groups were obtained, the estimates of β_0 and β_1 can be obtained and these were taken as their initial values. We then plugged in the estimates of β_0 and β_1 in the observed data log-likelihood function, which is then a function of γ only. Now, upon taking the first-order derivative of the observed data log-likelihood function with respect to γ, and after equating it to zero, we solved for γ to find an initial estimate of γ. In the simulation study, we observed that the above technique provides initial values which are reasonably close to the true parameter values. Now, in our simulation study, as we have considered random censoring, it is to be noted that the largest observed time in a group may not be censored, in which case the Kaplan-Meier estimate of the

Table 22.2 Estimates, bias, mean square error (MSE), and coverage probabilities (CP) for the Bernoulli cure rate model with exponential lifetime under different simulation settings

n	p	Parameter	Estimate (s.e.)	Bias	MSE	CP	
						90%	95%
400 (110, 120, 90, 80)	High	$\beta_0 = -1.192$	−1.192 (0.345)	−0.001	0.117	0.913	0.967
		$\beta_1 = 0.573$	0.577 (0.128)	0.005	0.016	0.915	0.956
		$\gamma = 1$	1.004 (0.099)	0.004	0.011	0.881	0.929
	Low	$\beta_0 = -0.038$	−0.055 (0.299)	−0.017	0.092	0.903	0.950
		$\beta_1 = 0.443$	0.454 (0.125)	0.011	0.016	0.901	0.949
		$\gamma = 1$	1.000 (0.069)	0.000	0.005	0.900	0.950
400 (110, 120, 90, 80)	High	$\beta_0 = -1.192$	−1.204 (0.345)	−0.013	0.121	0.913	0.958
		$\beta_1 = 0.573$	0.577 (0.128)	0.005	0.016	0.903	0.945
		$\gamma = 3$	3.029 (0.300)	0.029	0.096	0.893	0.947
	Low	$\beta_0 = -0.038$	−0.049 (0.299)	−0.012	0.100	0.888	0.943
		$\beta_1 = 0.443$	0.449 (0.124)	0.006	0.017	0.884	0.943
		$\gamma = 3$	3.003 (0.209)	0.003	0.040	0.916	0.960
600 (165, 180, 135, 120)	High	$\beta_0 = -1.192$	−1.195 (0.281)	−0.003	0.081	0.900	0.946
		$\beta_1 = 0.573$	0.577 (0.104)	0.004	0.011	0.895	0.949
		$\gamma = 1$	1.002 (0.081)	0.002	0.007	0.890	0.935
	Low	$\beta_0 = -0.038$	−0.045 (0.244)	−0.008	0.057	0.905	0.949
		$\beta_1 = 0.443$	0.449 (0.101)	0.006	0.010	0.911	0.960
		$\gamma = 1$	1.001 (0.057)	0.001	0.003	0.895	0.953

Note: In the table, n and p denote the sample size and the censoring proportion, respectively. "High" and "Low" correspond to the censoring proportions $(0.80, 0.65, 0.50, 0.35)$ and $(0.50, 0.40, 0.30, 0.20)$, respectively, for Groups 1–4.

cured proportion for that group becomes zero. To circumvent this problem, we ignored the largest observed time if it is not censored just for the purpose of determining the initial values.

Tables 22.2–22.4 show the results for the three special cases of COM-Poisson cure rate model. The results show that the EM algorithm converges to the true parameter values quite accurately. As expected, the standard error and the MSE of the estimates both decrease with an increase in the sample size and with a decrease in censoring proportion. The lifetime parameter γ does not seem to have any effect on the estimation of the regression parameters β_0 and β_1. The coverage probability of the confidence interval based on the asymptotic normality of the MLEs seems to be quite close to the nominal level. It should be mentioned that in the case of the geometric cure rate model, the EM algorithm sometimes shows a problem with convergence, irrespective of the sample size and level of censoring. This is because we link the parameter η in (22.3) to covariates $\boldsymbol{x}$ through the log-linear link function $\eta = \exp(\boldsymbol{x}'\boldsymbol{\beta})$ and there is no guarantee that in the subsequent steps of the EM algorithm η is less than one, which is a condition that must hold true for the geometric cure rate model, as mentioned earlier.

Similar findings were also observed when estimating the parameters of the COM-Poisson model with ϕ fixed at 0.5 but are not presented here for conciseness. However, for all practical purposes, ϕ is to be estimated from the data, for which we can employ the

Table 22.3 Estimates, bias, mean square error (MSE), and coverage probabilities (CP) for the Poisson cure rate model with exponential lifetime under different simulation settings

n	p	Parameter	Estimate (s.e.)	Bias	MSE	CP	
						90%	95%
400 (110, 120, 90, 80)	High	$\beta_0 = -1.232$	−1.234 (0.233)	−0.003	0.055	0.896	0.947
		$\beta_1 = 0.390$	0.393 (0.076)	0.004	0.006	0.903	0.949
		$\gamma = 1$	1.000 (0.124)	0.000	0.015	0.912	0.953
	Low	$\beta_0 = -0.330$	−0.332 (0.163)	−0.002	0.024	0.903	0.957
		$\beta_1 = 0.243$	0.243 (0.058)	0.000	0.003	0.911	0.958
		$\gamma = 1$	1.008 (0.094)	0.008	0.008	0.906	0.965
400 (110, 120, 90, 80)	High	$\beta_0 = -1.232$	−1.238 (0.234)	−0.006	0.052	0.911	0.955
		$\beta_1 = 0.390$	0.391 (0.076)	0.001	0.006	0.911	0.957
		$\gamma = 3$	3.029 (0.376)	0.029	0.136	0.907	0.956
	Low	$\beta_0 = -0.330$	−0.329 (0.163)	0.001	0.029	0.881	0.935
		$\beta_1 = 0.243$	0.244 (0.058)	0.002	0.004	0.876	0.931
		$\gamma = 3$	3.009 (0.281)	0.009	0.081	0.901	0.949
600 (165, 180, 135, 120)	High	$\beta_0 = -1.232$	−1.225 (0.190)	0.007	0.039	0.885	0.941
		$\beta_1 = 0.390$	0.387 (0.062)	−0.002	0.004	0.894	0.940
		$\gamma = 1$	1.003 (0.102)	0.003	0.011	0.889	0.945
	Low	$\beta_0 = -0.330$	−0.330 (0.133)	0.000	0.017	0.914	0.961
		$\beta_1 = 0.243$	0.242 (0.047)	0.000	0.002	0.913	0.955
		$\gamma = 1$	1.005 (0.076)	0.005	0.005	0.911	0.960

Note: In the table, n and p denote the sample size and the censoring proportion, respectively. "High" and "Low" correspond to the censoring proportions (0.80, 0.65, 0.50, 0.35) and (0.50, 0.40, 0.30, 0.20), respectively, for Groups 1–4.

Table 22.4 Estimates, bias, mean square error (MSE), and coverage probabilities (CP) for the geometric cure rate model with exponential lifetime under different simulation settings

n	p	Parameter	Estimate (s.e.)	Bias	MSE	CP 90%	CP 95%
400 (110, 120, 90, 80)	High	$\beta_0 = -1.304$	−1.299 (0.160)	0.004	0.026	0.889	0.946
		$\beta_1 = 0.254$	0.253 (0.043)	−0.001	0.002	0.899	0.950
		$\gamma = 1$	1.006 (0.162)	0.006	0.027	0.899	0.945
	Low	$\beta_0 = -0.627$	−0.625 (0.083)	0.002	0.006	0.916	0.953
		$\beta_1 = 0.116$	0.115 (0.022)	−0.001	0.000	0.910	0.953
		$\gamma = 1$	1.002 (0.137)	0.002	0.018	0.914	0.963
400 (110, 120, 90, 80)	High	$\beta_0 = -1.304$	−1.300 (0.160)	0.004	0.026	0.889	0.946
		$\beta_1 = 0.254$	0.253 (0.043)	−0.001	0.002	0.899	0.950
		$\gamma = 3$	3.018 (0.485)	0.018	0.239	0.898	0.945
	Low	$\beta_0 = -0.627$	−0.625 (0.083)	0.002	0.006	0.916	0.953
		$\beta_1 = 0.116$	0.115 (0.022)	−0.001	0.000	0.910	0.953
		$\gamma = 3$	3.006 (0.412)	0.006	0.161	0.914	0.963
600 (165, 180, 135, 120)	High	$\beta_0 = -1.304$	−1.301 (0.131)	0.003	0.018	0.894	0.941
		$\beta_1 = 0.254$	0.253 (0.035)	−0.001	0.001	0.898	0.955
		$\gamma = 1$	1.000 (0.131)	0.000	0.016	0.917	0.953
	Low	$\beta_0 = -0.627$	−0.625 (0.068)	0.002	0.004	0.907	0.950
		$\beta_1 = 0.116$	0.115 (0.018)	−0.001	0.000	0.909	0.949
		$\gamma = 1$	1.002 (0.112)	0.002	0.012	0.914	0.954

Note: In the table, n and p denote the sample size and the censoring proportion, respectively. "High" and "Low" correspond to the censoring proportions (0.80, 0.65, 0.50, 0.35) and (0.50, 0.40, 0.30, 0.20), respectively, for Groups 1–4.

profile likelihood technique. The corresponding results are shown in Tables 22.5, where the true value of ϕ is taken to be 0.5, but now estimated from the data along with other model parameters. The profile likelihood approach of estimating ϕ results in large bias and MSE of the estimate of ϕ. However, they both decrease with an increase in sample size and with a decrease in censoring proportion. The standard error and the MSE of the MLEs also show similar behavior. The standard errors of the estimates of β_0, β_1 and γ are found to be underestimated and consequently the coverage probabilities fall below the nominal level. This is mainly due to the imprecision in the estimate of ϕ by using the profile likelihood technique. The undercoverage seems to be more pronounced for β_1 and γ. [†] Now, to estimate ϕ by the profile likelihood approach, we chose the range of ϕ to be [0.05, 2], where the starting value of ϕ was taken to be 0.05 and in the interval [0.10, 1] ϕ was divided in steps of 0.1, whereas in the interval [1.05, 2] ϕ was divided in steps of 0.05. This accounted for 31 distinct choices of ϕ ranging from 0.05 to 2. It is clear that in order to get an estimate of ϕ for a given dataset, we must ensure that the EM algorithm converges for all chosen values of ϕ. However, for some datasets it was observed that for values of ϕ close to zero, the EM algorithm failed to converge and as such that dataset had to be discarded. This is because for values of ϕ close to

[†] Similar findings were also observed when $\phi = 2.0$, but are not presented here for conciseness.

Table 22.5 Estimates, bias, mean square error (MSE), and coverage probabilities (CP) for the COM–Poisson cure rate model with $\phi = 0.5$ and with exponential lifetime under different simulation settings

n	p	Parameter	Estimate (s.e.)	Bias	MSE	CP	
						90%	95%
400 (110, 120, 90, 80)	High	$\beta_0 = -1.262$	−1.278 (0.207)	−0.016	0.052	0.883	0.928
		$\beta_1 = 0.338$	0.352 (0.065)	0.015	0.017	0.518	0.613
		$\gamma = 1$	1.018 (0.137)	0.018	0.054	0.632	0.729
		$\phi = 0.5$	0.717 (–)	0.217	0.563	–	–
	Low	$\beta_0 = -0.449$	−0.449 (0.134)	0.000	0.025	0.860	0.913
		$\beta_1 = 0.195$	0.207 (0.046)	0.011	0.008	0.535	0.617
		$\gamma = 1$	1.021 (0.105)	0.021	0.054	0.508	0.592
		$\phi = 0.5$	0.662 (–)	0.162	0.360	–	–
400 (110, 120, 90, 80)	High	$\beta_0 = -1.262$	−1.278 (0.207)	−0.016	0.052	0.883	0.928
		$\beta_1 = 0.338$	0.352 (0.065)	0.015	0.017	0.518	0.613
		$\gamma = 3$	3.055 (0.411)	0.055	0.486	0.632	0.727
		$\phi = 0.5$	0.717 (–)	0.217	0.563	–	–
	Low	$\beta_0 = -0.449$	−0.449 (0.134)	0.000	0.025	0.858	0.913
		$\beta_1 = 0.195$	0.207 (0.046)	0.011	0.008	0.534	0.616
		$\gamma = 3$	3.061 (0.314)	0.061	0.489	0.507	0.591
		$\phi = 0.5$	0.661 (–)	0.161	0.360	–	–
600 (165, 180, 135, 120)	High	$\beta_0 = -1.262$	−1.270 (0.166)	−0.008	0.033	0.876	0.933
		$\beta_1 = 0.338$	0.342 (0.051)	0.005	0.013	0.485	0.571
		$\gamma = 1$	0.998 (0.111)	−0.002	0.043	0.559	0.668
		$\phi = 0.5$	0.632 (–)	0.132	0.414	–	–
	Low	$\beta_0 = -0.449$	−0.448 (0.107)	0.001	0.017	0.823	0.914
		$\beta_1 = 0.195$	0.200 (0.036)	0.004	0.006	0.552	0.616
		$\gamma = 1$	0.998 (0.084)	−0.002	0.037	0.519	0.602
		$\phi = 0.5$	0.591 (–)	0.091	0.228	–	–

Note: In the table, n and p denote the sample size and the censoring proportion, respectively. "High" and "Low" correspond to the censoring proportions (0.80, 0.65, 0.50, 0.35) and (0.50, 0.40, 0.30, 0.20), respectively, for Groups 1–4.

zero, the COM-Poisson distribution approaches the geometric distribution in which case, as mentioned earlier, sometimes a convergence problem is encountered.

22.7.2 Simulation Study: Model Discrimination

The flexibility of the COM-Poisson cure rate model and its inclusion of some of the well-known cure rate models in the literature due to Berkson and Gage (1952), Yakovlev and Tsodikov (1996) and Yin and Ibrahim (2005) often make it suitable as a tool to select a simple cure rate model within this family that provides an adequate fit. Therefore, it is of practical interest to explore the use of the COM-Poisson cure rate model to select a suitable distribution for the competing cause random variable M in (22.2) that provides an adequate fit to the data. This simulation study can be divided into two parts: one corresponding to the under-dispersed models ($\phi > 1$), where we investigate the performance of the likelihood ratio test in testing the null hypothesis that the number of competing causes relating to the event of interest can be described by either Bernoulli ($H_0 : \phi \to \infty$) or Poisson ($H_0 : \phi = 1$) distributions *versus* the alternative hypothesis that the number of competing causes can be described by a member of the COM-Poisson distribution other than the two distributions specified in the null hypothesis. The other part of the simulation study corresponds to the over-dispersed models ($\phi < 1$), where we investigate the performance of the likelihood ratio test in testing the null hypothesis that the number of competing causes relating to the event of interest can be described by either Poisson ($H_0 : \phi = 1$) or geometric ($H_0 : \phi = 0$) distributions *versus* the alternative hypothesis that the number of competing causes can be described by a member of the COM-Poisson distribution other than the two distributions specified in the null hypothesis. The likelihood ratio test statistic is defined as $\Lambda = -2(\hat{l}_0 - \hat{l})$, where $\hat{l}_0$ and $\hat{l}$ denote the restricted and unrestricted maximized log-likelihood functions, respectively. We note that under standard likelihood theory, the asymptotic null distribution of Λ is a central chi-square distribution with one degree of freedom. However, if the parameter lies on the boundary of the parameter space as in the case with Bernoulli and geometric models, the asymptotic null distribution of Λ is such that $P[\Lambda \le \lambda] = \frac{1}{2} + \frac{1}{2}P[\chi_1^2 \le \lambda]$, where χ_1^2 denotes a chi-square variable with one degree of freedom (Self and Liang 1987; Claeskens *et al.* 2008). Based on 1000

Table 22.6 Observed levels (in bold) and powers of the likelihood ratio test for under-dispersed cure rate models under different simulation settings

Fitted models	True COM-Poisson models				
	$\phi = 1$	$\phi = 1.2$	$\phi = 1.5$	$\phi = 1.8$	$\phi = 2$
	$n = 400$ (110, 120, 90, 80), $p =$ High, $\gamma = 1$				
Bernoulli	0.482	0.433	0.333	0.254	0.259
Poisson	**0.036**	0.015	0.016	0.013	0.013
	$n = 400$ (110, 120, 90, 80), $p =$ Low, $\gamma = 1$				
Bernoulli	0.837	0.752	0.670	0.587	0.540
Poisson	**0.041**	0.024	0.026	0.029	0.046
	$n = 400$ (110, 120, 90, 80), $p =$ High, $\gamma = 3$				
Bernoulli	0.482	0.433	0.333	0.254	0.259
Poisson	**0.037**	0.015	0.016	0.013	0.013
	$n = 400$ (110, 120, 90, 80), $p =$ Low, $\gamma = 3$				
Bernoulli	0.837	0.752	0.670	0.587	0.540
Poisson	**0.041**	0.024	0.026	0.029	0.046
	$n = 600$ (165, 180, 135, 120), $p =$ High, $\gamma = 1$				
Bernoulli	0.644	0.552	0.453	0.354	0.310
Poisson	**0.051**	0.039	0.023	0.018	0.013
	$n = 600$ (165, 180, 135, 120), $p =$ Low, $\gamma = 1$				
Bernoulli	0.936	0.887	0.815	0.745	0.687
Poisson	**0.059**	0.031	0.054	0.084	0.103

Note: In the table, n and p denote the sample size and the censoring proportion, respectively. "High" and "Low" correspond to the censoring proportions (0.80, 0.65, 0.50, 0.35) and (0.50, 0.40, 0.30, 0.20), respectively, for Groups 1–4.

Table 22.7 Observed levels (in bold) and powers of the likelihood ratio test for over-dispersed cure rate models under different simulation settings

Fitted models	True COM-Poisson models				
	$\phi = 1$	$\phi = 0.8$	$\phi = 0.5$	$\phi = 0.3$	$\phi = 0.05$
	$n = 400$ (110, 120, 90, 80), $p =$ High, $\gamma = 1$				
Poisson	**0.036**	0.047	0.092	0.168	0.417
Geometric	0.436	0.392	0.264	0.193	0.083
	$n = 400$ (110, 120, 90, 80), $p =$ Low, $\gamma = 1$				
Poisson	**0.041**	0.061	0.174	0.333	0.795
Geometric	0.757	0.668	0.501	0.312	0.049
	$n = 400$ (110, 120, 90, 80), $p =$ High, $\gamma = 3$				
Poisson	**0.037**	0.047	0.092	0.168	0.416
Geometric	0.436	0.391	0.264	0.193	0.083
	$n = 400$ (110, 120, 90, 80), $p =$ Low, $\gamma = 3$				
Poisson	**0.041**	0.061	0.175	0.333	0.795
Geometric	0.757	0.666	0.499	0.310	0.049
	$n = 600$ (165, 180, 135, 120), $p =$ High, $\gamma = 1$				
Poisson	**0.051**	0.059	0.139	0.244	0.543
Geometric	0.517	0.442	0.293	0.197	0.062
	$n = 600$ (165, 180, 135, 120), $p =$ Low, $\gamma = 1$				
Poisson	**0.059**	0.072	0.193	0.433	0.913
Geometric	0.851	0.796	0.595	0.352	0.052

Note: In the table, n and p denote the sample size and the censoring proportion, respectively. "High" and "Low" correspond to the censoring proportions (0.80, 0.65, 0.50, 0.35) and (0.50, 0.40, 0.30, 0.20), respectively, for Groups 1–4.

data sets in each situation and nominal significance level 0.05, we computed observed significance levels and powers of the likelihood ratio test corresponding to the under-dispersed and over-dispersed models and have presented them in Tables 22.6 and 22.7, respectively.

The numbers in bold face in Tables 22.6 and 22.7 are observed levels while other numbers are observed powers. These numbers were determined by the rejection rates of the null hypothesis. When the Poisson distribution ($\phi = 1$) is the true competing cause distribution, the test has good power to reject the Bernoulli distribution as the competing cause distribution (with the largest observed power being 94%) and a slightly less power to reject the geometric distribution as the competing cause distribution (with the largest observed power being 85%). From Table 22.6, when the COM-Poisson distribution ($\phi > 1$) is the true competing cause distribution, the test has reasonable power to reject the Bernoulli distribution as the competing cause distribution and the power of the test decreases as the true value of ϕ increases. However, the test does not appear to have much power to reject the Poisson distribution as the competing cause distribution. From Table 22.7, when the COM-Poisson distribution ($\phi < 1$) is the true competing cause distribution, the power of the test to reject the Poisson distribution as the competing cause distribution increases as the true value of ϕ decreases. This means that when the true model approaches the geometric model (ϕ close to 0), the test has the maximum power to reject the Poisson model, and the largest observed power in this case is 91%. As one would expect, when the true model is close to the geometric (say, $\phi = 0.05$), the power to reject the geometric model is very close to the nominal significance level. Furthermore, the power of the likelihood ratio test increases as the sample size increases and the censoring proportion decreases.

22.8 Weibull Lifetime Distribution

In this case, we assume the time-to-event W to follow the Weibull distribution with

$$S(w; \boldsymbol{\gamma}) = \exp(-(\gamma_2 w)^{1/\gamma_1}) \text{ and } f(w; \boldsymbol{\gamma}) = \frac{1}{\gamma_1 w}(\gamma_2 w)^{1/\gamma_1} S(w; \boldsymbol{\gamma}) \tag{22.9}$$

for $w > 0$, $\gamma_1 > 0$, $\gamma_2 > 0$. In order to circumvent the problem mentioned earlier with the use of the log-linear link function in case of the geometric cure rate model, we now relate the cure rate parameter p_0 to covariates $\boldsymbol{x}_i$ by the logistic link function $p_{0i} = \frac{1}{1+\exp(\boldsymbol{x}_i'\boldsymbol{\beta})}$, $i = 1, \ldots, n$. It is to be noted that for a fixed value of ϕ, p_0 is purely a function of η. Hence, given the link function for p_0, we can easily determine how η is linked with the covariates. Now, $p_0 = \frac{1}{Z(\eta,\phi)}$ and after linking it with the logistic link function, we have $H_\phi(\eta) = Z(\eta, \phi) = 1 + \exp(\boldsymbol{x}'\boldsymbol{\beta})$, for a fixed value of ϕ. This implies that $\eta = H_\phi^{-1}(1 + \exp(\boldsymbol{x}'\boldsymbol{\beta}))$. We note that $H_\phi(\eta)$ is a monotonically increasing function of η with $H_\phi(\eta)|_{\eta=0} = 1$ and $H_\phi(\eta)|_{\eta\to\infty} = \infty$. Hence, the link function for η is a valid one for any fixed value of ϕ. We also note that although the function $H_\phi(\eta)$ is explicit, the inverse function $H_\phi^{-1}(\cdot)$ is implicit and needs to be calculated numerically. However, an explicit form of $H_\phi^{-1}(\cdot)$ can be obtained in some special cases of the COM-Poisson model and are as presented in Table 22.8.

For the sake of conciseness, we present the expressions of the E-Step and M-Step and the expressions of the components of the observed information matrix for only the general case of COM-Poisson model in Appendix A2 and Appendix B2, respectively. The expressions for the three special cases of COM-Poisson model are available from the authors upon request.

22.8.1 *Simulation Study: Model Fitting*

We consider the same simulation setup as in the case of exponential lifetimes, except that here the lifetime W is generated from a Weibull distribution with shape $= \frac{1}{\gamma_1}$ and

Table 22.8 Expressions of η for the three special cases of the COM-Poisson cure rate model when using the logistic link function

Model	$H_\phi(\eta)$	$\eta = H_\phi^{-1}(1 + \exp(\boldsymbol{x}'\boldsymbol{\beta}))$
Bernoulli ($\phi \to \infty$)	$1 + \eta$	$\exp(\boldsymbol{x}'\boldsymbol{\beta})$
Poisson ($\phi = 1$)	$\exp(\eta)$	$\log\{1 + \exp(\boldsymbol{x}'\boldsymbol{\beta})\}$
geometric ($\phi = 0$)	$\frac{1}{1-\eta}$	$\frac{\exp(\boldsymbol{x}'\boldsymbol{\beta})}{1+\exp(\boldsymbol{x}'\boldsymbol{\beta})}$

scale $= \frac{1}{\gamma_2}$, for a specific choice of γ_1 and γ_2. The choice of the lifetime parameter (γ_1, γ_2) was obtained after equating the mean and variance of the underlying Weibull distribution in (22.9) to some fixed values. We considered two different choices for the mean $\frac{1}{\gamma_2}\Gamma(1+\gamma_1)$ as: 5 and 3.5, and two different choices for the variance $\frac{1}{\gamma_2^2}[\Gamma(1+2\gamma_1) - \{\Gamma(1+\gamma_1)\}^2]$ as: 3 and 1.5, where $\Gamma(\cdot)$ is the complete gamma function. The above choices of the mean and variance lead in total to four different choices of (γ_1, γ_2) : (0.316, 0.179), (0.215, 0.183), (0.470, 0.253), and (0.320, 0.256). This allows us to study the behavior of the model under the same mean but different variance and vice-versa. In this simulation study, we simulated data from different cure rate models and the estimation procedure was employed to estimate the unknown parameters of the model. We then determined the empirical bias and root mean square error (RMSE) of the estimates so as to examine the performance of the estimates. We also calculated the coverage probabilities of the confidence intervals based on the asymptotic normality of the MLEs for different nominal confidence levels. All simulations were performed using the R software and the results were based on 1000 Monte Carlo runs.

To find an initial guess of the parameters $\beta_0, \beta_1, \gamma_1$ and γ_2 so as to start the iterative procedure, we considered a discrete grid search on (β_0, β_1). For each possible choice of (β_0, β_1), we calculated the observed data log-likelihood function. For this purpose, we sampled a value of γ_1 and γ_2 independently from a set of values of γ_1 and γ_2 obtained by taking 20% deviation off the true values of γ_1 and γ_2, respectively. Finally, we took the value of $(\beta_0, \beta_1, \gamma_1, \gamma_2)$ for which the observed data log-likelihood function is the maximum as the initial value.

We present the results for the three special cases of the COM-Poisson cure rate model in Tables 22.9–22.11. The results show that the EM algorithm converges to the true parameter values quite accurately. As expected, the standard error and the RMSE of the estimates decrease with an increase in sample size and with a decrease in censoring proportion. On comparing models having the same mean but different variance for the lifetime distribution, the model having higher variance for the lifetime distribution is found to have a higher standard error and RMSE of the lifetime parameter $\boldsymbol{\gamma}$. Similarly, on comparing models having the same variance but different mean for the lifetime distribution, the standard error and RMSE of $\boldsymbol{\gamma}$ is found to be high in the model having a lower mean for the lifetime distribution. The coverage probabilities of the confidence intervals based on the asymptotic normality of the MLEs are found to be quite close to the nominal levels.

Similar findings were also observed when estimating the parameters of the COM-Poisson model with ϕ fixed at 0.5 but are not presented here for conciseness. However, as mentioned earlier, ϕ needs to be estimated from the data along with other model parameters, for which we can employ the profile likelihood approach. The corresponding results are shown in Table 22.12, where the true value of ϕ is taken to be 0.5. We again observe that the profile likelihood approach of estimating ϕ introduces large RMSE in the estimate of ϕ. However, it is seen to decrease with an increase in sample size and with a decrease in censoring proportion. The standard error of the MLEs also show similar behavior. The bias in the estimate of ϕ is large particularly when the censoring proportion is high. We also note that the bias and RMSE of the estimate of β_0 are high when compared to the bias and RMSE of the estimates of the other regression coefficient and lifetime parameters. The standard errors in the estimates of β_0, β_1, γ_1 and γ_2 are once

Table 22.9 Estimates, bias, root mean square error (RMSE), and coverage probabilities (CP) for the Bernoulli cure rate model with Weibull lifetime under different simulation settings

n	p	Parameter	Estimate (s.e.)	Bias	RMSE	CP	
						90%	95%
400 (110, 120, 90, 80)	High	$\beta_0 = -1.192$	−1.212 (0.326)	−0.021	0.336	0.891	0.941
		$\beta_1 = 0.573$	0.582 (0.123)	0.009	0.125	0.896	0.942
		$\gamma_1 = 0.316$	0.315 (0.019)	−0.001	0.019	0.892	0.942
		$\gamma_2 = 0.179$	0.179 (0.005)	0.000	0.005	0.893	0.944
	Low	$\beta_0 = -0.038$	−0.043 (0.288)	−0.006	0.291	0.899	0.956
		$\beta_1 = 0.443$	0.447 (0.120)	0.004	0.122	0.907	0.946
		$\gamma_1 = 0.316$	0.315 (0.015)	−0.001	0.015	0.893	0.944
		$\gamma_2 = 0.179$	0.179 (0.004)	0.000	0.004	0.901	0.954
400 (110, 120, 90, 80)	High	$\beta_0 = -1.192$	−1.218 (0.323)	−0.026	0.321	0.907	0.955
		$\beta_1 = 0.573$	0.584 (0.122)	0.012	0.121	0.905	0.955
		$\gamma_1 = 0.215$	0.214 (0.013)	−0.001	0.013	0.898	0.942
		$\gamma_2 = 0.183$	0.183 (0.003)	0.000	0.003	0.912	0.961
	Low	$\beta_0 = -0.038$	−0.031 (0.287)	0.007	0.288	0.911	0.953
		$\beta_1 = 0.443$	0.444 (0.120)	0.001	0.123	0.891	0.948
		$\gamma_1 = 0.215$	0.214 (0.010)	−0.001	0.010	0.896	0.946
		$\gamma_2 = 0.183$	0.183 (0.003)	0.000	0.003	0.881	0.940
400 (110, 120, 90, 80)	High	$\beta_0 = -1.192$	−1.203 (0.330)	−0.011	0.331	0.906	0.954
		$\beta_1 = 0.573$	0.578 (0.124)	0.005	0.127	0.902	0.950
		$\gamma_1 = 0.470$	0.467 (0.029)	−0.003	0.030	0.870	0.924
		$\gamma_2 = 0.253$	0.253 (0.010)	0.000	0.010	0.894	0.944
	Low	$\beta_0 = -0.038$	−0.044 (0.291)	−0.007	0.293	0.889	0.949
		$\beta_1 = 0.443$	0.450 (0.121)	0.007	0.123	0.897	0.954
		$\gamma_1 = 0.470$	0.469 (0.023)	−0.001	0.023	0.884	0.945
		$\gamma_2 = 0.253$	0.253 (0.008)	0.000	0.008	0.905	0.955
400 (110, 120, 90, 80)	High	$\beta_0 = -1.192$	−1.216 (0.326)	−0.025	0.332	0.897	0.948
		$\beta_1 = 0.573$	0.583 (0.123)	0.010	0.127	0.887	0.944
		$\gamma_1 = 0.320$	0.319 (0.019)	−0.001	0.019	0.908	0.955
		$\gamma_2 = 0.256$	0.256 (0.007)	0.000	0.007	0.908	0.947
	Low	$\beta_0 = -0.038$	−0.033 (0.289)	0.005	0.278	0.921	0.960
		$\beta_1 = 0.443$	0.445 (0.120)	0.002	0.117	0.918	0.958
		$\gamma_1 = 0.320$	0.319 (0.016)	−0.001	0.016	0.904	0.943
		$\gamma_2 = 0.256$	0.256 (0.005)	0.000	0.005	0.902	0.959
600 (165, 180, 135, 120)	High	$\beta_0 = -1.192$	−1.205 (0.265)	−0.013	0.275	0.879	0.940
		$\beta_1 = 0.573$	0.578 (0.100)	0.005	0.104	0.880	0.945
		$\gamma_1 = 0.316$	0.316 (0.016)	0.000	0.016	0.906	0.958
		$\gamma_2 = 0.179$	0.179 (0.004)	0.000	0.004	0.905	0.960
	Low	$\beta_0 = -0.038$	−0.056 (0.235)	−0.018	0.226	0.911	0.960
		$\beta_1 = 0.443$	0.453 (0.098)	0.009	0.094	0.915	0.963
		$\gamma_1 = 0.316$	0.315 (0.013)	−0.001	0.013	0.895	0.949
		$\gamma_2 = 0.179$	0.179 (0.003)	0.000	0.003	0.917	0.952

Note: In the table, n and p denote the sample size and the censoring proportion, respectively. "High" and "Low" correspond to the censoring proportions (0.80, 0.65, 0.50, 0.35) and (0.50, 0.40, 0.30, 0.20), respectively, for Groups 1–4.

Table 22.10 Estimates, bias, root mean square error (RMSE), and coverage probabilities (CP) for the Poisson cure rate model with Weibull lifetime under different simulation settings

n	p	Parameter	Estimate (s.e.)	Bias	RMSE	CP 90%	CP 95%
400 (110, 120, 90, 80)	High	$\beta_0 = -1.192$	−1.209 (0.310)	−0.018	0.305	0.905	0.957
		$\beta_1 = 0.573$	0.578 (0.115)	0.006	0.114	0.913	0.949
		$\gamma_1 = 0.316$	0.314 (0.018)	−0.002	0.019	0.888	0.935
		$\gamma_2 = 0.179$	0.179 (0.006)	0.000	0.006	0.888	0.942
	Low	$\beta_0 = -0.038$	−0.031 (0.268)	0.006	0.265	0.907	0.952
		$\beta_1 = 0.443$	0.443 (0.109)	0.000	0.107	0.905	0.950
		$\gamma_1 = 0.316$	0.314 (0.014)	−0.002	0.015	0.892	0.938
		$\gamma_2 = 0.179$	0.179 (0.005)	0.000	0.005	0.890	0.941
400 (110, 120, 90, 80)	High	$\beta_0 = -1.192$	−1.197 (0.310)	−0.005	0.321	0.896	0.944
		$\beta_1 = 0.573$	0.574 (0.115)	0.002	0.120	0.904	0.944
		$\gamma_1 = 0.215$	0.214 (0.012)	−0.001	0.013	0.901	0.949
		$\gamma_2 = 0.183$	0.183 (0.004)	0.000	0.004	0.889	0.939
	Low	$\beta_0 = -0.038$	−0.028 (0.267)	0.009	0.267	0.901	0.956
		$\beta_1 = 0.443$	0.439 (0.109)	−0.004	0.108	0.905	0.951
		$\gamma_1 = 0.215$	0.215 (0.010)	0.000	0.010	0.865	0.934
		$\gamma_2 = 0.183$	0.183 (0.003)	0.000	0.003	0.895	0.947
400 (110, 120, 90, 80)	High	$\beta_0 = -1.192$	−1.202 (0.310)	−0.010	0.314	0.905	0.947
		$\beta_1 = 0.573$	0.576 (0.115)	0.004	0.116	0.898	0.950
		$\gamma_1 = 0.470$	0.468 (0.028)	−0.002	0.029	0.879	0.937
		$\gamma_2 = 0.253$	0.253 (0.012)	0.000	0.013	0.894	0.947
	Low	$\beta_0 = -0.038$	−0.041 (0.268)	−0.003	0.262	0.912	0.955
		$\beta_1 = 0.443$	0.445 (0.109)	0.002	0.106	0.910	0.956
		$\gamma_1 = 0.470$	0.468 (0.022)	−0.002	0.022	0.895	0.937
		$\gamma_2 = 0.253$	0.254 (0.010)	0.001	0.010	0.881	0.940
400 (110, 120, 90, 80)	High	$\beta_0 = -1.192$	−1.190 (0.309)	0.002	0.310	0.900	0.949
		$\beta_1 = 0.573$	0.574 (0.115)	0.002	0.117	0.895	0.946
		$\gamma_1 = 0.320$	0.318 (0.019)	−0.002	0.019	0.895	0.937
		$\gamma_2 = 0.256$	0.256 (0.008)	0.000	0.008	0.890	0.945
	Low	$\beta_0 = -0.038$	−0.054 (0.267)	−0.017	0.282	0.888	0.939
		$\beta_1 = 0.443$	0.451 (0.109)	0.008	0.115	0.887	0.944
		$\gamma_1 = 0.320$	0.319 (0.015)	−0.001	0.015	0.891	0.944
		$\gamma_2 = 0.256$	0.256 (0.007)	0.000	0.007	0.905	0.954
600 (165, 180, 135, 120)	High	$\beta_0 = -1.192$	−1.194 (0.252)	−0.002	0.247	0.916	0.957
		$\beta_1 = 0.573$	0.573 (0.094)	0.000	0.094	0.904	0.949
		$\gamma_1 = 0.316$	0.315 (0.015)	−0.001	0.016	0.875	0.941
		$\gamma_2 = 0.179$	0.179 (0.005)	0.000	0.005	0.891	0.950
	Low	$\beta_0 = -0.038$	−0.031 (0.218)	0.006	0.213	0.911	0.953
		$\beta_1 = 0.443$	0.439 (0.089)	−0.004	0.087	0.910	0.953
		$\gamma_1 = 0.316$	0.316 (0.012)	0.000	0.012	0.885	0.943
		$\gamma_2 = 0.179$	0.179 (0.004)	0.000	0.004	0.889	0.934

Note: In the table, n and p denote the sample size and the censoring proportion, respectively. "High" and "Low" correspond to the censoring proportions (0.80, 0.65, 0.50, 0.35) and (0.50, 0.40, 0.30, 0.20), respectively, for Groups 1−4.

Table 22.11 Estimates, bias, root mean square error (RMSE), and coverage probabilities (CP) for the geometric cure rate model with Weibull lifetime under different simulation settings

n	p	Parameter	Estimate (s.e.)	Bias	RMSE	CP 90%	CP 95%
400 (110, 120, 90, 80)	High	$\beta_0=-1.192$	−1.207 (0.286)	−0.016	0.291	0.889	0.942
		$\beta_1=0.573$	0.579 (0.102)	0.007	0.105	0.903	0.945
		$\gamma_1=0.316$	0.314 (0.019)	−0.002	0.019	0.898	0.932
		$\gamma_2=0.179$	0.179 (0.007)	0.000	0.007	0.913	0.954
	Low	$\beta_0=-0.038$	−0.050 (0.231)	−0.013	0.229	0.911	0.953
		$\beta_1=0.443$	0.449 (0.087)	0.006	0.085	0.905	0.962
		$\gamma_1=0.316$	0.315 (0.015)	−0.001	0.015	0.908	0.957
		$\gamma_2=0.179$	0.179 (0.006)	0.000	0.006	0.898	0.951
400 (110, 120, 90, 80)	High	$\beta_0=-1.192$	−1.197 (0.286)	−0.005	0.288	0.893	0.945
		$\beta_1=0.573$	0.575 (0.103)	0.003	0.101	0.904	0.949
		$\gamma_1=0.215$	0.213 (0.013)	−0.002	0.013	0.887	0.934
		$\gamma_2=0.183$	0.183 (0.005)	0.000	0.005	0.908	0.949
	Low	$\beta_0=-0.038$	−0.043 (0.231)	−0.006	0.228	0.910	0.951
		$\beta_1=0.443$	0.446 (0.087)	0.003	0.087	0.899	0.948
		$\gamma_1=0.215$	0.215 (0.010)	0.000	0.010	0.900	0.947
		$\gamma_2=0.183$	0.183 (0.004)	0.000	0.004	0.911	0.950
400 (110, 120, 90, 80)	High	$\beta_0=-1.192$	−1.192 (0.284)	0.000	0.283	0.907	0.953
		$\beta_1=0.573$	0.574 (0.101)	0.001	0.100	0.913	0.953
		$\gamma_1=0.470$	0.467 (0.028)	−0.003	0.029	0.890	0.942
		$\gamma_2=0.253$	0.254 (0.016)	0.001	0.017	0.889	0.946
	Low	$\beta_0=-0.038$	−0.049 (0.231)	−0.011	0.227	0.901	0.951
		$\beta_1=0.443$	0.451 (0.087)	0.008	0.088	0.897	0.950
		$\gamma_1=0.470$	0.467 (0.022)	−0.003	0.023	0.880	0.940
		$\gamma_2=0.253$	0.253 (0.014)	0.000	0.014	0.895	0.946
400 (110, 120, 90, 80)	High	$\beta_0=-1.192$	−1.196 (0.285)	−0.005	0.283	0.910	0.957
		$\beta_1=0.573$	0.574 (0.102)	0.001	0.102	0.906	0.951
		$\gamma_1=0.320$	0.317 (0.019)	−0.003	0.019	0.886	0.933
		$\gamma_2=0.256$	0.257 (0.010)	0.001	0.010	0.913	0.954
	Low	$\beta_0=-0.038$	−0.051 (0.231)	−0.013	0.223	0.913	0.956
		$\beta_1=0.443$	0.448 (0.087)	0.005	0.086	0.905	0.959
		$\gamma_1=0.320$	0.319 (0.015)	−0.001	0.015	0.902	0.956
		$\gamma_2=0.256$	0.256 (0.009)	0.000	0.009	0.918	0.949
600 (165, 180, 135, 120)	High	$\beta_0=-1.192$	−1.208 (0.233)	−0.017	0.224	0.911	0.953
		$\beta_1=0.573$	0.581 (0.083)	0.009	0.077	0.924	0.963
		$\gamma_1=0.316$	0.314 (0.015)	−0.002	0.015	0.896	0.940
		$\gamma_2=0.179$	0.179 (0.006)	0.000	0.006	0.903	0.955
	Low	$\beta_0=-0.038$	−0.043 (0.189)	−0.006	0.188	0.896	0.949
		$\beta_1=0.443$	0.447 (0.071)	0.004	0.073	0.890	0.943
		$\gamma_1=0.316$	0.316 (0.012)	0.000	0.012	0.901	0.957
		$\gamma_2=0.179$	0.179 (0.005)	0.000	0.005	0.896	0.949

Note: In the table, n and p denote the sample size and the censoring proportion, respectively. "High" and "Low" correspond to the censoring proportions (0.80, 0.65, 0.50, 0.35) and (0.50, 0.40, 0.30, 0.20), respectively, for Groups 1–4.

Table 22.12 Estimates, bias, root mean square error (RMSE), and coverage probabilities (CP) for the COM-Poisson cure rate model with $\phi = 0.5$ and with Weibull lifetime under different simulation settings

n	p	Parameter	Estimate (s.e.)	Bias	RMSE	CP	
						90%	95%
400 (110, 120, 90, 80)	High	$\beta_0 = -1.192$	−1.183 (0.298)	−0.009	0.315	0.900	0.935
		$\beta_1 = 0.573$	0.571 (0.109)	−0.001	0.118	0.887	0.937
		$\gamma_1 = 0.316$	0.309 (0.018)	−0.007	0.026	0.744	0.828
		$\gamma_2 = 0.179$	0.178 (0.006)	−0.001	0.011	0.581	0.684
		$\phi = 0.5$	0.639 (–)	0.139	0.799	–	–
	Low	$\beta_0 = -0.038$	0.001 (0.247)	0.039	0.299	0.821	0.897
		$\beta_1 = 0.443$	0.419 (0.096)	−0.024	0.126	0.779	0.849
		$\gamma_1 = 0.316$	0.303 (0.014)	−0.013	0.029	0.498	0.583
		$\gamma_2 = 0.179$	0.175 (0.005)	−0.004	0.014	0.273	0.363
		$\phi = 0.5$	0.495 (–)	−0.005	0.717	–	–
400 (110, 120, 90, 80)	High	$\beta_0 = -1.192$	−1.121 (0.297)	0.071	0.302	0.893	0.938
		$\beta_1 = 0.573$	0.550 (0.109)	−0.023	0.115	0.883	0.930
		$\gamma_1 = 0.215$	0.211 (0.012)	−0.004	0.018	0.717	0.799
		$\gamma_2 = 0.183$	0.182 (0.004)	−0.001	0.007	0.566	0.680
		$\phi = 0.5$	0.636 (–)	0.136	0.787	–	–
	Low	$\beta_0 = -0.038$	0.002 (0.249)	0.040	0.293	0.837	0.911
		$\beta_1 = 0.443$	0.427 (0.098)	−0.016	0.125	0.784	0.868
		$\gamma_1 = 0.215$	0.207 (0.010)	−0.008	0.020	0.488	0.572
		$\gamma_2 = 0.183$	0.181 (0.004)	−0.002	0.010	0.259	0.347
		$\phi = 0.5$	0.587 (–)	0.087	0.772	–	–
400 (110, 120, 90, 80)	High	$\beta_0 = -1.192$	−1.199 (0.298)	−0.007	0.323	0.867	0.938
		$\beta_1 = 0.573$	0.573 (0.109)	0.001	0.119	0.853	0.929
		$\gamma_1 = 0.470$	0.458 (0.028)	−0.012	0.037	0.754	0.834
		$\gamma_2 = 0.253$	0.252 (0.013)	−0.001	0.025	0.555	0.668
		$\phi = 0.5$	0.644 (–)	0.144	0.813	–	–
	Low	$\beta_0 = -0.038$	0.008 (0.248)	0.045	0.293	0.840	0.911
		$\beta_1 = 0.443$	0.421 (0.097)	−0.022	0.125	0.782	0.861
		$\gamma_1 = 0.470$	0.452 (0.022)	−0.018	0.042	0.514	0.621
		$\gamma_2 = 0.253$	0.247 (0.011)	−0.006	0.030	0.288	0.368
		$\phi = 0.5$	0.549 (–)	0.049	0.744	–	–
400 (110, 120, 90, 80)	High	$\beta_0 = -1.192$	−1.180 (0.298)	0.011	0.318	0.872	0.936
		$\beta_1 = 0.573$	0.569 (0.109)	−0.004	0.117	0.873	0.939
		$\gamma_1 = 0.320$	0.313 (0.019)	−0.007	0.026	0.725	0.806
		$\gamma_2 = 0.256$	0.254 (0.009)	−0.002	0.016	0.557	0.660
		$\phi = 0.5$	0.597 (–)	0.097	0.774	–	–
	Low	$\beta_0 = -0.038$	0.011 (0.248)	0.048	0.311	0.808	0.886
		$\beta_1 = 0.443$	0.422 (0.097)	−0.021	0.130	0.753	0.841
		$\gamma_1 = 0.320$	0.307 (0.015)	−0.013	0.029	0.496	0.589
		$\gamma_2 = 0.256$	0.252 (0.007)	−0.004	0.020	0.288	0.374
		$\phi = 0.5$	0.533 (–)	0.033	0.731	–	–

Table 22.12 *(continued)*

n	p	Parameter	Estimate (s.e.)	Bias	RMSE	CP	
						90%	95%
600 (165, 180, 135, 120)	High	$\beta_0 =-1.192$	−1.145 (0.242)	0.047	0.275	0.845	0.912
		$\beta_1 = 0.573$	0.555 (0.088)	−0.017	0.100	0.832	0.910
		$\gamma_1 = 0.316$	0.310 (0.015)	−0.006	0.022	0.706	0.804
		$\gamma_2 = 0.179$	0.177 (0.005)	−0.002	0.010	0.460	0.540
		$\phi = 0.5$	0.547 (−)	0.047	0.723	−	−
	Low	$\beta_0 =-0.038$	0.023 (0.202)	0.060	0.261	0.776	0.873
		$\beta_1 = 0.443$	0.414 (0.078)	−0.029	0.106	0.739	0.842
		$\gamma_1 = 0.316$	0.302 (0.012)	−0.014	0.028	0.378	0.464
		$\gamma_2 = 0.179$	0.175 (0.004)	−0.004	0.013	0.199	0.276
		$\phi = 0.5$	0.469 (−)	−0.031	0.664	−	−

Note: In the table, n and p denote the sample size and the censoring proportion, respectively. "High" and "Low" correspond to the censoring proportions $(0.80, 0.65, 0.50, 0.35)$ and $(0.50, 0.40, 0.30, 0.20)$, respectively, for Groups 1–4.

again found to be underestimated and consequently the coverage probabilities fall below the nominal level. The under-coverage seems to be more pronounced for the lifetime parameters. Furthermore, we note that the under-coverage is more noticeable when the censoring proportion is low. It is interesting to note that the lifetime parameters (γ_1, γ_2) are quite robust to the cure rate model both in terms of its estimated value and the standard error of the estimates.

To estimate ϕ by the profile likelihood approach, we chose the range of ϕ to be [0, 2] and in this interval ϕ was divided in steps of 0.1. This accounted for 21 distinct choices of ϕ. For some datasets, it was observed that the EM algorithm failed to converge for some choices of ϕ and as such that dataset had to be discarded. Similar convergence problem was also noticed when estimating the parameters of the COM-Poisson cure rate model with ϕ fixed at 0.5. However, the number of samples rejected in this case was much less. But, no convergence problem was faced while estimating the parameters of the three special cases of the COM-Poisson cure rate model.

22.8.2 *Simulation Study: Model Discrimination*

Here, we use the COM-Poisson cure rate model for selecting a suitable competing cause distribution that provides an adequate fit to the data in two ways, either by using the likelihood-based method or by using the information-based criterion.

22.8.2.1 Likelihood-Based Method

In this simulation study, as done earlier in the case of exponential lifetimes, we investigate the performance of the likelihood ratio test in testing the null hypothesis that the competing cause random variable can be described by one of the Bernoulli ($H_0 : \phi \to \infty$), Poisson ($H_0 : \phi = 1$) and geometric ($H_0 : \phi = 0$) distributions *versus* an alternative hypothesis that the competing cause can be described by a member of the COM-Poisson other than the one specified in the null hypothesis. We restrict our study to the following parameter settings: Setting 1: $n = 400$ (110, 120, 90, 80), $p =$ Low, $(\gamma_1, \gamma_2) = (0.215, 0.183)$; Setting 2: $n = 600$ (165, 180, 135, 120), $p =$ Low, $(\gamma_1, \gamma_2) = (0.215, 0.183)$; Setting 3: $n = 400$ (110, 120, 90, 80), $p =$ High, $(\gamma_1, \gamma_2) = (0.215, 0.183)$, where n and p denote the sample size and the censoring proportion, respectively. "High" and "Low" correspond to the censoring proportions (0.80, 0.65, 0.50, 0.35) and (0.50, 0.40, 0.30, 0.20), respectively, for Groups 1–4. For each simulated data set, we calculated the likelihood ratio test statistic of the fitted Bernoulli, Poisson and geometric cure rate models *versus* the fitted COM-Poisson cure rate model. Based on 1000 data sets in each situation and nominal significance level of 0.05, we computed the observed significance levels and powers of the likelihood ratio test, and these results are reported in Table 22.13.

The numbers in bold represent the observed levels while other numbers represent the observed powers. These numbers were determined simply by the rejection rates of the null hypothesis. It is easy to see that the chi-square distribution provides only a reasonable approximation to the null distribution of the likelihood ratio test when testing for the Poisson distribution as the observed levels are not close to the nominal level in two of the three cases. However, the mixture chi-square distribution provides a good approximation to the null distribution of the likelihood ratio test when testing for the Bernoulli distribution. When testing for the geometric distribution, the observed levels are found to be much below the nominal level. To investigate the reason for this and to evaluate the performance of the mixture chi-square distribution, we constructed histograms of the likelihood ratio test statistic (Λ) values when testing for the geometric cure rate model (true competing cause distribution being geometric) and Bernoulli cure rate model (true competing cause distribution being Bernoulli). For this purpose, we considered Setting 1 above and the corresponding plots for the geometric and Bernoulli cases are presented in Figures 22.1 and 22.2, respectively. From Figure 22.1, we note that the likelihood ratio statistic takes the value zero 91% of times, which is significantly different from the

Table 22.13 Observed levels (in bold) and powers of the likelihood ratio test

Fitted model	True COM-Poisson model				
	$\phi = 0$	$\phi = 0.5$	$\phi = 1$	$\phi = 2$	$\phi \to \infty$
			Setting 1		
geometric	**0.013**	0.186	0.367	0.519	0.836
Poisson	0.746	0.223	**0.092**	0.038	0.076
Bernoulli	0.992	0.819	0.600	0.405	**0.070**
			Setting 2		
geometric	**0.007**	0.214	0.402	0.682	0.934
Poisson	0.883	0.301	**0.129**	0.045	0.287
Bernoulli	0.999	0.936	0.774	0.463	**0.051**
			Setting 3		
geometric	**0.018**	0.150	0.266	0.455	0.697
Poisson	0.373	0.081	**0.049**	0.013	0.007
Bernoulli	0.901	0.610	0.428	0.221	**0.057**

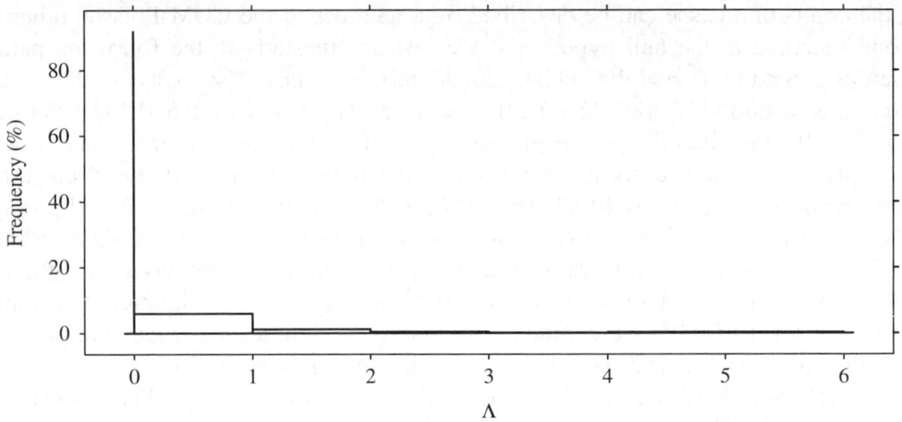

Figure 22.1 Λ values when testing for the geometric model (true model being geometric)

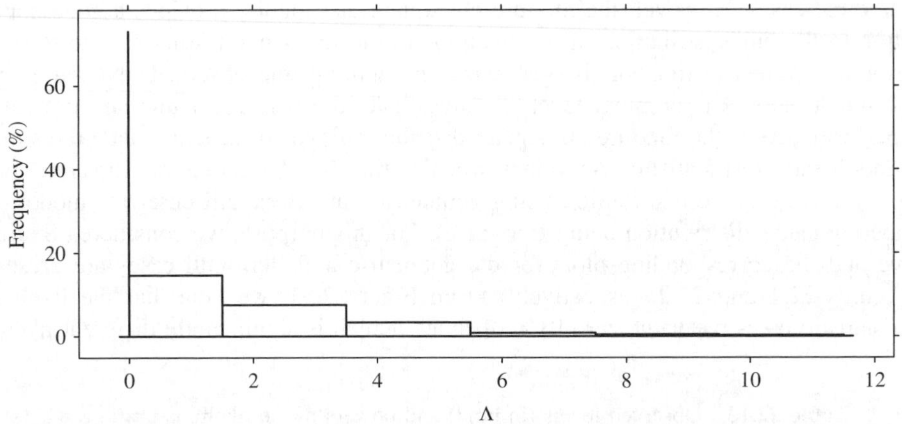

Figure 22.2 Λ values when testing for the Bernoulli model (true model being Bernoulli)

expected 50%, and this in turn explains why the observed level falls much below the nominal level when testing for the geometric cure rate model. However, Figure 22.2 represents a better mixture representation, in which case the likelihood ratio statistic takes the value zero 73% of times resulting in the observed level being close to the nominal level.

When the geometric distribution is the true competing cause distribution, the test has high power to reject the Bernoulli distribution as the competing cause distribution, and the largest observed power is 99%. The test also appears to have good power to reject the Poisson distribution as the competing cause distribution with the largest observed power being 88%. When the Poisson distribution is the true competing cause distribution, the test has moderate power to reject the Bernoulli distribution as the competing cause distribution (the largest observed power is 77%) and relatively less power to reject the geometric distribution as the competing cause distribution (the largest observed power is 40%). When the Bernoulli distribution is the true competing cause distribution, the test again has high power to reject the geometric distribution as the competing cause distribution (the largest observed power is 93%). However, the test does not appear to have good power to reject the Poisson distribution as the competing cause distribution, and the largest observed power, in this case, is only 28%. As expected, the power of the test to reject the wrong model increases with an increase in sample size and with a decrease in censoring proportion. Furthermore, as the true model approaches the Bernoulli distribution, the power of the test to reject the geometric distribution increases whereas the power of the test to reject the Bernoulli distribution decreases, as we would expect.

22.8.2.2 Information-Based Criterion

In this simulation study, we investigate the performance of the Akaike's information criterion (AIC) and the Bayesian information criterion (BIC), the two most widely used model selection criteria, in selecting one of the Bernoulli, Poisson and geometric distributions for the competing cause random variable, given a true competing cause distribution. We have AIC$=-2l+2k$ and BIC$=-2l+k\log(n)$, where l is the maximized log-likelihood value of the model, k is the number of parameters of the fitted model, and n is the sample size. Given a set of candidate models for the data, the preferred model is the one with the minimum AIC or BIC value. In particular, we considered the following true distributions for the competing cause random variable: Bernoulli, Poisson, geometric, COM-Poisson ($\phi = 0.5$) and COM-Poisson ($\phi = 2$). In this study, we considered the same parameter settings as mentioned earlier in Section 22.8.2.1. For each data set, we calculated the AIC and BIC values of the fitted Bernoulli, Poisson and geometric cure rate models. Based on 1000 data sets in each situation, we computed the observed selection rates for each of the two selection criteria used, and these results are reported in Table 22.14.

Table 22.14 Observed selection rates based on Akaike's information criterion*

Fitted model	True COM-Poisson model				
	$\phi = 0$	$\phi = 0.5$	$\phi = 1$	$\phi = 2$	$\phi \to \infty$
			Setting 1		
geometric	0.860	0.356	0.194	0.082	0.003
Poisson	0.140	0.596	0.632	0.555	0.166
Bernoulli	0.000	0.048	0.174	0.363	0.831
			Setting 2		
geometric	0.887	0.330	0.148	0.045	0.001
Poisson	0.113	0.635	0.749	0.608	0.137
Bernoulli	0.000	0.035	0.103	0.347	0.862
			Setting 3		
geometric	0.766	0.401	0.265	0.117	0.034
Poisson	0.220	0.478	0.496	0.452	0.224
Bernoulli	0.014	0.121	0.239	0.431	0.742

*In all cases, the results are the same for the Bayesian information criterion as well.

We first observe that both model selection criteria used result in the same selection rates since the models used all have the same number of parameters. Hence, in this case, inference can be drawn based on either the AIC or BIC. When the true competing cause distribution is either the geometric, Poisson or Bernoulli, the information criteria has high probability of selecting the correct model. Thus, the information criteria can distinguish between the Bernoulli, Poisson and geometric models with a high probability. Furthermore, as expected, this selection probability increases with an increase in sample size and a decrease in censoring proportion. When COM-Poisson ($\phi = 0.5$) is the true model, the selection rate of Poisson, as the correct model, is higher than that of geometric. This means that a COM-Poisson ($\phi = 0.5$) is closer to Poisson than geometric. When COM-Poisson ($\phi = 2$) is the true model, the selection rate of Poisson, as the correct model, is again higher than that of Bernoulli, which means that a COM-Poisson ($\phi = 2$) is closer to Poisson than Bernoulli. For model discrimination, we recommend using the information-based criterion as this method does not involve estimation of the parameter ϕ, unlike the likelihood-based method, which actually is quite time-consuming since the parameter ϕ needs to be estimated by the profile likelihood approach, in this case.

It will be of interest to examine the bias and MSE involved in estimating the cured proportions corresponding to four groups under the wrong model specification. For the bias, we may compute the ratio of the absolute bias of the cured proportion to that of the true cured proportion for each of the four groups and then add to get the total relative bias (TRB) in estimating the cured proportions. For the MSE, we may add the MSEs of the estimates of cured proportions corresponding to four groups to get the total MSE (TMSE), and then use it to determine the total relative efficiency (TRE) under model mis-specification, which is defined as the ratio of the TMSE of the cured proportion under true model specification to that under wrongly specified model. Based on 1000 data sets in each situation, we found the TRB when fitting different models for a given true model and these are presented in Table 22.15. The values in parentheses represent the TRE under model mis-specification. We note that model mis-specification in general leads to higher TRB of the cured proportion as compared to the true model specification. We also note that in certain situations, the TRB of the cured proportion under a wrongly fitted model is much higher than that under the correct model, which reveals that fitting a wrong model would lead to the estimation of cured proportion with a large bias. Similarly, it can also be seen that model mis-specification always leads to loss of efficiency. In certain situations, the TRE is considerably less than one, which implies significant loss of efficiency when fitting the wrong model. These findings strongly suggest that we should utilize the flexibility of the COM-Poisson model in selecting the most suitable competing cause distribution, thus avoiding the risk of estimating the cured proportion inefficiently.

When allowing the AIC to select a model from a set of candidate models, we note that a certain number of times the correct model is selected and the rest of the time the wrong model is selected. Of course, as seen in Table 22.14, the selection rate of the true model is high. In this case, for a given true model, we use the selection rates of candidate models as weights, and the TRB for a given true model is calculated as the weighted sum of TRBs of candidate models. In a similar way, one can also calculate the TMSE when allowing the AIC as the model selection criterion, and calculate the TRE for a given true model, which in this case is defined as the ratio of the TMSE when fitting the true model to that when using the AIC as a tool to select models. The corresponding results

Table 22.15 TRB (in %) (TRE) in estimation of cured proportions when fitting different models for a given true model

Fitted model	True model		
	Geometric	Poisson	Bernoulli
		Setting 1	
Geometric	0.837 (–)	36.486 (0.747)	66.838 (0.438)
Poisson	23.118 (0.692)	1.511 (–)	22.437 (0.913)
Bernoulli	1.583 (0.750)	1.628 (0.923)	1.002 (–)
		Setting 2	
Geometric	1.911 (–)	34.692 (0.672)	67.759 (0.333)
Poisson	23.569 (0.632)	1.330 (–)	23.727 (0.819)
Bernoulli	3.257 (0.744)	2.715 (0.880)	0.868 (–)
		Setting 3	
Geometric	0.829 (–)	17.298 (0.895)	33.484 (0.634)
Poisson	11.127 (0.832)	0.499 (–)	10.637 (0.964)
Bernoulli	4.628 (0.839)	2.337 (0.945)	0.576 (–)

Table 22.16 TRB (in %) and TRE when using the AIC as a model selection criterion between the three considered models

True model	TRB (%)	TRE
	Setting 1	
Geometric	5.810	0.900
Poisson	8.616	0.947
Bernoulli	3.912	1.007
	Setting 2	
Geometric	4.130	0.912
Poisson	8.623	0.922
Bernoulli	3.107	0.998
	Setting 3	
Geometric	4.375	0.933
Poisson	8.480	0.954
Bernoulli	3.474	0.987

are presented in Table 22.16. We note that for a given true model, the TRB values in some cases are less than that obtained when fitting the wrong model, as shown in Table 22.15. However, the TRE values for a given true model are always greater than those in Table 22.15. This in turn suggests that when we allow the AIC to select a model from a set of candidate models, although it selects the wrong model a certain proportion of times, the overall efficiency of the method is significantly greater than that obtained under wrong model specification. Furthermore, from Table 22.16, we note that the TRE values are all close to one, which further means that the loss of efficiency is negligible when using the AIC to select a working model.

22.9 Analysis of Cutaneous Melanoma Data

In this section, we illustrate an application of the proposed methodology to a real data set on cancer recurrence. The dataset, which is a part of an assay on cutaneous melanoma (a type of malignant cancer) for the evaluation of postoperative treatment performance with a high dose of interferon alpha-2b as a drug to prevent recurrence, includes 427 patients observed in the period 1991–1995 and followed until 1998. For 10 of the patients, the tumor thickness data (as covariate) were missing and hence were removed from our analysis. These data are taken from Ibrahim *et al.* (2001). The percentage of censored observations is 56%. The observed time refers to the time in years until the patient's death or the censoring time, and has mean = 3.18 and standard deviation = 1.69. In our application, we selected nodule category ($1 : n = 111; 2 : n = 137; 3 : n = 87; 4 : n = 82$) as the only covariate. As possible models, we considered the Bernoulli, Poisson, geometric and general COM-Poisson cure rate models and present the results of the data analysis separately for the exponential lifetime distribution with log-linear link function and the Weibull lifetime distribution with logistic link function.

22.9.1 *Exponential Lifetimes with Log-Linear Link Function*

For these data, the profile likelihood approach of estimating ϕ yields $\hat{\phi} = 0.05$. So, we first test the suitability of the geometric model for the number of causes M in (22.2). With the test procedure described in Section 22.7.2, we obtained $\Lambda = 1.366$ (p-value $= 0.121$), and we therefore accept the geometric model as a suitable model. In this way, the candidate models have the same number of parameters. Furthermore, since $\hat{\eta}_j = \exp(\hat{\beta}_0 + j\hat{\beta}_1) < 1$ for $j = 1, 2, 3, 4$, the distributions corresponding to the nodule categories are all legitimate.

For the Bernoulli, Poisson, and COM-Poisson (geometric) cure rate models, the maximized log-likelihood function values are -535.036, -533.333, and -533.269, respectively. Hence, the best fit is achieved with the COM-Poisson (geometric) cure rate model. We note that the maximized log-likelihood value for the Poisson and the geometric cure rate models are very close to each other. So, we also test for the suitability of the Poisson model for the number of causes M in (22.2). On following the test procedure described earlier in Section 22.7.2, we obtain $\Lambda = 1.494$ (p-value $= 0.222$), and so we accept the Poisson model as well. In order to explain why both the geometric and the Poisson models become acceptable for the data, we try to find an acceptable range of ϕ for the data. For this purpose, we consider testing for different values of ϕ and plot the values of ϕ against the corresponding test statistic Λ. The plot so obtained is presented in Figure 22.3. The test is rejected at $\alpha = 10\%$ level of significance if the value of the Λ statistic exceeds $\chi^2_{1,1-\alpha} = 2.71$. Now, for $\Lambda = 2.71$, the corresponding value of ϕ turns out to be 2.9 (from the plot). Hence, an admissible range of ϕ for the data becomes $(0, 2.9)$, which clearly explains why both geometric and Poisson models become acceptable for the data. On comparing the geometric and Poisson models, we adopt the geometric model as our working model based on the maximized log-likelihood values. Maximum likelihood estimates of the COM-Poisson (geometric) cure rate model parameters are presented in Table 22.17.

We finally deal with the estimation of the cure rate p_0. Estimates of the cure rate of patients stratified by nodule category are $\hat{p}_{0j} = 1 - \exp(\hat{\beta}_0 + j\hat{\beta}_1)$, $j = 1, 2, 3, 4$.

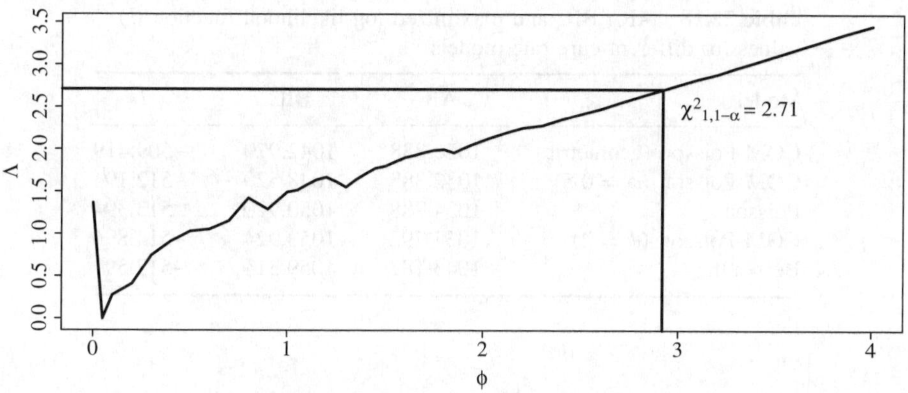

Figure 22.3 Values of ϕ against the corresponding Λ statistic

Table 22.17 Maximum likelihood estimates of the parameters of the COM-Poisson (geometric) cure rate model with exponential lifetimes

Parameter	Estimate	Standard error
β_0	−0.335	0.336
β_1	0.063	0.061
γ	0.035	0.042

By using the estimates in Table 22.17, we then obtain $\hat{p}_0 = 0.240$, 0.191, 0.138, and 0.082 corresponding to the four nodule categories (1–4).

22.9.2 *Weibull Lifetimes with Logistic Link Function*

In this case, we obtained $\hat{\phi} \approx 0$. So, we test the suitability of the geometric model for the number of causes M in (22.2). With the test procedure described earlier in Section 22.7.2, we obtained $\Lambda \approx 0$ (p-value=0.5), and we therefore adopt the geometric model as our working model. The p-values for testing the suitability of the Poisson and Bernoulli cure rate model came out to be 4.81×10^{-3} and 2.64×10^{-5}, respectively. These clearly show evidence against the Poisson and Bernoulli models. We note that the candidate models have the same number of parameters, and have reported the values of the maximized log-likelihood function, AIC and BIC for different cure rate models in Table 22.18. We observe that the AIC and BIC increases with the increase in COM-Poisson parameter ϕ, whereas the maximized log-likelihood value is monotonically decreasing with respect to ϕ. Hence, the best fit is achieved with the COM-Poisson (geometric) cure rate model, thus agreeing with the decision reached by the likelihood ratio test. Maximum likelihood estimates of the COM-Poisson (geometric) cure rate model parameters are presented in Table 22.19. We also considered testing for the exponential distribution ($H_0 : \gamma_1 = 1$)

Table 22.18 AIC, BIC and maximized log-likelihood function (l) values for different cure rate models

Model	AIC	BIC	l
COM-Poisson (geometric)	1026.838	1042.970	−509.419
COM-Poisson ($\phi = 0.5$)	1032.388	1048.520	−512.194
Poisson	1034.788	1050.920	−513.394
COM-Poisson ($\phi = 2$)	1037.792	1053.924	−514.896
Bernoulli	1043.182	1059.314	−517.591

Table 22.19 Maximum likelihood estimates of the parameters of the COM-Poisson (geometric) cure rate model with Weibull lifetimes

Parameter	Estimate	Standard error
β_0	−1.181	0.244
β_1	0.498	0.089
γ_1	0.528	0.034
γ_2	0.340	0.029

Table 22.20 Estimates, standard errors, and 95% confidence intervals (CI) of cure rate of patients stratified by nodule category

Nodule category	Cure rate	Estimate	Standard error	95% CI
1	p_{01}	0.664	0.039	(0.588, 0.740)
2	p_{02}	0.546	0.032	(0.483, 0.609)
3	p_{03}	0.422	0.033	(0.357, 0.487)
4	p_{04}	0.308	0.040	(0.230, 0.386)

in (22.9). The likelihood ratio test statistic in this case turned out to be 48.428 (p-value $= 3.4 \times 10^{-12}$), which clearly gives very strong evidence against the exponential distribution in (22.9).

Finally, we deal with the estimation of the cure rate p_0. Estimates of the cure rate of patients stratified by nodule category are $\hat{p}_{0j} = \frac{1}{1+\exp(\hat{\beta}_0+j\hat{\beta}_1)}$, $j = 1, 2, 3, 4$. We also calculated the standard errors and the 95% confidence intervals of the cure rate of patients corresponding to the four nodule categories $(1-4)$, and these are presented in Table 22.20. Observe that these cure rates are very different from those obtained under the assumption of exponential lifetimes which, as seen above, is not a reasonable assumption for these data.

22.10 Conclusion

In this chapter, we have discussed the implementation of the EM algorithm for the estimation of the parameters of the COM-Poisson cure rate model and its special cases when the data are right censored. Two distributions have been considered for the time-to-event and two different link functions have been used to relate the covariates to model parameters. An extensive Monte Carlo simulation study has been performed to evaluate the performance of the developed estimation procedures. The proposed methodology is also applied to analyze real data on cancer recurrence. The simulation study shows that the proposed methods work very well and that the estimates of the model parameters are close to their true values. However, the profile likelihood approach of estimating the parameter ϕ results in large RMSE in the estimate of ϕ. The profile likelihood approach also results in under-estimating the standard errors of the estimates due to which the coverage probabilities of the model parameters fall below the nominal levels. In this regard, a better method of estimation of ϕ would be good to develop.

It is also possible to consider other distributions, such as the lognormal and gamma, as the lifetime distribution and develop the associated likelihood inference based on the EM algorithm. However, one cannot perform a formal statistical test to determine which lifetime distribution among these provides the best fit to the data since they are not nested. Hence, it will be of interest to consider a wider class of lifetime distributions, such as the generalized gamma distribution considered by Balakrishnan and Peng (2006) in their work on frailty model. It would be useful to develop the EM algorithm for the determination of the MLEs of the parameters of different cure rate models based on right censored data, and also perform model discrimination within this general class of distributions to select a suitable lifetime distribution that provides the best fit to the data.

22.A1 Appendix A1: E-Step and M-Step Formulas for Exponential Lifetimes

Bernoulli cure rate model: The $Q(\boldsymbol{\theta}^*, \boldsymbol{\pi}^{(k)})$ function can be expressed as

$$Q(\boldsymbol{\theta}^*, \boldsymbol{\pi}^{(k)}) = Q_1(\boldsymbol{\beta}, \boldsymbol{\pi}^{(k)}) + Q_2(\gamma, \boldsymbol{\pi}^{(k)}),$$

where

$$Q_1(\boldsymbol{\beta}, \boldsymbol{\pi}^{(k)}) = \sum_{I_1} \boldsymbol{x}_i'\boldsymbol{\beta} + \sum_{I_0} \pi_i^{(k)} \boldsymbol{x}_i'\boldsymbol{\beta} - \sum_{I^*} \log(1 + \exp(\boldsymbol{x}_i'\boldsymbol{\beta})),$$

$$Q_2(\gamma, \boldsymbol{\pi}^{(k)}) = n_1 \log \gamma - \gamma \left(\sum_{I_1} t_i + \sum_{I_0} \pi_i^{(k)} t_i \right),$$

$$\pi_i^{(k)} = \frac{\exp(\boldsymbol{x}_i'\boldsymbol{\beta}^{(k)} - \gamma^{(k)} t_i)}{1 + \exp(\boldsymbol{x}_i'\boldsymbol{\beta}^{(k)} - \gamma^{(k)} t_i)}$$

for censored cases only. We see that the function $Q(\boldsymbol{\theta}^*, \boldsymbol{\pi}^{(k)})$ can be separated into two parts, one containing the parameter $\boldsymbol{\beta}$ and the other containing the parameter γ. Hence, if $\boldsymbol{\beta}^{(k)}$ and $\gamma^{(k)}$ are the estimates of $\boldsymbol{\beta}$ and γ, respectively, at the k-th iteration step, then $\boldsymbol{\beta}^{(k+1)}$ can be obtained by maximizing $Q_1(\boldsymbol{\beta}, \boldsymbol{\pi}^{(k)})$ with respect to $\boldsymbol{\beta}$ and $\gamma^{(k+1)}$ can be obtained by maximizing $Q_2(\gamma, \boldsymbol{\pi}^{(k)})$ with respect to γ, separately. The first- and second-order derivatives of the function $Q_1(\boldsymbol{\beta}, \boldsymbol{\pi}^{(k)})$ with respect to $\boldsymbol{\beta}$ are as follows:

$$\frac{\partial Q_1}{\partial \beta_l} = \sum_{I_1} x_{il} + \sum_{I_0} \pi_i^{(k)} x_{il} - \sum_{I^*} x_{il} \frac{\exp(\boldsymbol{x}_i'\boldsymbol{\beta})}{1+\exp(\boldsymbol{x}_i'\boldsymbol{\beta})},$$

$$\frac{\partial^2 Q_1}{\partial \beta_l \partial \beta_{l'}} = -\sum_{I^*} x_{il} x_{il'} \frac{\exp(\boldsymbol{x}_i'\boldsymbol{\beta})}{(1+\exp(\boldsymbol{x}_i'\boldsymbol{\beta}))^2}$$

for $l, l' = 0, 1, \ldots, k$, and $x_{i0} = 1 \forall i = 1, 2, \ldots, n$. Using the above derivatives, we can then apply the one-step Newton-Raphson method for finding $\boldsymbol{\beta}^{(k+1)}$. To find $\gamma^{(k+1)}$, we equate the first-order derivative of the function $Q_2(\gamma)$ with respect to γ to zero, which is given by

$$\frac{n_1}{\gamma} - \sum_{I_1} t_i - \sum_{I_0} \pi_i^{(k)} t_i = 0. \tag{22.10}$$

We solve (22.10) for γ to get $\gamma^{(k+1)}$ as

$$\gamma^{(k+1)} = \frac{n_1}{\sum_{I_1} t_i - \sum_{I_0} \pi_i^{(k)} t_i}.$$

We also note that

$$\frac{\partial^2 Q_2}{\partial \gamma^2} = -\frac{n_1}{\gamma^2} < 0.$$

Poisson cure rate model: The $Q(\boldsymbol{\theta}^*, \boldsymbol{\pi}^{(k)})$ function can be expressed as

$$Q(\boldsymbol{\theta}^*, \boldsymbol{\pi}^{(k)}) = n_1 \log \gamma + \sum_{I_1} \{\boldsymbol{x}_i'\boldsymbol{\beta} - \gamma t_i + \exp(\boldsymbol{x}_i'\boldsymbol{\beta} - \gamma t_i)\}$$
$$+ \sum_{I_0} \pi_i^{(k)} \log(\exp(\exp(\boldsymbol{x}_i'\boldsymbol{\beta} - \gamma t_i)) - 1) - \sum_{I^*} \exp(\boldsymbol{x}_i'\boldsymbol{\beta}),$$

where

$$\pi_i^{(k)} = \frac{\exp(\exp(\boldsymbol{x}_i'\boldsymbol{\beta}^{(k)} - \gamma^{(k)} t_i)) - 1}{\exp(\exp(\boldsymbol{x}_i'\boldsymbol{\beta}^{(k)} - \gamma^{(k)} t_i))}$$

for censored cases only. The required first- and second-order derivatives of the function $Q(\boldsymbol{\theta}^*, \boldsymbol{\pi}^{(k)})$ with respect to $\boldsymbol{\beta}$ and γ to carry out the maximization step are as follows:

$$\frac{\partial Q}{\partial \beta_l} = \sum_{I_1} x_{il}\{1 + \exp(\boldsymbol{x}_i'\boldsymbol{\beta} - \gamma t_i)\} + \sum_{I_0} \pi_i^{(k)} x_{il} A_i(\boldsymbol{\beta}, \gamma) - \sum_{I^*} x_{il} \exp(\boldsymbol{x}_i'\boldsymbol{\beta}),$$

$$\frac{\partial Q}{\partial \gamma} = \frac{n_1}{\gamma} - \sum_{I_1} t_i\{1 + \exp(\boldsymbol{x}_i'\boldsymbol{\beta} - \gamma t_i)\} - \sum_{I_0} \pi_i^{(k)} t_i A_i(\boldsymbol{\beta}, \gamma),$$

$$\frac{\partial^2 Q}{\partial \beta_l \partial \beta_{l'}} = \sum_{I_1} x_{il} x_{il'} \exp(\boldsymbol{x}_i'\boldsymbol{\beta} - \gamma t_i) + \sum_{I_0} \pi_i^{(k)} x_{il} x_{il'} A_i(\boldsymbol{\beta}, \gamma) B_i(\boldsymbol{\beta}, \gamma)$$

$$- \sum_{I^*} x_{il} x_{il'} \exp(\boldsymbol{x}_i' \boldsymbol{\beta}),$$

$$\frac{\partial^2 Q}{\partial \gamma^2} = -\frac{n_1}{\gamma^2} + \sum_{I_1} t_i^2 \exp(\boldsymbol{x}_i' \boldsymbol{\beta} - \gamma t_i) + \sum_{I_0} \pi_i^{(k)} t_i^2 A_i(\boldsymbol{\beta}, \gamma) B_i(\boldsymbol{\beta}, \gamma),$$

$$\frac{\partial^2 Q}{\partial \beta_l \partial \gamma} = -\sum_{I_1} x_{il} t_i \exp(\boldsymbol{x}_i' \boldsymbol{\beta} - \gamma t_i) - \sum_{I_0} \pi_i^{(k)} x_{il} t_i A_i(\boldsymbol{\beta}, \gamma) B_i(\boldsymbol{\beta}, \gamma)$$

for $l, l' = 0, 1, \ldots, k$, and $x_{i0} = 1 \forall i = 1, 2, \ldots, n$, where

$$A_i(\boldsymbol{\beta}, \gamma) = \frac{\exp(\exp(\boldsymbol{x}_i' \boldsymbol{\beta} - \gamma t_i) + \boldsymbol{x}_i' \boldsymbol{\beta} - \gamma t_i)}{\exp(\exp(\boldsymbol{x}_i' \boldsymbol{\beta} - \gamma t_i)) - 1},$$

$$B_i(\boldsymbol{\beta}, \gamma) = \frac{\exp(\exp(\boldsymbol{x}_i' \boldsymbol{\beta} - \gamma t_i)) - \exp(\boldsymbol{x}_i' \boldsymbol{\beta} - \gamma t_i) - 1}{\exp(\exp(\boldsymbol{x}_i' \boldsymbol{\beta} - \gamma t_i)) - 1}$$

for $i \in I_0$.

Geometric cure rate model: The $Q(\boldsymbol{\theta}^*, \boldsymbol{\pi}^{(k)})$ function can be expressed as

$$Q(\boldsymbol{\theta}^*, \boldsymbol{\pi}^{(k)}) = n_1 \log \gamma + \sum_{I_1} \{\boldsymbol{x}_i' \boldsymbol{\beta} - \gamma t_i - 2 \log(1 - \exp(\boldsymbol{x}_i' \boldsymbol{\beta} - \gamma t_i))\}$$

$$+ \sum_{I_0} \pi_i^{(k)} \{\boldsymbol{x}_i' \boldsymbol{\beta} - \gamma t_i - \log(1 - \exp(\boldsymbol{x}_i' \boldsymbol{\beta} - \gamma t_i))\} + \sum_{I^*} \log(1 - \exp(\boldsymbol{x}_i' \boldsymbol{\beta})),$$

where

$$\pi_i^{(k)} = \exp(\boldsymbol{x}_i' \boldsymbol{\beta}^{(k)} - \gamma^{(k)} t_i)$$

for censored cases only. The required first- and second-order derivatives of the function $Q(\boldsymbol{\theta}^*, \boldsymbol{\pi}^{(k)})$ with respect to $\boldsymbol{\beta}$ and γ to carry out the maximization step are as follows:

$$\frac{\partial Q}{\partial \beta_l} = \sum_{I_1} x_{il} \frac{1 + \exp(\boldsymbol{x}_i' \boldsymbol{\beta} - \gamma t_i)}{1 - \exp(\boldsymbol{x}_i' \boldsymbol{\beta} - \gamma t_i)} + \sum_{I_0} \pi_i^{(k)} x_{il} \frac{1}{1 - \exp(\boldsymbol{x}_i' \boldsymbol{\beta} - \gamma t_i)}$$

$$- \sum_{I^*} x_{il} \frac{\exp(\boldsymbol{x}_i' \boldsymbol{\beta})}{1 - \exp(\boldsymbol{x}_i' \boldsymbol{\beta})},$$

$$\frac{\partial Q}{\partial \gamma} = \frac{n_1}{\gamma} - \sum_{I_1} t_i \frac{1 + \exp(\boldsymbol{x}_i' \boldsymbol{\beta} - \gamma t_i)}{1 - \exp(\boldsymbol{x}_i' \boldsymbol{\beta} - \gamma t_i)} - \sum_{I_0} \pi_i^{(k)} t_i \frac{1}{1 - \exp(\boldsymbol{x}_i' \boldsymbol{\beta} - \gamma t_i)},$$

$$\frac{\partial^2 Q}{\partial \beta_l \partial \beta_{l'}} = 2 \sum_{I_1} x_{il} x_{il'} \frac{\exp(\boldsymbol{x}_i' \boldsymbol{\beta} - \gamma t_i)}{(1 - \exp(\boldsymbol{x}_i' \boldsymbol{\beta} - \gamma t_i))^2} + \sum_{I_0} \pi_i^{(k)} x_{il} x_{il'} \frac{\exp(\boldsymbol{x}_i' \boldsymbol{\beta} - \gamma t_i)}{(1 - \exp(\boldsymbol{x}_i' \boldsymbol{\beta} - \gamma t_i))^2}$$

$$- \sum_{I^*} x_{il} x_{il'} \frac{\exp(\boldsymbol{x}_i' \boldsymbol{\beta})}{(1 - \exp(\boldsymbol{x}_i' \boldsymbol{\beta}))^2},$$

$$\frac{\partial^2 Q}{\partial \gamma^2} = -\frac{n_1}{\gamma^2} + 2\sum_{I_1} t_i^2 \frac{\exp(\boldsymbol{x}_i'\boldsymbol{\beta} - \gamma t_i)}{(1-\exp(\boldsymbol{x}_i'\boldsymbol{\beta} - \gamma t_i))^2} + \sum_{I_0} \pi_i^{(k)} t_i^2 \frac{\exp(\boldsymbol{x}_i'\boldsymbol{\beta} - \gamma t_i)}{(1-\exp(\boldsymbol{x}_i'\boldsymbol{\beta} - \gamma t_i))^2},$$

$$\frac{\partial^2 Q}{\partial \beta_l \partial \gamma} = -2\sum_{I_1} x_{il} t_i \frac{\exp(\boldsymbol{x}_i'\boldsymbol{\beta} - \gamma t_i)}{(1-\exp(\boldsymbol{x}_i'\boldsymbol{\beta} - \gamma t_i))^2} - \sum_{I_0} \pi_i^{(k)} x_{il} t_i \frac{\exp(\boldsymbol{x}_i'\boldsymbol{\beta} - \gamma t_i)}{(1-\exp(\boldsymbol{x}_i'\boldsymbol{\beta} - \gamma t_i))^2}$$

for $l, l' = 0, 1, \ldots, k$, and $x_{i0} = 1 \forall i = 1, 2, \ldots, n$.

General case of COM-Poisson cure rate model: The $Q(\boldsymbol{\theta}^*, \boldsymbol{\pi}^{(k)})$ function, for a fixed value of ϕ, can be expressed as

$$Q(\boldsymbol{\theta}^*, \boldsymbol{\pi}^{(k)}) = n_1 \log \gamma + \sum_{I_1} \log z_2(\boldsymbol{\beta}, \gamma, \phi; \boldsymbol{x}_i, t_i) + \sum_{I_0} \pi_i^{(k)} \log z_1(\boldsymbol{\beta}, \gamma, \phi; \boldsymbol{x}_i, t_i)$$
$$- \sum_{I^*} \log z(\boldsymbol{\beta}, \phi; \boldsymbol{x}_i),$$

where

$$z(\boldsymbol{\beta}, \phi; \boldsymbol{x}_i) = \sum_{j=0}^{\infty} \frac{(\exp(\boldsymbol{x}_i'\boldsymbol{\beta}))^j}{(j!)^\phi},$$

$$z_1(\boldsymbol{\beta}, \gamma, \phi; \boldsymbol{x}_i, t_i) = \sum_{j=1}^{\infty} \frac{(\exp(\boldsymbol{x}_i'\boldsymbol{\beta} - \gamma t_i))^j}{(j!)^\phi},$$

$$z_2(\boldsymbol{\beta}, \gamma, \phi; \boldsymbol{x}_i, t_i) = \sum_{j=1}^{\infty} \frac{j(\exp(\boldsymbol{x}_i'\boldsymbol{\beta} - \gamma t_i))^j}{(j!)^\phi},$$

$$\pi_i^{(k)} = \frac{z_1(\boldsymbol{\beta}^{(k)}, \gamma^{(k)}, \phi; \boldsymbol{x}_i, t_i)}{1 + z_1(\boldsymbol{\beta}^{(k)}, \gamma^{(k)}, \phi; \boldsymbol{x}_i, t_i)}$$

for censored cases only. The required first- and second-order derivatives of the function $Q(\boldsymbol{\theta}^*, \boldsymbol{\pi}^{(k)})$ with respect to $\boldsymbol{\beta}$ and γ to carry out the maximization step are as follows:

$$\frac{\partial Q}{\partial \beta_l} = \sum_{I_1} x_{il} \frac{z_{21}(\boldsymbol{\beta}, \gamma, \phi; \boldsymbol{x}_i, t_i)}{z_2(\boldsymbol{\beta}, \gamma, \phi; \boldsymbol{x}_i, t_i)} + \sum_{I_0} \pi_i^{(k)} x_{il} \frac{z_2(\boldsymbol{\beta}, \gamma, \phi; \boldsymbol{x}_i, t_i)}{z_1(\boldsymbol{\beta}, \gamma, \phi; \boldsymbol{x}_i, t_i)}$$
$$- \sum_{I^*} x_{il} \frac{z_{01}(\boldsymbol{\beta}, \phi; \boldsymbol{x}_i)}{z(\boldsymbol{\beta}, \phi; \boldsymbol{x}_i)},$$

$$\frac{\partial Q}{\partial \gamma} = \frac{n_1}{\gamma} - \sum_{I_1} t_i \frac{z_{21}(\boldsymbol{\beta}, \gamma, \phi; \boldsymbol{x}_i, t_i)}{z_2(\boldsymbol{\beta}, \gamma, \phi; \boldsymbol{x}_i, t_i)} - \sum_{I_0} \pi_i^{(k)} t_i \frac{z_2(\boldsymbol{\beta}, \gamma, \phi; \boldsymbol{x}_i, t_i)}{z_1(\boldsymbol{\beta}, \gamma, \phi; \boldsymbol{x}_i, t_i)},$$

$$\frac{\partial^2 Q}{\partial \beta_l \partial \beta_{l'}} = \sum_{I_1} x_{il} x_{il'} \frac{z_2(\boldsymbol{\beta}, \gamma, \phi; \boldsymbol{x}_i, t_i) z_{22}(\boldsymbol{\beta}, \gamma, \phi; \boldsymbol{x}_i, t_i) - z_{21}^2(\boldsymbol{\beta}, \gamma, \phi; \boldsymbol{x}_i, t_i)}{z_2^2(\boldsymbol{\beta}, \gamma, \phi; \boldsymbol{x}_i, t_i)}$$
$$+ \sum_{I_0} \pi_i^{(k)} x_{il} x_{il'} \frac{z_1(\boldsymbol{\beta}, \gamma, \phi; \boldsymbol{x}_i, t_i) z_{21}(\boldsymbol{\beta}, \gamma, \phi; \boldsymbol{x}_i, t_i) - z_2^2(\boldsymbol{\beta}, \gamma, \phi; \boldsymbol{x}_i, t_i)}{z_1^2(\boldsymbol{\beta}, \gamma, \phi; \boldsymbol{x}_i, t_i)}$$

$$
\begin{aligned}
&-\sum_{I^*} x_{il}x_{il'}\frac{z(\boldsymbol{\beta},\phi;\boldsymbol{x}_i)z_{02}(\boldsymbol{\beta},\phi;\boldsymbol{x}_i)-z_{01}^2(\boldsymbol{\beta},\phi;\boldsymbol{x}_i)}{z^2(\boldsymbol{\beta},\phi;\boldsymbol{x}_i)},\\
\frac{\partial^2 Q}{\partial\gamma^2} &= -\frac{n_1}{\gamma^2}+\sum_{I_1} t_i^2\frac{z_2(\boldsymbol{\beta},\gamma,\phi;\boldsymbol{x}_i,t_i)z_{22}(\boldsymbol{\beta},\gamma,\phi;\boldsymbol{x}_i,t_i)-z_{21}^2(\boldsymbol{\beta},\gamma,\phi;\boldsymbol{x}_i,t_i)}{z_2^2(\boldsymbol{\beta},\gamma,\phi;\boldsymbol{x}_i,t_i)}\\
&\quad+\sum_{I_0}\pi_i^{(k)}t_i^2\frac{z_1(\boldsymbol{\beta},\gamma,\phi;\boldsymbol{x}_i,t_i)z_{21}(\boldsymbol{\beta},\gamma,\phi;\boldsymbol{x}_i,t_i)-z_2^2(\boldsymbol{\beta},\gamma,\phi;\boldsymbol{x}_i,t_i)}{z_1^2(\boldsymbol{\beta},\gamma,\phi;\boldsymbol{x}_i,t_i)},\\
\frac{\partial^2 Q}{\partial\beta_l\partial\gamma} &= -\sum_{I_1} x_{il}t_i\frac{z_2(\boldsymbol{\beta},\gamma,\phi;\boldsymbol{x}_i,t_i)z_{22}(\boldsymbol{\beta},\gamma,\phi;\boldsymbol{x}_i,t_i)-z_{21}^2(\boldsymbol{\beta},\gamma,\phi;\boldsymbol{x}_i,t_i)}{z_2^2(\boldsymbol{\beta},\gamma,\phi;\boldsymbol{x}_i,t_i)}\\
&\quad-\sum_{I_0}\pi_i^{(k)}x_{il}t_i\frac{z_1(\boldsymbol{\beta},\gamma,\phi;\boldsymbol{x}_i,t_i)z_{21}(\boldsymbol{\beta},\gamma,\phi;\boldsymbol{x}_i,t_i)-z_2^2(\boldsymbol{\beta},\gamma,\phi;\boldsymbol{x}_i,t_i)}{z_1^2(\boldsymbol{\beta},\gamma,\phi;\boldsymbol{x}_i,t_i)}
\end{aligned}
$$

for $l, l' = 0, 1, \ldots, k$, and $x_{i0} = 1 \forall i = 1, 2, \ldots, n$, where

$$
\begin{aligned}
z_{21}(\boldsymbol{\beta},\gamma,\phi;\boldsymbol{x}_i,t_i) &= \sum_{j=1}^{\infty}\frac{j^2(\exp(\boldsymbol{x}_i'\boldsymbol{\beta}-\gamma t_i))^j}{(j!)^\phi},\\
z_{22}(\boldsymbol{\beta},\gamma,\phi;\boldsymbol{x}_i,t_i) &= \sum_{j=1}^{\infty}\frac{j^3(\exp(\boldsymbol{x}_i'\boldsymbol{\beta}-\gamma t_i))^j}{(j!)^\phi},\\
z_{01}(\boldsymbol{\beta},\phi;\boldsymbol{x}_i) &= \sum_{j=1}^{\infty} j\frac{(\exp(\boldsymbol{x}_i'\boldsymbol{\beta}))^j}{(j!)^\phi},\\
z_{02}(\boldsymbol{\beta},\phi;\boldsymbol{x}_i) &= \sum_{j=1}^{\infty} j^2\frac{(\exp(\boldsymbol{x}_i'\boldsymbol{\beta}))^j}{(j!)^\phi}.
\end{aligned}
$$

The quantities $z_i, z_{1i}, z_{2i}, z_{21i}, z_{22i}, z_{01i}$ and z_{02i} can all be computed by truncating the numerical series for each i. Let S_n denote the n-th order partial sum of the infinite series. Then, the infinite series can be computed with n terms if $S_{n+1} - S_n \le \epsilon$, where ϵ is a pre-fixed tolerance value set. In our case, we chose $\epsilon = 0.01$ for computational ease and efficiency.

22.A2 Appendix A2: E-Step and M-Step Formulas for Weibull Lifetimes

General case of COM-Poisson cure rate model: The $Q(\boldsymbol{\theta}^*, \boldsymbol{\pi}^{(k)})$ function, for a fixed value of ϕ, can be expressed as

$$
Q(\boldsymbol{\theta}^*,\boldsymbol{\pi}^{(k)}) = -n_1\log\gamma_1+\frac{n_1}{\gamma_1}\log\gamma_2+\left(\frac{1}{\gamma_1}-1\right)\sum_{I_1}\log t_i+\sum_{I_1}\log z_{2i}^*
$$

$$+\sum_{I_0} \pi_i^{(k)} \log z_{1i}^* - \sum_{I^*} \log\{1+\exp(\boldsymbol{x}_i'\boldsymbol{\beta})\},$$

where

$$z_1^* = z_1^*(\boldsymbol{\theta};\boldsymbol{x},t) = \sum_{j=1}^{\infty} \frac{\{H_\phi^{-1}(1+\exp(\boldsymbol{x}'\boldsymbol{\beta}))\exp(-(\gamma_2 t)^{1/\gamma_1})\}^j}{(j!)^\phi},$$

$$z_2^* = z_2^*(\boldsymbol{\theta};\boldsymbol{x},t) = \sum_{j=1}^{\infty} \frac{j\{H_\phi^{-1}(1+\exp(\boldsymbol{x}'\boldsymbol{\beta}))\exp(-(\gamma_2 t)^{1/\gamma_1})\}^j}{(j!)^\phi},$$

$$\pi_i^{(k)} = \left.\frac{z_1^*(\boldsymbol{\theta};\boldsymbol{x}_i,t_i)}{1+z_1^*(\boldsymbol{\theta};\boldsymbol{x}_i,t_i)}\right|_{\boldsymbol{\theta}^*=\boldsymbol{\theta}^{*(k)}}$$

for the i-th censored observation. The required first- and second-order derivatives of the $Q(\boldsymbol{\theta}^*, \boldsymbol{\pi}^{(k)})$ function with respect to $\boldsymbol{\beta}$ and $\boldsymbol{\gamma}$ are as follows:

$$\frac{\partial Q}{\partial \beta_l} = \sum_{I_1} x_{il} \frac{\exp(\boldsymbol{x}_i'\boldsymbol{\beta}) z_{21i}^*}{z_{2i}^* z_{01i}^*} + \sum_{I_0} \pi_i^{(k)} x_{il} \frac{\exp(\boldsymbol{x}_i'\boldsymbol{\beta}) z_{2i}^*}{z_{1i}^* z_{01i}^*} - \sum_{I^*} x_{il} \frac{\exp(\boldsymbol{x}_i'\boldsymbol{\beta})}{1+\exp(\boldsymbol{x}_i'\boldsymbol{\beta})},$$

$$\frac{\partial Q}{\partial \gamma_1} = -\frac{n_1}{\gamma_1} - \frac{n_1}{\gamma_1^2}\log\gamma_2 - \frac{1}{\gamma_1^2}\sum_{I_1}\log t_i + \frac{1}{\gamma_1^2}\sum_{I_1}\frac{G_{Ci}(\boldsymbol{\gamma}) z_{21i}^*}{z_{2i}^*}$$
$$+\frac{1}{\gamma_1^2}\sum_{I_0}\pi_i^{(k)}\frac{G_{Ci}(\boldsymbol{\gamma}) z_{2i}^*}{z_{1i}^*},$$

$$\frac{\partial Q}{\partial \gamma_2} = \frac{n_1}{\gamma_1\gamma_2} - \frac{1}{\gamma_1\gamma_2}\sum_{I_1}\frac{(\gamma_2 t_i)^{1/\gamma_1} z_{21i}^*}{z_{2i}^*} - \frac{1}{\gamma_1\gamma_2}\sum_{I_0}\pi_i^{(k)}\frac{(\gamma_2 t_i)^{1/\gamma_1} z_{2i}^*}{z_{1i}^*},$$

$$\frac{\partial^2 Q}{\partial\beta_l\partial\beta_{l'}} = \sum_{I_1} x_{il}x_{il'}\frac{\exp(\boldsymbol{x}_i'\boldsymbol{\beta})}{(z_{2i}^* z_{01i}^*)^2}\left[z_{2i}^*\left\{\exp(\boldsymbol{x}_i'\boldsymbol{\beta}) z_{22i}^* + z_{01i}^* z_{21i}^*\right\}\right.$$
$$\left. - \exp(\boldsymbol{x}_i'\boldsymbol{\beta}) z_{21i}^* \left\{\frac{z_{2i}^* z_{02i}^*}{z_{01i}^*} + z_{21i}^*\right\}\right]$$
$$+\sum_{I_0}\pi_i^{(k)} x_{il}x_{il'}\frac{\exp(\boldsymbol{x}_i'\boldsymbol{\beta})}{(z_{1i}^* z_{01i}^*)^2}\left[z_{1i}^*\left\{\exp(\boldsymbol{x}_i'\boldsymbol{\beta}) z_{21i}^* + z_{01i}^* z_{2i}^*\right\}\right.$$
$$\left. - \exp(\boldsymbol{x}_i'\boldsymbol{\beta}) z_{2i}^* \left\{\frac{z_{1i}^* z_{02i}^*}{z_{01i}^*} + z_{2i}^*\right\}\right] - \sum_{I^*} x_{il}x_{il'}\frac{\exp(\boldsymbol{x}_i'\boldsymbol{\beta})}{(1+\exp(\boldsymbol{x}_i'\boldsymbol{\beta}))^2},$$

$$\frac{\partial^2 Q}{\partial\beta_l\partial\gamma_1} = \frac{1}{\gamma_1^2}\sum_{I_1} x_{il}\frac{G_{Ci}(\boldsymbol{\gamma})\exp(\boldsymbol{x}_i'\boldsymbol{\beta})}{z_{01i}^* {z_{2i}^*}^2}\{z_{2i}^* z_{22i}^* - {z_{21i}^*}^2\}$$
$$+\frac{1}{\gamma_1^2}\sum_{I_0}\pi_i^{(k)} x_{il}\frac{G_{Ci}(\boldsymbol{\gamma})\exp(\boldsymbol{x}_i'\boldsymbol{\beta})}{z_{01i}^* {z_{1i}^*}^2}\{z_{1i}^* z_{21i}^* - {z_{2i}^*}^2\},$$

$$\frac{\partial^2 Q}{\partial \beta_l \partial \gamma_2} = -\frac{1}{\gamma_1 \gamma_2} \sum_{I_1} x_{il} \frac{(\gamma_2 t_i)^{1/\gamma_1} \exp(\boldsymbol{x}'_i \boldsymbol{\beta})}{z^*_{01i} {z^*_{2i}}^2} \{z^*_{2i} z^*_{22i} - {z^*_{21i}}^2\}$$

$$- \frac{1}{\gamma_1 \gamma_2} \sum_{I_0} \pi_i^{(k)} x_{il} \frac{(\gamma_2 t_i)^{1/\gamma_1} \exp(\boldsymbol{x}'_i \boldsymbol{\beta})}{z^*_{01i} {z^*_{1i}}^2} \{z^*_{1i} z^*_{21i} - {z^*_{2i}}^2\},$$

$$\frac{\partial^2 Q}{\partial {\gamma_1}^2} = \frac{n_1}{{\gamma_1}^2} + \frac{2n_1}{{\gamma_1}^3} \log \gamma_2 + \frac{2}{{\gamma_1}^3} \sum_{I_1} \log t_i$$

$$+ \frac{1}{\gamma_1^4} \sum_{I_1} \frac{G_{Ci}(\boldsymbol{\gamma})}{{z^*_{2i}}^2} [\log(\gamma_2 t_i) z^*_{2i} \{(\gamma_2 t_i)^{1/\gamma_1} z^*_{22i} - z^*_{21i}\}$$

$$- z^*_{21i} \{G_{Ci}(\boldsymbol{\gamma}) z^*_{21i} + 2\gamma_1 z^*_{2i}\}]$$

$$+ \frac{1}{\gamma_1^4} \sum_{I_0} \pi_i^{(k)} \frac{G_{Ci}(\boldsymbol{\gamma})}{{z^*_{1i}}^2} [\log(\gamma_2 t_i) z^*_{1i} \{(\gamma_2 t_i)^{1/\gamma_1} z^*_{21i} - z^*_{2i}\}$$

$$- z^*_{2i} \{G_{Ci}(\boldsymbol{\gamma}) z^*_{2i} + 2\gamma_1 z^*_{1i}\}],$$

$$\frac{\partial^2 Q}{\partial {\gamma_2}^2} = -\frac{n_1}{\gamma_1 {\gamma_2}^2} + \frac{1}{\gamma_1 {\gamma_2}^2} \sum_{I_1} \frac{(\gamma_2 t_i)^{1/\gamma_1}}{{z^*_{2i}}^2}$$

$$\times \left[z^*_{2i} \left\{ \frac{1}{\gamma_1} (\gamma_2 t_i)^{1/\gamma_1} z^*_{22i} - \left(\frac{1}{\gamma_1} - 1 \right) z^*_{21i} \right\} - \frac{1}{\gamma_1} (\gamma_2 t_i)^{1/\gamma_1} {z^*_{21i}}^2 \right]$$

$$+ \frac{1}{\gamma_1 {\gamma_2}^2} \sum_{I_0} \pi_i^{(k)} \frac{(\gamma_2 t_i)^{1/\gamma_1}}{{z^*_{1i}}^2} \left[z^*_{1i} \left\{ \frac{1}{\gamma_1} (\gamma_2 t_i)^{1/\gamma_1} z^*_{21i} - \left(\frac{1}{\gamma_1} - 1 \right) z^*_{2i} \right\} \right.$$

$$\left. - \frac{1}{\gamma_1} (\gamma_2 t_i)^{1/\gamma_1} {z^*_{2i}}^2 \right],$$

$$\frac{\partial^2 Q}{\partial \gamma_1 \partial \gamma_2} = -\frac{n_1}{\gamma_2 {\gamma_1}^2} - \frac{1}{\gamma_2 {\gamma_1}^2} \sum_{I_1} \frac{(\gamma_2 t_i)^{1/\gamma_1}}{{z^*_{2i}}^2}$$

$$\times \left[z^*_{2i} \left\{ \frac{1}{\gamma_1} G_{Ci}(\boldsymbol{\gamma}) z^*_{22i} - z^*_{21i} - \frac{1}{\gamma_1} \log(\gamma_2 t_i) z^*_{21i} \right\} - \frac{1}{\gamma_1} G_{Ci}(\boldsymbol{\gamma}) {z^*_{21i}}^2 \right]$$

$$- \frac{1}{\gamma_2 {\gamma_1}^2} \sum_{I_0} \pi_i^{(k)} \frac{(\gamma_2 t_i)^{1/\gamma_1}}{{z^*_{1i}}^2} \left[z^*_{1i} \left\{ \frac{1}{\gamma_1} G_{Ci}(\boldsymbol{\gamma}) z^*_{21i} - z^*_{2i} - \frac{1}{\gamma_1} \log(\gamma_2 t_i) z^*_{2i} \right\} \right.$$

$$\left. - \frac{1}{\gamma_1} G_{Ci}(\boldsymbol{\gamma}) {z^*_{2i}}^2 \right]$$

for $l, l' = 0, 1, \ldots, k$, and $x_{i0} = 1 \forall i = 1, 2, \ldots, n$, where

$$z^*_{21} = z^*_{21}(\boldsymbol{\theta}; \boldsymbol{x}, t) = \sum_{j=1}^{\infty} \frac{j^2 \{H_\phi^{-1}(1 + \exp(\boldsymbol{x}'\boldsymbol{\beta})) \exp(-(\gamma_2 t)^{1/\gamma_1})\}^j}{(j!)^\phi},$$

$$z_{22}^* = z_{22}^*(\boldsymbol{\theta}; \boldsymbol{x}, t) = \sum_{j=1}^{\infty} \frac{j^3\{H_\phi^{-1}(1+\exp(\boldsymbol{x}'\boldsymbol{\beta}))\exp(-(\gamma_2 t)^{1/\gamma_1})\}^j}{(j!)^\phi},$$

$$z_{01}^* = z_{01}^*(\boldsymbol{\theta}_1; \boldsymbol{x}) = \sum_{j=1}^{\infty} \frac{j\{H_\phi^{-1}(1+\exp(\boldsymbol{x}'\boldsymbol{\beta}))\}^j}{(j!)^\phi},$$

$$z_{02}^* = z_{02}^*(\boldsymbol{\theta}_1; \boldsymbol{x}) = \sum_{j=1}^{\infty} \frac{j^2\{H_\phi^{-1}(1+\exp(\boldsymbol{x}'\boldsymbol{\beta}))\}^j}{(j!)^\phi},$$

$$G_{Ci}(\boldsymbol{\gamma}) = (\gamma_2 t_i)^{1/\gamma_1}\log(\gamma_2 t_i) \tag{22.11}$$

for $i = 1, 2, \ldots, n$. The quantities $z_{1i}^*, z_{2i}^*, z_{21i}^*, z_{22i}^*, z_{01i}^*$ and z_{02i}^* are all computed as described in Appendix A1.

22.B1 Appendix B1: Observed Information Matrix for Exponential Lifetimes

Bernoulli cure rate model: The observed information matrix has its components as

$$-\frac{\partial^2 l}{\partial\beta_l\partial\beta_{l'}} = -\sum_{I_0} x_{il}x_{il'}w_i(1-w_i) + \sum_{I^*} x_{il}x_{il'}\frac{\exp(\boldsymbol{x}_i'\boldsymbol{\beta})}{(1+\exp(\boldsymbol{x}_i'\boldsymbol{\beta}))^2},$$

$$-\frac{\partial^2 l}{\partial\gamma^2} = \frac{n_1}{\gamma^2} - \sum_{I_0} t_i^2 w_i(1-w_i),$$

$$-\frac{\partial^2 l}{\partial\beta_l\partial\gamma} = \sum_{I_0} x_{il}t_i w_i(1-w_i)$$

for $l, l' = 0, 1, \ldots, k$, and $x_{i0} = 1 \forall i = 1, 2, \ldots, n$, where

$$w_i = \frac{\exp(\boldsymbol{x}_i'\boldsymbol{\beta} - \gamma t_i)}{1+\exp(\boldsymbol{x}_i'\boldsymbol{\beta} - \gamma t_i)}$$

for $i \in I_0$.

Poisson cure rate model: The observed information matrix has its components as

$$-\frac{\partial^2 l}{\partial\beta_l\partial\beta_{l'}} = \sum_{I^*} x_{il}x_{il'}\exp(\boldsymbol{x}_i'\boldsymbol{\beta})\{1-\exp(-\gamma t_i)\},$$

$$-\frac{\partial^2 l}{\partial\gamma^2} = \frac{n_1}{\gamma^2} - \sum_{I^*} t_i^2\exp(\boldsymbol{x}_i'\boldsymbol{\beta} - \gamma t_i),$$

$$-\frac{\partial^2 l}{\partial\beta_l\partial\gamma} = \sum_{I^*} x_{il}t_i\exp(\boldsymbol{x}_i'\boldsymbol{\beta} - \gamma t_i)$$

for $l, l' = 0, 1, \ldots, k$, and $x_{i0} = 1 \forall i = 1, 2, \ldots, n$.

Geometric cure rate model: The observed information matrix has its components as

$$-\frac{\partial^2 l}{\partial\beta_l\partial\beta_{l'}} = -2\sum_{I_1} x_{il}x_{il'}\frac{\exp(\boldsymbol{x}_i'\boldsymbol{\beta}-\gamma t_i)}{(1-\exp(\boldsymbol{x}_i'\boldsymbol{\beta}-\gamma t_i))^2} - \sum_{I_0} x_{il}x_{il'}\frac{\exp(\boldsymbol{x}_i'\boldsymbol{\beta}-\gamma t_i)}{(1-\exp(\boldsymbol{x}_i'\boldsymbol{\beta}-\gamma t_i))^2}$$

$$+\sum_{I^*} x_{il}x_{il'}\frac{\exp(\boldsymbol{x}_i'\boldsymbol{\beta})}{(1-\exp(\boldsymbol{x}_i'\boldsymbol{\beta}))^2},$$

$$-\frac{\partial^2 l}{\partial\gamma^2} = \frac{n_1}{\gamma^2} - 2\sum_{I_1} t_i^2\frac{\exp(\boldsymbol{x}_i'\boldsymbol{\beta}-\gamma t_i)}{(1-\exp(\boldsymbol{x}_i'\boldsymbol{\beta}-\gamma t_i))^2} - \sum_{I_0} t_i^2\frac{\exp(\boldsymbol{x}_i'\boldsymbol{\beta}-\gamma t_i)}{(1-\exp(\boldsymbol{x}_i'\boldsymbol{\beta}-\gamma t_i))^2},$$

$$-\frac{\partial^2 l}{\partial\beta_l\partial\gamma} = 2\sum_{I_1} x_{il}t_i\frac{\exp(\boldsymbol{x}_i'\boldsymbol{\beta}-\gamma t_i)}{(1-\exp(\boldsymbol{x}_i'\boldsymbol{\beta}-\gamma t_i))^2} + \sum_{I_0} x_{il}t_i\frac{\exp(\boldsymbol{x}_i'\boldsymbol{\beta}-\gamma t_i)}{(1-\exp(\boldsymbol{x}_i'\boldsymbol{\beta}-\gamma t_i))^2}$$

for $l, l' = 0, 1, \ldots, k$, and $x_{i0} = 1 \forall i = 1, 2, \ldots, n$.

COM-Poisson cure rate model: The observed information matrix, for a fixed value of ϕ, has its components as

$$-\frac{\partial^2 l}{\partial\beta_l\partial\beta_{l'}} = -\sum_{I_1} x_{il}x_{il'}\frac{z_2(\boldsymbol{\beta},\gamma,\phi;\boldsymbol{x}_i,t_i)z_{22}(\boldsymbol{\beta},\gamma,\phi;\boldsymbol{x}_i,t_i) - z_{21}^2(\boldsymbol{\beta},\gamma,\phi;\boldsymbol{x}_i,t_i)}{z_2^2(\boldsymbol{\beta},\gamma,\phi;\boldsymbol{x}_i,t_i)}$$

$$-\sum_{I_0} x_{il}x_{il'}\frac{(1+z_1(\boldsymbol{\beta},\gamma,\phi;\boldsymbol{x}_i,t_i))z_{21}(\boldsymbol{\beta},\gamma,\phi;\boldsymbol{x}_i,t_i) - z_2^2(\boldsymbol{\beta},\gamma,\phi;\boldsymbol{x}_i,t_i)}{(1+z_1(\boldsymbol{\beta},\gamma,\phi;\boldsymbol{x}_i,t_i))^2}$$

$$+\sum_{I^*} x_{il}x_{il'}\frac{z(\boldsymbol{\beta},\phi;\boldsymbol{x}_i)z_{02}(\boldsymbol{\beta},\phi;\boldsymbol{x}_i) - z_{01}^2(\boldsymbol{\beta},\phi;\boldsymbol{x}_i)}{z^2(\boldsymbol{\beta},\phi;\boldsymbol{x}_i)},$$

$$-\frac{\partial^2 l}{\partial\gamma^2} = \frac{n_1}{\gamma^2} - \sum_{I_1} t_i^2\frac{z_2(\boldsymbol{\beta},\gamma,\phi;\boldsymbol{x}_i,t_i)z_{22}(\boldsymbol{\beta},\gamma,\phi;\boldsymbol{x}_i,t_i) - z_{21}^2(\boldsymbol{\beta},\gamma,\phi;\boldsymbol{x}_i,t_i)}{z_2^2(\boldsymbol{\beta},\gamma,\phi;\boldsymbol{x}_i,t_i)}$$

$$-\sum_{I_0} t_i^2\frac{(1+z_1(\boldsymbol{\beta},\gamma,\phi;\boldsymbol{x}_i,t_i))z_{21}(\boldsymbol{\beta},\gamma,\phi;\boldsymbol{x}_i,t_i) - z_2^2(\boldsymbol{\beta},\gamma,\phi;\boldsymbol{x}_i,t_i)}{(1+z_1(\boldsymbol{\beta},\gamma,\phi;\boldsymbol{x}_i,t_i))^2},$$

$$-\frac{\partial^2 l}{\partial\beta_l\partial\gamma} = \sum_{I_1} x_{il}t_i\frac{z_2(\boldsymbol{\beta},\gamma,\phi;\boldsymbol{x}_i,t_i)z_{22}(\boldsymbol{\beta},\gamma,\phi;\boldsymbol{x}_i,t_i) - z_{21}^2(\boldsymbol{\beta},\gamma,\phi;\boldsymbol{x}_i,t_i)}{z_2^2(\boldsymbol{\beta},\gamma,\phi;\boldsymbol{x}_i,t_i)}$$

$$+\sum_{I_0} x_{il}t_i\frac{(1+z_1(\boldsymbol{\beta},\gamma,\phi;\boldsymbol{x}_i,t_i))z_{21}(\boldsymbol{\beta},\gamma,\phi;\boldsymbol{x}_i,t_i) - z_2^2(\boldsymbol{\beta},\gamma,\phi;\boldsymbol{x}_i,t_i)}{(1+z_1(\boldsymbol{\beta},\gamma,\phi;\boldsymbol{x}_i,t_i))^2}$$

for $l, l' = 0, 1, \ldots, k$, and $x_{i0} = 1 \forall i = 1, 2, \ldots, n$.

22.B2 Appendix B2: Observed Information Matrix for Weibull Lifetimes

COM-Poisson cure rate model: The observed information matrix, for a fixed value of ϕ, has its components as

$$
\begin{aligned}
-\frac{\partial^2 l}{\partial \beta_l \partial \beta_{l'}} &= -\sum_{I_1} x_{il} x_{il'} \frac{\exp(\boldsymbol{x}'_i \boldsymbol{\beta})}{(z^*_{2i} z^*_{01i})^2} [z^*_{2i}\{\exp(\boldsymbol{x}'_i\boldsymbol{\beta}) z^*_{22i} + z^*_{01i} z^*_{21i}\} \\
&\quad - \exp(\boldsymbol{x}'_i\boldsymbol{\beta}) z^*_{21i} \left\{ \frac{z^*_{2i} z^*_{02i}}{z^*_{01i}} + z^*_{21i} \right\} \Bigg] - \sum_{I_0} x_{il} x_{il'} \frac{\exp(\boldsymbol{x}'_i\boldsymbol{\beta})}{((1+z^*_{1i}) z^*_{01i})^2} \\
&\quad \times \left[(1+z^*_{1i}) z^*_{01i} \left\{ \frac{\exp(\boldsymbol{x}'_i\boldsymbol{\beta}) z^*_{21i}}{z^*_{01i}} + z^*_{2i} \right\} - \exp(\boldsymbol{x}'_i\boldsymbol{\beta}) z^*_{2i} \left\{ \frac{(1+z^*_{1i}) z^*_{02i}}{z^*_{01i}} + z^*_{2i} \right\} \right] \\
&\quad + \sum_{I^*} x_{il} x_{il'} \frac{\exp(\boldsymbol{x}'_i\boldsymbol{\beta})}{(1+\exp(\boldsymbol{x}'_i\boldsymbol{\beta}))^2}, \\
-\frac{\partial^2 l}{\partial \beta_l \partial \gamma_1} &= -\frac{1}{\gamma_1^2} \sum_{I_1} x_{il} \frac{G_{Ci}(\boldsymbol{\gamma}) \exp(\boldsymbol{x}'_i\boldsymbol{\beta})}{z^*_{01i} {z^*_{2i}}^2} \{ z^*_{2i} z^*_{22i} - {z^*_{21i}}^2 \} \\
&\quad - \frac{1}{\gamma_1^2} \sum_{I_0} x_{il} \frac{G_{Ci}(\boldsymbol{\gamma}) \exp(\boldsymbol{x}'_i\boldsymbol{\beta})}{(1+z^*_{1i})^2 z^*_{01i}} \{ (1+z^*_{1i}) z^*_{21i} - {z^*_{2i}}^2 \}, \\
-\frac{\partial^2 l}{\partial \beta_l \partial \gamma_2} &= \frac{1}{\gamma_1\gamma_2} \sum_{I_1} x_{il} \frac{(\gamma_2 t_i)^{1/\gamma_1} \exp(\boldsymbol{x}'_i\boldsymbol{\beta})}{z^*_{01i} {z^*_{2i}}^2} \{ z^*_{2i} z^*_{22i} - {z^*_{21i}}^2 \} \\
&\quad + \frac{1}{\gamma_1\gamma_2} \sum_{I_0} x_{il} \frac{(\gamma_2 t_i)^{1/\gamma_1} \exp(\boldsymbol{x}'_i\boldsymbol{\beta})}{(1+z^*_{1i})^2 z^*_{01i}} \{ (1+z^*_{1i}) z^*_{21i} - {z^*_{2i}}^2 \}, \\
-\frac{\partial^2 l}{\partial {\gamma_1}^2} &= -\frac{n_1}{{\gamma_1}^2} - \frac{2n_1}{{\gamma_1}^3} \log\gamma_2 - \frac{2}{{\gamma_1}^3} \sum_{I_1} \log t_i - \frac{1}{\gamma_1^4} \sum_{I_1} \frac{G_{Ci}(\boldsymbol{\gamma})}{z^{*2}_{2i}} \\
&\quad \times [\log(\gamma_2 t_i) z^*_{2i} \{ (\gamma_2 t_i)^{1/\gamma_1} z^*_{22i} - z^*_{21i} \} - z^*_{21i} \{ G_{Ci}(\boldsymbol{\gamma}) z^*_{21i} + 2\gamma_1 z^*_{2i} \}] \\
&\quad - \frac{1}{\gamma_1^4} \sum_{I_0} \frac{G_{Ci}(\boldsymbol{\gamma})}{(1+z^*_{1i})^2} [\log(\gamma_2 t_i)(1+z^*_{1i}) \{ (\gamma_2 t_i)^{1/\gamma_1} z^*_{21i} - z^*_{2i} \} \\
&\quad - z^*_{2i} \{ G_{Ci}(\boldsymbol{\gamma}) z^*_{2i} + 2\gamma_1 (1+z^*_{1i}) \}], \\
-\frac{\partial^2 l}{\partial {\gamma_2}^2} &= \frac{n_1}{\gamma_1 {\gamma_2}^2} - \frac{1}{\gamma_1 {\gamma_2}^2} \sum_{I_1} \frac{(\gamma_2 t_i)^{1/\gamma_1}}{{z^*_{2i}}^2} \left[z^*_{2i} \left\{ \frac{1}{\gamma_1} (\gamma_2 t_i)^{1/\gamma_1} z^*_{22i} - \left(\frac{1}{\gamma_1} - 1 \right) z^*_{21i} \right\} \right. \\
&\quad \left. - \frac{1}{\gamma_1} (\gamma_2 t_i)^{1/\gamma_1} {z^*_{21i}}^2 \right]
\end{aligned}
$$

$$
\begin{aligned}
&+\frac{1}{\gamma_1{\gamma_2}^2}\sum_{I_0}\frac{(\gamma_2 t_i)^{1/\gamma_1}}{(1+z_{1i}^*)^2}\left[\frac{1}{\gamma_1}(\gamma_2 t_i)^{1/\gamma_1}{z_{2i}^*}^2+(1+z_{1i}^*)\right. \\
&\left.\times\left\{\left(\frac{1}{\gamma_1}-1\right)z_{2i}^*-\frac{1}{\gamma_1}(\gamma_2 t_i)^{1/\gamma_1}z_{21i}^*\right\}\right],
\end{aligned}
$$

$$
\begin{aligned}
-\frac{\partial^2 l}{\partial\gamma_1\partial\gamma_2}=&\frac{n_1}{\gamma_2{\gamma_1}^2}+\frac{1}{\gamma_2{\gamma_1}^2}\sum_{I_1}\frac{(\gamma_2 t_i)^{1/\gamma_1}}{{z_{2i}^*}^2}\left[z_{2i}^*\left\{\frac{1}{\gamma_1}G_{Ci}(\boldsymbol{\gamma})z_{22i}^*-z_{21i}^*-\frac{1}{\gamma_1}\log(\gamma_2 t_i)z_{21i}^*\right\}\right. \\
&\left.-\frac{1}{\gamma_1}G_{Ci}(\boldsymbol{\gamma}){z_{21i}^*}^2\right]-\frac{1}{\gamma_2{\gamma_1}^2}\sum_{I_0}\frac{(\gamma_2 t_i)^{1/\gamma_1}}{(1+z_{1i}^*)^2}\left[(1+z_{1i}^*)\left\{z_{2i}^*+\frac{1}{\gamma_1}\log(\gamma_2 t_i)z_{2i}^*\right.\right. \\
&\left.\left.-\frac{1}{\gamma_1}G_{Ci}(\boldsymbol{\gamma})z_{21i}^*\right\}+\frac{1}{\gamma_1}G_{Ci}(\boldsymbol{\gamma}){z_{2i}^*}^2\right].
\end{aligned}
$$

The above are defined for $l, l' = 0, 1, \ldots, k$, $x_{i0} = 1 \forall i = 1, 2, \ldots, n$, where $G_{Ci}(\boldsymbol{\gamma})$ is as defined in (22.11).

References

Balakrishnan, N. and A. P. Basu. 1995. *The Exponential Distribution: Theory, Methods and Applications*. Leiden, The Netherlands: Gordon and Breach Publishers.

Balakrishnan, N. and S. Pal. 2013. Lognormal liftetimes and likelihood-based inference for flexible cure rate models based on COM-Poisson family. *Computational Statistics and Data Analysis* 67: 41–67.

Balakrishnan, N. and S. Pal. 2012. EM algorithm-based likelihood estimation for some cure rate models. *Journal of Statistical Theory and Practice* 6 (4): 698–724.

Balakrishnan, N. and Y. Peng. 2006. Generalized gamma frailty model. *Statistics in Medicine* 25 (16): 2797–2816.

Berkson, J. and R. P. Gage. 1952. Survival cure for cancer patients following treatment. *Journal of the American Statistical Association* 47 (259): 501–515.

Boatwright, P., S. Borle, J. B. Kadane, T. P. Minka and G. Shmueli. 2006. Conjugate analysis of the Conway-Maxwell-Poisson distribution. *Bayesian Analysis* 1 (2): 363–374.

Claeskens, G., R. Nguti and P. Janssen. 2008. One-sided tests in shared frailty models. *Test* 17 (1): 69–82.

Conway, R. W. and W. L. Maxwell. 1961. A queuing model with state dependent services rates. *The Journal of Industrial Engineering* 12: 132–136.

Dempster, A. P., N. M. Laird and D. B. Rubin. 1977. Maximum likelihood from incomplete data via the EM algorithm. *Journal of the Royal Statistical Society* B39 (1): 1–38.

Hoggart, C. J. and J. E. Griffin. 2001. A Bayesian partition model for customer attrition. In *Bayesian Methods with Applications to Science, Policy, and Official Statistics,* ed. E. I. George. Selected Papers from ISBA 2000. Proceedings of the Sixth World Meeting of the International Society for Bayesian Analysis. 222–232. International Society for Bayesian Analysis, Crete Greece.

Ibrahim, J. G., M. -H. Chen and D. Sinha. 2001. *Bayesian Survival Analysis*. New York: Springer.

Johnson, N. L., S. Kotz and N. Balakrishnan. 1994. *Continuous Univariate Distributions*. Vol. 1. New York: John Wiley & Sons.

Kokonendji, C. C., T. S. Kiesse and N. Balakrishnan. 2009. Semiparametric estimation for count data through weighted distributions. *Journal of Statistical Planning and Inference* 139 (10): 3625–3638.

Kokonendji, C. C., D. Mizere and N. Balakrishnan. 2008. Connections of the Poisson weight function to overdispersion and underdispersion. *Journal of Statistical Planning and Inference* 138 (5): 1287–1296.

Lange, K. 1995. A gradient algorithm locally equivalent to the EM algorithm. *Journal of the Royal Statistical Society* B57 (2): 425–437.

Louis, T. A. 1982. Finding the observed information matrix when using the EM algorithm. *Journal of the Royal Statistical Society* B44 (2): 226–233.

McLachlan, G. J. and T. Krishnan. 2008. *The EM Algorithm and Extensions*. Hoboken, New Jersey: Wiley.

Meeker, W. Q. and L. A. Escobar. 1998. *Statistical Methods for Reliability Data*. New York: Wiley.

Rodrigues, J., M. de Castro, V. G. Cancho and N. Balakrishnan. 2009. COM-Poisson cure rate survival models and an application to a cutaneous melanoma data. *Journal of Statistical Planning and Inference* 139 (10): 3605–3611.

Self, S. G. and K. -Y. Liang. 1987. Asymptotic properties of maximum likelihood estimators and likelihood ratio tests under nonstandard conditions. *Journal of the American Statistical Association* 82 (398): 605–610.

Shmueli, G., T. P. Minka, J. B. Kadane, S. Borle and P. Boatwright. 2005. A useful distribution for fitting discrete data: revival of the Conway-Maxwell-Poisson distribution. *Journal of the Royal Statistical Society* C54 (1): 127–142.

Sy, J. P. and J. M. G. Taylor. 2000. Estimation in a Cox proportional hazards cure model. *Biometrics* 56 (1): 227–236.

Yakovlev, A. Yu. and A. D. Tsodikov. 1996. *Stochastic Models of Tumor Latency and their Biostatistical Applications*. Singapore: World Scientific.

Yakovlev, A. Yu., A. D. Tsodikov and L. Bass. 1993. A stochastic model of hormesis. *Mathematical Biosciences* 116 (2): 197–219.

Yin, G. and J. G. Ibrahim. 2005. Cure rate models: a unified approach. *The Canadian Journal of Statistics* 33 (4): 559–570.

23

Exponential Expansions for Perturbed Discrete Time Renewal Equations

Dmitrii Silvestrov and Mikael Petersson
Department of Mathematics, Stockholm University, Sweden

23.1 Introduction

Let $q^{(\varepsilon)}(n), n = 0, 1, \ldots$, be a sequence of real numbers and let $f^{(\varepsilon)}(n), n = 0, 1, \ldots$, be a discrete probability distribution which may be improper but not concentrated at zero, depending on a parameter $\varepsilon \geq 0$. Consider, for every $\varepsilon \geq 0$ a discrete time renewal equation,

$$x^{(\varepsilon)}(n) = q^{(\varepsilon)}(n) + \sum_{k=0}^{n} x^{(\varepsilon)}(n-k) f^{(\varepsilon)}(k), \quad n = 0, 1, \ldots \tag{23.1}$$

As known, the renewal equation (23.1) has a unique solution.

Suppose that the distribution $f^{(\varepsilon)}(n)$ and the function $q^{(\varepsilon)}(n)$ converge in some natural sense to $f^{(0)}(n)$ and $q^{(0)}(n)$, respectively, as $\varepsilon \to 0$. Then for $\varepsilon > 0$, Equation (23.1) can be interpreted as a perturbed version of the renewal equation obtained by setting $\varepsilon = 0$.

In the non-perturbed case, the classical renewal theorem given in Erdos *et al.* (1949) states that if $f^{(0)}(n)$ is a non-periodic, proper distribution with finite expectation $m_1^{(0)} = \sum_{k=0}^{\infty} k f^{(0)}(k) < \infty$ and $\sum_{k=0}^{\infty} |q^{(0)}(k)| < \infty$, then

$$x^{(0)}(n) \to \frac{1}{m_1^{(0)}} \sum_{k=0}^{\infty} q^{(0)}(k) \quad \text{as } n \to \infty. \tag{23.2}$$

Applied Reliability Engineering and Risk Analysis: Probabilistic Models and Statistical Inference, First Edition.
Edited by Ilia B. Frenkel, Alex Karagrigoriou, Anatoly Lisnianski and Andre Kleyner.

This result can also be found in Feller (1950) together with an introduction to discrete time renewal theory.

This theorem plays an important role in ergodic theorems for discrete time Markov type processes and queuing and reliability applications, due to the well-known fact that the distribution at moment n for such processes usually satisfies a renewal equation.

In the present chapter, we present results about the asymptotic behavior of the solution $x^{(\varepsilon)}(n)$ of a perturbed discrete time renewal equation as $\varepsilon \to 0$ and $n \to \infty$ simultaneously. We consider two cases of so-called quasi-stationary and pseudo-stationary asymptotics, where the limiting distribution $f^{(0)}(k)$ may be, respectively, improper or proper. We improve the asymptotic relation (23.2) to the much more advanced form of an exponential asymptotic expansion. We also illustrate theoretical results by examples related to queuing systems and risk processes.

In the pseudo-stationary case, our results generalize results given in Englund and Silvestrov (1997). At the same time, our results are discrete time analogues of results given in the book by Gyllenberg and Silvestrov (2008) for perturbed continuous time renewal equations, where one can also find an extended bibliography of works in the area.

Some examples of more recent works related to perturbed renewal equations in continuous time and applications to risk theory are Kartashov (2009), Blanchet and Zwart (2010), and Ni (2010).

Since the results in this chapter have applications to discrete time Markov chains, we also mention that some results on asymptotic expansions for perturbed Markov chains can be found in, for example, Kartashov (1988, 1996), Latouche (1988), Hassin and Haviv (1992), Khasminskii *et al.* 1996, Yin and Zhang (2003), Altman *et al.* (2004) and Yin and Nguyen (2009).

23.2 Asymptotic Results

This section presents asymptotic results for the perturbed renewal equation in discrete time with a possibly improper limiting distribution. Without loss of generality it can be assumed that $n = n^{(\varepsilon)}$ is a function of ε such that $n^{(\varepsilon)} \to \infty$ as $\varepsilon \to 0$. Let $f^{(\varepsilon)}$ denote the defect of the distribution $f^{(\varepsilon)}(n)$,

$$f^{(\varepsilon)} = 1 - \sum_{n=0}^{\infty} f^{(\varepsilon)}(n).$$

Furthermore, let $\phi^{(\varepsilon)}(\rho)$ be the moment generating function for the distribution $f^{(\varepsilon)}(n)$,

$$\phi^{(\varepsilon)}(\rho) = \sum_{n=0}^{\infty} e^{\rho n} f^{(\varepsilon)}(n), \quad \rho \geq 0. \tag{23.3}$$

We assume that the following conditions on the probability distributions $f^{(\varepsilon)}(n)$ hold.

A: (a) $f^{(\varepsilon)}(n) \to f^{(0)}(n)$ as $\varepsilon \to 0$, $n = 0, 1, \ldots$, where the limiting distribution is non-periodic and not concentrated in zero.

(b) $f^{(\varepsilon)} \to f^{(0)} \in [0, 1)$ as $\varepsilon \to 0$.

B: There exists $\delta_c > 0$ such that
(a) $\overline{\lim}_{\varepsilon\to 0}\phi^{(\varepsilon)}(\delta_c) < \infty$,
(b) $\phi^{(0)}(\delta_c) > 1$.

We define the characteristic equation of the probability distribution $f^{(\varepsilon)}(n)$ by

$$\phi^{(\varepsilon)}(\rho) = 1. \tag{23.4}$$

As will be shown in Theorem 23.1, there exists a unique non-negative root $\rho^{(\varepsilon)}$ of equation (23.4) for sufficiently small ε and, moreover, $\rho^{(\varepsilon)} \to \rho^{(0)}$ as $\varepsilon \to 0$, if conditions **A** and **B** hold.

Thus, under conditions **A** and **B**, the following conditions on the functions $q^{(\varepsilon)}(n)$ are well-founded:

C: (a) $\overline{\lim}_{\varepsilon\to 0}|q^{(\varepsilon)}(n)| < \infty$, for $n = 0, 1, \ldots$
(b) $\sum_{n=0}^{\infty} e^{\rho^{(\varepsilon)}n} q^{(\varepsilon)}(n) \to \sum_{n=0}^{\infty} e^{\rho^{(0)}n} q^{(0)}(n)$ as $\varepsilon \to 0$.
(c) $\overline{\lim}_{\varepsilon\to 0}\sum_{n=0}^{\infty} e^{(\rho^{(0)}+\delta)n}|q^{(\varepsilon)}(n)| < \infty$, for some $\delta > 0$.

Henceforth we will use the following notations.

$$\tilde{m}_1^{(\varepsilon)} = \sum_{n=0}^{\infty} n e^{\rho^{(\varepsilon)}n} f^{(\varepsilon)}(n), \quad \tilde{x}^{(\varepsilon)}(\infty) = \frac{1}{\tilde{m}_1^{(\varepsilon)}}\sum_{n=0}^{\infty} e^{\rho^{(\varepsilon)}n} q^{(\varepsilon)}(n).$$

Theorem 23.1. Assume that conditions **A** and **B** hold. Then

(i) There exists a unique nonnegative root $\rho^{(\varepsilon)}$ of the equation (23.4) for all sufficiently small ε and $\rho^{(\varepsilon)} \to \rho^{(0)} < \delta_c$ as $\varepsilon \to 0$.
(ii) If, in addition, condition **C** is satisfied, then for any non-negative integer-valued function $n^{(\varepsilon)} \to \infty$ as $\varepsilon \to 0$, the following asymptotic relation holds.

$$\frac{x^{(\varepsilon)}(n^{(\varepsilon)})}{\exp\,(-\rho^{(\varepsilon)}n^{(\varepsilon)})} \to \tilde{x}^{(0)}(\infty) \quad \text{as } \varepsilon \to 0.$$

If the limiting distribution is proper, that is $f^{(0)} = 0$, then $\rho^{(0)} = 0$ and Theorem 23.1 reduce to the corresponding result in Englund and Silvestrov (1997).

In particular, the asymptotic result in Theorem 23.1 is good in the sense that the limit does not depend on in which way $n^{(\varepsilon)}$ tends to infinity. However, a drawback is that the relation involves $\rho^{(\varepsilon)}$, which is only given implicitly as the solution to a nonlinear equation. Next we present how some additional perturbation conditions guarantee that $\rho^{(\varepsilon)}$ can be expanded in an asymptotic power series with respect to ε. These conditions require that certain asymptotical expansions hold for the following mixed power-exponential moment generating functions,

$$\phi^{(\varepsilon)}(\rho, r) = \sum_{n=0}^{\infty} n^r e^{\rho n} f^{(\varepsilon)}(n), \quad \rho \geq 0, \quad r = 0, 1, \ldots \tag{23.5}$$

Before proceeding it is useful to note that if r is a nonnegative integer, then $\phi^{(\varepsilon)}(\rho^{(0)}, r)$ and $\phi^{(\varepsilon)}(\rho^{(\varepsilon)}, r)$ are finite for all ε small enough. Indeed, if $0 < \beta < \delta_c$ then $c_r = \sup_{n \geq 0} n^r e^{-(\delta_c - \beta)n} < \infty$ and

$$\phi^{(\varepsilon)}(\beta, r) \leq c_r \sum_{n=0}^{\infty} e^{\delta_c n} f^{(\varepsilon)}(n) = c_r \phi^{(\varepsilon)}(\delta_c).$$

It follows from condition **B** that the left-hand side is finite for all sufficiently small ε, say, $\varepsilon \leq \varepsilon_1$, for some $\varepsilon_1 > 0$. Since $\rho^{(\varepsilon)} \to \rho^{(0)} < \delta_c$, we can choose $\beta < \delta_c$ and $\varepsilon_2 = \varepsilon_2(\beta) > 0$ such that $\rho^{(\varepsilon)} < \beta$ for all $\varepsilon \leq \varepsilon_2$. From this it follows that if $\varepsilon_0 = \min\{\varepsilon_1, \varepsilon_2\}$, then, for all $0 \leq \varepsilon \leq \varepsilon_0$,

$$\phi^{(\varepsilon)}(\rho^{(0)}, r) < \infty, \quad \phi^{(\varepsilon)}(\rho^{(\varepsilon)}, r) < \infty.$$

We assume that the following perturbation condition holds for some positive integer k:

$\mathbf{P^{(k)}}$: $\phi^{(\varepsilon)}(\rho^{(0)}, r) = \phi^{(0)}(\rho^{(0)}, r) + b_{1,r}\varepsilon + \cdots + b_{k-r,r}\varepsilon^r + o(\varepsilon^{k-r})$ for $r = 0, \ldots, k$, where $|b_{n,r}| < \infty$, $n = 1, \ldots, k - r$, $r = 0, \ldots, k$.

Note that if $f^{(0)} = 0$, then $\phi^{(\varepsilon)}(\rho^{(0)}, 0) = 1 - f^{(\varepsilon)}$ and $\phi^{(\varepsilon)}(\rho^{(0)}, r) = \sum_{n=0}^{\infty} n^r f^{(\varepsilon)}(n)$ for positive integers r, so in this case $\mathbf{P^{(k)}}$ reduces to conditions on the defect and the usual moments.

Henceforth we will denote $b_{0,r} = \phi^{(0)}(\rho^{(0)}, r)$.

The following theorem is the main result of the chapter.

Theorem 23.2. Suppose that conditions **A**, **B** and $\mathbf{P^{(k)}}$ hold. Then

(i) The root $\rho^{(\varepsilon)}$ of equation (23.4) has the asymptotical expansion

$$\rho^{(\varepsilon)} = \rho^{(0)} + a_1\varepsilon + \cdots + a_k\varepsilon^k + o(\varepsilon^k). \tag{23.6}$$

The coefficients $a_1, \ldots, a_k$ are given by the recurrence formulas $a_1 = -b_{1,0}/b_{0,1}$ and for $n = 2, \ldots, k$,

$$a_n = -\frac{1}{b_{0,1}}\left(b_{n,0} + \sum_{q=1}^{n-1} b_{n-q,1}a_q + \sum_{m=2}^{n}\sum_{q=m}^{n} b_{n-q,m} \cdot \sum_{n_1, \ldots, n_{q-1} \in D_{m,q}} \prod_{p=1}^{q-1} \frac{a_p^{n_p}}{n_p!}\right), \tag{23.7}$$

where $D_{m,q}$ is the set of all nonnegative integer solutions to the system

$$n_1 + \cdots + n_{q-1} = m, \quad n_1 + \cdots + (q-1)n_{q-1} = q. \tag{23.8}$$

(ii) If $b_{l,0} = 0, l = 1, \ldots, r$, for some $1 \leq r \leq k$, then $a_1, \ldots, a_r = 0$. If $b_{l,0} = 0, l = 1, \ldots, r-1$, and $b_{r,0} < 0$ hold for some $1 \leq r \leq k$, then $a_1, \ldots, a_{r-1} = 0$ and $a_r > 0$.

(iii) If, in addition, condition **C** holds and $n^{(\varepsilon)} \to \infty$ as $\varepsilon \to 0$ in such a way that $\varepsilon^r n^{(\varepsilon)} \to \lambda_r \in [0, \infty)$ for some $1 \leq r \leq k$, then

$$\frac{x^{(\varepsilon)}(n^{(\varepsilon)})}{\exp\left(-(\rho^{(0)} + a_1\varepsilon + \cdots + a_{r-1}\varepsilon^{r-1})n^{(\varepsilon)}\right)} \to \frac{\tilde{x}^{(0)}(\infty)}{e^{\lambda_r a_r}} \quad \text{as } \varepsilon \to 0.$$

In order to prove Theorems 23.1 and 23.2, we shall use the following result from Englund and Silvestrov (1997), which is a modification of an earlier result by Kalashnikov (1978).

Let us impose the following conditions on the distributions $f^{(\varepsilon)}(n)$ and functions $q^{(\varepsilon)}(n)$:

A′: (a) $f^{(\varepsilon)}(n) \to f^{(0)}(n)$ as $\varepsilon \to 0$, $n = 0, 1, \ldots$, where the limiting distribution is proper, non-periodic and not concentrated in zero.

(b) $m_1^{(\varepsilon)} \to m_1^{(0)} < \infty$ as $\varepsilon \to 0$.

C′: (a) $\overline{\lim}_{\varepsilon \to 0} |q^{(\varepsilon)}(n)| < \infty$, for $n = 0, 1, \ldots$

(b) $\lim_{n\to\infty} \overline{\lim}_{\varepsilon\to 0} \sum_{k=n}^{\infty} |q^{(\varepsilon)}(k)| = 0$.

(c) $\sum_{n=0}^{\infty} q^{(\varepsilon)}(n) \to \sum_{n=0}^{\infty} q^{(0)}(n)$ as $\varepsilon \to 0$.

Lemma 23.1. Assume that **A**′ and **C**′ hold. If $n^{(\varepsilon)} \to \infty$ in such a way that $f^{(\varepsilon)} n^{(\varepsilon)} \to \lambda \in [0, \infty]$ as $\varepsilon \to 0$, then

$$x^{(\varepsilon)}(n^{(\varepsilon)}) \to \exp\left(-\lambda/m_1^{(0)}\right) \frac{1}{m_1^{(0)}} \sum_{k=0}^{\infty} q^{(0)}(k) \quad \text{as } \varepsilon \to 0.$$

23.3 Proofs

Proof of Theorem 23.1. Condition **B** implies that $\phi^{(\varepsilon)}(\delta_c) < \infty$ for ε sufficiently small, say $\varepsilon \leq \varepsilon_1$. From condition **A** it follows that there exists ε_2 such that the distributions $f^{(\varepsilon)}(n)$ are not concentrated at zero for $\varepsilon \leq \varepsilon_2$. From this it can be concluded that for every fixed $\varepsilon \leq \min\{\varepsilon_1, \varepsilon_2\}$, the function $\phi^{(\varepsilon)}(\rho)$ is nonnegative, continuous and strictly increasing on the interval $[0, \delta_c]$. Since $\phi^{(0)}(0) = 1 - f^{(0)} \leq 1$ and $\phi^{(0)}(\delta_c) > 1$ it follows that there exists a unique root $\rho^{(0)} \in [0, \delta_c)$ to the equation $\phi^{(0)}(\rho) = 1$. For any $\beta \in (\rho^{(0)}, \delta_c]$,

$$\begin{aligned}\underline{\lim}_{\varepsilon\to 0} \phi^{(\varepsilon)}(\beta) &\geq \lim_{N\to\infty} \underline{\lim}_{\varepsilon\to 0} \sum_{n=0}^{N} e^{\beta n} f^{(\varepsilon)}(n) \\ &= \lim_{N\to\infty} \sum_{n=0}^{N} e^{\beta n} f^{(0)}(n) = \phi^{(0)}(\beta) > 1.\end{aligned}$$

For any such β we can choose $\varepsilon_3 = \varepsilon_3(\beta)$ such that $\phi^{(\varepsilon)}(\beta) > 1$ for $\varepsilon \leq \varepsilon_3$. Let $\varepsilon_0(\beta) = \{\varepsilon_1, \varepsilon_2, \varepsilon_3(\beta)\}$. Since $\phi^{(\varepsilon)}(0) = 1 - f^{(\varepsilon)} \leq 1$, it follows that there exists a unique root

$\rho^{(\varepsilon)} \in [0, \beta)$ to the characteristic equation (23.4) for $\varepsilon \leq \varepsilon_0(\beta)$. To show that $\rho^{(\varepsilon)} \to \rho^{(0)}$, first note that since β can be chosen arbitrarily close from above to $\rho^{(0)}$ we have

$$\overline{\lim_{\varepsilon\to 0}}\rho^{(\varepsilon)} \leq \rho^{(0)}.$$

Using that, β can be chosen such that $\rho^{(\varepsilon)} < \beta < \delta_c$ for ε small enough yields

$$\begin{aligned}
&\lim_{N\to\infty} \overline{\lim_{\varepsilon\to 0}} \sum_{n=N+1}^{\infty} e^{\rho^{(\varepsilon)}n} f^{(\varepsilon)}(n) \\
&\leq \lim_{N\to\infty} \overline{\lim_{\varepsilon\to 0}} \sum_{n=N+1}^{\infty} e^{\beta n} f^{(\varepsilon)}(n) \\
&\leq \lim_{N\to\infty} e^{-(\delta_c-\beta)(N+1)} \overline{\lim_{\varepsilon\to 0}} \phi^{(\varepsilon)}(\delta_c) = 0.
\end{aligned} \tag{23.9}$$

Assume that there exists a number $\gamma > 0$ and a subsequence $\varepsilon_k \to 0$ as $k \to \infty$ such that $\rho^{(\varepsilon_k)} \leq \rho^{(0)} - \gamma$ for all k. Then

$$\begin{aligned}
&\lim_{N\to\infty} \overline{\lim_{k\to\infty}} \sum_{n=0}^{N} e^{\rho^{(\varepsilon_k)}n} f^{(\varepsilon_k)}(n) \\
&\leq \lim_{N\to\infty} \overline{\lim_{k\to\infty}} \sum_{n=0}^{N} e^{(\rho^{(0)}-\gamma)n} f^{(\varepsilon_k)}(n) \\
&\leq \lim_{N\to\infty} \sum_{n=0}^{N} e^{(\rho^{(0)}-\gamma)n} f^{(0)}(n) = \phi^{(0)}(\rho^{(0)} - \gamma) < 1.
\end{aligned} \tag{23.10}$$

It follows from (23.9) and (23.10) that if such subsequence exists, then

$$\begin{aligned}
1 &= \overline{\lim_{k\to\infty}} \phi^{(\varepsilon_k)}(\rho^{(\varepsilon_k)}) \\
&= \lim_{N\to\infty} \overline{\lim_{k\to\infty}} \left(\sum_{n=0}^{N} e^{\rho^{(\varepsilon_k)}n} f^{(\varepsilon_k)}(n) + \sum_{n=N+1}^{\infty} e^{\rho^{(\varepsilon_k)}n} f^{(\varepsilon_k)}(n) \right) < 1.
\end{aligned}$$

This contradiction implies that $\underline{\lim}_{\varepsilon\to 0}\rho^{(\varepsilon)} \geq \rho^{(0)}$ and hence the first part of the theorem is proved.

For the proof of the second part of the theorem, set

$$\tilde{x}^{(\varepsilon)}(n) = e^{\rho^{(\varepsilon)}n} x^{(\varepsilon)}(n), \quad \tilde{q}^{(\varepsilon)}(n) = e^{\rho^{(\varepsilon)}n} q^{(\varepsilon)}(n), \quad \tilde{f}^{(\varepsilon)}(n) = e^{\rho^{(\varepsilon)}n} f^{(\varepsilon)}(n),$$

Multiplying both sides of the renewal equation (23.1) by $e^{\rho^{(\varepsilon)}n}$ yields

$$\tilde{x}^{(\varepsilon)}(n) = \tilde{q}^{(\varepsilon)}(n) + \sum_{k=0}^{n} \tilde{x}^{(\varepsilon)}(n-k) \tilde{f}^{(\varepsilon)}(k), \quad n = 0, 1, \ldots$$

Next step is to show that if $f^{(\varepsilon)}(n)$ and $q^{(\varepsilon)}(n)$ satisfy conditions **A**, **B** and **C**, then $\tilde{f}^{(\varepsilon)}(n)$ and $\tilde{q}^{(\varepsilon)}(n)$ satisfy conditions **A**$'$ and **C**$'$. It follows in a direct way that **A**$'$(**a**) holds for $\tilde{f}^{(\varepsilon)}(n)$. In particular, the distributions $\tilde{f}^{(\varepsilon)}(n)$ are proper since the definition of $\rho^{(\varepsilon)}$ as the root of the characteristic equation (23.4) implies that

$$\sum_{n=0}^{\infty} \tilde{f}^{(\varepsilon)}(n) = \sum_{n=0}^{\infty} e^{\rho^{(\varepsilon)} n} f^{(\varepsilon)}(n) = 1.$$

In order to prove that the expectations $\tilde{m}_1^{(\varepsilon)}$ of the distributions $\tilde{f}^{(\varepsilon)}(n)$ satisfy **A**$'$(**b**), first note that it follows from the first part of the proof that there exists $\gamma > 0$ such that $\rho^{(\varepsilon)} < \rho^{(0)} + \gamma < \delta_c$ for sufficiently small ε. Using this, we get

$$\begin{aligned} &\lim_{N\to\infty} \overline{\lim_{\varepsilon\to 0}} \sum_{n=N+1}^{\infty} n \tilde{f}^{(\varepsilon)}(n) \\ &\leq \lim_{N\to\infty} \overline{\lim_{\varepsilon\to 0}} \sum_{n=N+1}^{\infty} n e^{(\rho^{(0)}+\gamma)n} f^{(\varepsilon)}(n) \\ &\leq \lim_{N\to\infty} (N+1) e^{-(\delta_c-\rho^{(0)}-\gamma)(N+1)} \overline{\lim_{\varepsilon\to 0}} \phi^{(\varepsilon)}(\delta_c) = 0. \end{aligned} \tag{23.11}$$

Relation (23.11) and the fact that $\tilde{f}^{(\varepsilon)}(n)$ satisfies **A**$'$(**a**) give

$$\begin{aligned} \lim_{\varepsilon\to 0} \tilde{m}_1^{(\varepsilon)} &= \lim_{\varepsilon\to 0} \sum_{n=0}^{\infty} n \tilde{f}^{(\varepsilon)}(n) = \lim_{N\to\infty} \lim_{\varepsilon\to 0} \sum_{n=0}^{N} n \tilde{f}^{(\varepsilon)}(n) \\ &= \lim_{N\to\infty} \sum_{n=0}^{N} n \tilde{f}^{(0)}(n) = \tilde{m}_1^{(0)}. \end{aligned}$$

Moreover, if **C** holds, then the functions $\tilde{q}^{(\varepsilon)}(n)$ satisfy condition **C**$'$. The result now follows from Lemma 23.1.

Proof of Theorem 23.2. Let $\Delta^{(\varepsilon)} = \rho^{(\varepsilon)} - \rho^{(0)}$. It follows from the Taylor expansion of the exponential function that for every $n = 0, 1, \ldots,$

$$e^{\rho^{(\varepsilon)} n} = e^{\rho^{(0)} n} \left(\sum_{r=0}^{k} \frac{n^r (\Delta^{(\varepsilon)})^r}{r!} + \frac{n^{k+1} (\Delta^{(\varepsilon)})^{k+1}}{(k+1)!} e^{|\Delta^{(\varepsilon)}| n} \theta_{k+1}^{(\varepsilon)}(n) \right), \tag{23.12}$$

where $0 \leq \theta_{k+1}^{(\varepsilon)}(n) \leq 1$. Since $\rho^{(\varepsilon)} \to \rho^{(0)} < \delta_c$, there exist $\beta < \delta_c$ and $\varepsilon_1 = \varepsilon_1(\beta)$ such that

$$\rho^{(\varepsilon)} \leq \rho^{(0)} + |\Delta^{(\varepsilon)}| < \beta, \quad \varepsilon \leq \varepsilon_1.$$

It follows from the discussion in Section 23.2 that for this β, there exists $\varepsilon_2 = \varepsilon_2(\beta)$ such that

$$\phi^{(\varepsilon)}(\rho^{(0)}, r) < \infty, \quad r = 0, 1, \ldots, \quad \varepsilon \leq \varepsilon_2.$$

Furthermore, the Cramér type condition **B** implies that

$$\overline{\lim_{\varepsilon\to 0}} \sum_{n=0}^{\infty} n^{k+1} e^{\beta n} f^{(\varepsilon)}(n) \leq \overline{\lim_{\varepsilon\to 0}} c_{k+1} \sum_{n=0}^{\infty} e^{\delta_c n} f^{(\varepsilon)}(n) < \infty,$$

where $c_{k+1} = \sup_{n\geq 0} n^{k+1} e^{-(\delta_c - \beta)n} < \infty$. From this it can be concluded that there exists $\varepsilon_3 = \varepsilon_3(\beta)$ such that

$$M_{k+1} = \frac{1}{(k+1)!} \sup_{\varepsilon \leq \varepsilon_3} \sum_{n=0}^{\infty} n^{k+1} e^{\beta n} f^{(\varepsilon)}(n) < \infty. \tag{23.13}$$

Set $\varepsilon_4 = \varepsilon_4(\beta) = \min\{\varepsilon_1, \varepsilon_2, \varepsilon_3\}$. By first multiplying both sides in (23.12) with $f^{(\varepsilon)}(n)$ and then summing both sides with respect to n, using the notation introduced in Equation (23.5) and the fact that $\rho^{(\varepsilon)}$ is the solution of the characteristic equation (23.4), we obtain for $\varepsilon \leq \varepsilon_4$,

$$1 = \sum_{r=0}^{k} \frac{(\Delta^{(\varepsilon)})^r}{r!} \phi^{(\varepsilon)}(\rho^{(0)}, r) + \frac{(\Delta^{(\varepsilon)})^{k+1}}{(k+1)!} \sum_{n=0}^{\infty} n^{k+1} e^{(\rho^{(0)} + |\Delta^{(\varepsilon)}|)n} \theta_{k+1}^{(\varepsilon)}(n) f^{(\varepsilon)}(n)$$

It follows from Equation (23.13) that the last term on the right-hand side can be written as $(\Delta^{(\varepsilon)})^{k+1} M_{k+1} \theta_{k+1}^{(\varepsilon)}$, where $M_{k+1} < \infty$ and $\theta_{k+1}^{(\varepsilon)} \in [0, 1]$. Hence, we have shown that for $\varepsilon \leq \varepsilon_4$,

$$\sum_{r=1}^{k} \frac{(\Delta^{(\varepsilon)})^r}{r!} \phi^{(\varepsilon)}(\rho^{(0)}, r) + (\Delta^{(\varepsilon)})^{k+1} M_{k+1} \theta_{k+1}^{(\varepsilon)} = 1 - \phi^{(\varepsilon)}(\rho^{(0)}, 0). \tag{23.14}$$

Putting $k = 1$ in Equation (23.14) and dividing by ε gives

$$\frac{\Delta^{(\varepsilon)}}{\varepsilon} (\phi^{(\varepsilon)}(\rho^{(0)}, 1) + \Delta^{(\varepsilon)} M_2 \theta_2^{(\varepsilon)}) = \frac{1 - \phi^{(\varepsilon)}(\rho^{(0)}, 0)}{\varepsilon}, \quad 0 < \varepsilon \leq \varepsilon_4. \tag{23.15}$$

It follows from the perturbation condition $\mathbf{P^{(k)}}$ that $\phi^{(\varepsilon)}(\rho^{(0)}, 1) \to b_{0,1} \in (0, \infty)$ and $(1 - \phi^{(\varepsilon)}(\rho^{(0)}, 0))/\varepsilon \to b_{1,0} \in [0, \infty)$ as $\varepsilon \to 0$. From this and (23.15) it can be concluded that $\Delta^{(\varepsilon)}/\varepsilon \to -b_{1,0}/b_{0,1}$ as $\varepsilon \to 0$, which means that $\Delta^{(\varepsilon)}$ can be represented as

$$\Delta^{(\varepsilon)} = a_1 \varepsilon + \Delta_1^{(\varepsilon)}, \tag{23.16}$$

where $a_1 = -b_{1,0}/b_{0,1}$ and $\Delta_1^{(\varepsilon)}$ is of order $o(\varepsilon)$.

This proves part (**i**) in the case $k = 1$.

If the perturbation condition holds for some $k \geq 2$ we can continue and substitute (23.16) and $\mathbf{P^{(k)}}$ into (23.14) in the case when $k = 2$. This yields

$$\begin{aligned}(a_1\varepsilon + \Delta_1^{(\varepsilon)})(b_{0,1} + b_{1,1}\varepsilon + o(\varepsilon)) + \frac{1}{2}(a_1\varepsilon + \Delta_1^{(\varepsilon)})^2(b_{0,2} + o(1)) \\ + (a_1\varepsilon + \Delta_1^{(\varepsilon)})^3 M_3 \theta_3^{(\varepsilon)} = -b_{1,0}\varepsilon - b_{2,0}\varepsilon^2 + o(\varepsilon^2).\end{aligned} \tag{23.17}$$

Dividing both sides of this equation by $b_{0,1}\varepsilon^2$, using the identity $a_1 = -b_{1,0}/b_{0,1}$ and letting ε tend to zero it can be concluded that

$$\frac{\Delta_1^{(\varepsilon)}}{\varepsilon^2} \to -\frac{1}{b_{0,1}}\left(b_{2,0} + b_{1,1}a_1 + \frac{1}{2}b_{0,2}a_1^2\right), \quad \text{as } \varepsilon \to 0.$$

Using this and equation (23.16), it is found that $\Delta^{(\varepsilon)}$ can be written in the form

$$\Delta^{(\varepsilon)} = a_1\varepsilon + a_2\varepsilon^2 + \Delta_2^{(\varepsilon)},$$

where

$$a_1 = -\frac{b_{1,0}}{b_{0,1}}, \quad a_2 = -\frac{1}{b_{0,1}}\left(b_{2,0} + b_{1,1}a_1 + \frac{1}{2}b_{0,2}a_1^2\right),$$

and the remainder term $\Delta_2^{(\varepsilon)}$ is order $o(\varepsilon^2)$.

This proves part (**i**) in the case $k = 2$.

Continuing in this way, the result is obtained for any positive integer k. However, once we know that the asymptotic expansion (23.6) exists, the coefficients $a_1, \ldots, a_k$ can be obtained in a simpler way. Consider the following formal equation.

$$\begin{aligned}(b_{0,1} &+ b_{1,1}\varepsilon + \cdots)(a_1\varepsilon + a_2\varepsilon^2 + \cdots)/1! \\ &+ (b_{0,2} + b_{1,2}\varepsilon + \cdots)(a_1\varepsilon + a_2\varepsilon^2 + \cdots)^2/2! + \cdots \\ &= -(b_{1,0}\varepsilon + b_{2,0}\varepsilon^2 + \cdots).\end{aligned} \tag{23.18}$$

By equating the coefficients of ε^n in the left- and in the right-hand sides of (23.18) for $n = 1, \ldots, k$, the formula (23.7) for calculating the coefficients $a_1, \ldots, a_k$ is obtained. Let $\alpha_{q,m}$ denote the coefficient of ε^q in the expansion of $(a_1\varepsilon + a_2\varepsilon^2 + \cdots)^m/m!$. Using this notation, the left-hand side of Equation (23.18) is

$$\begin{aligned}(b_{0,1} &+ b_{1,1}\varepsilon + \cdots)(a_1\varepsilon + a_2\varepsilon^2 + \cdots) \\ &+ (b_{0,2} + b_{1,2}\varepsilon + \cdots)(\alpha_{2,2}\varepsilon^2 + \alpha_{3,2}\varepsilon^3 + \cdots) \\ &+ (b_{0,3} + b_{1,3}\varepsilon + \cdots)(\alpha_{3,3}\varepsilon^3 + \alpha_{4,3}\varepsilon^4 + \cdots) + \cdots\end{aligned} \tag{23.19}$$

The contribution from the first summand in (23.19) to the coefficient of ε^n is

$$b_{n-1,1}a_1 + b_{n-2,1}a_2 + \cdots + b_{0,1}a_n = b_{0,1}a_n + \sum_{q=1}^{n-1} b_{n-q,1}a_q. \tag{23.20}$$

If $2 \le m \le n$, then the contribution from the mth summand in (23.19) to the coefficient of ε^n is

$$b_{n-m,m}\alpha_{m,m} + b_{n-m-1,m}\alpha_{m+1,m} + \cdots + b_{0,m}\alpha_{n,m} = \sum_{q=m}^{n} b_{n-q,m}\alpha_{q,m}. \tag{23.21}$$

No further contributions are given by the rest of the summands in (23.19). Using (23.19), (23.20) and (23.21) we see that equating the coefficients of ε^n in the left- and in the right-hand sides of (23.18) yields $b_{0,1}a_1 = -b_{1,0}$ if $n = 1$ and

$$b_{0,1}a_n + \sum_{q=1}^{n-1} b_{n-q,1}a_q + \sum_{m=2}^{n}\sum_{q=m}^{n} b_{n-q,m}\alpha_{q,m} = -b_{n,0}, \quad \text{for } n = 2, \dots, k.$$

To determine $\alpha_{q,m}$ it is sufficient to expand $(a_1\varepsilon + \cdots + a_{q-1}\varepsilon^{q-1})^m/m!$ and find the coefficient in front of ε^q. Using the multinomial theorem we obtain

$$\alpha_{q,m} = \sum_{n_1,\dots,n_{q-1}\in D_{m,q}} \prod_{p=1}^{q-1} \frac{a_p^{n_p}}{n_p!},$$

where $D_{m,q}$ is the set of all non-negative integer solutions to the system (23.8).

This concludes the proof of part (**i**).

Part (**ii**) follows directly from the recurrence formula (23.7).

Finally, let us prove part (**iii**). Note that it follows from Theorem 23.1 that

$$\frac{x^{(\varepsilon)}(n^{(\varepsilon)})}{\exp\,(-\rho^{(\varepsilon)}n^{(\varepsilon)})} \to \tilde{x}^{(0)}(\infty), \quad \text{as } \varepsilon \to 0. \tag{23.22}$$

If $\varepsilon^r n^{(\varepsilon)} \to \lambda_r \in [0, \infty)$ for some $1 \le r \le k$, it follows from part (**i**) that

$$\frac{\exp\,(-\rho^{(\varepsilon)}n^{(\varepsilon)})}{\exp\,(-(\rho^{(0)} + a_1\varepsilon + \cdots + a_{r-1}\varepsilon^{r-1})n^{(\varepsilon)})} \to e^{-\lambda_r a_r}, \quad \text{as } \varepsilon \to 0. \tag{23.23}$$

Part (**iii**) now follows from (23.22) and (23.23).

23.4 Discrete Time Regenerative Processes

Let, for every $\varepsilon \ge 0$, $Z_n^{(\varepsilon)}, n = 0, 1, \dots$, be a discrete time regenerative process with a phase space $\mathcal{X}$, with a σ-field of measurable subsets $\mathcal{B}_{\mathcal{X}}$ and regeneration times $0 < \tau_1^{(\varepsilon)} < \tau_2^{(\varepsilon)} < \cdots$, and let $\tau^{(\varepsilon)}$ be a regenerative stopping time, which regenerates together with the process $Z_n^{(\varepsilon)}$.

This means that the probability $P^{(\varepsilon)}(n, A) = \mathsf{P}\{Z_n^{(\varepsilon)} \in A, \tau^{(\varepsilon)} > n\}$ satisfies the following renewal equation,

$$P^{(\varepsilon)}(n, A) = q^{(\varepsilon)}(n, A) + \sum_{k=0}^{n} P^{(\varepsilon)}(n - k, A) f^{(\varepsilon)}(k), \quad n = 0, 1, \dots, \tag{23.24}$$

where

$$q^{(\varepsilon)}(n, A) = \mathsf{P}\{Z_n^{(\varepsilon)} \in A, \tau^{(\varepsilon)} \wedge \tau_1^{(\varepsilon)} > n\}, \quad n = 0, 1, \dots,$$

and

$$f^{(\varepsilon)}(n) = \mathsf{P}\{\tau_1^{(\varepsilon)} = n, \tau^{(\varepsilon)} > \tau_1^{(\varepsilon)}\}, \quad n = 0, 1, \dots$$

A typical example of a regenerative stopping time is $\tau^{(\varepsilon)} = \min(n \ge 1 : Z_n^{(\varepsilon)} \in D^{(\varepsilon)})$, the first hitting time of the process $Z_n^{(\varepsilon)}$ into some set $D^{(\varepsilon)} \in \mathcal{B}_{\mathcal{X}}$.

The results presented in Theorems 23.1 and 23.2 can be reformulated for the perturbed renewal equation (23.24).

Note that $\rho^{(0)} > 0$ if and only if the stopping probability $f^{(0)} = \mathsf{P}\{\tau^{(0)} \leq \tau_1^{(0)}\} > 0$, that is, the limiting distribution $f^{(0)}(k)$ is improper. In this case, Theorems 23.1 and 23.2 describe the so-called quasi-stationary asymptotics for the probabilities $P^{(\varepsilon)}(n, A)$.

Also, $\rho^{(0)} = 0$ if and only if the stopping probability $f^{(0)} = \mathsf{P}\{\tau^{(0)} \leq \tau_1^{(0)}\} = 0$, that is, the limiting distribution $f^{(0)}(k)$ is proper. In this case, Theorems 23.1 and 23.2 describe the so-called pseudo-stationary asymptotics for the probabilities $P^{(\varepsilon)}(n, A)$.

One can find additional comments concerning quasi- and pseudo-stationary phenomena in the book by Gyllenberg and Silvestrov (2008).

23.5 Queuing and Risk Applications

Suppose that a queuing system from the beginning contains u customers. The input flow of customers in the system is described by a sequence of i.i.d. random variables $V_n, n = 1, 2, \dots$, taking values 1 and 0 with probabilities q and $1 - q$, respectively. If $V_n = 1$, then one customer goes in the system at moment n. Such input flow is a discrete time analogue of a Poisson flow. The output flow is described by a sequence of i.i.d. random variables $W_n, n = 1, 2, \dots$, taking values $0, 1, \dots$, with the corresponding probabilities $h(0), h(1), \dots$. If $W_n = k$ then k customers go out from the system at moment n. We also assume that the random sequences $V_n, n = 1, 2, \dots$, and $W_n, n = 1, 2, \dots$, are independent.

The total number of customers in the system at moment n is given by the following relation,

$$Z_{u,n} = u + \sum_{k=1}^{n} V_k - \sum_{k=1}^{n} W_k, \quad n = 0, 1, \dots$$

Note that we formally admit the situation, where the random variables Z_n take negative values.

Usual assumptions are also that: **(a)** $q > 0$; **(b)** $\mathsf{P}\{W_1 > 1\} = 1 - h(0) - h(1) > 0$; **(c)** $\mathsf{E}W_1 = \sum_{x=0}^{\infty} xh(x) < \infty$.

It is possible and convenient to interpret the process $Z_{u,n}$ as a discrete time risk process. It can be done by defining a sequence of non-negative i.i.d. random variables $X_n = W_n - V_n + 1, n = 1, 2, \dots$, and representing the process $Z_{u,n}$ in the following form,

$$Z_{u,n} = u + n - \sum_{k=1}^{n} X_k, \quad n = 0, 1, \dots$$

In this case, $Z_{u,n}$ may be interpreted as a capital of an insurance company (counted in units equivalent to expected premium per time unit) at moment n, and the random variable X_n as claims at moment n, counted in the same units.

An object of our interest is the following probability,

$$\psi(u) = \mathsf{P}\{\min_{n\geq 0} Z_{u,n} < 0\}, \quad u = 0, 1, \dots, \tag{23.25}$$

which can be interpreted as a ruin probability for the infinite time horizon.

Let us denote,

$$p = \mathsf{P}\{X_1 > 0\}, \quad \mu = \sum_{u=0}^{\infty} u g(u), \quad g(u) = \mathsf{P}\{X_1 = u/X_1 > 0\}, \quad u = 0, 1, \ldots$$

It is not difficult to show that $\psi(u) = 1$ if $\mathsf{E}X_1 = \alpha = p\mu > 1$. Let us, therefore, assume that **(d)** $\mathsf{E}X_1 = \alpha = p\mu \leq 1$.

In this case, the ruin probability $\psi(u)$ satisfies the following discrete time renewal equation,

$$\psi(u) = q(u) + \sum_{k=0}^{u} \psi(u-k) f(k), \quad u = 0, 1, \ldots, \tag{23.26}$$

where

$$G(u) = \sum_{k=0}^{u} g(k), \quad f(u) = \alpha \frac{1 - G(u)}{\mu}, \quad q(u) = \sum_{k=u+1}^{\infty} f(k), \quad u = 0, 1, \ldots$$

Just, in order to make the chapter understandable, let us briefly show the way of getting the renewal equation (23.26). It repeats the method of finding a similar continuous time renewal equation for ruin probabilities, given, for example in Feller (1966) and Grandell (1991).

Let $\phi(u) = 1 - \psi(u)$. By conditioning on X_1 we obtain, for any $u = 0, 1, \ldots,$

$$\begin{aligned}
\phi(u) &= \sum_{x=0}^{u+1} \phi(u+1-x)\mathsf{P}\{X_1 = x\} \\
&= (1-p)\phi(u+1) + p\sum_{x=1}^{u+1} \phi(u+1-x)g(x).
\end{aligned}$$

Rearranging this gives, for any $u = 0, 1, \ldots,$

$$\phi(u+1) - \phi(u) = p\phi(u+1) - p\sum_{x=1}^{u+1} \phi(u+1-x)g(x).$$

Using this it follows that for any $u = 1, 2, \ldots,$

$$\begin{aligned}
\phi(u) - \phi(0) &= \sum_{t=0}^{u-1} (\phi(t+1) - \phi(t)) \\
&= p\sum_{t=1}^{u} \phi(t) - p\sum_{t=1}^{u}\sum_{x=1}^{t} \phi(t-x)g(x) \\
&= -p\phi(0) + p\sum_{t=0}^{u} \phi(u-t)(1 - G(t)).
\end{aligned}$$

The left-hand side is equal to the right-hand side also for $u = 0$, thus,

$$\phi(u) = (1-p)\phi(0) + p\sum_{t=0}^{u}\phi(u-t)(1-G(t)), \quad u = 0, 1, \ldots \tag{23.27}$$

It follows from the strong law of large numbers that $n^{-1}Z_{u,n} \to (1-\alpha)$ almost surely as $n \to \infty$. In the case $\alpha < 1$, this implies that for almost every $\omega \in \Omega$, there exists a positive integer $N(\omega)$ such that $Z_{u,n}(\omega) > 0$ for all $n \geq N(\omega)$. This yields that $\min_{n\geq 0} Z_{u,n}$ is almost surely finite and, therefore, we can conclude that $\phi(u) \to 1$ as $u \to \infty$. From this and (23.27) it follows by monotone convergence that $1 = (1-p)\phi(0) + \alpha$. Solving for $\phi(0)$, putting this into Equation (23.27) and rearranging we obtain the renewal equation (23.26) for $\psi(u)$.

In the case, where the parameters of the above risk process depend on a perturbation parameter $\varepsilon \geq 0$, the results presented in Theorems 23.1 and 23.2 can be applied to the renewal equation (23.26) and yield exponential asymptotic expansions for the ruin probability $\psi^{(\varepsilon)}(u)$ which, in this case, also depends on the perturbation parameter ε.

It is worth noting that in the above model, $\rho^{(0)} > 0$ if and only if the limiting parameter $\alpha^{(0)} < 1$. In this case, the quasi-stationary asymptotics for $\psi^{(\varepsilon)}(u)$ given by Theorems 23.1 and 23.2 generalize the classical Cramér-Lundberg approximation for these ruin probabilities to the more advanced form of an exponential asymptotic expansion. Also, $\rho^{(0)} = 0$ if and only if the limiting parameter $\alpha^{(0)} = 1$. In this case, the pseudo-stationary asymptotics for $\psi^{(\varepsilon)}(u)$ given in these theorems generalizes the so-called diffusion approximation for these ruin probabilities.

References

Altman, E., K. E. Avrachenkov and R. Nunes-Queija. 2004. Perturbation analysis for denumerable Markov chains with application to queueing models. *Advances in Applied Probability* 36 (3): 839–853.

Blanchet, J. and B. Zwart. 2010. Asymptotic expansions of defective renewal equations with applications to perturbed risk models and processors sharing queues. *Mathematical Methods of Operations Research* 72 (2): 311–326.

Englund, E. and D. S. Silvestrov. 1997. Mixed large deviation and ergodic theorems for regenerative processes with discrete time. In *Proceedings of the Second Scandinavian - Ukrainian Conference in Mathematical Statistics*, ed. P. Jagers, G. Kulldorff, N. Portenko, and D. Silvestrov. *Theory of Stochastic Processes* 3(19) (1-2): 164–176.

Erdos, P., W. Feller and H. Pollard. 1949. A theorem on power series. *Bulletin of the American Mathematical Society* 55: 201–204.

Feller, W. 1950, 1957, 1968. *An Introduction to Probability Theory and Its Applications*, Vol. I. New York: John Wiley & Sons.

Feller, W. 1966, 1971. *An Introduction to Probability Theory and Its Applications*, Vol. II. New York: Wiley.

Grandell, J. 1991. *Aspects of Risk Theory*, New York: Springer.

Gyllenberg, M. and D. S. Silvestrov. 2008. *Quasi-Stationary Phenomena in Nonlinearly Perturbed Stochastic Systems*, Berlin: Walter de Gruyter.

Hassin, R. and M. Haviv. 1992. Mean passage times and nearly uncoupled Markov chain. *SIAM Journal on Discrete Mathematics* 5 (3): 386–397.

Kalashnikov, V. V. 1978. *Qualitative Analysis of the Behaviour of Complex Systems by the Method of Test Functions*, Moskow: Nauka.

Kartashov, N. V. 2009. Inhomogeneous perturbations of a renewal equation and the Cramer-Lundberg theorem for a risk process with variable premium rates. *Theory of Probability and Mathematical Statistics* 78: 61–73.

Kartashov, N. V. 1988. Asymptotic expansions and inequalities in stability theorems for general Markov chains with relatively bounded perturbations. *Journal of Mathematical Sciences* 40 (4): 509–518.

Kartashov, N. V. 1996. *Strong Stable Markov Chains*, VSP: Utrecht, The Netherlands and TBiMC Scientific Publishers: Kiev, Ukraine.

Khasminskii, R. Z., G. Yin and Q. Zhang. 1996. Singularly perturbed Markov chains: quasi-stationary distribution and asymptotic expansion. In *Proceedings of Dynamic Systems and Applications*, ed. G.S. Ladde and M. Sambandhan, Vol. 2, 301–308. Atlanta, GA, 1996, Dynamic, Atlanta, GA.

Latouche, G. 1988. Perturbation analysis of a phase-type queue with weakly correlated arrivals. *Advances in Applied Probability* 20: 896–912.

Ni, Y. 2010. Perturbed renewal equations with multivariate non-polynomial perturbations. In *Proceedings of the International Symposium on Stochastic Models in Reliability Engineering, Life Science and Operations Management*, ed. I. Frenkel, I. Gertsbakh, L. Khvatskin, Z. Laslo and A. Lisnianski, 754–763, Beer Sheva, Israel.

Yin, G. and D. T. Nguyen. 2009. Asymptotic expansions of backward equations for two-time-scale Markov chains in continuous time. *Acta Mathematicae Applicatae Sinica (Eng. Ser.)* 25 (3): 457–476.

Yin, G. and Q. Zhang. 2003. Discrete-time singularly perturbed Markov chains. In *Stochastic Modelling and Optimization*, ed. D.D. Yao, H. Zhang, X.U. Zhou, 1–42, New York: Springer.

24

On Generalized Extreme Shock Models under Renewal Shock Processes

Ji Hwan Cha[1] and Maxim Finkelstein[2]
[1]*Department of Statistics, Ewha Womans University, Korea*
[2]*Department of Mathematical Statistics, University of the Free State, South Africa*

24.1 Introduction

Shock models are widely used in many different areas such as reliability, structure and infrastructure engineering, insurance, credit risk, etc. (see, e.g. Nakagawa 2007, for the comprehensive list of references on the subject). Along with meaningful mathematical properties, they have significant practical importance and a wide range of applications.

Two major types of shock models are extensively studied in the literature: cumulative shock models (systems fail because of some cumulative effect) and extreme shock models (systems fail due to one single "large" shock). In the latter case, only the impact of the current, possibly fatal shock is usually taken into account: a system fails with some probability and survives with the complementary probability. The most important assumption in the extreme shock models is that this probability does not depend on the corresponding histories of the shock process and system operation. However, this is often not the case in practice and the history-dependent settings should be considered for a more adequate stochastic description of shock models.

Some generalizations (to name a few) of traditional models have been considered in the literature. For instance, Sumita and Shanthikumar (1985) studied the cumulative shock model when the arrival time and the corresponding shock magnitude are correlated;

Applied Reliability Engineering and Risk Analysis: Probabilistic Models and Statistical Inference, First Edition.
Edited by Ilia B. Frenkel, Alex Karagrigoriou, Anatoly Lisnianski and Andre Kleyner.

Gut and Hüsler (2005) deal with the consequences of only latest shocks and define the corresponding "influence window"; Mallor and Omey (2001) study systems that fail when k consecutive shocks (a run) with critical magnitude (above a certain level) occur; in Cha and Finkelstein (2009), the critical value of accumulated damage that defines the failure of a system is an exponentially distributed random variable and, therefore, a probabilistic description of the cumulative and the extreme shock models appears to be mathematically equivalent in this specific case.

Our approach can be considered a meaningful generalization of the classical extreme shock model. We assume that the probability of failure on each shock's occurrence "depends" not only on the history (or its elements) of the shock process but also on the operational history of a system as well (see Cha and Finkelstein 2011, for some specific models of this kind). It should be noted that the conventional models usually deal with the NHPP of shocks, which makes the corresponding stochastic analysis fairly simple (see Section 24.2.2). However, in many instances the external shock processes are more likely to follow the renewal process pattern and the derivations and final results for this setting (as usually found in problems with renewal processes) are much more cumbersome and can often be expressed only in the form of various infinite sums. This will be illustrated by our results in Section 24.3.

The chapter is organized as follows. In Section 24.2, we first present a solution (in terms of Laplace transforms) for a history-independent shock model when the underlying shock process is a renewal process. Then we state a fairly general result for history-dependent shock models with arbitrary underlying shock processes. In Section 24.3, we apply this result to the specific settings with renewal shock process. Finally, some concluding remarks are given in Section 24.4.

24.2 Generalized Extreme Shock Models

24.2.1 "Classical" Extreme Shock Model for Renewal Process of Shocks

Consider a system subject to a renewal process of shocks. Denote the inter-arrival times by X_i, $i = 1, 2, \ldots,$ its Cdf (the governing distribution of the renewal process), the corresponding pdf, and the failure rate function by$F_X(x)$ $(\overline{F}_X(x) \equiv 1 - F_X(x))$, $f_X(x)$, and $\lambda_X(x)$, respectively. Let our system be, for simplicity, "absolutely reliable" in the absence of shocks. Assume that each shock (regardless of its number) results in its failure with probability $p(t)$ and is harmless (i.e., has no effect) with probability $q(t) = 1 - p(t)$, "independently of everything else," where t is the time elapsed since the last shock occurrence. This setting is often referred to as the *extreme shock model* (see, e.g. Gut and Hüsler 2005). Denote by T_S the time to termination (failure) of the process and by $\overline{F}_S(t)$ the corresponding survival function, respectively. Then, the standard renewal-type reasoning results in the following equation:

$$\overline{F}_S(t) = 1 - F_X(t) + \int_0^t f_X(x)q(x)\overline{F}_S(t - x)dx, \tag{24.1}$$

where $f_X(t)$ is the density that corresponds to $F_X(t)$. Indeed, the first term on the right-hand side is the probability that there is no shock in $(0, t]$. The integrand defines the probability that the first shock has occurred in $(x, x + dx]$, it is harmless to the system (the probability of this event is $1 - p(x)$), and then the system has survived in $(x, t]$. Define $g(t) \equiv f_X(t)q(x)$. Letting $\overline{F}_S{}^*(s)$, $g^*(s)$ and $f_X^*(s)$ be the Laplace transforms of $\overline{F}_S(t)$, $g(t)$ and $f_X(x)$, respectively. Applying the Laplace transform to both sides of (24.1) results in

$$\overline{F}_S{}^*(s) = \frac{1 - f_X^*(s)}{s} + g^*(s)\overline{F}_S{}^*(s).$$

From this,

$$\overline{F}_S{}^*(s) = \frac{1}{s}\left(\frac{1 - f_X^*(s)}{1 - g^*(s)}\right)$$

and the survival function can be obtained by inversion of this Laplace transform. Note that a closed form of $\overline{F}_S(t)$ can be derived by the inversion procedure only for limited cases. For instance, when $f_X(t) = \lambda \exp\{\lambda t\}$ and $q(t) \equiv q$, the survival function, $\overline{F}_X(t)$ can be obviously obtained from its Laplace transform as $\exp\{-p\lambda t\}$ with the corresponding failure rate λt. However, it is well known (see, e.g. Finkelstein 2008, and references therein) that when the underlying shock process is the NHPP, the survival probability for a classical extreme shock model can be obtained for general $\lambda(t)$ and $p(t)$ in a simple and meaningful form

$$\overline{F}_S(t) = \exp\left(-\int_0^t p(u)\,\nu(u)du\right) \tag{24.2}$$

Thus, the renewal case to be considered in this chapter is far more computationally complex even in a well-defined history-independent case.

As mentioned before, the crucial assumptions for deriving $\overline{F}_S(t)$ is that the probability $p(t)$ does not depend on the history of the shock process. A careful probabilistic reasoning is required to generalize this classical extreme shock model to the history-dependent case and this is done in Section 24.2.2 for a rather general setting.

24.2.2 *History-Dependent Extreme Shock Model*

In this section, in line with Cha and Finkelstein (2012), we consider the problem in a far more generality when both $p(t)$ and the shock process depend on the history. After the general discussion, in Section 24.3 we will consider several meaningful specific cases.

Let both the counting process of shocks and the failure probability of a system on each shock depend now on the history of the shock process up to t, $H_{t-} \equiv \{N(s), 0 \le s < t\}$, and on some other random history process up to t, Φ_{t-}. The latter can describe, for example, the evolving operational characteristics of our system (e.g. covariates such as the total resource of the system, accumulated amount of wear, the state of the system, etc.). We also assume that the initial shock process $N(t), t \ge 0$ is *external* for the system in the sense that $N(t), t \ge 0$ may influence the system's performance but is not influenced by it (Fleming and Harrington 1991).

Let $N(t), t \ge 0$, denote an orderly point process of shocks with arrival times $T_i, i = 1, 2, \ldots$ affecting our system and let shocks be the only cause of its failure. Denote by $t-$, as usual, the time instant just prior t. For example, $N(t-)$ denotes the number of shocks in $[0, t)$. Then the conditional probability of failure in the infinitesimal interval of time can be formally written as

$$\begin{aligned}
&P[T_S \in [t, t+dt)|T_S \geq t, H_{t-}, \Phi_{(t+dt)-}] \\
&\quad = P[T_S \in [t, t+dt)|T_S \geq t, H_{t-}, \Phi_{(t+dt)-}, N((t+dt)-) - N(t-) = 1] \\
&\qquad \times P[N((t+dt)-) - N(t-) = 1|T_S \geq t, H_{t-}, \Phi_{(t+dt)-}] \\
&\qquad + P[T_S \in [t, t+dt)|T_S \geq t, H_{t-}, \Phi_{(t+dt)-}, N((t+dt)-) - N(t-) = 0] \\
&\qquad \times P[N((t+dt)-) - N(t-) = 0|T_S \geq t, H_{t-}, \Phi_{(t+dt)-}] \\
&\quad = P[T_S \in [t, t+dt)|T_S \geq t, H_{t-}, \Phi_{(t+dt)-}, N((t+dt)-) - N(t-) = 1] \\
&\qquad \times P[N((t+dt)-) - N(t-) = 1|H_{t-}], \qquad (24.3)
\end{aligned}$$

where

$$P[N((t+dt)-) - N(t-) = 1|T_S \geq t, H_{t-}, \Phi_{(t+dt)-}]$$

reduces to

$$P[N((t+dt)-) - N(t-) = 1|H_{t-}]$$

as H_{t-} completely defines the corresponding conditional probability due to our assumption that the initial shock process is external and also due to the fact that the event "$T_S \geq t$" does not give any additional information given the full history H_{t-}. Accordingly, from (24.3),

$$\begin{aligned}
&P[T_S \in [t, t+dt)|T_S \geq t, H_{t-}, \Phi_{(t+dt)-}] \\
&\quad = P[T_S \in [t, t+dt)|T_S \geq t, H_{t-}, \Phi_{(t+dt)-}, N((t+dt)-) - N(t-) = 1] \cdot \nu(t|H_{t-})dt, \qquad (24.4)
\end{aligned}$$

where $\nu(t|H_{t-})$ is the conditional intensity function (CIF) (Cox and Isham 1980; Anderson *et al.* 1993) defined by

$$\nu(t|H_{t-}) \equiv \lim_{\Delta t \to 0} \frac{P[N((t+\Delta t)-) - N(t-) = 1|H_{t-}]}{\Delta t}.$$

Therefore, from (24.4), the only way for

$$P[T_S \in [t, t+dt)|T_S \geq t, H_{t-}, \Phi_{(t+dt)-}, N((t+dt)-) - N(t-) = 1] \cdot \nu(t|H_{t-}) \qquad (24.5)$$

to become the full-fledged failure rate that corresponds to the lifetime T_S and therefore, for the ordinary exponential representation to hold, is when there is no dependence on H_{t-} and $\Phi_{(t+dt)-}$ in both multipliers in (24.5). However, when there is found this dependence in at least one of the multipliers, the failure rate function of the lifetime T_S can be derived by the following proposition.

Proposition 24.1. *The unconditional failure rate function of T_S is*

$$\begin{aligned}
\lambda_S(t) = E[P[T_S \in [t, t+dt)|T_S \geq t, H_{t-}, \Phi_{(t+dt)-},] \\
N((t+dt)-) - N(t-) = 1] \cdot \nu(t|H_{t-})], \qquad (24.6)
\end{aligned}$$

where the expectation is with respect to the joint conditional distribution of

$$(H_{t-}, \Phi_{(t+dt)-}|T_S \geq t), t \geq 0.$$

Obtaining the explicit expressions for the conditional failure rate in (24.6) is usually possible only for the relatively simple settings to be considered in what follows. However, it should be pointed out that practical implementation of Proposition 24.1 is meaningful and not straightforward , which will be illustrated in the next section.

24.3 Specific Models

In this section, we consider several models where the theoretical message of the previous section can be applied in different ways. We focus our discussion on the renewal process of shocks. As in Section 24.2, denote the i.i.d. inter-arrival times by X_i, $i = 1, 2, \ldots$ and the Cdf, the pdf, and the failure rate function by $F_X(x)$ $(\overline{F}_X(x) \equiv 1 - F_X(x))$, $f_X(x)$, and $\lambda_X(x)$, respectively. Denote also the arrival times by $T_i, i = 1, 2, \ldots$.

24.3.1 Stress-Strength Model

Let the ith shock be characterized by its magnitude (stress) S_i, $i = 1, 2, \ldots$ Assume that S_i are i.i.d. random variables with the Cdf $F_S(s)$ $(\overline{F}_S(s) \equiv 1 - F_S(s))$ and the corresponding pdf $f_S(s)$. Let first, the strength of the system be given by a constant u. Then the system survives the ith shock if $S_i \leq u$; and it fails if $S_i > u$, 'independently of everything else'. Obviously, equation (1) can be applied with $p(t)$ replaced by $P(S_i > u)$, which is time-independent in this case.

In practice, it is more likely that the strength of a system is a random variable. Denote it by U with the Cdf $G_U(u)$ $(\overline{G}_U(u) \equiv 1 - G_U(u))$ and the pdf $g_U(u)$. It is clear that in this case we cannot apply equation (1) as there is no renewal point now. Indeed, after the first survived shock, the system's survival probability on the next shock is different from $P(S_i > u)$, as the survived system should have a different (updated) strength distribution. However, we still can proceed by deriving $\lambda_S(t)$ in a direct way using Proposition 24.1. In this model, $\Phi_{(t+dt)-} = \{S_1, S_2, \ldots, S_{N((t+dt)-)}; U\}$ and

$$P[T_S \in [t, t+dt)|T_S \geq t, H_{t-}, \Phi_{(t+dt)-}, N((t+dt)-) - N(t-) = 1] = I(S_{N(t-)+1} > U),$$

where the conditional distribution of $U|T_S \geq t$ does depend on the event $\{T_S \geq t\}$ and $I(\cdot)$ is the corresponding indicator. On the other hand, as shocks arrive in accordance with the renewal process, its history which determines the conditional intensity is just the time elapsed since the last renewal, that is

$$\nu(t|H_{t-}) = \lambda_X(t - T_{N(t-)}).$$

Therefore, from (24.6),

$$\lambda_S(t) = E[I(S_{N(t-)+1} > U)\lambda(t - T_{N(t-)})], t \geq 0, \tag{24.7}$$

where the expectation is with respect to the joint conditional distribution

$$(T_{N(t-)}, U|T_S \geq t).$$

Observe that

$$\begin{aligned}P(T_S \geq t|N(t-) &= n, T_1 = t_1, T_2 = t_2, \ldots, T_{N(t-)} = t_n, U = u) \\ &= P(u \geq \max\{S_1, S_2, \ldots, S_n\}) = (F_S(u))^n.\end{aligned}$$

As the joint distribution of the event $\{N(t-) = n, T_1 = t_1, T_2 = t_2, \ldots, T_n = t_n\}$ is given by

$$\left(\prod_{j=1}^{n} f_X\left(t_j - t_{j-1}\right)\right) \overline{F}_X(t - t_n), \quad 0 = t_0 < t_1 < t_2 < \cdots < t_n < t,$$

the joint distribution of $(T_S \geq t, T_1 = t_1, T_2 = t_2, \ldots, T_{N(t-)} = t_n, U = u)$ is

$$\begin{aligned}f(T_S \geq t, N(t-) &= n, T_1 = t_1, T_2 = t_2, \ldots, T_{N(t-)} = t_n, U = u) \\ &= (F_S(u))^n \left(\prod_{j=1}^{n} f_X\left(t_j - t_{j-1}\right)\right) \overline{F}_X(t - t_n) \cdot g_U(u),\end{aligned}$$

where $t_0 \equiv 0$. In the following, it would be more convenient to use the notation y instead of t_n. Thus, the joint distribution of $(T_S \geq t, T_{N(t-)} = y, U = u)$ is

$$\begin{aligned}&f(T_S \geq t, T_{N(t-)} = y, U = u) \\ &= \sum_{n=0}^{\infty} \int_0^y \int_0^{t_{n-1}} \cdots \int_0^{t_2} (F_S(u))^n \left(\prod_{j=1}^{n-1} f_X\left(t_j - t_{j-1}\right)\right) f_X(y - t_{n-1}) \\ &\quad \overline{F}_X(t - y) dt_1 dt_2 \cdots dt_{n-1} \cdot g_U(u) \\ &= \sum_{n=0}^{\infty} (F_S(u))^n \cdot \overline{F}_X(t - y) f_X^{(n)}(y) \cdot g_U(u),\end{aligned}$$

where $f_X^{(n)}(y)$ denotes the corresponding convolution. Therefore, the conditional distribution $(T_{N(t-)} = y, U = u|T_S \geq t)$ is

$$\frac{\sum_{n=0}^{\infty} (F_S(u))^n \cdot \overline{F}_X(t - y) f_X^{(n)}(y) \cdot g_U(u)}{\int_0^{\infty} \int_0^t \sum_{n=0}^{\infty} (F_S(r))^n \cdot \overline{F}_X(t - z) f_X^{(n)}(z) \cdot g_U(r) dz dr}.$$

Finally, from (24.7),

$$\begin{aligned}\lambda_S(t) &= E[I(S_{N(t-)+1} > U) \cdot \lambda_X(t - T_{N(t-)})|T_S \geq t] \\ &= E[E[I(S_{N(t-)+1} > U) \cdot \lambda_X(t - T_{N(t-)})|T_S \geq t, S_{N(t-)+1}]],\end{aligned}$$

where the outermost expectation is with respect to the distribution of $(S_{N(t-)+1}|T_S \geq t)=_d S_{N(t-)+1}$. Therefore,

$$\lambda_S(t) = \int_0^\infty \frac{\int_0^s \int_0^t \lambda_X(t-y) \sum_{n=0}^\infty (F_S(u))^n \cdot \overline{F}_X(t-y) f_X^{(n)}(y) \cdot g_U(u) dy du}{\int_0^\infty \int_0^t \sum_{n=0}^\infty (F_S(r))^n \cdot \overline{F}_X(t-z) f_X^{(n)}(z) \cdot g_U(r) dz dr} f_S(s) ds, t \geq 0.$$

24.3.2 Model A in Cha and Finkelstein (2011)

We revisit now Model A of Cha and Finkelstein (2011), where the survival probability on each shock was initially defined as a decreasing function of the total number of shocks, $n(t)$ (history) that the system has experienced in $[0, t)$. The specific multiplicative form of this function was considered: $q(t)\rho(n(t))$, where $\rho(\cdot)$ is a decreasing function of its argument and $q(t)$ is the corresponding history-independent survival probability. As the approach of this chapter relied on the properties of the NHPP of shocks, it cannot be used for the renewal process of shocks. Therefore, in accordance with Proposition 24.1, we will now obtain the corresponding failure rate *directly*.

In this case,

$$P[T_S \in [t, t+dt)|T_S \geq t, H_{t-}, N((t+dt)-) - N(t-) = 1] = 1 - \rho(N(t-)+1)q(t),$$

where the conditional probability in this case does not involve the process $\Phi_{(t+dt)-}$. Therefore, (24.6) simplifies to

$$\lambda_S(t) = E[(1 - \rho(N(t-)+1)q(t))\lambda(t - T_{N(t-)})|T_S \geq t], t \geq 0. \tag{24.8}$$

We will now derive the joint conditional distribution of $(T_{N(t-)}, N(t-)|T_S \geq t)$. It is clear that

$$P(T_S \geq t|N(t-), T_1, T_2, \ldots, T_{N(t-)}) = \prod_{i=1}^{N(t-)} q(T_i)\rho(i).$$

The joint distribution of $f(T_S \geq t, T_{N(t-)} = y, N(t-) = n)$ is given by

$$\begin{aligned} &f(T_S \geq t, T_{N(t-)} = y, N(t-) = n) \\ &= \left\{\prod_{i=1}^n \rho(i)\right\} \cdot \int_0^y \int_0^{t_{n-1}} \cdots \int_0^{t_2} q(y) f_X(y - t_{n-1}) \\ &\quad \times \left(\prod_{j=1}^{n-1} q(t_j) f_X(t_j - t_{j-1})\right) \overline{F}_X(t-y) dt_1 dt_2 \cdots dt_{n-1}. \end{aligned}$$

Therefore,

$$f(T_{N(t-)} = y, N(t-) = n | T_S \geq t)$$

$$= \frac{\left\{\prod_{i=1}^{n} \rho(i)\right\} \cdot \int_0^y \int_0^{t_{n-1}} \cdots \int_0^{t_2} q(y) f_X(y - t_{n-1}) \left(\prod_{j=1}^{n-1} q(t_j) f_X(t_j - t_{j-1})\right) \times \overline{F}_X(t - y) dt_1 dt_2 \cdots dt_{n-1}}{\sum_{n=0}^{\infty} \int_0^t \left\{\prod_{i=1}^{n} \rho(i)\right\} \cdot \int_0^{t_n} \int_0^{t_{n-1}} \cdots \int_0^{t_2} q(t_n) f_X(t_n - t_{n-1}) \left(\prod_{j=1}^{n-1} q(t_j) f_X(t_j - t_{j-1})\right) \times \overline{F}_X(t - t_n) dt_1 dt_2 \cdots dt_{n-1} dt_n},$$

$n = 0, 1, 2, \ldots,$ where $\prod_{i=1}^{n} (\cdot)_i \equiv 1$ for $n = 0$. Finally, from (24.8),

$$\lambda_S(t) =$$

$$\int_0^t \frac{\lambda_X(t - y) \sum_{n=0}^{\infty} \left\{\prod_{i=1}^{n} \rho(i)\right\} \cdot \int_0^y \int_0^{t_{n-1}} \cdots \int_0^{t_2} q(y) f_X(y - t_{n-1}) \left(\prod_{j=1}^{n-1} q(t_j) f_X(t_j - t_{j-1})\right) \overline{F}_X(t - y) dt_1 dt_2 \cdots dt_{n-1}}{\sum_{n=0}^{\infty} \int_0^t \left\{\prod_{i=1}^{n} \rho(i)\right\} \cdot \int_0^{t_n} \int_0^{t_{n-1}} \cdots \int_0^{t_2} q(z) f_X(t_n - t_{n-1}) \left(\prod_{j=1}^{n-1} q(t_j) f_X(t_j - t_{j-1})\right) \overline{F}_X(t - t_n) dt_1 dt_2 \cdots dt_{n-1} dt_n} dy$$

$$- \int_0^t \frac{q(t) \lambda_X(t - y) \sum_{n=0}^{\infty} \left\{\prod_{i=1}^{n+1} \rho(i)\right\} \cdot \int_0^y \int_0^{t_{n-1}} \cdots \int_0^{t_2} q(y) f_X(y - t_{n-1}) \left(\prod_{j=1}^{n-1} q(t_j) f_X(t_j - t_{j-1})\right) \overline{F}_X(t - y) dt_1 dt_2 \cdots dt_{n-1}}{\sum_{n=0}^{\infty} \int_0^t \left\{\prod_{i=1}^{n} \rho(i)\right\} \cdot \int_0^{t_n} \int_0^{t_{n-1}} \cdots \int_0^{t_2} q(t_n) f_X(t_n - t_{n-1}) \left(\prod_{j=1}^{n-1} q(t_j) f_X(t_j - t_{j-1})\right) \overline{F}_X(t - z) dt_1 dt_2 \cdots dt_{n-1} dt_n} dy.$$

24.3.3 State-Dependent Shock Model

It is natural to discuss the case when the failure probability on each shock's occurrence depends on the state of the system at that time. The stress-strength model of Section 24.3.1 can be viewed as a simple continuous-state model of this kind, because the random strength, in fact, defines the state of our system. Let us now discuss the following discrete-state model with the renewal process of shocks. The forthcoming approach can be generalized to the more complex state-dependent settings. However, the results are rather cumbersome even for the case below.

The system has $m + 1$ states, denoted as $0, 1, \ldots, m$. It starts operation at state 0 (the best state) and sequentially increases its state ,staying in each of them exponentially distributed (with parameter η) random time. Thus, the state of our system changes in accordance with the described Markov chain with the finite number of states.

Let $S(t)$ be the state of the system at time t. Assume that the failure probability upon a shock at time t is $p(t, S(t))$ $(q(t, S(t)) \equiv 1 - p(t, S(t)))$, where $p(t, s)$ is increasing in s for all fixed t. Obviously, as in the previous models, $\lambda_S(t)$ cannot be defined by just as $E[p(t, S(t))]\nu(t)$ and Proposition 24.1 should be applied. Note that, in this case,

$\Phi_{(t+dt)-} = \{S(u), 0 \le u < t + dt\}$. Therefore,

$$P[T_S \in [t, t+dt) | T_S \ge t, H_{t-}, \Phi_{(t+dt)-}, N((t+dt)-) - N(t-) = 1]$$
$$= p(t, S(T_{N(t-)+1})),$$

where $t \le T_{N(t-)+1} < t + dt$. It can easily be shown for $T_{N(t-)+1}$ in this interval that

$$\begin{aligned} &P\left(S\left(T_{N(t-)+1}\right) = j | T_S \ge t\right) \\ &= \sum_{k=0}^{m} P(S(T_{N(t-)+1}) = j | T_S \ge t, S(t-) = k) P(S(t-) = k | T_S \ge t) \\ &= \sum_{k=0}^{j} P(S(T_{N(t-)+1}) = j | T_S \ge t, S(t-) = k) P(S(t-) = k | T_S \ge t) \\ &= P(S(T_{N(t-)+1}) = j | T_S \ge t, S(t-) = j) P(S(t-) = j | T_S \ge t) \\ &= P(S(T_{N(t-)}) = j | T_S \ge t), \; j = 0, 1, .., m, \end{aligned} \tag{24.9}$$

where, for notational sake, the corresponding additive terms $O(dt)$ are omitted. This is intuitively obvious, as in the limit ($dt \to 0$) the probability of the state change in this interval is vanishing (the processes are orderly). Note that (24.9) is equivalent to

$$\begin{aligned} &\int_0^t \left[f\left(S\left(T_{N(t-)+1}\right) = j, T_{N(t-)} = y | T_S \ge t\right)\right. \\ &\quad \left. - f\left(S\left(T_{N(t-)}\right) = j, T_{N(t-)} = y | T_S \ge t\right)\right] dy = O(dt). \end{aligned}$$

Then, as $p(t, j)\lambda_X(t - y) \le M$ for all $j = 0, 1, \ldots m, 0 \le y \le t$ for some M (i.e., bounded by a certain value M), for $T_{N(t-)+1}$ in the interval $t \le T_{N(t-)+1} < t + dt$, we have

$$\begin{aligned} &\int_0^t p(t, j)\lambda_X(t - y)\left[f\left(S\left(T_{N(t-)+1}\right) = j, T_{N(t-)} = y | T_S \ge t\right)\right. \\ &\quad \left. - f\left(S\left(T_{N(t-)}\right) = j, T_{N(t-)} = y | T_S \ge t\right)\right] dy \\ &\le \int_0^t M\left[f\left(S\left(T_{N(t-)+1}\right) = j, T_{N(t-)} = y | T_S \ge t\right)\right. \\ &\quad \left. - f\left(S\left(T_{N(t-)}\right) = j, T_{N(t-)} = y | T_S \ge t\right)\right] dy \\ &= M\int_0^t \left[f\left(S\left(T_{N(t-)+1}\right) = j, T_{N(t-)} = y | T_S \ge t\right)\right. \\ &\quad \left. - f\left(S\left(T_{N(t-)}\right) = j, T_{N(t-)} = y | T_S \ge t\right)\right] dy = O(dt)] \end{aligned}$$

for all $j = 0, 1, \ldots m$. Accordingly,

$$\begin{aligned} &E[p(t, S(T_{N(t-)+1}))\lambda_X(t - T_{N(t-)}) | T_S \ge t] \\ &\quad = E[p(t, S(T_{N(t-)}))\lambda_X(t - T_{N(t-)}) | T_S \ge t] + O(dt), \end{aligned}$$

and based on (24.4) and (24.6), $\lambda_S(t)$can be obtained (exactly) as

$$\lambda_S(t) = E[p(t, S(T_{N(t-)}))\lambda_X(t - T_{N(t-)})|T_S \geq t], t \geq 0. \tag{24.10}$$

Observe that the conditional distribution

$$(T_S \geq t|N(t-), T_1, T_2, \ldots, T_{N(t-)}; S(T_1), S(T_2), \ldots, S(T_{N(t-)}))$$

is given by

$$P\left(T_S \geq t|N(t-) = n, T_1 = t_1, T_2 = t_2, \ldots, T_{N(t-)} = t_n;\right.$$
$$\left. S(T_1) = i_1, S(T_2) = i_2, \ldots, S(T_{N(t-)}) = i_n\right) = \prod_{i=1}^{n} q(t_i, i_i).$$

On the other hand, the joint conditional distribution of

$$\left(S\left(T_1\right) = i_1, S(T_2) = i_2, \ldots, S(T_{N(t-)}) = i_n|N(t-) = n,\right.$$
$$\left. T_1 = t_1, T_2 = t_2, \ldots, T_{N(t-)} = t_n\right)$$

is

$$\frac{(\eta t_1)^{i_1}}{i_1!}\exp\{-\eta t_1\}\cdot\frac{(\eta(t_2 - t_1))^{i_2-i_1}}{(i_2 - i_1)!}\exp\{-\eta(t_2 - t_1)\}\cdots$$
$$\times\frac{(\eta(t_n - t_{n-1}))^{i_n-i_{n-1}}}{(i_n - i_{n-1})!}\exp\{-\eta(t_n - t_{n-1})\}$$
$$= \prod_{k=1}^{n}\frac{(\eta(t_k - t_{k-1}))^{i_k-i_{k-1}}}{(i_k - i_{k-1})!}\exp\{-\eta(t_k - t_{k-1})\}$$
$$\equiv g(i_1, i_2, \ldots, i_n, t_1, t_2, \ldots, t_n),$$

where $i_0 = t_0 \equiv 0$. Then, following the same procedure as that found in Section 24.3.2, the joint distribution of $(T_S \geq t, S(T_{N(t-)}), T_{N(t-)})$ can be obtained as

$$f\left(T_S \geq t, S\left(T_{N(t-)}\right) = j, T_{N(t-)} = y\right)$$
$$= \sum_{n=0}^{\infty}\int_0^y\int_0^{t_{n-2}}\cdots\int_0^{t_2}\sum_{0\leq i_1\leq i_2\leq\cdots\leq i_{n-1}\leq j} q(y, j)\left(\prod_{i=1}^{n-1} q\left(t_i, i_i\right)\right)$$
$$\times g\left(i_1, i_2, \ldots, i_{n-1}, j, t_1, t_2, \ldots, y\right)$$
$$\times f_X(t_1)f_X(t_2 - t_1)\cdots f_X(y - t_{n-1})\overline{F}_X(t - y)dt_1dt_2\cdots dt_{n-1}.$$

Thus

$$f(S(T_{N(t-)}) = j, T_{N(t-)} = y|T_S \geq t)$$
$$= \left(\sum_{n=0}^{\infty}\int_0^y\int_0^{t_{n-2}}\cdots\int_0^{t_2}\sum_{0\leq i_1\leq i_2\leq\cdots\leq i_{n-1}\leq j} q\left(y, j\right)\left(\prod_{i=1}^{n-1} q\left(t_i, i_i\right)\right)\right.$$

$$
\times g(i_1, i_2, \ldots, i_{n-1}, j, t_1, t_2, \ldots, y)
$$

$$
\left. \times f_X(t_1) f_X(t_2 - t_1) \cdots f_X(y - t_{n-1}) \overline{F}_X(t - y) dt_1 dt_2 \cdots dt_{n-1} \right)
$$

$$
\times \left(\sum_{n=0}^{\infty} \int_0^t \int_0^{t_n} \cdots \int_0^{t_2} \sum_{0 \le i_1 \le i_2 \le \cdots \le i_{n-1} \le i_n \le m} \left(\prod_{i=1}^{n} q(t_i, i_i) \right) g(i_1, i_2, \ldots, i_n, t_1, t_2, \ldots, t_n) \right.
$$

$$
\left. \times f_X(t_1) f_X(t_2 - t_1) \cdots f_X(t_n - t_{n-1}) \overline{F}_X(t - t_n) dt_1 dt_2 \cdots dt_n \right)^{-1},
$$

$j = 0, 1, .., m, 0 \le y \le t.$

Therefore, finally, from (24.10):

$$
\lambda_S(t) = \sum_{j=0}^{m} \int_0^t p(t, j) \lambda_X(t - y) f(S(T_{N(t-)}) = j, T_{N(t-)} = y | T_S \ge t) dy.
$$

24.4 Conclusion

In this chapter, we have discussed the generalization of the classical extreme shock model to the history-dependent case when the external shocks arrive according to a renewal process. We have discussed an important proposition which provides a new perspective on the generalized extreme shock models. Based on this general theoretical result, several history-dependent extreme shock models have been considered and analyzed.

It should be noted that the conventional models usually deal with the NHPP of shocks, which makes the corresponding stochastic analysis fairly simple. However, in many instances the external shock processes are more likely to follow the renewal process pattern and the derivations and final results for this setting (as usually in problems with renewal processes) are much more complex and can often be expressed only in the form of various infinite sums. Therefore, the analytical results of Section 24.3 are rather cumbersome, but the corresponding probabilistic reasoning clearly follows the message of Proposition 24.1.

Acknowledgements

This work was supported by the National Research Foundation of Korea (NRF) grant funded by the Korea government (MEST) (No. 2011-0017338). The work of the second author was supported by the NRF (National Research Foundation of South Africa) grant FA2006040700002.

References

Anderson, P.K., O. Borgan, R.D. Gill, and N. Keiding. 1993. *Statistical Models Based on Counting Processes*. New York: Springer-Verlag.

Cha, J.H. and M. Finkelstein. 2009. On terminating shock process with independent wear increments. *Journal of Applied Probability* 46 (2): 353–362.

Cha, J.H. and M. Finkelstein. 2011. On new classes of extreme shock models and some generalizations. *Journal of Applied Probability* 48 (1): 258–270.

Cha, J.H. and M. Finkelstein. 2012. Information-based thinning of point processes and its application to shocks models. *Journal of Statistical Planning and Inference* 142 (8): 2345–2350.

Cox, D.R. and V. Isham. 1980. *Point Processes*. London: Chapman & Hall/CRC.

Finkelstein, M. 2008. *Failure Rate Modelling for Reliability and Risk*. London: Springer.

Fleming, T.R. and D.P. Harrington. 1991. *Counting Processes and Survival Analysis*. New York: John Wiley & Sons.

Gut, A. and J. Hüsler. 2005. Realistic variation of shock models. *Statistics and Probability Letters* 74 (2): 187–204.

Mallor, F. and E. Omey. 2001. Shocks, runs and random sums. *Journal of Applied Probability* 38: 438-448.

Nakagawa, T. 2007. *Shocks and Damage Models in Reliability Theory*. London: Springer.

Sumita, U. and J. Shanthikumar. 1985. A class of correlated cumulative shocks models. *Advances in Applied Probability* 17: 347–366.

Part Five

Systemability, Physics-of-Failure and Reliability Demonstration

25

Systemability Theory and its Applications

Hoang Pham
Department of Industrial and Systems Engineering, Rutgers, The State University of New Jersey, USA

25.1 Introduction

As more products become increasingly smaller in physical terms but larger in terms of complexity, the functionality of the entire system becomes more important and more complicated. In the midst of these challenges, developing reliable systems and products is of greater importance. Therefore, it is imperative for the developers to explore techniques that can effectively quantify the reliability of complex systems subject to global operating environments and conditions. Reliability is the probability of a product successfully performing its function for a specified period of time under specified conditions of operation.

The successful operation of a product will not only depend on the use of its applications but also depend on the operating conditions and the users. The effect of the unknown operating environment is often ignored in reliability practice. In other words, engineers and developers often assume that the controlled testing (or laboratory) environment and the field environment are the same. Therefore, the reliability calculation of the entire system may not truly represent the product's usefulness in actual experience according to the above assumption. In fact, most of the systems which operate in the field environments are different from the controlled-testing environment (Pham 2010a).

Recent authors (Cox 1972; Currit and Singpurwalla 1988; Pham and Xie 2002; Pham 2005a, 2005b, 2010a; Battini *et al.* 2007, 2008; Sohn *et al.* 2007; Persona *et al.* 2009, 2010; Sgarbossa and Pham 2010; Wang and Pham 2011, 2012) have proposed approaches to estimate the impact of operating environments on the failure rate. Cox (1972) first introduced the concept of proportional hazard rate by incorporating the environmental

Applied Reliability Engineering and Risk Analysis: Probabilistic Models and Statistical Inference, First Edition.
Edited by Ilia B. Frenkel, Alex Karagrigoriou, Anatoly Lisnianski and Andre Kleyner.

factors and the operating conditions into the hazard rate. The effect of the unknown operating environment, which is commonly ignored, has recently received a lot of attention from researchers (Currit and Singpurwalla 1988; Pham 2005a, 2005b, 2010b; Battini *et al.* 2007, 2008; Sohn *et al.* 2007; Persona *et al.* 2009, 2010; Sgarbossa and Pham 2010; Wang and Pham 2011, 2012). The introduction of various covariates has been studied by many researchers (Currit and Singpurwalla 1988; Pham 2002, 2005a, 2005b, 2010b; Battini *et al.* 2007, 2008; Persona *et al.* 2009, 2010; Sohn *et al.* 2007; Sgarbossa and Pham 2010; Wang and Pham 2011, 2012) in order to develop methods by considering the uncertainty of operating environments in the modeling. For example, a harsh environment suggests the random variable η that represents the uncertainty of operating environment $\eta > 1$, whereas in the case of over-stress, testing in the laboratory would suggest $\eta < 1$. The case $\eta = 1$ corresponds to ignoring the difference between the two environments or if the two environments are "theoretically" the same.

Pham (2005a) introduced the concept of systemability by taking into account the uncertainty of operating environments in the reliability function. *Systemability* (Pham 2005a) is defined as the probability that the system will perform its intended function for a specified mission time subject to a random variable that represents the uncertainty of the operating environments. In general, the exact systemability function (see Equation 25.2, Pham 2005a) would require great care in computations in order to obtain the systemability values. In this chapter, we also discuss an approximation systemability function in the field environments using the Taylor series expansion. Numerical comparisons for systems with and without considering the field environments as well as systemability approximation are given to illustrate the results.

The following notation is used:

- $h(t)$ system hazard rate function
- $R_c(t)$ system reliability function
- $R_s(t)$ systemability function
- $f(t)$ system failure time density function
- λ Intensity parameter of Weibull distribution
- γ shape parameter of Weibull distribution
- η random variable that represents the system's operating environments with distribution function G
- $G(\eta)$ cumulative distribution function of η
- α shape parameter of Gamma distribution
- β scale parameter of Gamma distribution

25.2 Systemability Measures

Reliability is the probability that a system will be successful in the interval from time 0 to time t under specified conditions of operation. Mathematically, reliability function is given by

$$R(t) = \int_t^\infty f(s)ds = e^{-\int_0^t h(s)ds} \tag{25.1}$$

where $f(s)$ and $h(s)$ are, respectively, the failure time density and failure rate function.

The underlying assumption of traditional reliability is that the operating environments and the laboratory-testing environments are the same. The operating environments are often unknown due to the uncertainties of environments in the field (Pham 2010a). Pham (2005a) introduced the concept of systemability by taking into account the randomness of the operating environments in a mathematical reliability function. We now discuss it briefly.

Definition (Pham 2005a). *Systemability* is defined as the probability that the system will perform its intended function for a specified mission time under the random operating environments.

In a mathematical form, the *systemabililty* function is given by

$$R_s(t) = \int_\eta e^{-\eta \int_0^t h(s)ds} dG(\eta) \tag{25.2}$$

where η is a random variable that represents the system operating environments with a distribution function G.

It is worth noting that systemability is, in summary, the reliability of a system in operating environments. This new function captures the uncertainty of complex operating environments of systems in terms of the system failure rate. It also would reflect the reliability estimation of the system in the field.

If we assume that η has a gamma distribution with parameters α and β, that is, $\eta \sim gamma(\alpha, \beta)$ where the pdf of η is

$$f_\eta(x) = \frac{\beta^\alpha x^{\alpha-1} e^{-\beta x}}{\Gamma(\alpha)} \quad \text{for } \alpha, \beta > 0;\ x \geq 0 \tag{25.3}$$

then the systemability function of the system in Equation (25.2) is given by

$$R_s(t) = \left[\frac{\beta}{\beta + \int_0^t h(s)ds} \right]^\alpha \tag{25.4}$$

25.3 Systemability Analysis of *k*-out-of-*n* Systems

This subsection presents several systemability results and variances of some system configurations such as series, parallel, and *k*-out-of-*n* systems (Pham 2005a). Consider the following assumptions:

1. A system consists of n independent and identical components where the system is subject to the uncertainty of operating environments η.
2. Each component lifetime is assumed to follow the Weibull density function.

$$\text{Component hazard rate } h(t) = \lambda\gamma t^{\gamma-1}$$

$$\text{Component reliability } R(t) = e^{-\lambda t^\gamma} \qquad \text{for } t > 0.$$

In a k-out-of-n configuration, the system will operate if at least k out of n components are operating. As for the uncertainty environments of gamma random variable, that is, $\eta \sim gamma(\alpha, \beta)$, the systemability of k-out-of-n systems is as follows (Pham 2005a):

$$R_s(t) = \sum_{j=k}^{n} \binom{n}{j} \sum_{l=0}^{n-j} \binom{n-j}{l} (-1)^l \left[\frac{\beta}{\beta + \lambda (j+l) t^{\gamma}} \right]^{\alpha} \tag{25.5}$$

The variance of the k-out-of-n systems systemability function is given by (Pham 2005a):

$$Var(R_s(t)) = \sum_{i=k}^{n} \binom{n}{i} \sum_{j=k}^{n} \binom{n}{j} \sum_{l=0}^{2n-i-j} \binom{2n-i-j}{l} (-1)^l \left(\frac{\beta}{\beta + (i+j+l)\lambda t^{\gamma}} \right)^2 - \left(\sum_{j=k}^{n} \binom{n}{j} \sum_{l=0}^{n-j} \binom{n-j}{l} (-1)^l \left(\frac{\beta}{\beta + (j+l)\lambda t^{\gamma}} \right)^2 \right)^2. \tag{25.6}$$

Example Consider a k-out-of-5 system where $\lambda = 0.0001$, $\gamma = 1.5$, and $\eta \sim gamma$ (α, β). The systemability of k-out-of-5 system configurations is given by

$$R_s(t) = \sum_{j=k}^{5} \binom{5}{j} \sum_{l=0}^{5-j} \binom{5-j}{l} (-1)^l \left[\frac{\beta}{\beta + \lambda (j+l) t^{\gamma}} \right]^{\alpha} \tag{25.7}$$

The reliability of k-out-of-5 systems is

$$R_c(t) = \sum_{\mathrm{i=k}}^{5} \binom{5}{\mathrm{i}} (e^{-\lambda t^{\gamma}})^{\mathrm{i}} (1 - e^{-\lambda t^{\gamma}})^{5\text{-}\mathrm{i}}. \tag{25.8}$$

Figures 25.1 and 25.2 show the reliability (notation: $R_{kn}(t)$) and systemability ($R_{kn}(t|\eta)$) functions of a 3-out-of-5 system for $\alpha = 2, \beta = 3$ and for $\alpha = 2, \beta = 1$, respectively. Similarly, Figures 25.3 and 25.4 show the reliability and systemability functions of a parallel system (here $k = 1$) for $\alpha = 2, \beta = 3$ and for $\alpha = 2, \beta = 1$, respectively.

25.3.1 Variance of Systemability Calculations

Assume $\lambda = 0.00001$, $\gamma = 1.5$, $n = 3$, $k = 2$, and $\eta \sim gamma(\alpha, \beta)$, Figures 25.5 shows the systemability and its confidence intervals of a 2-out-of-3 system for $\alpha = 2, \beta = 1$.

25.4 Systemability Function Approximation

We now describe the systemability function approximation using the Taylor series. Assume that X has a pdf $g_1(x)$ where the mean is μ and the variance is σ^2, and that X be around μ, that is, $X \sim g_1(\mu, \sigma^2)$. Let f_1 be a function with derivatives of all orders throughout some open interval containing μ. Then,

$$f_1(x) = \sum_{\mathrm{k=0}}^{\infty} \frac{(x-\mu)^k}{k!} f_1^{(k)}(\mu) \tag{25.9}$$

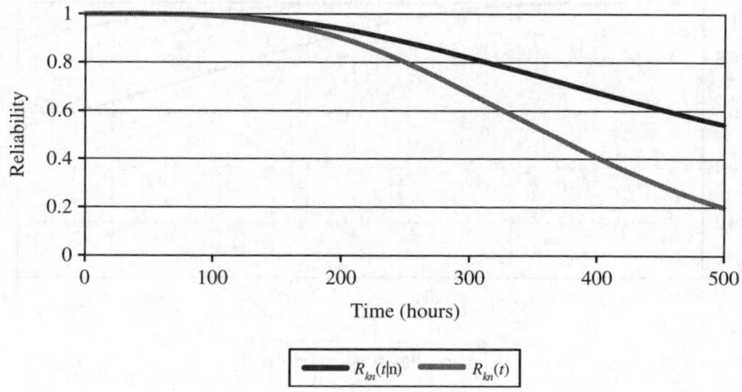

Figure 25.1 Comparisons of *3*-out-of-*5* system reliability vs. systemability functions for $\alpha = 2$ and $\beta = 3$

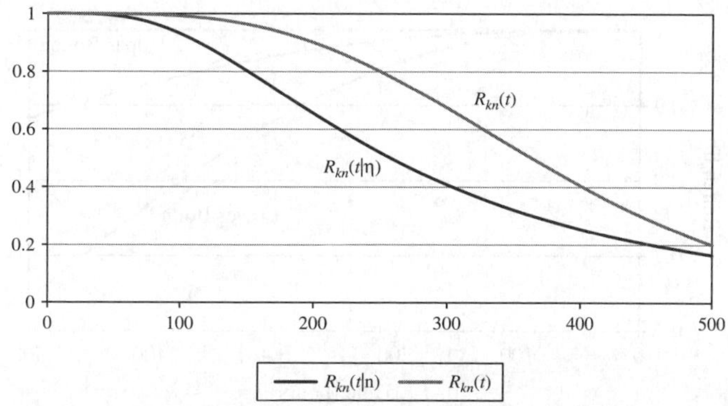

Figure 25.2 Comparisons of *3*-out-of-*5* system reliability vs. systemability functions for $\alpha = 2$ and $\beta = 1$

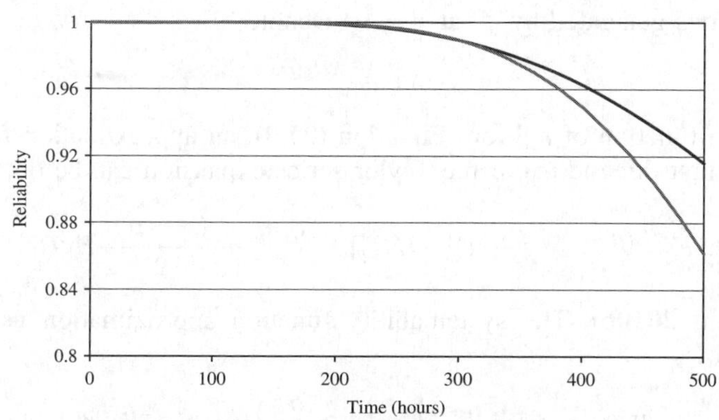

Figure 25.3 Comparisons of parallel system reliability vs. systemability function for $\alpha = 2$ and $\beta = 3$

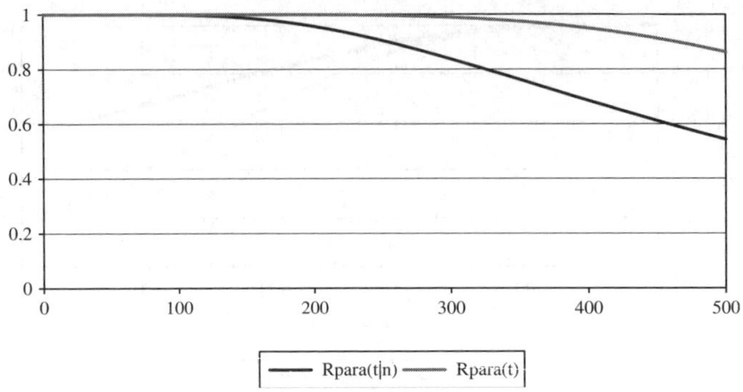

Figure 25.4 Comparisons of parallel system reliability vs. systemability functions for $\alpha = 2$ and $\beta = 1$

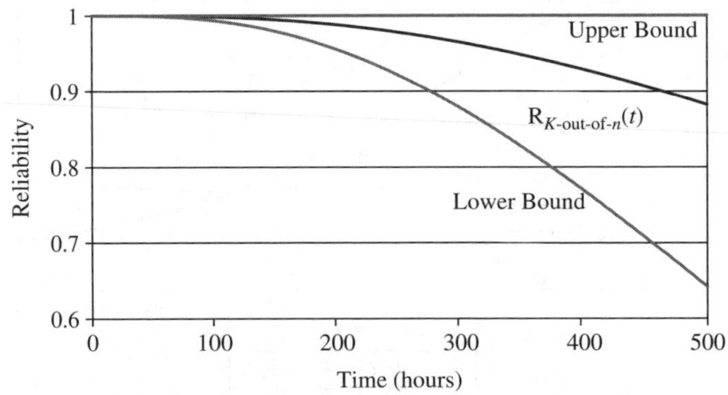

Figure 25.5 A *2*-out-of-*3* systemability and its 95% confidence interval where $\alpha = 2$, $\beta = 1$

is the Taylor series generated by f_1 at $x = \mu$. Denote

$$f_1(x) = e^{-H(t)x}$$

where $H(t)$ is a function of t. From Equation (25.9), an approximation for the function $f_1(x)$ using the first three terms of the Taylor series expansion can be obtained

$$f_1(x) \simeq e^{-H(t)\mu} + (x-\mu)[-H(t)]\, e^{-H(t)\mu} + \frac{(x-\mu)^2}{2!} H^2(t)\, e^{-H(t)\mu}$$

Theorem (Pham 2010b). The systemability function approximation using the Taylor series is as follows

$$R_s(t) = E[e^{-H(t)x}] = \left[1 + \frac{\sigma^2}{2!} H^2(t)\right] e^{-H(t)\mu} \tag{25.10}$$

where $H(t) = \displaystyle\int_0^t h(s)ds$ and μ and σ^2 are the mean and variance of the random variable η, respectively.

25.5 Systemability with Loglog Distribution

We first present numerical examples for a system's operating environments and then compare the results where the failure time of a system follows loglog distribution (Pham 2002). Note that the failure rate of the loglog distribution has the V-tub-shaped whereas Weibull distribution can obtain a bath-tub-shaped. As for the bath-tub-shaped, after the infant mortality period, the useful life of the system begins. During its useful life, the system fails as a constant rate. This period is then followed by a period of wearing out when the system starts slowly to decrease with the onset of the wearing out. For the V-tub-shaped, after the infant mortality period, the system starts to experience at a relatively low increasing rate, but not a constant, and then increasingly more failures due to aging.

25.5.1 Loglog Distribution

The loglog hazard rate and cumulative hazard functions are, respectively (Pham 2002):

$$h(t) = b \ln a \cdot t^{b-1} a^{t^b}, \qquad t > 0, a > 1, b > 0$$

and

$$H(t) = a^{t^b} - 1 \tag{25.11}$$

The loglog reliability function is

$$R(t) = e^{1-a^{t^b}} \tag{25.12}$$

Figure 25.6 describes the failure rate curves for various values of a and b.

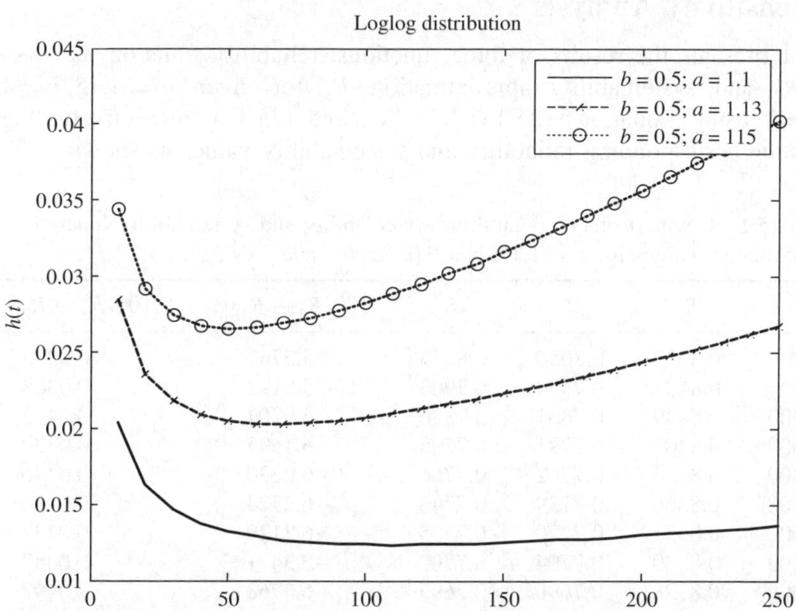

Figure 25.6 Failure rate h(t) for various values of a and $b = 0.5$

25.5.1.1 System Reliability Function $[R_1(t)]$

The system reliability (without considering the uncertainty of the operating environments) $R_1(t)$ is

$$R_1(t) = e^{1-a^{t^b}} \tag{25.13}$$

25.5.1.2 Systemability Function $[R_2(t)]$

Suppose a random variable η is governed by the gamma distribution, that is, $\eta \sim gamma\,(\alpha, \beta)$ with parameters α and β where the pdf is given in Equation (25.3), then from Equation (25.11), the systemability function, $R_2(t)$, is

$$R_2(t) = \left(\frac{\beta}{\beta + a^{t^b} - 1}\right)^{\alpha} \tag{25.14}$$

25.5.1.3 Systemability Function Approximation $[R_3(t)]$

Similarly, from the theorem above and Equation (25.11) we obtain

$$R_3(t) = \left(1 + \frac{\alpha\left(a^{t^b} - 1\right)^2}{2\beta^2}\right) e^{-\frac{\alpha(a^{t^b}-1)}{\beta}} \tag{25.15}$$

25.6 Sensitivity Analysis

Table 25.1 presents the results of three functions: reliability function R_1, systemability funtion R_2 and systemability approximation R_3 for given $a = 1.15, b = 0.05, \alpha = 3$ and $\beta = 2$ using Equations (25.13), (25.14), and (25.15), respectively. Figure 25.7 illustrates the corresponding reliability and systemability values as shown in Table 25.1.

Table 25.1 Comparisons of reliability, systemability and systemability function approximation values for $a = 1.15, b = 0.05, \alpha = 3$ and $\beta = 2$

t	R_1	R_2	R_3	$100(R_2 - R_1)\%$	$\lvert 100(R_3 - R_2)\%\rvert$
1.0000	0.8607	0.8050	0.8053	−5.5747	0.0293
5.0000	0.8491	0.7899	0.7903	−5.9192	0.0368
10.0000	0.8439	0.7831	0.7836	−6.0704	0.0405
15.0000	0.8407	0.7791	0.7795	−6.1595	0.0429
20.0000	0.8384	0.7762	0.7766	−6.2230	0.0446
25.0000	0.8366	0.7739	0.7743	−6.2724	0.0460
30.0000	0.8351	0.7720	0.7725	−6.3129	0.0472
35.0000	0.8339	0.7704	0.7709	−6.3471	0.0483
40.0000	0.8328	0.7690	0.7695	−6.3768	0.0492
45.0000	0.8318	0.7677	0.7682	−6.4031	0.0500
50.0000	0.8309	0.7666	0.7671	−6.4266	0.0507

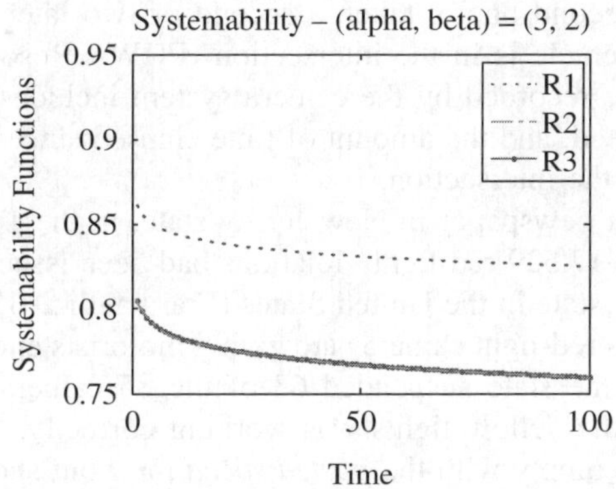

Figure 25.7 Reliability, systemability and systemability function approximation curves for $a = 1.15, b = 0.05, \alpha = 3$ and $\beta = 2$

Table 25.2 Reliability, systemability and systemability function approximation for various values a, b, α, β

a	b	α	β	R_1 and R_2	R_2 and R_3	$\lvert(R_3 - R_2)100\%\rvert \leq$
1.15	0.05	3	2	$R_1 > R_2$	$R_2 \leq R_3$	0.08%
1.13	0.05	3	2	$R_1 > R_2$	$R_2 \leq R_3$	0.07%
1.15	0.05	2	3	$R_1 < R_2$	$R_2 \leq R_3$	0.03%
1.13	0.05	2	3	$R_1 < R_2$	$R_2 \leq R_3$	0.02%
1.10	0.05	3	2	$R_1 > R_2$	$R_2 \leq R_3$	0.03%
1.10	0.05	2	3	$R_1 < R_2$	$R_2 \leq R_3$	0.01%

From Table 25.2, we observe that the estimated errors of the systemability in operating environments between the exact systemability function (Equation 25.14) and the systemability approximation from Equation (25.15) are significantly small and less than 0.1%, which again is very encouraging.

25.7 Applications: Red Light Camera Systems

A red light camera system is a traffic enforcement camera that captures an image of a vehicle which has entered an intersection against a red traffic light (FHWA-SA-05-002 (2005)). In general, the camera is triggered when a vehicle enters the intersection after the traffic light has turned red. Today these cameras are used worldwide in many countries including Canada, the United Kingdom, Singapore and the United States. Red light camera systems usually employ two closely spaced inductive loops embedded in the pavement just before the limit line, in order to measure the speed of vehicles. Using the speed measured, the system predicts if a particular vehicle will not be able to stop before entering the intersection, and takes two photographs of the event. Typically, the first photo is taken of the front of the vehicle just before it enters the intersection, with the light showing red, and the second photo, taken a second or two later, is taken of the rear of the vehicle when the vehicle is in the intersection (FHWA Resource Center 2012). The information that may be recorded by the camera system includes: the date and time; the location; the vehicle speed, and the amount of time elapsed since the light turned red and the vehicle passed into the intersection.

According to a recent newspaper in New Jersey state, in the 11 months from May 26, 2011 to April 30, 2012, 17,529 red-light violations had been issued at an intersection of a township in New Jersey state in the United States (Frassinelli 2012). Currently the state is investigating whether its red-light cameras are giving motorists enough time to get through intersections. Recently the state suspended 63 of the 85 cameras amid questions about whether the timing of the yellow lights was working correctly. "Yellow lights on those systems were timed to comply with the *posted speed limit* but should have been timed to coincide with the speed at which 85 percent of the drivers approach the intersections," NJ Department of Transportation officials said.

The question here is whether the increases of these recent red-light violations and red-light system errors are due to software errors, hardware errors of the camera systems, or depends on a local traffic condition such as posted speed limit and traffic flow, the location, environment, or a combination of all above. The reliability modeling of software and hardware systems can be found in (Pham 2010a).

Consider a moderately harsh environment given at these particular setting where the vehicles are in the given intersections, the loglog distribution discussed in Section 25.5 seems to be applicable in describing time to failure for some types of camera devices. For this purpose we have considered applying the loglog distribution in this application. Consider that the time to failure of a red-light camera system follows the loglog distribution with parameters:

$$a = 1.15 \text{ and } b = 0.05.$$

Given the two parameters from gamma distribution are: $\alpha = 2.42$ and $\beta = 2.2$. Table 25.3 shows the reliability values for all three functions: reliability function R_1, systemability function R_2 and systemability approximation R_3 for given $a = 1.15, b = 0.05, \alpha = 2.42$

Table 25.3 Comparisons of reliability, systemability and systemability function approximation values for $a = 1.15, b = 0.05, \alpha = 2.42$ and $\beta = 2.2$

t	R_1	R_2	R_3	$100(R_2 - R_1)\%$	$\lvert 100(R_3 - R_2)\%\rvert$
1.0000	0.8607	0.8525	0.8527	−0.8239	0.0194
5.0000	0.8491	0.8407	0.8409	−0.8434	0.0245
10.0000	0.8439	0.8354	0.8356	−0.8499	0.0271
15.0000	0.8407	0.8322	0.8324	−0.8530	0.0288
20.0000	0.8384	0.8298	0.8301	−0.8550	0.0300
25.0000	0.8366	0.8280	0.8283	−0.8564	0.0310
30.0000	0.8351	0.8265	0.8269	−0.8574	0.0318
35.0000	0.8339	0.8253	0.8256	−0.8581	0.0325
40.0000	0.8328	0.8242	0.8245	−0.8587	0.0332
45.0000	0.8318	0.8232	0.8235	−0.8592	0.0337
50.0000	0.8309	0.8223	0.8226	−0.8596	0.0343

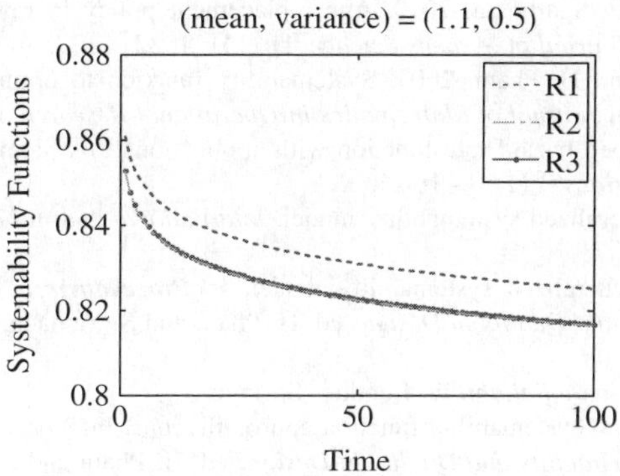

Figure 25.8 Reliability, systemability and systemability function approximation curves for $a = 1.15, b = 0.05, \alpha = 2.42$ and $\beta = 2.2$

and $\beta = 2.2$ using Equations (25.13), (25.14), and (25.15), respectively. Figure 25.8 illustrates the corresponding reliability and system values as shown in Table 25.3. We observe in Table 25.3 that the estimated errors of the system reliability in operating environments between the exact systemability function (Equation 25.14) and the systemability approximation from Equation (25.15) are significantly less than 0.1%, which is, in practice, very encouraging. Therefore, it is worth the effort to further study other critical applications such as complex security systems and safety monitoring control systems.

25.8 Conclusion

Growing competition has increased the need for all developers and designers to ensure the systemability of their products in the field while enabling them to be produced at the lowest possible cost. Systemability is the ability of a system to perform its required functions subject to the uncertainty of the user's environments and conditions.

References

Battini, D., Faccio, M., Persona, A. and F. Sgarbossa. 2007. Reliability in random environment: systemability and its application. In *Proceedings of 13th ISSAT International Conference on Reliability and Quality in Design*, ed. H. Pham and T. Nakagawa, August 2–4, 2007, Seattle, Washington, USA.

Battini, D., Faccio, M., Persona, A. and F. Sgarbossa. 2008. Reliability of motorcycle components using systemability approach. In *Proceedings of 14th ISSAT International Conference on Reliability and Quality in Design*, ed. H. Pham and T. Nakagawa, August 6–9, 2008, Orlando, FL.

Cox, D. R. 1972. Regression models and life-tables. *Journal of the Royal Statistical Society* B-34 (2): 187–220.

Currit, A. and N. D. Singpurwalla. 1988. On the reliability function of a system of components sharing a common environment. *Journal of Applied Probability* 25 (4): 763–771.

FHWA-SA-05-002. 2005. *Red Light Camera Systems Operational Guidelines*. United States Department of Transportation, Federal Highway Administration, FHWA Office of Safety, Washington, DC.

FHWA Resource Center. 2012. United States Department of Transportation, Federal Highway Administration, FHWA Office of Safety, Washington, DC. http://www.fhwa.dot.gov/resourcecenter/teams/safety/safe_mrt_redlight.cfm.

Frassinelli, M. 2012. Seeing red: Cherry Hill faces first class-action suit over red-light cameras. *The Star-Ledger Newspaper*. June 30, 2012. New Jersey, USA.

Persona, A., H. Pham and F. Sgarbossa. 2010. Age replacement policy in random environment using systemability. *International Journal of Systems Science* 41 (11): 1383–1397.

Persona, A., F. Sgarbossa and H. Pham. 2009. Systemability function to optimization reliability in random environment. *International Journal of Mathematics in Operational Research* 1 (3): 397–417.

Pham, H. 2002. A Vtub-shaped hazard rate function with applications to system safety. *International Journal of Reliability and Applications* 3 (1): 1–16.

Pham, H. 2005a. A new generalized systemability model. *International Journal of Performability Engineering* 1 (2): 145–155.

Pham, H. 2005b. A new generalized systemability model. In *Proceedings of the 11th ISSAT International Conference on Reliability and Quality in Design*, ed. H. Pham and S. Yamada, August 4–6, 2005, St. Louis, Missouri.

Pham, H. 2010a. *System Software Reliability*. London: Springer.

Pham, H. 2010b. Mathematical systemability function approximations. In *Proceedings of the 16th ISSAT International Conference on Reliability and Quality in Design*, ed. H. Pham and S. Yamada, August 5–7, 2010, Washington, DC.

Pham, H. and M. Xie. 2002. A generalized surveillance model with applications to systems' safety. *IEEE Transactions on Systems, Man, and Cybernetics* C-32 (4): 485–492.

Sgarbossa, F. and H. Pham. 2010. A cost analysis of systems subject to random field environments and reliability. *IEEE Transactions on Systems, Man, and Cybernetics* C-40 (4): 429–437.

Sohn, S. Y., I. S. Chang and T. H. Moon. 2007. Random effects Weibull regression model for occupational lifetime. *European Journal of Operational Research* 179 (1): 124–131.

Wang, Y. and H. Pham. 2011. A multi-objective optimization of imperfect preventive maintenance policy for dependent competing risk systems with hidden failure. *IEEE Transactions on Reliability* 60 (4): 770–781.

Wang, Y. and H. Pham. 2012. Modeling the dependent competing risks with multiple degradation processes and random shock using time-varying copulas. *IEEE Transactions on Reliability* 61 (1): 13–22.

26

Physics-of-Failure based Reliability Engineering

Pedro O. Quintero[1] and Michael Pecht[2]

[1]*Department of Mechanical Engineering, University of Puerto Rico at Mayagüez, Puerto Rico*

[2]*Center for Advanced Life Cycle Engineering (CALCE), Department of Mechanical Engineering, University of Maryland, USA*

26.1 Introduction

Reliability is the ability of a product to perform its intended function without failure and within specified performance limits, for a specified time, in its life cycle application conditions (the environment). Development of a reliable product is not a matter of chance or good fortune; rather, it is a rational consequence of conscious, systematic, and rigorous efforts conducted through the entire life cycle of the product: from design to scrap. Early approaches to reliability prediction relied on statistical analysis of field data characterized by a wide spectrum of applications and therefore a significant variation in environments. Owing to the randomness of the failure data, the assumption of a constant failure rate became the foundation of the classical approach, with Mil-Hdbk-217 becoming the epitome of classical prediction methodology in electronics. As a consequence of the assumption of a constant failure rate, however, the classical approach does not provide an accurate estimate of life when we consider the ever-increasing nature of technological progress and the competitiveness of today's market.

The electronics industry is rapidly progressing in miniaturization and complexity, and new failure modes are arising, not only due to the higher stress-strength ratios associated with miniaturization, but also due to the advent of new materials and manufacturing processes (Matic and Sruk 2008). These requirements demand new reliability methodologies

Applied Reliability Engineering and Risk Analysis: Probabilistic Models and Statistical Inference, First Edition.
Edited by Ilia B. Frenkel, Alex Karagrigoriou, Anatoly Lisnianski and Andre Kleyner.

to assess the reliability of components as accurately as possible in up-front steps within the development process. Debates in the literature (Knowles 1993; Pecht 1996; Talmor and Arueti 1997; Goe and Graves 2006) about the appropriateness of applying the procedures based on the classical concept have generated the need for another approach that overcomes the pitfalls of the constant hazard rate approach. Thus, acceptable product reliability can only be ensured through robust product designs, capable and controlled processes, and qualified components and materials from vendors whose processes are also capable of meeting quality requirements. Quantitative understanding and modeling of the relevant failure mechanisms guide product design and manufacturing and the planning of test specifications and tolerances.

Reliability is an expectation that must be met in the competitive and global business environment where time-to-profit is a key metric for success. Time-to-market pressure does not allow companies to stumble with unreliable products without serious financial consequences. As a result, companies must ensure reliability, but at the same time cannot afford to design and test products regarding their reliability over a long period of time. It has been acknowledged that there is a gap between reliability theory and its application to industrial fields (Codier 1969); therefore, a myriad of reliability prediction methods have been discussed in the literature (O'Connor and Head 1985; Bowles 1992; Leonard and Pecht 1993; Pecht and Nash 1994; Stadterman *et al.* 1995; Johnson and Gullo 2000), with physics-of-failure (PoF) as the most promising approach. The physics-of-failure approach focuses on the key words in the definition of reliability: time and environment.

Several comparisons between the physics-of-failure and classical approaches have been published in both the quantitative and qualitative domains (Matic and Sruk 2008). Mortin *et al.* (1995) made a comparison between the constant hazard rate and the hazard rate specific for the electromigration failure mechanism. The discrepancy between an assumed constant hazard rate and the corresponding real hazard rate has the potential to inhibit, for example, the support activities determined by logistic and maintenance requirements based on the assumed hazard rates, and also the design process itself as a result of over-designing or under-designing it. As a consequence, tremendous life cycle costs could be imposed due to incorrect assumptions about the hazard rate.

By analyzing test and field data, Weil *et al.* (1993) showed that when high volume suppliers who employed the best reliability and quality practices manufactured plastic encapsulated microcircuits (PEMs), the PEMs had considerably lower failure rates than the failure rates predicted by Mil-Hdbk-217. Cushing *et al.* (1993) compared the classical and PoF approaches from model development to final cost analysis. This chapter favors the PoF approach because it targets a broader perspective of reliability assessment. A key factor in favor of the PoF approach is that whenever test and field data are inconsistent with reliability assessments, the reliability is not well understood, and therefore additional analysis is required. Cushing *et al.* (1993) pointed to the ignorance of the classical approach regarding the dominant failure mechanism in determining reliability, while the PoF approach includes reliability analysis of contributing failure mechanisms. Foucher *et al.* (2002) provided a non-subjective comparison between the PoF and classical approaches. When the most significant characteristics, such as accuracy, ease of customization, traceability, and ability to evolve, were compared, the PoF approach was reported to be the superior approach in reliability engineering. These comparisons between the PoF and the classical approach distinguished the PoF as the more effective reliability approach for use in concurrent engineering practice, which involves methods for both designing-in reliability and performance improvement.

Physics-of-failure (PoF) encompasses the identification and understanding of the failure mechanisms that degrade products and ultimately produce failure. Stipan *et al.* (2000) noted:

> The PoF concepts pioneered by Pecht *et al.* have provided the contemporary electronics community with perhaps the most unified approach in dealing with electronics reliability. The key philosophy is to develop a fundamental understanding of the underlying mechanisms that govern the degradation of electronics hardware. Then, by assigning rate constants or maximum limits to these various failure mechanisms, insight is obtained with regard to the anticipated lifetime of the electronics.

PoF is an approach to the design and development of reliable products and the prevention of failure based on the knowledge of failure mechanisms (JEDEC 2008). PoF can be used in reliability engineering applications including design-for-reliability (DfR), reliability prediction, test planning, and prognostics. To be used in the design and testing of a product, PoF requires that the product and its anticipated life cycle loading profile be sufficiently defined to identify potential failure sites and failure mechanisms. Failure mechanisms are the processes by which physical, chemical, mechanical, or any other processes induce product failures (Hu *et al.* 1993). Whether a specific failure mechanism will be active in a product and cause failure will depend on the environmental and the operating loading conditions. In addition to loading conditions, failure mechanisms depend on product architecture design, materials (or components), and all production (assembly) processes.

In PoF, the root cause of an individual failure mechanism is studied and corrected to achieve some predetermined lifetime. In root cause analysis, a failure site is a physical location within a product where a failure mechanism occurs and causes a failure of the product. These sites undergo changes that result in the product becoming inoperable.

To predict reliability, the time-to-failure at identified failure sites for the anticipated use conditions must be determined. Numerical and/or analytical models that are based on failure mechanisms and forecast time to failure are used to make PoF-based reliability predictions. These models may be referred to as failure models or PoF models. PoF models provide the different stress–time relationships that describe the failure mechanisms (JEDEC 2008; Dasgupta and Pecht 1991). Bernstein *et al.* (2006) presented a comprehensive review of failure mechanisms and reliability models for state-of-the-art MOSFETs. In their chapter, the PoF approach is discussed as a valuable tool for predicting circuit reliability and failure rate from a system point of view. In general, inputs to these failure models include specific product geometry, material information, and stress information. Stress information needs to include stress levels as well as the duration or frequency of the application of stress. The time-to-failure predicted by a failure model generally represents time to specific percentage of failure depending on how a model is developed and validated. Since the inputs to all the models have known or expected levels of uncertainty associated with them, simulation of those uncertainties allows for the development of a series of possible times-to-failure and the statistical distribution that represents the failure probability over time. Using these distribution parameters, a confidence interval can be associated with the estimated times to failure and other reliability parameters (Ryu and Chang 2005).

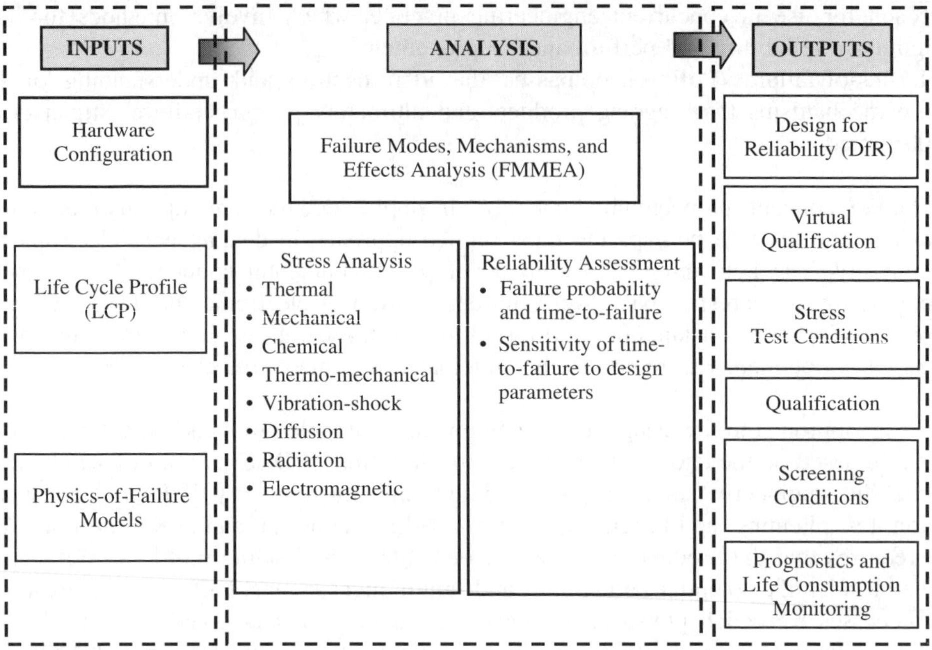

Figure 26.1 A flowchart of the physics-of-failure methodology

A PoF methodology for reliability engineering is shown in Figure 26.1. With input information, PoF analysis is conducted, and then the outputs of PoF analysis can help various activities in the product development process. A more detailed description and discussion are provided in the following sections.

PoF-based reliability assessment relies on an analysis of physical design with regard to the loads which the design is expected to endure. The identification of failure mechanisms that act on the product and manifest failure allows designers to address potential design weaknesses. As a result, the PoF-based reliability assessment allows for value-added engineering activities. Testing based on PoF-based reliability assessment provides feedback and confirmation that the physical product is sufficiently understood. Identification of critical parameters that influence the failures allows for the design and use conditions to be modified in order to reduce the negative impact of these parameters on the performance of the system.

Identification of failure mechanisms as the basis of reliability assessment and qualification test design for a system has been accepted by the EIA/JEDEC and SEMATECH. Intel's product qualification method (Mencinger 2000), which defines environmental, lifetime, and manufacturing use conditions based on target market segments, determines probable stresses with reliability implications, estimates stress levels for modeling and the modes of testing, defines accelerated stress conditions necessary to identify failure mechanisms, and determines final stress conditions for testing, following PoF principles.

IEEE Standard 1413 (1998) identifies the framework for the reliability prediction process for electronic systems (products) and equipment. An IEEE 1413-compliant reliability prediction report must include the reasons why the reliability predictions were performed, the intended use of the reliability prediction results, warnings as to how the reliability prediction results must not be used, and identification of where precautions are necessary. In order to assist the user in the selection and use of a particular reliability prediction methodology complying with IEEE 1413, a list of criteria is provided in IEEE 1413.1-2002 (2003).

26.2 Physics-of-Failure-based Reliability Assessment

The process to achieve reliability requirements can be divided into four fundamental steps, as discussed by Snook *et al.* (2003): (1) understand the application environmental extremes, life cycle environment, and life and reliability requirements; (2) ensure that each component part and assembly can survive and function over the whole application environment; (3) ensure that every part's construction and assembly technology has the capability to meet the life requirement subjected to the life cycle environment; and (4) assess the probability of failure within the desired life and review and amend the design and manufacturing methods to minimize this probability. The information requirements, analysis processes, and applications for PoF-based reliability assessment are presented in this section. The process is shown in Figure 26.2 as a flowchart.

26.2.1 Information Requirements

To conduct a PoF-based reliability assessment, the design and life cycle profile (LCP) of the product under analysis must be defined. The various types of information required for a PoF-based reliability assessment are discussed in the following sections.

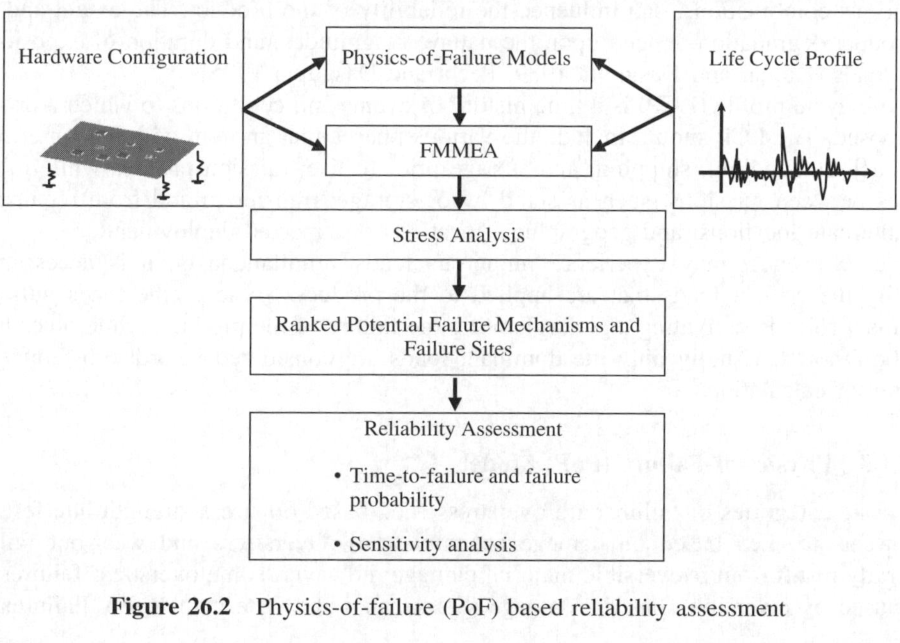

Figure 26.2 Physics-of-failure (PoF) based reliability assessment

26.2.1.1 Hardware Configuration

As discussed by Tilgner (2009), the design of a product identifies the materials, components, manufacturing processes, and connectivity required to realize the functional requirements of the product. The hardware configuration of the product describes the design of the components and subassemblies and the product architecture. It may also include the effects of the manufacturing processes on the final product in the form of tolerances on the dimensions and material properties.

For electronic products, designs will likely include electronic parts, printed circuit boards, connectors, and enclosures. The part geometry and structure, the sub-component geometry, and the connection methods can also be used as inputs for the PoF methodology as design information depending on the level of analysis involved.

Materials used to construct a product influence the stress at potential failure sites within the product due to external and internal loads and the process of damage accumulation. To determine the extent to which materials influence stress and damage, their physical properties also need to be inputs to PoF-based failure models. For example, a failure in a solder joint may be driven by stress arising from repeated temperature excursions through a fatigue failure mechanism. In this case, the coefficients of thermal expansion of materials are needed to determine the cyclic stress state. Properties of common materials used in electronic products can be found in references (Pecht *et al.* 1995; Stipan *et al.* 2000).

26.2.1.2 Life Cycle Profile (LCP)

Loads applied to a product during its life cycle from its environment and operation degrade the product and cause failures. In PoF-based reliability assessment, all loads that can degrade and fail the product should be characterized. The life cycle loads include thermal, mechanical, chemical, physical, and operational loading conditions. These loads, in various combinations, can influence the reliability of the product. The extent and rate of product degradation depend upon the nature, magnitudes, and duration of exposure to such loads (Kumar and Dasgupta 1989; Pecht and Dasgupta 1995).

A life cycle profile (LCP) is a time history of events and conditions to which a product is exposed. The LCP should include the various phases that an item will encounter in its life, such as handling, shipping, and storage prior to use; mission profiles while in use; phases between missions, such as stand by or storage; transfer to and from repair sites and alternate locations; and geographical locations of expected deployment.

Since a product may experience numerous loads simultaneously, it is necessary to identify the critical loads that are applied to the product. Some of the loads will play dominant roles in activating and accelerating the failure of the product, while other loads can be ignored. Usually only the dominant loads are considered in order to reduce the amount of calculation.

26.2.1.3 Physics-of-Failure (PoF) Models

The basic categories of failures are overstress (i.e., based on stress-strength interference) and wear-out (i.e., based on damage accumulation). Overstress and wear-out failures generally result from irreversible material damage; however, some overstress failures can be caused by reversible material damage (Stadterman and Osterman 1999). Failures can

Table 26.1 Examples of overstress and wear-out failure mechanisms for electronic products (Snook *et al*. 2003)

Failure mechanisms			
Overstress (environmental extreme)		Wear-out (life)	
Mechanical	Yield, fracture, delamination	Mechanical	Fatigue, creep
Thermal	Softening due to glass transition	Thermal	Diffusion voiding
Electrical	Electrostatic discharge, dielectric breakdown	Electrical	Electro-migration, conductive filament
Radiation	Single-event upsets	Radiation	Embrittlement, charge tapping
Chemical	Etching	Chemical	Corrosion, dendrites, intermetallic growth

also be broadly categorized by the nature of the loads that trigger or accelerate the failure mechanisms. Examples of both overstress and wear-out failure types for electronic assemblies are given in Table 26.1.

Failure models are used as tools to assess failure propensity. There are two categories of models: empirical models and PoF models. PoF methodology needs inputs from PoF-based models rather than empirical models. Many models, such as the Arrhenius model, the Coffin-Mason model, and the Basquin model, predict the time-to-failure of design elements (i.e., failure sites). All failure models have associated assumptions that help users determine the applicability (or lack thereof) of failure models to specific design and age conditions (Tilgner 2009).

In PoF models, the stresses and the various stress parameters and their relationships to materials, geometry, and product life are considered. Each potential failure mechanism is represented by one or more of the prevalent models. For electronic products, there are many PoF models describing the behaviors of components such as printed circuit boards, interconnections, and metallizations under various conditions, such as temperature cycling, vibration, humidity, and corrosion (Dasgupta and Pecht 1991). A PoF model should provide repeatable results, be sensitive to the variables and interactions that are causing degradations and failures, and predict the behavior of the product over the entire domain of its operational environment. This type of model allows for the development of accelerated tests and also allows for the transformation of the accelerated test results to application conditions.

Many PoF models are available in the literature, such as the strain-range-based model for solder joint fatigue (Osterman *et al*. 2006), which describes temperature-cycle-induced solder interconnect fatigue; Black's model and its variations (Clement 2001), which describe electromigration in semiconductor device metallizations; the Fowler-Nordheim model (Lee *et al*. 1988), which describes time-dependent dielectric breakdown due to tunneling in gate oxide devices; and the Rudra model (Rudra and Jennings 2004), which describes conductive filament formation in printed circuit boards. Models applicable to electronic products are available in references (Li and Dasgupta 1994; Lall *et al*. 1997; Pecht 1999). If no models are available or if the models are found not to be applicable to specified failure sites and loads, then new models can be developed by using controlled experiments that identify the design and environmental factors governing failure and the mathematical relationship linking those factors to the time-to-failure. New failure mechanisms or variations of known failure mechanisms in products usually arise with the introduction of new materials and technologies. As a result, research into the failure of new materials and technologies is critical for evaluating the life expectancies of products.

26.2.2 Failure Modes, Mechanisms, and Effects Analysis (FMMEA)

Failure site identification is critical to PoF-based reliability assessment. Failure modes, mechanisms, and effects analysis (FMMEA) is an essential step in conducting a PoF evaluation. FMMEA utilizes the basic steps in developing a traditional failure modes and effects analysis (FMEA) in combination with knowledge of PoF to identify failure sites. It then uses a life cycle profile to identify active stresses and select the potential failure mechanisms. Knowledge of load type, level, and frequency, combined with a known failure site, is used to prioritize failure mechanisms according to their severity and likelihood of occurrence (Bauernschub and Lall 1995). Figure 26.3 is a schematic diagram of FMMEA.

FMMEA prioritizes failure mechanisms based on their potential occurrence, failure time, and severity to provide guidelines for determining the major operational stresses and environmental and operational parameters that must be accounted for in the design or be controlled. The LCP is used to evaluate failure susceptibility. If certain environmental and operating conditions are nonexistent or generate a stress level that is below the trigger conditions for a failure mechanism, the failure mechanisms that are exclusively dependent on those environmental and operating conditions are said to have a low occurrence. Quality levels of products also affect the possible occurrence level for a failure mechanism. Severity ratings are obtained from the failure modes and sites associated with

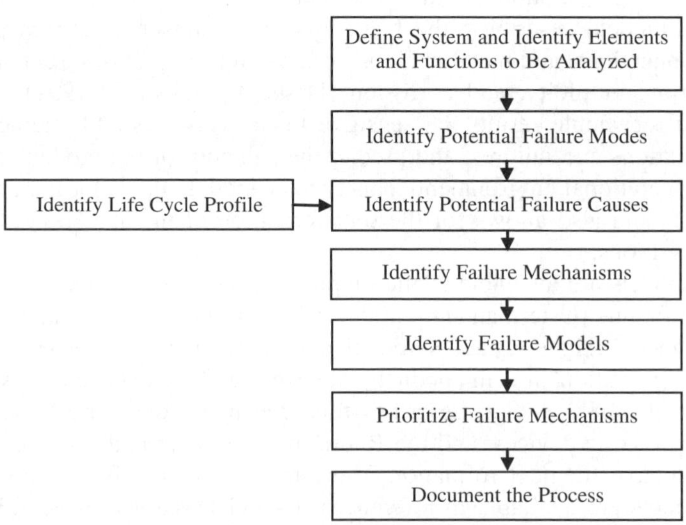

Figure 26.3 FMMEA methodology

the mechanism and not from the mechanism itself. The same failure mechanism can result in a small change in some electrical parameters at one site and shut down the system in another site. The damage will be at more serious in the latter case. The high-priority failure mechanisms identified through a combination of occurrence statistics, failure time, and severity are the critical mechanisms. Each critical failure mechanism has one or more associated causes, sites, and modes in an FMMEA result.

26.2.3 Stress Analysis

The identification of failure mechanisms, sites, and models in the FMMEA process allows for the quantification of time to and probability of failure. However, to evaluate the failures using failure models, it is often necessary to evaluate the stress levels acting at the failure sites. Stress analysis is used to convert the loads on the system to stresses at the potential failure sites. The product's life cycle loads and design information are inputs to the stress analysis process. Stress can be evaluated through physical testing. However, simulation techniques are generally required to determine the stress levels within a product over its anticipated life cycle. For electronic products, thermal analysis, thermo-mechanical analysis, and vibration analysis are often performed to determine the inputs for the relevant failure models (Pecht *et al.* 1998; Tilgner 2009).

26.2.4 Reliability Assessment

With the failure mechanisms, failure models, stress inputs, failure modes, and failure sites, the reliability of a product can be estimated and reported in terms of time-to-failure of the identified failure sites with a certain probability. Most failure models define time to failure under specific loading conditions. The stress status at potential failure sites identified from the FMEA is obtained by stress analysis. By inputting the stress status at the potential failure sites into the PoF models for the potential failure mechanisms, the time-to-failure and failure probability can be calculated. Thus, the potential failure mechanisms associated with failure sites can be ranked according to their time-to-failure, failure probability, and severity. The ranked failure mechanisms associated with failure sites can be used for a reliability assessment.

In general, time-to-failure data is obtained as a distribution for each failure site and failure mechanism. This distribution for time-to-failure is achieved by considering the input parameters for the failure models as distributions. In reality, all dimensional and material properties are distributed around a nominal value as a result of variations in manufacturing. The same is true for the environmental loads. The PoF-based approach allows for the utilization of statistical distributions of input parameters into the reliability assessment. With the time-to-failure distribution on each site known, reliability can be evaluated with different metrics such as hazard rate, warranty return rate, or mean-time-to-failure (MTTF).

In addition to evaluating time-to-failure, the use of failure models allows for the examination of time-to-failure dependency on materials, geometry, and life cycle profile. By considering the impact of identified material and product geometries and loading conditions, the most influential parameters can be identified. This information can be used to improve design through closer attention to critical design parameters.

26.3 Uses of Physics-of-Failure

The first aspect of the PoF-based reliability assessment process is the identification of possible failure mechanisms and sites. The PoF-based reliability assessment process can be used to support various product development activities. These activities include the ranking of potential failures, DfR, product qualification, accelerated testing, and prognostics implementation. Higher-risk failure mechanisms and failure sites are given a high priority in these activities.

26.3.1 Design-for-Reliability (DfR)

Design-for-reliability (DfR) is a systematic, streamlined, and concurrent engineering program in which reliability engineering is woven into the total development cycle. The process encompasses a variety of tools and practices and describes the overall order of deployment that an organization needs to follow in order to design reliability into its products. PoF can help an organization to examine the effect of variations in design parameters on a product's reliability. The PoF-based reliability assessment process can be used to examine the impact of proposed design changes as well as to compare competing designs. Reliability assessments may be performed on multiple product designs and manufacturing variables to compare their estimated reliability. The sensitivity analysis in PoF assessment also provides a way to obtain reliability as a function of product attributes. Hence, implementing PoF into the design makes the process of trade-off analysis efficient and effective. Typically, the PoF approach outputs the time-to-failure and the ranked list of failure mechanisms associated with failure modes and failure sites. The ranked list allows designers to concentrate on the locations and mechanisms that are most likely to cause product failure. The PoF methodology can be used to guide design improvement by identifying the drivers for the dominant failure mechanisms during each life cycle phase. Design trade-offs can then be evaluated by determining the sensitivity of the dominant degradation mechanisms to the drivers of the mechanisms.

Virtual qualification is a useful tool in DfR. It assesses whether a part or system can meet its reliability goals under the anticipated life cycle profiles based on its materials, geometry, and operating characteristics. The technique involves the application of simulation software to model physical hardware to determine the probability of the system's meeting its desired life goals (Cushing *et al.* 1993; Larson and Newel 1997; Cunningham *et al.* 2000). Virtual qualification is applied in the design stage and thus helps in the implementation of the reliability assessment process in the design phase (Caruso and Dasgupta 1998). This methodology takes advantage of advances in computer-aided engineering software, permitting components and systems to be qualified based on analysis of the susceptibility of their designs to failure using the critical failure mechanisms and the failure models associated with them (McCluskey *et al.* 1997).

26.3.2 Stress Testing Conditions

The application of stress is usually for the purpose of ruggedizing the design and manufacturing process of the device through systematic step-stress testing and increasing the stress margins by corrective action (reliability enhancement testing). The application of stress is also used to conduct compressed/accelerated life tests in the laboratory to verify in-service reliability (qualification testing). The stress conditions are determined from the LCP of the product, FMMEA, experience from previous similar products, the application requirements, and also the time-to-market. Target values and tolerances have to be met by the product or product elements under the test conditions selected with respect to the LCP. Critical failure mechanisms, identified by FMMEA, associated with failure sites and failure modes are considered in the selection of stress conditions. Proper stresses and their levels can be determined from the stresses associated with the inputs to the PoF models for the identified failure mechanisms (Pecht and Dasgupta 1995; Rothman *et al.* 1995).

26.3.3 Qualification

The aim of qualification is to verify whether a product meets or exceeds the reliability and quality requirements of its intended application. Qualification plays an important role in the process of product development. PoF can assist in the qualification of products in terms of qualification requirement definitions, especially qualification test conditions. Qualification requirements are determined based on insights from the potential failure modes and failure mechanisms. Failure analysis is also performed, when applicable, to verify the potential failure modes and failure mechanisms. Quality and reliability assessments are performed on the qualification testing results to give answers with confidence. This approach to qualification ensures that it successfully addresses the failure mechanisms applicable to the product's specific design, manufacture, and application conditions.

26.3.4 Screening Conditions

A screening process is used to eliminate defective products with a certain level of tolerance after production. Products are subjected, as part of an end-of-line inspection at the manufacturing site, to a set of defined stress tests to expose the defective products. To be effective, the screening conditions must identify defective products while not introducing defects into the products or reducing their life expectancy below acceptable levels. By providing an understanding of the failure mechanisms that affect the product, PoF allows for the identification of design parameters that affect product life and the stress conditions that can be used to precipitate failure. This information is used to select appropriate stress tests to be used as screens (Foucher *et al.* 2006). PoF is used to define stress conditions that will produce failure in defective products with design parameters below a desired threshold. More importantly, PoF identifies the damage induced on products subjected to a screen, which is critical information for determining if the screen will compromise the reliability of the surviving products.

26.3.5 Prognostics and Health Management (PHM)

Prognostics is the process of predicting the remaining useful life of a product by assessing its degradation from its expected state of health relative to its expected future use conditions. PoF-based PHM permits in-situ assessment of system reliability under actual application conditions. This method uses sensor data with models that enable in-situ assessment of the deviation or degradation of a product from its expected normal operating condition and the prediction of its future state of reliability. PoF helps PHM to identify the components which are critical to system reliability, provides an estimate of damage for given loading conditions and failure mechanisms, enables remaining life estimation for different loading conditions, and is suitable for obtaining remaining useful life even in non-operating conditions. PoF-based PHM provides previously unknown information on life cycle environmental and operational conditions. PoF-based PHM can prevent premature failures and provide information on remaining life. PoF-based PHM provides a cost-effective solution to improve the reliability of electronic products and systems.

26.4 Conclusion

The physics-of-failure (PoF)-based methodology is the preferred methodology for reliability assessment since it utilizes the knowledge of how failures actually occur in a product through degradation methods and can be used to evaluate a product's reliability based on its design and life cycle. PoF ranks potential failure mechanisms and assists in the design for reliability, selection of stress test conditions, product qualification and screening, and performing prognostics and health management. Currently, PoF as a reliability assessment methodology is widely accepted by industry, professional societies, and the US government organizations, including the IEEE, EIA/JEDEC, and SEMATECH, which have been developing standards to incorporate PoF in the product development and evaluation process. The US Army has also been working with research institutions to develop PoF and support the use of PoF and PHM in military and defense applications.

References

Bauernschub, R. and P. Lall. 1995. Addressing defect related reliability and screening levels through physics-of-failure analysis. *Advances in Electronic Packaging* 10–12: 635-645.

Bernstein, J.B., M. Gurfinkel, X. Li, J. Walters, Y. Shapira and M. Talmor. 2006. Electronic circuit reliability modeling. *Microelectronics Reliability* 46 (12): 1957–1979.

Bowles, J.B. 1992. A survey of reliability prediction procedures for microelectronic devices. *IEEE Transactions on Reliability* 41 (1): 2–12.

Caruso, H. and A. Dasgupta. 1998. A fundamental overview of analytical accelerated testing models. *Journal of the Institute of Environmental Sciences* 41 (1): 16–30.

Clement, J.J. 2001. Electromigration modeling for integrated circuit interconnect reliability analysis. *IEEE Transactions on Device and Materials Reliability.* 1 (1): 33–42.

Codier, E.O. 1969. Reliability prediction: help or hoax? In *Annual Symposium on Reliability*, 383–390.

Cunningham, J., R. Valentin, C. Hillman, A. Dasgupta and M. Osterman. 2001. A Demonstration of virtual qualification for the design of electronic hardware. In *Proceedings of the Institute of Environmental Sciences and Technology Meeting*, April 24, 2001. Rolling Meadows, IL: IEST.

Cushing, M.J., D.E. Mortin, T.J. Stadterman and A. Malhorta. 1993. Comparison of electronics-reliability assessment approaches. *IEEE Transactions on Reliability* 42 (4): 542–546.

Dasgupta, A. and M. Pecht. 1991. Material failure mechanisms and damage models. *IEEE Transactions on Reliability* 40 (5): 531–536.

Evans, J., Cushing, M. and R. Bauernschub. 1993. Physics of failure (POF) approach to addressing device reliability in accelerated testing of MCMs. In *Proceedings of the IEEE Multi-Chip Module Conference* (MCMC-95). 14–25. 31 Jan.–2 Feb. 1995.

Foucher, A., J. Tomas, F. Mounsi and M. Jerremias. 2006. Life margin assessment with Physics of Failure tools applications for BGA packages. *Microelectronics Reliability* 46 (5–6): 1013–1018.

Foucher, B., J. Boullie, B. Meslet and D. Das. 2002. A review of reliability prediction methods for electronic devices. *Microelectronics Reliability* 42 (8): 1155–1162.

Goe, A. and R.J. Graves. 2006. Electronic system reliability: collating prediction models. *IEEE Transaction on Device and Materials Reliability* 6 (2): 258–265.

Hu, J., D. Barker, A. Dasgupta and A. Arora. 1993. Role of failure mechanism identification in accelerated testing. *Journal of IES* 36 (4): 39–45.

IEEE Standard 1413-1998. 1998. IEEE Standard Methodology for Reliability Prediction and Assessment for Electronic Systems and Equipment. December 1998.

IEEE Standard 1413.1-2002. 2003. IEEE Guide for Selecting and Using Reliability Predictions Based on IEEE 1413. February 2003.

JEDEC. 2008. Reliability Qualification of Semiconductor Devices Based on Physics of Failure Risk and Opportunity Assessment, JEP 148. April 2008. Solid State Technology Association.

Johnson, B.G. and L. Gullo. 2000. Improvements in reliability assessment and prediction methodology. In *Proceedings of the Annual Reliability Maintainability Symposium* (RAMS), pp. 181–187.

Knowles, I. 1993. Is it time for a new approach? *IEEE Transactions on Reliability* 42 (1): 2–3.

Kumar, U. and A. Dasgupta. 1989. Guidelines for physics-of-failure based accelerated stress testing. In *Proceedings of the Annual Reliability and Maintainability Symposium*. 345–357. January 1922, 1998.

Lall, P., M. Pecht and E. Hakim. 1997. *Influence of Temperature on Microelectronics System Reliability: A Physics of Failure Approach*. Boca Raton, FL: CRC Press.

Larson, T. and J. Newel. 1997. Test philosophies for the new millennium. *Journal of the Institute of Environmental Sciences* 40 (3): 22–27.

Lee, J.C., I.-C. Chen and C. Hu. 1988. Modeling and characterization of gate oxide reliability. *IEEE Transactions on Electron Device* 35 (12): 2268–2278.

Leonard, C.T. and M. Pecht. 1993. How failure prediction methodology affects electronic equipment design. *Quality and Reliability International* 6 (4): 243–249.

Li, J. and A. Dasgupta. 1994. Failure mechanism models for material aging due to interdiffusion. *IEEE Transactions on Reliability* 43 (1): 2–10.

Matic, Z. and V. Sruk. 2008. The Physics-of-failure approach in reliability engineering. In *Proceedings of the 30th International Conference on Information Technology Interfaces* June 23–26, 2008. 745–750. Cavtat, Croatia.

McCluskey, P., M. Pecht and S. Azarm. 1997. Reducing time-to-market using virtual qualification. In *Proceedings of the Institute of Environmental Sciences Conference,* pp. 148–152.

Mencinger, N.P. 2000. A mechanism-based methodology for processor package reliability assessments. *Intel Technology Journal* Q3: 1–8.

Mortin, D.E., J.G. Krolewski and M.J. Cushing. 1995. Consideration of component failure mechanisms in the reliability assessment of electronic equipment: addressing the constant failure rate assumption. In *Proceedings of the Annual Reliability and Maintainability Symposium*, January 16–19, 1995, Washington, DC, pp. 54–59.

O'Connor, P.D.T. and M.G. Head. 1985. Reliability prediction for microelectronic systems. *Reliability Engineering* 10 (3): 129–140.

Osterman, M., A. Dasgupta and B. Han. 2006. A strain range based model for life assessment of Pb-free SAC solder interconnects. In *Proceedings of the 56th Electronic Component and Technology Conference*. 884–890. May 30–June 2, 2006.

Pecht, M. 1996. Why the traditional reliability prediction models do not work: Is there an alternative? *Electronic Cooling* 2 (1): 10–12.

Pecht, M., R. Agarwal, P. McCluskey, T. Dishongh, S. Javadpour and R. Mahajan. 1998. *Electronic Packaging Materials and their Properties*. Boca Raton, FL: CRC Press.

Pecht, M. and A. Dasgupta. Physics of failure: An approach to reliable product development. In *Proceedings of the 1995 International Integrated Reliability Workshop*. 1–4. October 22–25, 1995. Lake Tahoe, CA.

Pecht, M. and F. Nash. 1994. Predicting the reliability of electronic equipment. *Proceedings of the IEEE* 82 (7): 992–1004.

Pecht, M., L. Nguyen and E. Hakim. 1995. *Plastic Encapsulated Microelectronics: Materials, Processes, Quality, Reliability, and Applications*. New York: John Wiley & Sons.

Pecht, M., R. Radojcic and G. Rao. 1999. *Guidebook for Managing Silicon Chip Reliability*. Boca Raton, FL: CRC Press.

Rothman, T.P., A. Dasgupta and M.C. Binder. 1995. Physics-of-failure case study for accelerated testing for electronic packaging. In *Proceedings of the Annual Technical meeting of the Institute of Environmental Sciences*, pp. 63–71.

Rudra B. and Jennings D. 2004. Failure-mechanism models for conductive-filament formation. *IEEE Transactions on Reliability* 43 (3): 354–360.

Ryu, D. and S. Chang. 2005. Novel concepts for reliability technology. *Microelectronics Reliability* 45 (3-4): 611–622.

Snook, I., J. Marshall and R. Newman, 2003. Physics of failure as an integrated part of design for reliability. In *Proceedings of the IEEE Annual Reliability and Maintainability Symposium*, pp. 46–54.

Stadterman, T., M. Cushing, B. Hum, A. Malhorta and M. Pecht. 1995. The transition from statistical-field failure based models to physics-of-failure based models for reliability assessment of electronic packages. In *Proceedings of the International Electronic Packaging Conference* (InterPACK'95), March 26–30, 1995, Maui, HI, pp. 619–625. American Society of Mechanical Engineers.

Stadterman, T. and M. Osterman, 1999. Reliability and performance of advanced PWB assemblies. In *High Performance Printed Circuit Boards*, ed. C. A. Harper. New York: McGraw-Hill.

Stipan, P., B. Beihoff and M. Shaw. 2000. Electronics package reliability and failure analysis: a micromechanics-based approach. In *Electronic Packaging Handbook*, ed. G. T. Blackwell. Boca Raton, FL: CRC Press.

Talmor, M. and S. Arueti. 1997. Reliability prediction: The turn-over point. In *Proceedings of the Annual Reliability and Maintainability Symposium*. 254–262. January 13–16, 1997. Philadelphia, PA.

Tilgner, R. 2009. Physics of failure for interconnect structures: an essay. *Microsystem Technology* 15: 129–138.

Weil, L., M. Pecht and E. Hakim. 1993. Reliability evaluation of plastic encapsulated parts. *IEEE Transactions on Reliability* 42 (4): 536–540.

27

Accelerated Testing: Effect of Variance in Field Environmental Conditions on the Demonstrated Reliability

Andre Kleyner
Delphi Electronics & Safety, USA

27.1 Introduction

Product validation and reliability demonstration by test are important parts of product development and are commonly included in technical documentation specifying certain levels of reliability needed to be demonstrated by product testing. The most common ways of demonstrating reliability include success-based testing and test-to-failure.

Success-based (also referred as *success run*) testing is probably the most common technique utilized in the industry, where survival of a product at the end of a test is expected. Under those conditions the product is subjected to a test, often accelerated, representing an equivalent of one mission life (test to a bogey), which is expected to be completed without failure by all the units in the test sample.

Success run test statistics are most often based on the binomial distribution shown by (27.1) (see, e.g. O'Connor and Kleyner 2012):

$$C = 1 - \sum_{i=0}^{k} \frac{N!}{i!(N-i)!} R^{N-i}(1-R)^{i}, \tag{27.1}$$

where R is reliability, C is confidence level, N is total number of test samples and k is number of failures experienced during the test.

Applied Reliability Engineering and Risk Analysis: Probabilistic Models and Statistical Inference, First Edition. Edited by Ilia B. Frenkel, Alex Karagrigoriou, Anatoly Lisnianski and Andre Kleyner.

In the case of no failures ($k = 0$), Equation (27.1) turns into a simple equation for success run testing:

$$C = 1 - R^N \tag{27.2}$$

For example, according to (27.2), testing 22 samples would be sufficient to demonstrate 90% reliability with 90% confidence or alternatively 97% reliability with 50% confidence, which is a common requirement in the automotive industry. It is important to note that according to (27.2) the demonstrated reliability R approaching 100% would require a sample size N approaching infinity.

$$N = \frac{\ln(1 - C)}{\ln R} \tag{27.3}$$

Reliability demonstration through test to failure is usually done by testing a test sample until at least several failures are observed and then fitting a statistical distribution (most often the two- or three-parameter Weibull) into the collected life data. The three-parameter Weibull reliability (occasionally referred in this book as Weibull-Gnedenko) can be described by:

$$R(t) = \exp\left[-\left(\frac{t - \gamma}{\eta}\right)^{\beta}\right] \tag{27.4}$$

where β is Weibull slope (shape parameter), η - Characteristic Life (scale parameter) and γ is Minimum life (location parameter).

More on test-to-failure and life data analysis can be found in ReliaSoft (2012) and various other sources. Test to failure provides more flexibility and information regarding the product's design margins, but usually takes longer (sometimes $2\times$ to $6\times$ more) than success-based testing, thus extending a product development cycle.

The main limitation of the reliability demonstration testing (either success-based or test-to-failure) is in the number of test samples (see, e.g. (27.3) limiting the number of test samples). Therefore, the reliability numbers demonstrated by either test are relatively low compared to what is typically required by product specifications on field reliability (often 99% and higher). This problem often causes disputes and misunderstanding in customer–supplier relationships, which sometimes even requires mediation. This issue, however, can be addressed by taking into account variation in the field stress and/or usage conditions (e.g. driving mileage, vibration level, temperature exposure, number of on/off cycles, etc.) and if the test simulates more severe conditions than the average (which is often the case), then the actual field population reliability will be higher than that demonstrated by the test.

27.2 Accelerated Testing and Field Stress Variation

The acceleration factor (*AF*) is defined as a ratio of the expected life in unaccelerated (field) conditions to the expected life under accelerated (test) conditions (see, e.g. Nelson 2004, or O'Connor and Kleyner 2012, Chapter 13):

$$AF = \frac{L_{Field}}{L_{Test}} \tag{27.5}$$

where L_{Field} and L_{Test} are the expected product lives in field and test conditions respectively.

The acceleration factor relationship with reliability function can be modeled as:

$$R_{Field}(t) = R_{Test}\left(\frac{t}{AF}\right) \tag{27.6}$$

where R_{Field} is Reliability under field conditions and R_{Test} is Reliability under test conditions.

Accelerated testing is usually done based on either an increased usage rate (e.g. increased number of loading cycles, extended number of miles driven, increased duty cycle, etc.) or based on an increased stress level (e.g. elevated temperature, vibration, voltage, etc.). In either case, if the total product population has a variation in the usage rate or the stress level, the acceleration factor will also vary. This chapter will focus on the cases where a test is accelerated by more severe stress conditions, although the approach to the increased usage rate analysis will be very similar.

The acceleration factor AF is a function of multiple stress variables. However in many practical applications one dominant stress factor S can often be identified. In those cases the acceleration factor would depend on that dominant stress variable, i.e. $AF = AF(S)$, and thus would vary for each unit in the population depending on the stress level experienced by that particular unit. Therefore, the reliability for the whole population of products can be calculated based on averaging over the stress value varying in the field population of units.

$$R_{Field}(t) = \int_0^{S_{\max}} R_{Test}\left(\frac{t}{AF(S)}\right) f(S)dS \tag{27.7}$$

where $S_{\max}$ is maximum possible stress value achieved in the population of products and $f(S)$ is the probability density function (pdf) of the stress distribution over the population, and $f(S)$ requires the knowledge (or at least an educated estimate) of how the stress is distributed in the total population of units.

Considering that both stress values and field failures are random variables we can re-write (27.7) in a parametric form as:

$$R_{Field}(t) = \int_0^{S_{\max}} R_{Test}\left(\frac{t; \alpha_1, \ldots, \alpha_q}{AF(S)}\right) f(S; \beta_1, \ldots, \beta_p)dS \tag{27.8}$$

where:

$\alpha_1 \ldots \alpha_q$ = vector of failure distribution parameters (dimension q)
$\beta_1 \ldots \beta_p$ = vector of stress distribution parameters (dimension p)

Parameters α_i can be estimated based on previously collected reliability data; and β_i will characterize the distribution of the field stress variable and can be based on field environment studies.

27.3 Case Study: Reliability Demonstration Using Temperature Cycling Test

In the consumer and automotive electronics industries temperature cycling (TC) became a very common procedure used in reliability demonstration testing. It simulates the temperature changes caused by an internal heating during on/off cycles, by changes in the ambient temperature caused by diurnal cycles and by external heat sources, such as engines in the cases of electronics that is mounted under the hood of a vehicle.

During TC (low cycle fatigue) the strain range caused by the materials mismatch of the coefficients of thermal expansion is proportional to the temperature excursion $\Delta T = T_{Max} - T_{Min}$, thus making ΔT the dominant stress factor for this type of reliability testing. Therefore, using the Coffin-Manson model (see, e.g. O'Connor and Kleyner 2012, Chapter 13), the acceleration factor can be modeled by:

$$AF = \left(\frac{\Delta T_{Test}}{\Delta T_{Field}}\right)^{m} \tag{27.9}$$

where ΔT_{Test} is the temperature range experienced during the test, ΔT_{Field} is the temperature range experienced in the field and m is Coffin-Manson exponent ($m = 2.65$ is often used for lead-free solder fatigue, see, for example, Arnold *et al.* 2008).

The test is designed to simulate 10 years field life by performing a continuous thermal cycling. The available 14 product samples are tested with the expectation of surviving a number (to be determined) of [−40; 125]°C thermal cycles. These test conditions simulate the environment of the electronics mounted under the hood where operating temperatures vary based on the field climate conditions and driving durations.

In the cases of thermal cycling the acceleration factor is typically applied to the number of thermal cycles rather than overall temperature exposure time. Therefore, the number of field cycles, assuming two cold cycles per day is 10yr × 365days/yr × 2cycles/day = 7300 cycles.

Analysis of the automotive field data suggested $\Delta T_{Field} = 70°C$, which corresponds to the 99th percentile user, thus applying equations (27.5) and (27.9), the test duration can be calculated as:

$$7300\left(\frac{70}{125-(-40)}\right)^{2.65} = 752 \text{ cycles}$$

Therefore, 752 cycles under [−40; 125]°C conditions would demonstrate according to (27.2) $R = 0.95$ with 50% confidence.

Now, considering the fact that the test has been designed for the conditions simulating the 99th percentile stress $\Delta T_{Field} = 70°C$, let us calculate the reliability demonstrated for the total product population using the approach discussed in Section 27.2.

Let us assume that field failures follow the two-parameter Weibull distribution (27.4) with no minimum life, $\gamma = 0$:

$$R(t) = \exp\left[-\left(\frac{t}{\eta}\right)^{\beta}\right] \tag{27.10}$$

For an accelerated test, Weibull reliability according to (27.6) becomes:

$$R_{Field}(t) = R_{Test}\left(\frac{t}{AF}\right) = \exp\left[-\left(\frac{t}{\eta AF}\right)^{\beta}\right] \tag{27.11}$$

While ΔT_{Test} in (27.9) is known and defined by the test specifications, the exact value of ΔT_{Field} is unknown and needs to be modeled as a random variable, designated further as ΔT. Its value is subject to variation based on a region's climate and the individual drive times. It would range from low ΔT of short driving in mild climates to high ΔT of extended driving in extreme climates. It has been observed that distributions of field usage variables, such as mileage, temperature, vibration, etc. are often skewed to the right and typically can be modeled by the likes of the lognormal distribution, commonly designated as $LN(\mu, \sigma)$.

Therefore, combining (27.8) with (27.9) and (27.11) will allow us to calculate the reliability of field population:

$$R(t) = \int_0^{\Delta T_{\max}} \exp\left(-\left(\frac{t}{\eta_0\left(\frac{\Delta T_{Test}}{\Delta T}\right)^m}\right)^{\beta_0}\right) f(\Delta T; \mu, \sigma) d(\Delta T) \tag{27.12}$$

where η_0 and β_0 are the Weibull distribution parameters obtained as a result of the reliability demonstration testing and $f(\Delta T, \mu, \sigma)$ represents the pdf of the random variable ΔT (in our case, $LN(\mu, \sigma)$).

In the cases where a reliability demonstration was conducted via success run testing (no failures occurred and no life data analysis performed), the values of η_0 and β_0 can be estimated based on success test equation (27.2) and engineering assumptions about the Weibull slope value β_0. Since we have only one equation, it cannot be solved for both η_0 and β_0, thus one of the parameters needs to be assumed:

$$R(t_{Life}) = \exp\left[-\left(\frac{t_{Test}}{\eta_0}\right)^{\beta_0}\right] = (1 - C)^{\frac{1}{N}} \tag{27.13}$$

where t_{Life} is the mission life of the product and t_{Test} is the test time corresponding to one mission life in the field (test to a bogey).

Therefore,

$$\eta_0 = t_{Test}\left[-\frac{1}{N}\ln(1 - C)\right]^{-\frac{1}{\beta_0}} \tag{27.14}$$

Based on (27.14) our case study data produces $\eta_0 = 2504$ cycles, assuming $\beta_0 = 2.5$ a generic value for lead-free solder fatigue failures.

Field studies for these types of automotive electronics show the 50th percentile temperature $\Delta T_{50} = 40°\text{C}$ and as mentioned before, the 99th percentile $\Delta T_{99} = 70°\text{C}$ with the maximum value of $\Delta T_{\max} = 100°\text{C}$. The lognormal distribution fitting this data would have the following parameters: $LN(3.69, 0.24)$ truncated at 100.

Therefore, substituting all the numerical values into (27.12) produces a numerical solution:

$$R(7300\ cycles) = \int_0^{100} \exp\left(-\left(\frac{7300}{2503\left(\frac{165}{\Delta T}\right)^{2.65}}\right)^{2.5}\right) f(\Delta T; 3.69, 0.24) d(\Delta T) = 0.996$$

where $f(\Delta T; 3.69, 0.24)$ is the pdf of $LN(3.69, 0.24)$. As shown, the mission life reliability $R(t_{Life}) = 99.6\%$ is significantly higher than the originally demonstrated value of 95%, thus shifting the "true" demonstrated population reliability into the more desirable region above 99%.

27.4 Conclusion

A variation in field stress conditions may significantly affect field reliability demonstrated by test. When a reliability demonstration test (either accelerated or conducted at the field stress level) is designed based on the stress conditions corresponding to a higher severity user, the actual field reliability will be higher than that demonstrated by test. The higher the percentile stress severity conditions for the test, the higher the difference between the actual and the demonstrated reliability.

Calculation of the demonstrated field reliability requires two key pieces of information. First is an acceleration model expressed as a function of stress variable(s) and the second is a statistical model of the stress distribution over the field population. The mathematical approach presented in this chapter can be applied to both types of accelerated testing: increased stress level and increased usage rate. This mathematical approach can also be used during the accelerated test planning phase. It can also be applied in reverse in order to determine the test stress level needed to demonstrate required reliability in the cases where the sample size and test duration have already been predetermined, for example, due to resource allocations.

References

Arnold, J., N. Blattau, C. Hillman and K. Sweatman. 2008. Reliability testing of Ni-Modified SnCu and SAC305: accelerated thermal cycling. In *Proceedings of the IPC/JEDEC Lead Free Conference*, March 2008.

Nelson, W. 2004. *Accelerated Testing: Statistical Models, Test Plans, and Data Analysis*. Hoboken, NJ: John Wiley & Sons.

O'Connor, P. and A. Kleyner. 2012. *Practical Reliability Engineering*. Chichester: John Wiley & Sons.

ReliaSoft. 2012. *Life Data Analysis (Weibull Analysis)*. Available at http://www.weibull.com/basics/lifedata.htm.

Index

Applied Reliability Engineering and Risk Analysis: Probabilistic Models and Statistical Inference, First Edition.
Edited by Ilia B. Frenkel, Alex Karagrigoriou, Anatoly Lisnianski and Andre Kleyner.